Studies in Systems, Decision and Control

Volume 605

The series "Studies in Systems, Decision and Control" (SSDC) covers both new developments and advances, as well as the state of the art, in the various areas of broadly perceived systems, decision making and control–quickly, up to date and with a high quality. The intent is to cover the theory, applications, and perspectives on the state of the art and future developments relevant to systems, decision making, control, complex processes and related areas, as embedded in the fields of engineering, computer science, physics, economics, social and life sciences, as well as the paradigms and methodologies behind them. The series contains monographs, textbooks, lecture notes and edited volumes in systems, decision making and control spanning the areas of Cyber-Physical Systems, Autonomous Systems, Sensor Networks, Control Systems, Energy Systems, Automotive Systems, Biological Systems, Vehicular Networking and Connected Vehicles, Aerospace Systems, Automation, Manufacturing, Smart Grids, Nonlinear Systems, Power Systems, Robotics, Social Systems, Economic Systems and other. Of particular value to both the contributors and the readership are the short publication timeframe and the worldwide distribution and exposure which enable both a wide and rapid dissemination of research output.

Indexed by SCOPUS, DBLP, WTI Frankfurt eG, zbMATH, SCImago.

All books published in the series are submitted for consideration in Web of Science.

Le Van Hien

Two-Dimensional Time-Delay Systems

Analysis and Control

Springer

Le Van Hien
Hanoi National University of Education
Hanoi, Vietnam

ISSN 2198-4182 ISSN 2198-4190 (electronic)
Studies in Systems, Decision and Control
ISBN 978-3-031-95677-5 ISBN 978-3-031-95678-2 (eBook)
https://doi.org/10.1007/978-3-031-95678-2

This Springer imprint is published by the registered company Springer Nature Switzerland AG
The registered company address is: Gewerbestrasse 11, 6330 Cham, Switzerland

To my family
To my wife, Truong Thi Hong,
and to our lovely children, Le Thi Nguyet Ha,
Le Thi Ngan Giang

Preface

Qualitative theory and, in particular, stability theory are essential issues in the systems and control theory of dynamical systems. Recently, the study on stability, performance analysis and applications in control of two-dimensional (2-D) dynamical systems has become one the most active research topics. On one hand, this area has been found in various practical applications in control engineering, digital filtering, image processing and many other fields. On the other hand, the development of this theory also raises much challenge and difficulty due to the lack of efficient tools. It is because that, according to major differences in their structures and the nature of the dynamic propagation characteristics, established methodologies for one-dimensional (1-D) dynamical systems are typically inappropriate and inapplicable for 2-D systems. This urges to develop suitable and efficient schemes for analysis and synthesis problems of 2-D systems.

In the past more 10 years, fundamental problems in the systems and control theory of 2-D systems have attracted significant research attention from both academia and engineers. Many results concerning qualitative studies and control of 2-D systems have been reported in the literature. These results, however, have just been developed for 2-D systems with special structures, which will narrow the applicability to practical engineering especially for 2-D complex systems such as singular systems, 2-D systems with positivity constraints or 2-D systems with stochastic and/or deterministic jumping parameters. Some other aspects in the analysis and synthesis of 2-D systems are still under construction and development.

This book provides a systematic study of the author and his team with fundamental concepts, useful techniques and utilizable schemes developed for various classes of 2-D time-delay systems and complements the systems and control theory of 2-D systems with delays. The information presented in this book is subject to current state of the art research in this field. The book will serve as a useful reference to researchers in this area to understand the fundamentals involved the theory of 2-D time-delay systems and to gather most recent advancements in the area of stability/performance analysis and control of 2-D time-delay systems. The text can be accessible to readers with basic knowledge of matrix analysis and linear algebra, mathematical analysis, differential/difference equations, and control theory. Except the introductory chapter, the remaining of the book is organized and divided into ten chapters. Each subsequent chapter is concerned with particular problems. More specifically, main content of this book includes

- Stability and stabilization of 2-D systems with state delays.
- Asymptotic estimation of reachable set for 2-D time-delay systems.
- Stability analysis and observers design for positive 2-D systems.
- Performance analysis and l_1-gain control of 2-D positive systems with delays.
- Peak-to-peak performance analysis of 2-D positive time-delay systems.
- Stability of 2-D time-delay Roesser systems with Markovian switchings.
- Analysis and control of 2-D Roesser systems with stochastic noises.
- Stability and stabilization of 2-D Takagi–Sugeno fuzzy systems with time-varying delays.
- Stability and stabilization via event-triggered control of 2-D singular systems with delays.
- Finite-region stability and dissipativity analysis of 2-D singular systems with delays.

The aforementioned problems have been solved by the author during the past decade. The obtained results are presented with detail proofs and illustrations. The book contains 94 figures and around 300 compiled references, which is very useful to postgraduate students and academic researchers to look insights in the systems and control theory of 2-D time-delay systems. This book can also be useful for specialized final year undergraduate courses in control systems engineering and applied mathematics.

Hanoi, Vietnam
April 2025

Le Van Hien
hienlv@hnue.edu.vn

Acknowledgements The author is grateful to Prof. Hieu Trinh, School of Engineering, Deakin University, Australia, who worked with me on this subject of two-dimensional time-delay systems. In particular, the unconditional help and support, the encouragement, and invaluable discussion from Prof. Hieu Trinh while I was visiting and working at Deakin University made an enormous contribution to my scientific development in the field of systems and control theory. I also thank my former supervisor, Prof. Vu Ngoc Phat, Institute of Mathematics, Vietnam Academy of Science and Technology, my Ph.D. students, Le Huy Vu, Mai Thi Hong, Nguyen Thi Lan Huong, and my colleagues at the Faculty of Mathematics and Informatics, Hanoi National University of Education, for their indirect help and support.

This book would not be formable without my research achievement from sabbatical years accorded by the Faculty of Mathematics and Informatics, Hanoi National University of Education. Let me take this opportunity to thank all my colleagues from the Division of Mathematical Analysis, who shared my annual pedagogic service. I also would like to thank institutional organizations, which provided me with excellent support, Hanoi National University of Education, Deakin University, Vietnam Institute for Advanced Study in Mathematics, and the National Foundation for Science and Technology Development (NAFOSTED) of Vietnam.

Competing Interests The author has no competing interests to declare that are relevant to the content of this manuscript.

Contents

Acronyms

1-D	One-dimensional
2-D	Two-dimensional
BMIs	Bilinear matrix inequalities
EPSS	Energy-to-peak stochastic stability
ETC	Event-triggered control
EUB	Ellipsoidal ultimate bound
F-M	Fornasini–Marchesini
FMLSS	Fornasini–Marchesini local state-space model
FRB	Finite-region boundedness
FTB	Finite-time boundedness
FTS	Finite-time stability
GAS	Global asymptotic stability
GES	Global exponential stability
LKF	Lyapunov–Krasovskii functional
LMIs	Linear matrix inequalities
LP	Linear programming
LTI	Linear time-invariant
MIMO	Multiple-input multiple-output
MJSs	Markov jump systems
MNSs	Multiplicative noisy systems
NoDv	Number of decision variable
RSE	Reachable set estimation
SFCs	State-feedback controllers
SFRB	Singular finite-region boundedness
SIMO	Single-input multiple-output
SISO	Single-input single-output
SOFCs	Static output-feedback controllers
SSFCs	Synchronous state-feedback controllers
TPs	Transition probabilities
T–S	Takagi–Sugeno
WSIs	Weighted summation inequalities

$\mathbb{Z}$	The set of integers
$\mathbb{N}$	The set of positive integers
$[n]$	The set $\{1, 2, \ldots, n\}$ for an $n \in \mathbb{N}$
$\mathbb{N}_0$	$\mathbb{N} \cup \{0\}$
$\mathbb{Z}[a, b]$	The set of integers between a and b
$\mathbb{Z}[a, b] \times [c, d]$	$\{(p, q) : p \in \mathbb{Z}[a, b], q \in \mathbb{Z}[c, d]\}$
$l_{\mathbb{Z}}(a, b)$	The length $b - a + 1$ of $\mathbb{Z}[a, b]$
$\mathbb{R}$	The set of real numbers
$\mathbb{R}^n$	The n-dimensional Euclidean vector space
$\mathbf{1}_n$	The vector in $\mathbb{R}^n$ with all entries equal one
$x \preceq y$	If $x_i \leq y_i$ for all $i \in [n]$
$x \prec y$	If $x_i < y_i$ for all $i \in [n]$
$\mathbb{R}^n_+$	Positive orthant $\{x \in \mathbb{R}^n : x \succeq 0\}$
$\Vert . \Vert$	The Euclidean vector-norm
$\Vert . \Vert_\infty$	The max vector-norm
$\Vert . \Vert_1$	The sum vector-norm
$\Vert f \Vert_{l_1}$	The l_1-norm $\sum_{i,j=0}^{\infty} \Vert f(i, j) \Vert_1$ of $f : \mathbb{Z}_+ \times \mathbb{Z}_+ \to \mathbb{R}^n$
$l_1(\mathbb{R}^n)$	The space $\left\{f : \mathbb{Z}_+ \times \mathbb{Z}_+ \to \mathbb{R}^n : \Vert f \Vert_{l_1} < \infty\right\}$
$\Vert w \Vert_{l_2}$	The l_2-norm $\sqrt{\sum_{i,j=0}^{\infty} w^\top(i, j) w(i, j)}$ of sequence $w(i, j)$
$l_2(\mathbb{R}^n)$	The space of $\mathbb{R}^n$-valued square-summable sequences
$\Vert f \Vert_{l_\infty}$	The l_∞-norm $\sup_{k,l \in \mathbb{Z}_+} \Vert f(k, l) \Vert_\infty$ of $f : \mathbb{Z}_+ \times \mathbb{Z}_+ \to \mathbb{R}^n$
$l_\infty(\mathbb{R}^n)$	The space $\left\{f : \mathbb{Z}_+ \times \mathbb{Z}_+ \to \mathbb{R}^n : \Vert f \Vert_{l_\infty} < \infty\right\}$
$\mathbb{R}^{n \times m}$	The set of $n \times m$ real matrices
I_n	Identity matrix of n dimension
$A^\top$	The transpose of matrix A
$A^\perp$	Null-space of matrix A
A^{-1}	Inverse matrix of A
$A \succeq 0$	Matrix A is nonnegative (i.e. $A = (a_{ij})$ and $a_{ij} \geq 0$ for all i, j)
$A \succ 0$	Matrix A is positive (i.e. $a_{ij} > 0$ for all i, j)
$A \succ B$	The matrix $A - B$ is nonnegative
$A \succ B$	The matrix $A - B$ is positive
$\lvert A \rvert = (\lvert a_{ij} \rvert)$	The absolute matrix of $A = (a_{ij}) \in \mathbb{R}^{m \times n}$
$\Vert A \Vert_1$	The induced 1-norm of matrix A
$\Vert A \Vert_\infty$	The induced max-norm of matrix A
$\mathrm{Sym}\{.\}$	Symmetry operator
$\mathrm{diag}(\cdot)$	The diagonal block-matrix
$\mathrm{col}\{.\}$	Column block-vector or matrix
D_χ	Diagonal matrix $\mathrm{diag}(\chi_1, \ldots, \chi_n)$ of $\chi = (\chi_i) \in \mathbb{R}^n$
$\lambda_{\max}(A)$	Maximal real part of eigenvalues of matrix A
$\lambda_{\min}(A)$	Minimal real part of eigenvalues of matrix A
$\lambda(A)$	The set of all eigenvalues of matrix A
$\rho(A)$	The spectral radius $\max\{\lvert\lambda\rvert : \lambda \in \lambda(A)\}$ of A
$\sigma(A)$	The spectral abscissa of matrix A

$\rho(E, A)$	Generalized spectral radius of the pair (E, A) defined by $\max_{\lambda \in \{z \mid \det(zE-A)=0\}} \lvert\lambda\rvert$
$\lVert A \rVert$	The induced norm $\sqrt{\lambda_{\max}(A^\top A)}$ of matrix A
$M \geq 0$	Semi-positive definite matrix (i.e. $x^\top M x \geq 0$ for all x)
$M > 0$	Positive definite matrix (i.e. $x^\top M x > 0$ for all $x \neq 0$)
$A \geq B$	The matrix $A - B$ is semi-positive definite
$A > B$	The matrix $A - B$ is positive definite
$\mathbb{S}_n^+$	The set $\{M \in \mathbb{R}^{n \times n} : M > 0\}$
$\mathbb{D}_n^+(p, q)$	The set $\{\mathrm{diag}(M_p, M_q) : M_p \in \mathbb{S}_p^+, M_q \in \mathbb{S}_q^+\}$
$\lfloor x \rfloor$	The floor function $\lfloor x \rfloor = \max\{m \in \mathbb{Z} \mid m < x\}$
$*$	The terms induced by symmetry
$\alpha \vee \beta$	$\max\{\alpha, \beta\}$ for scalars α, β
$\alpha \wedge \beta$	$\min\{\alpha, \beta\}$
$\mathbb{I}(\alpha, \beta)$	The diagonal block matrix $\mathrm{diag}(\alpha I_{n_h}, \beta I_{n_v})$
$\mathbb{E}[.]$	The mathematical expectation operator
$\mathbb{E}[X \mid Y]$	The conditional expectation of X with respect to Y

Chapter 1
Introduction and Preview

Abstract This chapter gives a brief introduction to the study of two-dimension (2-D) systems, mathematical models that lead to the representation using 2-D systems as well as some basic models of 2-D systems. We also present some preliminary results on the stability and stabilization problems of a basic model of 2-D systems. Finally, we give a preview of the main content of each chapter presented in this monograph.

1.1 Brief Introduction

Two-dimensional (2-D) systems form a category of dynamical systems, where the dynamic propagation occurs in each of the two independent directions [1]. This type of systems can be used to model a large number of physical processes and engineering systems [1, 2]. Typical examples of 2-D systems can be found in finite approximations of partial differential equations [3] which appear in heat exchanger or gas filtration models [4, 5]. Moreover, 2-D system models are also taken into various areas of practical applications in circuit analysis, image processing, multi-dimensional digital filtering, thermal processes, seismographic data transmissions, repetitive processes or iterative learning control [6–14]. According to a wide range of applications and interesting features, the theory of 2-D systems has received significant research attention in the past few decades. In particular, the problems of performance analysis and design have been extensively studied and developed for various types of 2-D systems [15–22].

There are numerous state-space models of 2-D systems that can be used in modeling practical phenomena. Among those, Roesser model [23] and Fornasini–Marchesini local state-space model (FMLSS) [24] are the most widely studied in the literature. Unlike the FMLSS model, in Roesser model, two independent state vectors, called the horizontal and vertical state vectors, are used to describe dynamics of 2-D systems which prevails in modeling control processes governed by partial differential equations [25]. In very special cases, some state-space models of 2-D systems can be mutually transformed to each other. However, for complex 2-D systems, they are often studied separately due to the nature of their special structures and potential applications [19, 20].

L. V. Hien, *Two-Dimensional Time-Delay Systems*, Studies in Systems, Decision and Control 605, https://doi.org/10.1007/978-3-031-95678-2_1

On the other hand, time-delay phenomena are inevitably encountered in practical systems and industrial processes due to many technical reasons [26, 27]. It is because that the reaction of practical engineering systems to exogenous signals is never instantaneous and always infected by time delays [28]. For example, in networked control or multi-agent systems, the data transmission processing through low-rate channel or the network congestion often induces communication delays, a key factor affecting the consensus the system [29]. Many other models of time-delay systems can also be found in chemical processes, transmission lines, or telecommunication networks.

It is well-known that, in the presence of time-delay, the system performance and behavior becomes unpredictable, chaotic or even unstable. In addition, according to deep and major differences in their analytical characteristics, deal with the analysis and control problems of time-delay systems is typically much more challenging than nominal ones [30]. Thus, with a long history, the systems and control theory of one-dimensional (1-D) time-delay systems is still one of the most vibrant research topics in the past few decades [31–37]. It has been well recognized from the aforementioned literatures that the Lyapunov–Krasovskii functional (LKF) method and its variants are widely used to derive tractable stability conditions for time-delay systems [35, 36]. Such stability conditions are derived using some types of linear matrix inequalities (LMIs), which can be solved by using convex algorithms and many available computational tools. For some particular types of time-delay systems, for example, delayed systems with positivity constraints on states and/or outputs, variants of LKF method via the so-called co-positive Lyapunov functions [38] (a type of linear state-dependent functionals) are developed together the use of monotone properties induced by the positivity [39–42].

Most important problems in the systems and control theory have been extensively studied and developed for one-dimensional (1-D) time-delay systems. However, despite of interesting theoretical properties and a wide range of practical applications, this theory for 2-D systems has just been exploited and broader extended to various classes of complex dynamical systems in the past two decades. For a few references, we refer the reader to recent works for various problems in analysis and control of 2-D systems [43–62].

Due to substantial differences in the theory of 1-D and 2-D systems, existing methodologies established for conventional 1-D systems cannot be directly adaptable to analysis of 2-D processes since such an approach ignores its inherent to 2-D systems structure. In addition, a 2-D framework typically has more complex structures and available preliminaries associated to 2-D systems are insufficient. This makes the analysis and synthesis of 2-D systems much more complicated than those of 1-D systems.

In this book, we present a systematic study with fundamental concepts, effective techniques and schemes developed for various classes of 2-D time-delay systems. More specifically, we complement the systems and control theory for 2-D time-delay systems with popular problems such as

- Stability and performance analysis.
- Controller design under state/output feedback and event-triggered control schemes.
- State estimation and reachable set bounding.
- Input-output attenuation analysis subject to exogenous disturbances and noises.
- Analysis and control in finite-time regions.

1.2 Motivation Examples of 2-D Systems

In this section, we present some numerical examples that lead to some models of 2-D systems.

Example 1.1 Consider a schematic control system as given in Fig. 1.1. This model can be used to describe various thermal processes in chemical reactors, heat exchanger or pipe furnaces which is described by the following Darboux-type equation [1]

$$\begin{aligned} \frac{\partial T(x,t)}{\partial x} + \frac{\partial T(x,t)}{\partial t} &= -aT(x,t) + bu(x,t), \\ z(x,t) &= cT(x,t), \end{aligned} \tag{1.1}$$

where $T(x,t)$ is the unknown function representing, for example, temperature at space $x \in [0,d]$ and time t, $u = u(x,t)$ is the control input, $z(x,t)$ is the measurement output, and a, b, c are real constants.

We now discretize equation (1.1) as follows. For given increments Δx, Δt, by a simple forward Euler approximation

$$\frac{\partial T(x,t)}{\partial x} \approx \frac{T(i,j) - T(i-1,j)}{\Delta x}, \quad \frac{\partial T(x,t)}{\partial t} \approx \frac{T(i,j+1) - T(i,j)}{\Delta t},$$

and $T(i,j) = T(i\Delta x, j\Delta t)$, $z(i,j) = z(i\Delta x, j\Delta t)$, $u(i,j) = u(i\Delta x, j\Delta t)$, Eq. (1.1) can be written in the form

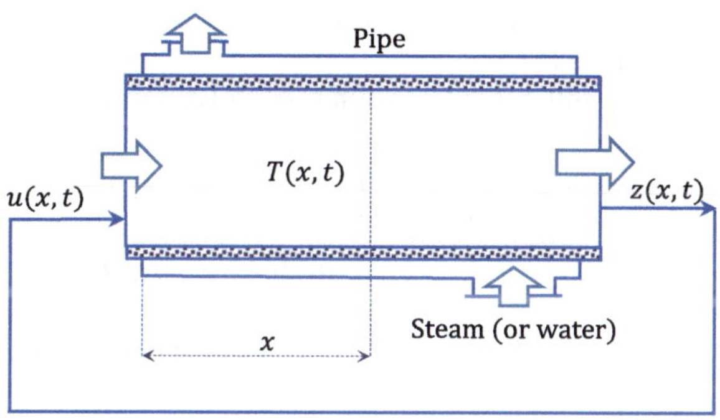

Fig. 1.1 A schematic heat exchanger control

$$
\begin{aligned}
T(i, j+1) &= \left(1 - \frac{\Delta t}{\Delta x} - a\Delta t\right) T(i, j) + \frac{\Delta t}{\Delta x} T(i-1, j) + b\Delta t u(i, j), \\
z(i, j) &= cT(i, j).
\end{aligned}
\tag{1.2}
$$

A discrete-time static output feedback controller is designed in the form

$$
u(i, j) = kz(i, j) = kcT(i, j).
$$

We define the two state variables $x^h(i, j) = T(i-1, j)$ and $x^v(i, j) = T(i, j)$. Then, the closed-loop system of (1.2) can be obtained as

$$
\begin{bmatrix} x^h(i+1, j) \\ x^v(i, j+1) \end{bmatrix} = \begin{bmatrix} 0 & 1 \\ \frac{\Delta t}{\Delta x} & 1 - \frac{\Delta t}{\Delta x} - a\Delta t + bkc\Delta t \end{bmatrix} \begin{bmatrix} x^h(i, j) \\ x^v(i, j) \end{bmatrix}. \tag{1.3}
$$

Equation (1.3) represents a 2-D system in the Roesser model. On the other hand, from (1.2), if we define $x(i, j) = \begin{bmatrix} T(i-1, j) \\ T(i, j) \end{bmatrix}$, then we have

$$
x(i+1, j+1) = A_1 x(i, j+1) + A_2 x(i+1, j), \tag{1.4}
$$

where

$$
A_1 = \begin{bmatrix} 0 & 1 \\ 0 & 0 \end{bmatrix}, \quad A_2 = \begin{bmatrix} 0 & 0 \\ \frac{\Delta t}{\Delta x} & 1 - \frac{\Delta t}{\Delta x} - a\Delta t + bkc\Delta t \end{bmatrix}.
$$

Equation (1.4) gives a 2-D system in the second Fornasini–Marchesini model.

Example 1.2 In this example, we consider a delayed control system described by the following partial differential equation

$$
\begin{aligned}
\frac{\partial T(x, t)}{\partial x} + \frac{\partial T(x, t)}{\partial t} &= -a_0 T(x, t) - a_1 T(x - d_x, t) + bu(x, t), \\
T_{\text{out}} &= cT(x, t),
\end{aligned}
\tag{1.5}
$$

where, similar to (1.1), $T(x, t)$ represents an unknown function, $u_{\text{in}} = u(x, t)$ is the control input, T_{out} is the measurement output, d_x is a quantity involving space delay, and a_0, a_1, b, c are real constants. The schematic model of system (1.5) is presented as in Fig. 1.2.

By a similar discretization used in Example 1.1, system (1.5) can be written as

$$
\begin{aligned}
T(i, j+1) = {} & \left(1 - \frac{\Delta t}{\Delta x} - a_0\Delta t\right) T(i, j) + \frac{\Delta t}{\Delta x} T(i-1, j) \\
& - a_1 \Delta t T(i - 1 - d_s(i), j) + b\Delta t u(i, j),
\end{aligned}
\tag{1.6}
$$

where $d_s(i) = \lfloor \frac{d_x}{\Delta x} \rfloor$ and $\lfloor . \rfloor$ denotes the floor function.

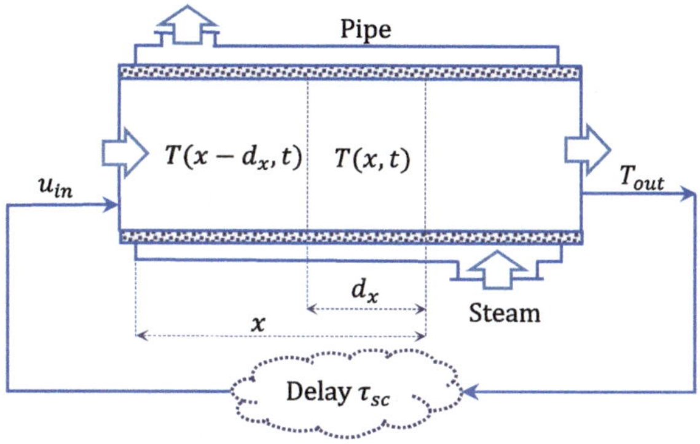

Fig. 1.2 A schematic delayed control system in heat exchanger

In control engineering, due to many practical reasons, for example, the finite speed of data processing through a low-rate communication channel or sensor technology, the impact of time-delay cannot be ignored. Assume that the communication delay occurs in the signal processing. Then, a delayed static output-feedback for system (1.6) is designed in the form

$$u(i, j) = kcT(i, j - d_c(j)).$$

We define the state variables $x^h(i, j) = T(i-1, j)$ and $x^v(i, j) = T(i, j)$. Then, the closed-loop system of (1.6) can be recast into the following system

$$\begin{bmatrix} x^h(i+1, j) \\ x^v(i, j+1) \end{bmatrix} = A \begin{bmatrix} x^h(i, j) \\ x^v(i, j) \end{bmatrix} + A_d \begin{bmatrix} x^h(i-d_h(i), j) \\ x^v(i, j-d_v(j)) \end{bmatrix}, \tag{1.7}$$

where the system matrices A and A_d are given by

$$A = \begin{bmatrix} 0 & 1 \\ \frac{\Delta t}{\Delta x} & 1 - \frac{\Delta t}{\Delta x} - a_0 \Delta t \end{bmatrix}, \quad A_d = \begin{bmatrix} 0 & 0 \\ -a_1 \Delta t & bkc\Delta t \end{bmatrix}.$$

System (1.7) represents a 2-D system with directional delays in the Roesser model.

Example 1.3 Consider the following control system described by a Darboux-type equation

$$\begin{aligned} \frac{\partial^2 w(x,t)}{\partial x \partial t} &= a_0 w(x,t) + a_1 \frac{\partial w(x,t)}{\partial t} + a_2 \frac{\partial w(x,t)}{\partial x} \\ &\quad + a_d w(x, t-\tau(t)) + bu(x,t), \end{aligned} \tag{1.8}$$

where $w(x, t)$ is an unknown function of variable x and time t, $\tau(t)$ represents a time-dependent delay, $u(x, t)$ is the control input, a_0, a_1, a_2, a_d, and b are real coefficients.

We denote

$$x^h(x,t) = \frac{\partial w(x,t)}{\partial t} - a_2 w(x,t), \quad x^v(x,t) = w(x,t),$$

then Eq. (1.8) can be written in the form

$$\begin{bmatrix} \frac{\partial x^h}{\partial x} \\ \frac{\partial x^v}{\partial t} \end{bmatrix} = \begin{bmatrix} a_1 & a_0 + a_1 a_2 \\ 1 & a_2 \end{bmatrix} \begin{bmatrix} x^h \\ x^v \end{bmatrix} + \begin{bmatrix} 0 & a_d \\ 0 & 0 \end{bmatrix} \begin{bmatrix} 0 \\ x_d^v \end{bmatrix} + \begin{bmatrix} b \\ 0 \end{bmatrix} u, \tag{1.9}$$

where $x_d^v(x,t) = w(x, t - \tau(t))$.

For given increments Δx, Δt, we define the state variables

$$\begin{aligned} x^h(i,j) &= x^h(i\Delta x, j\Delta t), \quad x^v(i,j) = x^v(i\Delta x, j\Delta t), \\ u(i,j) &= u(i\Delta x, j\Delta t), \quad d_v(j) = \lfloor \tau(j\Delta t) \rfloor. \end{aligned}$$

Then, by a simple Euler approximation scheme, Eq. (1.9) can be written as

$$\begin{bmatrix} x^h(i+1,j) \\ x^v(i,j+1) \end{bmatrix} = A \begin{bmatrix} x^h(i,j) \\ x^v(i,j) \end{bmatrix} + A_d \begin{bmatrix} x^h(i - d_h(i), j) \\ x^v(i, j - d_v(j)) \end{bmatrix} + Bu(i,j), \tag{1.10}$$

where the system matrices are given by

$$A = \begin{bmatrix} 1 + a_1\Delta x & (a_0 + a_1 a_2)\Delta x \\ \Delta t & 1 + a_2 \Delta t \end{bmatrix}, \quad A_d = \begin{bmatrix} 0 & a_d \Delta x \\ 0 & 0 \end{bmatrix}, \quad B = \begin{bmatrix} b\Delta x \\ 0 \end{bmatrix}.$$

Once again, we obtain a Roesser model of 2-D control systems with state delays.

Example 1.4 Consider a thermal process described by the following Darboux-type equation

$$\begin{aligned} \frac{\partial w(s,t)}{\partial x} + \frac{\partial w(x,t)}{\partial t} &= a_0 w(x,t) + a_1 w(x - d_x, t) + a_2 w(x, t - d_t) \\ &\quad + c_1 \int_0^{x_f} w(x - s, t) ds + c_2 \int_0^{t_f} w(x, t - \theta) d\theta, \end{aligned} \tag{1.11}$$

where $w(x,t)$ represents the unknown function at space x and time t, a_0, a_1, a_2, and c_1, c_2 are real scalars, d_x, d_t, and x_f, t_f represent delay terms.

For given increments Δx and Δt, denote $w(i,j) = w(i\Delta x, j\Delta t)$. The derivatives $\frac{\partial w(x,t)}{\partial x}$ and $\frac{\partial w(x,t)}{\partial t}$ will be approximated using a backward difference scheme as

$$\begin{aligned} \frac{\partial w(x,t)}{\partial x} &\simeq \frac{w(i,j) - w(i-1,j)}{\Delta x}, \\ \frac{\partial w(x,t)}{\partial t} &\simeq \frac{w(i,j) - w(i,j-1)}{\Delta t}. \end{aligned}$$

Define $x^h(i, j) = w(i-1, j)$ and $x^v(i, j) = w(i, j)$. Then, the Eq. (1.11) can be transformed into the following 2-D system

$$\begin{bmatrix} 1 & 0 \\ 0 & 0 \end{bmatrix} \begin{bmatrix} x^h(i+1, j) \\ x^v(i, j+1) \end{bmatrix} = A \begin{bmatrix} x^h(i, i) \\ x^v(i, j) \end{bmatrix} + A_d \begin{bmatrix} x^h(i - d_h(i), j) \\ x^v(i, j - d_v(j)) \end{bmatrix} + A_s \begin{bmatrix} \sum_{l=1}^{s_h} x^h(i-l, j) \\ \sum_{l=1}^{s_v} x^v(i, j-l) \end{bmatrix}, \tag{1.12}$$

where $d_h(i) = \lfloor d_x/\Delta x \rfloor$, $d_v(j) = \lfloor d_t/\Delta t \rfloor$, $s_h = \lfloor x_f/\Delta x \rfloor$ and $s_v = \lfloor t_f/\Delta t \rfloor$.

System (1.12) represents a 2-D singular system with mixed discrete- and distributed-delays.

1.3 Basic Models of 2-D Systems

In this section, we introduce some basic models of 2-D systems and the relation between them.

1.3.1 Roesser Model

In the existing literature, the Roesser model of 2-D systems is widely studied due to its natural structure, which can be easily adaptive to various practical systems. The Roesser model of a 2-D system without delays is given by

$$\begin{aligned} \begin{bmatrix} x^h(i+1, j) \\ x^v(i, j+1) \end{bmatrix} &= \begin{bmatrix} A_{11} & A_{12} \\ A_{21} & A_{22} \end{bmatrix} \begin{bmatrix} x^h(i, j) \\ x^v(i, j) \end{bmatrix} + \begin{bmatrix} B_1 \\ B_2 \end{bmatrix} u(i, j), \\ y(i, j) &= \begin{bmatrix} C_1 & C_2 \end{bmatrix} \begin{bmatrix} x^h(i, j) \\ x^v(i, j) \end{bmatrix} + Du(i, j), \end{aligned} \tag{1.13}$$

where $x^h(i, j) \in \mathbb{R}^{n_h}$ and $x^v(i, j) \in \mathbb{R}^{n_v}$ represent the horizontal and vertical state vectors, respectively, $u(i, j) \in \mathbb{R}^m$ is the control input, $y(i, j) \in \mathbb{R}^{n_o}$ is the measurement output vector, and $A_{11}, A_{12}, A_{21}, A_{22}, B_1, B_2, C_1, C_2, D$ are given matrices of appropriate dimensions.

The initial condition of system (1.13) is determined by sequences $\phi(j)$ and $\psi(i)$ (also known as boundary valued sequences)

$$x^h(0, j) = \phi(j), \quad j \geq 0, \text{ and } x^v(i, 0) = \psi(i), \quad i \geq 0. \tag{1.14}$$

In many existing works, the sequences ϕ and ψ are assumed to have finite support, that is, there exist positive integers T_h and T_v such that

$$\phi(j) = 0 \text{ for } j \geq T_h \text{ and } \psi(i) = 0 \text{ for } i \geq T_v,$$

or, more general, ϕ and ψ belong to the spaces $l_2(\mathbb{R}^{n_h})$ and $l_2(\mathbb{R}^{n_v})$, that is

$$X_0 = \lim_{T\to\infty} \sum_{k=0}^{T} \left(\phi^2(k) + \psi^2(k)\right) < \infty.$$

1.3.2 Attasi Model

Attasi model of 2-D systems can be used in various applications in image processing or digital data processing [6]. This model is given as

$$\begin{aligned} x(i+1, j+1) &= A_1 x(i, j+1) + A_2 x(i+1, j) - A_2 A_1 x(i, j) + B_0 u(i, j), \\ y(i, j) &= C x(i, j), \end{aligned} \tag{1.15}$$

where A_1, A_2, B_0, and C are given real matrices of appropriate dimensions and $A_1 A_2 = A_2 A_1$.

1.3.3 Fornasini–Marchesini Models

Fornasini–Marchesini (F-M) models are widely used in the area of control and signal processing [8]. The first F-M model is described by the following system

$$\begin{aligned} x(i+1, j+1) &= A_0 x(i, j) + A_1 x(i, j+1) + A_2 x(i+1, j) + B u(i, j), \\ y(i, j) &= C x(i, j), \end{aligned} \tag{1.16}$$

and the second F-M model is given by

$$\begin{aligned} x(i+1, j+1) &= A_1 x(i, j+1) + A_2 x(i+1, j) \\ &\quad + B_1 u(i, j+1) + B_2 u(i+1, j), \\ y(i, j) &= C x(i, j). \end{aligned} \tag{1.17}$$

The general F-M model is given as

$$\begin{aligned} x(i+1, j+1) &= A_0 x(i, j) + A_1 x(i, j+1) + A_2 x(i+1, j) + B_0 u(i, j) \\ &\quad + B_1 u(i, j+1) + B_2 u(i+1, j), \\ y(i, j) &= C x(i, j). \end{aligned} \tag{1.18}$$

1.3.4 Transformations of Basic 2-D Models

It is clear that Attasi model is a special case of the first F-M model for $A_0 = -A_1 A_2$ and $B_0 = B$. In addition, by using the state transformation

$$x^h(i, j) = x(i, j+1) - A_2 x(i, j) \text{ and } x^v(i, j) = x(i, j). \tag{1.19}$$

From (1.15), we have

$$\begin{aligned} x^h(i+1, j) &= x(i+1, j+1) - A_2 x(i+1, j) \\ &= A_1 x(i, j+1) - A_2 A_1 x(i, j) + B_0 u(i, j) \\ &= A_1 x^h(i, j) + B_0 u(i, j) \end{aligned}$$

and

$$x^v(i, j+1) = x(i, j+1) = x^h(i, j) + A_2 x^v(i, j).$$

Therefore, Eq. (1.15) can always be written in the form

$$\begin{aligned} \begin{bmatrix} x^h(i+1, j) \\ x^v(i, j+1) \end{bmatrix} &= \begin{bmatrix} A_1 & 0_n \\ I_n & A_2 \end{bmatrix} \begin{bmatrix} x^h(i, j) \\ x^v(i, j) \end{bmatrix} + \begin{bmatrix} B_0 \\ 0_{n\times m} \end{bmatrix} u(i, j), \\ y(i, j) &= \begin{bmatrix} 0_{n_o\times n} & C \end{bmatrix} \begin{bmatrix} x^h(i, j) \\ x^v(i, j) \end{bmatrix}, \end{aligned} \tag{1.20}$$

which is clearly a special case of Roesser model (1.13).

On the other hand, base on the transformation Eqs. (1.19) and (1.16) can be represented in the form

$$\begin{aligned} \begin{bmatrix} x^h(i+1, j) \\ x^v(i, j+1) \end{bmatrix} &= \begin{bmatrix} A_1 & A_0 + A_1 A_2 \\ I_n & A_2 \end{bmatrix} \begin{bmatrix} x^h(i, j) \\ x^v(i, j) \end{bmatrix} + \begin{bmatrix} B \\ 0_{n\times m} \end{bmatrix} u(i, j), \\ y(i, j) &= \begin{bmatrix} 0_{n_o\times n} & C \end{bmatrix} \begin{bmatrix} x^h(i, j) \\ x^v(i, j) \end{bmatrix} + Du(i, j). \end{aligned} \tag{1.21}$$

Thus, by the state transformation (1.19), the first F-M model can be recast into the Roesser model (1.13).

Now, we define the state vector

$$\hat{x}(i, j) = \begin{bmatrix} x^\top(i, j) & x^\top(i, j-1) & u^\top(i, j-1) \end{bmatrix}^\top .$$

Then, system (1.16) becomes

$$
\begin{aligned}
\hat{x}(i+1, j+1) &= \left[\begin{array}{c|c} \begin{matrix} A_2 \\ I_n \end{matrix} & 0_{2n} \\ \hline 0_n & 0_{n\times 2n} \end{array}\right] \hat{x}(i+1, j) + \left[\begin{array}{cc|c} A_1 & A_0 & B \\ \hline \multicolumn{2}{c|}{0_{2n}} & 0_{2n\times m} \end{array}\right] \hat{x}(i, j+1) \\
&\quad + \left[\begin{array}{c} 0_{2n\times n} \\ \hline I_n \end{array}\right] u(i, j+1) + 0_{3n\times m} u(i, j+1), \\
y(i, j) &= \begin{bmatrix} C & 0_{n_o\times 2n} \end{bmatrix} \hat{x}(i, j).
\end{aligned} \tag{1.22}
$$

Therefore, the first F-M model can be embedded into the second F-M model.

Finally, consider the Roesser model (1.13). We define the state vector $x(i, j) = \begin{bmatrix} x^h(i, j) \\ x^v(i, j) \end{bmatrix}$ then system (1.13) can be transformed into the following model

$$
\begin{aligned}
x(i+1, j+1) &= \begin{bmatrix} A_{11} & A_{12} \\ 0_{n_2\times n_1} & 0_{n_2\times n_2} \end{bmatrix} x(i, j+1) + \begin{bmatrix} 0_{n_1\times n_2} & 0_{n_2\times n_2} \\ A_{21} & A_{22} \end{bmatrix} x(i+1, j) \\
&\quad \begin{bmatrix} B_1 \\ 0_{n_2\times m} \end{bmatrix} u(i, j+1) + \begin{bmatrix} 0_{n_1\times m} \\ B_2 \end{bmatrix} u(i+1, j), \\
y(i, j) &= Cx(i, j),
\end{aligned} \tag{1.23}
$$

where $C = \begin{bmatrix} C_1 & C_2 \end{bmatrix}$. This shows that the Roesser model (1.13) can be represented as a second F-M model.

Figure 1.3 below shows a relation between basic models of 2-D systems.

It can be seen that, in some sense, basic models of 2-D systems can be mutually recast into other models. In addition, the second F-M model can be regarded as one of the most general models. However, it should be emphasized that the use of a model that is transformed from another one might not be adequate from numerical point of view as it might give ill-conditioned problems. For example, in the second F-M model (1.23) transformed from Roesser model (1.13), the size of the resulting matrices considerably increases by adding zero rows. Thus, the computational burden gets much higher. Moreover, for 2-D systems with more complex structures, for

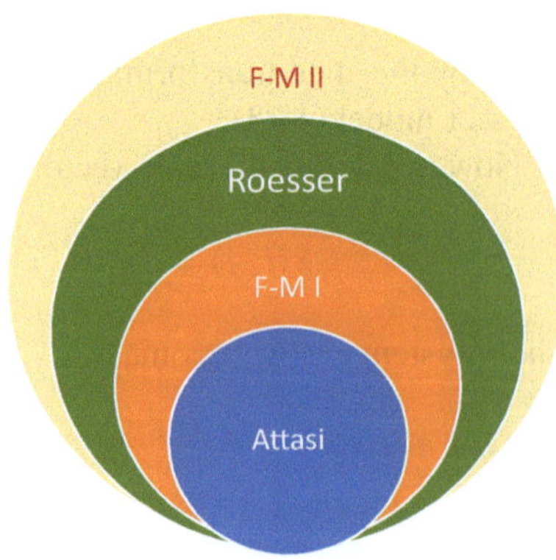

Fig. 1.3 Relation between basic models of 2-D systems

instance, 2-D systems in the presence of time-delays, uncertainties, and exogenous deterministic and/or stochastic disturbances, the aforementioned models cannot be transformed to other ones. Thus, it is generally worth working always with a model that resembles more the original 2-D problem. In particular, 2-D systems described by Roesser model and F-M models are often studied separately due to the nature of their applications and special structures.

1.4 Stability and Stabilization of Basic 2-D Roesser Systems

1.4.1 Stability Conditions

Consider the following 2-D Roesser system

$$\begin{bmatrix} x^h(i+1,j) \\ x^v(i,j+1) \end{bmatrix} = \underbrace{\begin{bmatrix} A_{11} & A_{12} \\ A_{21} & A_{22} \end{bmatrix}}_{\mathscr{A}} \begin{bmatrix} x^h(i,j) \\ x^v(i,j) \end{bmatrix}, \quad i, j \in \mathbb{N}_0, \tag{1.24}$$

where $x^h(i,j) \in \mathbb{R}^{n_h}$ and $x^v(i,j) \in \mathbb{R}^{n_v}$ are the horizontal and vertical state vectors, respectively, $\mathscr{A} \in \mathbb{R}^{n \times n}$, $n = n_h + n_v$, is a given real matrix.

Let $x(i,j) = \begin{bmatrix} x^h(i,j) \\ x^v(i,j) \end{bmatrix}$ denote the system state vector. The initial condition of (1.24) is determined by sequences $\phi \in l_2(\mathbb{R}^{n_h})$, $\psi \in l_2(\mathbb{R}^{n_v})$ as

$$\begin{aligned} & x^h(0,j) = \phi(j), \quad j \geq 0, \quad x^v(i,0) = \psi(i), \quad i \geq 0, \\ & \|x_0^h\|_2^2 = \sum_{j=0}^{\infty} \|\phi(j)\|^2 < \infty, \quad \|x_0^v\|_2^2 = \sum_{i=0}^{\infty} \|\psi(i)\|^2 < \infty. \end{aligned} \tag{1.25}$$

Definition 1.1 System (1.24) is said to be globally asymptotically stable (GAS) if any solution $x(i,j)$ of (1.24) with initial condition (1.25) satisfies

$$\lim_{r \to \infty} X_r = \lim_{r \to \infty} \left\{ \sup_{i+j=r} \|x(i,j)\| \right\} = 0.$$

The characteristic polynomial of system (1.24) is given by

$$C(z_1, z_2) = \det \begin{bmatrix} I_{n_h} - z_1 A_{11} & -z_1 A_{12} \\ -z_2 A_{21} & I_{n_v} - z_2 A_{22} \end{bmatrix}.$$

The following theorem gives a stability criterion for system (1.24) based on the characteristic polynomial.

Theorem 1.1 (see [1]) *The 2-D Roesser system* (1.24) *is GAS if and only if* $C(z_1, z_2) \neq 0$ *for all* (z_1, z_2) *belongs to the close disk* $\overline{U} \triangleq \{(z_1, z_2) \in \mathbb{C}^2 : |z_1| \leq 1, |z_2| \leq 1\}$.

Based on the criterion derived in Theorem 1.1, we have the following result.

Theorem 1.2 *System* (1.24) *is GAS if there exists a symmetric positive definite matrix* $P = \text{diag}(P^h, P^v) \in \mathbb{S}_n^+$ *that satisfies the following LMI condition*

$$\mathscr{A}^\top P \mathscr{A} - P < 0. \tag{1.26}$$

Proof By virtue of the contrary arguments, assume that there is a point $(z_1, z_2) \in \overline{U}$ such that

$$\begin{vmatrix} I_{n_h} - z_1 A_{11} & -z_1 A_{12} \\ -z_2 A_{21} & I_{n_v} - z_2 A_{22} \end{vmatrix} = 0.$$

Then there exists a nonzero vector $x \in \mathbb{R}^{n \times n}$ that satisfies

$$\begin{bmatrix} I_{n_h} - z_1 A_{11} & -z_1 A_{12} \\ -z_2 A_{21} & I_{n_v} - z_2 A_{22} \end{bmatrix} x = 0.$$

Thus,

$$x = \mathbb{I}(z_1, z_2) \mathscr{A} x.$$

Denote $Q = P - \mathscr{A}^\top P \mathscr{A} > 0$ and let x^*, z^* be the transpose conjugate vectors of x and $z = (z_1, z_2)$. Then, we have

$$\begin{aligned} x^* P x &= x^* \mathscr{A}^\top \mathbb{I}(z_1^*, z_2^*) P \mathbb{I}(z_1, z_2) \mathscr{A} x \\ &= x^* \mathscr{A}^\top \mathbb{I}(|z_1|^2, |z_2|^2) P \mathscr{A} x \\ &= x^* P x - x^* Q x - x^* \mathscr{A}^\top \mathbb{I}(1 - |z_1|^2, 1 - |z_2|^2) P \mathscr{A} x \\ &\leq x^* P x - x^* Q x, \end{aligned} \tag{1.27}$$

where, for scalars α, β, $\mathbb{I}(\alpha, \beta) = \text{diag}(\alpha I_{n_h}, \beta I_{n_v})$.

The last inequality in (1.27) implies that $x^* Q x = 0$ and hence $x = 0$ since $Q > 0$. This clearly yields a contradiction with $x \neq 0$. The proof is completed. □

Proposition 1.1 (A 2-D Lyapunov function method) *Assume that there exist functionals* $V^h(x^h)$, $V^v(x^v)$, *and positive scalars* ρ, $\underline{\lambda}_h$, $\overline{\lambda}_h$, $\underline{\lambda}_v$, *and* $\overline{\lambda}_v$ *that satisfy the following conditions*

(i) *For any* $x^h \in \mathbb{R}^{n_v}$ *and* $x^v \in \mathbb{R}^{n_v}$, *we have*

$$\underline{\lambda}_h \|x^h\|^2 \leq V^h(x^h) \leq \overline{\lambda}_h \|x^h\|^2, \quad \underline{\lambda}_v \|x^v\|^2 \leq V^v(x^v) \leq \overline{\lambda}_v \|x^v\|^2;$$

(ii) *The partial difference of the functional* $V(x) = V^h(x^h) + V^v(x^v)$ *along any state trajectory* $x(i, j) = \left[x^{h\top}(i, j)\ x^{v\top}(i, j)\right]$ *satisfies*

$$\begin{aligned}\Delta V(x(i, j)) &\triangleq \Delta_h(V^h(x^h(i, j))) + \Delta_v(V^v(x^v(i, j))) \\ &\leq -\rho\|x(i, j)\|^2,\end{aligned} \tag{1.28}$$

where the directional differences $\Delta_h(V^h(x^h(i, j)))$ *and* $\Delta_v(V^v(x^v(i, j)))$ *are defined as*

$$\begin{aligned}\Delta_h(V^h(x^h(i, j))) &= V^h(x^h(i+1, j)) - V^h(x^h(i, j)), \\ \Delta_v(V^v(x^v(i, j))) &= V^v(x^v(i, j+1)) - V^v(x^v(i, j)).\end{aligned}$$

Then, system (1.24) *is GAS.*

Proof For any positive integers T, K, by taking summation both sides of (1.28), we then obtain

$$\begin{aligned}\sum_{j=0}^{K}\left[V^h(x^h(T, j) - V^h(x^h(0, j))\right] + \sum_{i=0}^{T}\left[V^v(x^v(i, K)) - V^v(x^v(i, 0))\right] \\ \leq -\rho\sum_{i=0}^{T-1}\sum_{j=0}^{K-1}\|x(i, j)\|^2.\end{aligned} \tag{1.29}$$

Cho $T, K \to \infty$, from (1.29) and condition (i), we can get

$$\sum_{i=0}^{\infty}\sum_{j=0}^{\infty}\|x(i, j)\|^2 \leq \frac{\overline{\lambda}_h\|\phi\|_{l_2}^2 + \overline{\lambda}_v\|\psi\|_{l_2}^2}{\rho}.$$

It can be deduced from the last estimation that system (1.24) is GAS. The proof is completed. □

Remark 1.1 The LMI-based stability condition derived in (1.26) can be formulated by using the 2-D Lypunov function $V(x) = V^h(x^h) + V^v(x^v)$, where

$$V^h(x^h) = x^{h\top}P^h x^h, \quad V^v(x^v) = x^{v\top}P^v x^v.$$

Indeed, let $\mathscr{A}_1 = \left[A_{11}\ \ A_{12}\right]$ and $\mathscr{A}_2 = \left[A_{21}\ \ A_{22}\right]$ then, from (1.24), we have

$$x^h(i+1, j) = \mathscr{A}_1 x(i, j) \text{ and } x^v(i, j+1) = \mathscr{A}_2 x(i, j).$$

Therefore,

$$\begin{aligned}\Delta V(x(i,j)) &= x^{\top}(i,j)\left(\mathscr{A}_1^{\top}P^h\mathscr{A}_1 + \mathscr{A}_2^{\top}P^v\mathscr{A}_2\right)x(i,j)\\ &\quad - x^{h\top}(i,j)P^hx^h(i,j) - x^{v\top}(i,j)P^vx^v(i,j)\\ &= x^{\top}(i,j)\left(\begin{bmatrix}\mathscr{A}_1\\ \mathscr{A}_2\end{bmatrix}^{\top} P \begin{bmatrix}\mathscr{A}_1\\ \mathscr{A}_2\end{bmatrix}\right)x(i,j) - x^{\top}(i,j)Px(i,j)\\ &= x^{\top}(i,j)\left(\mathscr{A}^{\top}P\mathscr{A} - P\right)x(i,j).\end{aligned} \tag{1.30}$$

The last inequality in (1.30) shows that condition is satisfied if and only if

$$\mathscr{A}^{\top}P\mathscr{A} - P < 0$$

which coincides with the LMI-based condition (1.28). According to Proposition 1.1, system (1.24) is GAS. The proof is completed.

1.4.2 Stabilization of 2-D Roesser Systems

Consider a 2-D system as (1.24) with control input, which is represented as follows

$$\begin{bmatrix}x^h(i+1,j)\\ x^v(i,j+1)\end{bmatrix} = \begin{bmatrix}A_{11} & A_{12}\\ A_{21} & A_{22}\end{bmatrix}\begin{bmatrix}x^h(i,j)\\ x^v(i,j)\end{bmatrix} + \begin{bmatrix}B_1\\ B_2\end{bmatrix}u(i,j), \tag{1.31}$$

where $u(i,j) \in \mathbb{R}^{n_u}$ is the control input vector and $\mathscr{B} = \begin{bmatrix}B_1\\ B_2\end{bmatrix} \in \mathbb{R}^{n \times n_u}$ is a given matrix.

We illustrate the stabilization problem of system (1.31) via state-feedback controllers (SFCs). More specifically, we design an SFC in the form

$$u(i,j) = Kx(i,j) \tag{1.32}$$

that makes the resulting closed-loop system stable, where $K \in \mathbb{R}^{n_u \times n}$ is the controller gain matrix which will be determined. By incorporating the controller (1.32), the closed-loop system (1.31) is obtained as

$$\begin{bmatrix}x^h(i+1,j)\\ x^v(i,j+1)\end{bmatrix} = \mathscr{A}_c\begin{bmatrix}x^h(i,j)\\ x^v(i,j)\end{bmatrix}, \tag{1.33}$$

where $\mathscr{A}_c = \mathscr{A} + \mathscr{B}K$.

According to condition (1.26), the closed-loop system (1.33) is GAS if there exists a matrix $P = \mathrm{diag}(P^h, P^v) \in \mathbb{S}_+^n$ such that

$$(\mathscr{A} + \mathscr{B}K)^\top P (\mathscr{A} + \mathscr{B}K) - P < 0. \tag{1.34}$$

Denote $X = P^{-1}$. By pre- and post-multiplying with X, condition (1.34) holds if and only if

$$X (\mathscr{A} + \mathscr{B}K)^\top X^{-1} (\mathscr{A} + \mathscr{B}K) X - X < 0. \tag{1.35}$$

By the Schur complement lemma [see, Chap. 2, Lemma 2.7 for more details], condition (1.35) is equivalent to the following one

$$\begin{bmatrix} -X & X\mathscr{A}^\top + XK^\top\mathscr{B}^\top \\ \mathscr{A}X + \mathscr{B}KX & -X \end{bmatrix} < 0. \tag{1.36}$$

Finally, by changing the variable $KX = Z \in \mathbb{R}^{n_u \times n}$, condition (1.36) is reduced to the following LMI

$$\begin{bmatrix} -X & X\mathscr{A}^\top + Z^\top\mathscr{B}^\top \\ \mathscr{A}X + \mathscr{B}Z & -X \end{bmatrix} < 0. \tag{1.37}$$

In summary, we have the following result.

Theorem 1.3 *The control system* (1.31) *is stabilizable via SFCs if there exist a symmetric positive definite matrix* $X \in \mathbb{S}_+^n$ *and a real matrix* $Z \in \mathbb{R}^{n_u \times n}$ *that satisfy the LMI condition* (1.37). *A stabilizing controller of the SFC* (1.32) *is obtained as*

$$K = ZX^{-1},$$

where (X, Z) *is a feasible solution of* (1.37).

1.5 Preview of Chapters

The remaining of the monograph is composed of ten chapters, Chaps. 2–11. Each subsequent chapter is concerned with particular problems. A summary of each chapter in this monograph is given below.

Chapter 2 investigates the problems of stability analysis and controller design for some classes of discrete-time 2-D systems with directional state delays. An analysis scheme extended from the conventional LKF method is first developed. The proposed scheme is then utilized in combination with some 2-D summation inequalities to derive LMI-based delay-dependent stability conditions. The obtained results are employed to address the stabilization problem of 2-D time-delay control systems via SFCs and event-triggered state-feedback controllers.

Chapter 3 deals with the asymptotic estimation problem of the reachable set of 2-D time-delay systems with exogenous disturbances. The directional time-varying delays are confined within certain ranges and the input disturbances are norm bounded. Based on a newly proposed scheme, and by utilizing some 2-D weighted summation inequalities, delay-dependent tractable conditions are derived to guarantee the existence of an ellipsoid that attracts exponentially all system states in the presence of bounded disturbances. An application to exponential stability of 2-D systems with delays is also presented.

In the first part of Chap. 4, based on the monotone dynamical systems theory, necessary and sufficient delay-independent exponential stability conditions are formulated for 2-D nonlinear systems in the Roesser model with time-varying delays. Applications to the stability analysis of positive 2-D systems with delays are also obtained. The second part of Chap. 4 focuses on the design problem of positive observers for 2-D positive linear systems with delays. Extended Luenberger-type observers that generate nonnegative asymptotic estimations of state vectors are constructed and a reduced-order functional state observer that is an asymptotic approximation of a given linear functional of the state vector is also taken into consideration for 2-D positive time-delay systems. Necessary and sufficient design conditions are derived using linear programming (LP) approach.

Chapter 5 addresses the problem of l_1-gain control for positive 2-D systems with directional delays. Exact value characterizations of l_1-induced norm of the input-output operators are obtained using 2-D Z-transform approach. The obtained l_1-gain characterizations are then utilized to establish necessary and sufficient conditions in LP settings for l_1-induced performance with a prescribed attenuation level. Finally, by utilizing a linear convex optimization approach, necessary and sufficient conditions for the existence of an SFC that makes the closed-loop system positive, stable and has prescribed l_1-induced performance are formulated.

In Chap. 6, the problems of peak-to-peak performance analysis and control are studied for 2-D positive linear systems with interval time-varying delays. Novel comparison techniques involving interval state estimations subject to peak values of exogenous disturbances are presented and a characterization of l_∞-induced norm of the input-output operator is formulated. The obtained l_∞-gain characterization is then utilized to establish necessary and sufficient LP-based conditions for l_∞-induced performance with a prescribed attenuation level. Applications to l_∞-gain control via static output-feedback controllers (SOFCs) are also discussed.

Chapter 7 is devoted to the stability problem of 2-D Markov jump Roesser systems with interval delays and uncertain transition rates. More precisely, the transition probabilities of the jumping process driven by a Markov chain are assumed to be not exactly known but can be estimated by prescribed ranges. An analysis scheme for the stochastic stability property is first developed. The proposed scheme is then utilized in combination with weighted 2-D summation inequalities to derive delay-dependent stochastic stability conditions.

Chapter 8 investigates the problems of stochastic stability, robust stabilization via synchronous state-feedback, and observer-based control for some classes of 2-D Roesser systems with state delays and multiplicative stochastic noises. An analysis scheme is first presented and utilized to derive delay-dependent stability conditions. Applications to the design problem of an observer-based control for 2-D systems with exogenous disturbances and random packet dropouts is then presented. Extending to 2-D systems with multiplicative noises, the stabilization problem via synchronous SFCs is also discussed for 2-D Roesser systems driven by inhomogeneous Markov chains.

Chapter 9 studies the stability and stabilization problems via fuzzy rule-based SFCs for 2-D Takagi–Sugeno fuzzy systems in the Roesser model with interval state delays. An augmented fuzzy parameter-dependent LKF candidate is constructed and employed to establish delay-dependent LMI-based stability conditions. On the basis of the obtained stability results, the design problem of a stabilizing rule-based SFC is addressed.

Chapter 10 considers the stability and stabilization problems via SFCs and event-triggered SFCs for 2-D singular Roesser systems with state delays. By utilizing a free matrix equations technique, delay-dependent conditions that guarantee the regularity, causallity and internal stability of 2-D singular Roesser systems with time-varying delays are first derived. The obtained stability conditions are then employed to address the design problem of event-triggered SFCs for 2-D singular Roesser systems with delays.

In Chap. 11, the problems of stability and dissipativity analysis in finite-region are developed for 2-D singular Roesser systems with mixed time-varying delays. Based on a 2-D LKF method via weighted LKF candidates, and by utilizing zero-type free matrix equations, delay-dependent conditions are derived in LMI settings to guarantee the finite-region stability and dissipativity with respect to quadratic supplied rate functions.

References

1. Kaczorek, T.: Two-Dimensional Linear Systems. Springer, Berlin (1985)
2. Rogers, E., Galkowski, K., Paszke, W., Moore, K.L., Bauer, P.H., Hladowski, L., Dabkowski, P.: Multidimensional control systems: case studies in design and evaluation. Multidim Syst. Sig. Process **26**, 895–939 (2015)
3. Chen, C.W., Tsai, J.S., Shieh, L.S.: Modeling of variable coefficient Roesser model for systems described by second-order partial differential equation. Circuit Syst. Sig. Process **22**, 423–463 (2003)
4. Marszałek, W., Kekkerist, G.T.: Heat exchangers and linear image processing theory. Int. J. Heat Mass Trans. **32**, 2363–2374 (1989)
5. Bors, D., Walczak, S.: Application of 2D systems to investigation of a process of gas filtration. Multidim Syst. Sig. Process **23**, 119–130 (2012)
6. Lu, W.S.: Two-Dimensional Digital Filters. Marcel-Dekker, New York (1992)
7. Kurek, J.E., Zaremba, M.B.: Iterative learning control synthesis based on 2-D system theory. IEEE Trans. Autom. Control **38**, 121–125 (1993)
8. Du, C., Xie, L.: H_∞ Control and Filtering of Two-Dimensional Systems. Springer-Verlag, Berlin (2002)
9. Bose, N.K., Buchberger, B., Guiver, J.P.: Multidimensional Systems Theory and Applications. Springer, Berlin (2003)

10. Bristow, D.A., Tharayil, M., Alleyne, A.G.: A survey of iterative learning control: A learning based method for high performance tracking control. IEEE Control Syst. Magaz. **26**, 96–114 (2006)
11. Ahn, H.S., Chen, Y.Q., Moore, K.L.: Iterative learning control: Brief survey and categorization. IEEE Trans. Syst. Man Cybern: Appl. Rev. **37**, 1099–1121 (2007)
12. Freeman, C.T., Rogers, E., Hughes, A.M., Burridge, J.H., Meadmore, K.L.: Iterative learning control in healthcare electrical stimulation and robotic-assisted upper limb stroke rehabilitation. IEEE Control Syst. Magaz. **32**, 18–43 (2012)
13. Katz, E., Bar-Ness, Y.: Two-dimensional (2-D) spatial domain modulation methods for unipolar pixelated optical wireless communication systems. J. Lightwave Technol. **33**, 4233–4239 (2015)
14. Hoelzle, D.J., Barton, K.L.: On spatial iterative learning control via 2-D convolution: Stability analysis and computational efficiency. IEEE Trans. Control Syst. Technol. **24**, 1504–1512 (2016)
15. Li, X., Gao, H.: Robust finite frequency H_∞ filtering for uncertain 2-D Roesser systems. Automatica **48**, 1163–1170 (2012)
16. Ahn, C.K.: Overflow oscillation elimination of 2-D digital filters in the Roesser model with Wiener process noise. IEEE Sig. Process Lett. **21**, 1302–1305 (2014)
17. Wang, Z., Shang, H.: Observer based fault detection for two dimensional systems described by Roesser models. Multidim. Syst. Sig. Process **26**, 753–775 (2015)
18. Ahn, C.K., Shi, P., Basin, M.V.: Two-dimensional dissipative control and filtering for Roesser model. IEEE Trans. Autom. Control **60**, 1745–1759 (2015)
19. Ahn, C.K., Kar, H.: Expected power bound for two-dimensional digital filters in the Fornasini-Marchesini local state-space model. IEEE Sig. Process Lett. **22**, 1065–1069 (2015)
20. Bachelier, O., Paszke, W., Yeganefar, N., Mehdi, D., Cherifi, A.: LMI stability conditions for 2D Roesser models. IEEE Trans. Autom. Control **61**, 766–770 (2016)
21. Hien, L.V., Trinh, H.: Switching design for suboptimal guaranteed cost control of 2-D nonlinear switched systems in the Roesser model. Nonlinear Anal. Hybrid Syst. **24**, 45–57 (2017)
22. Shi, S., Fei, Z., Sun, W., Yang, X.: Stabilization of 2-D switched systems with all modes unstable via switching signal regulation. IEEE Trans. Autom. Control **63**, 857–863 (2018)
23. Roesser, R.P.: A discrete state-space model for linear image processing. IEEE Trans. Autom. Control **20**, 1–10 (1975)
24. Fornasini, E., Marchesini, G.: State-space realization theory of two-dimensional filters. IEEE Trans. Autom. Control **21**, 484–492 (1976)
25. Yang, R., Xie, L., Zhang, C.: H_2 and mixed H_2/H_∞ control of two-dimensional systems in Roesser model. Automatica **42**, 1507–1514 (2006)
26. Sipahi, R., Niculescu, S.-I., Abdallah, C.T., Michiels, W., Gu, K.: Stability and stabilization of systems with time delay. IEEE Control Syst. Magaz. **31**, 38–65 (2011)
27. Witrant, E., Fridman, E., Sename, O., Dugard, L.: Recent Results on Time-Delay Systems: Analysis and Control. Springer, Basel (2016)
28. Pham, T.N., Trinh, H., Hien, L.V., Wong, K.P.: Integration of electric vehicles for load frequency output feedback H_∞ control of smart grids. IET Gen. Trans. Distrib. **10**, 3341–3352 (2016)
29. Zhang, H., Yue, D., Zhao, W., Hu, S., Dou, C.: Distributed optimal consensus control for multiagent systems with input delay. IEEE Trans. Cybern. **48**, 1747–1759 (2018)
30. Li, X., Gao, H., Gu, K.: Delay-independent stability analysis of linear time-delay systems based on frequency discretization. Automatica **70**, 288–294 (2016)
31. Tian, Y., Wang, Z.: Composite slack-matrix-based integral inequality and its application to stability analysis of time-delay systems. Appl. Math. Lett. **120**, 107252 (2021)
32. Zeng, H.B., He, Y., Teo, K.L.: Monotone-delay-interval-based Lyapunov functionals for stability analysis of systems with a periodically varying delay. Automatica **138**, 110030 (2022)
33. Zhang, C.K., Chen, W.H., Zhu, C., He, Y., Wu, M.: Stability analysis of discrete-time systems with time-varying delay via a delay-dependent matrix- separation-based inequality. Automatica **156**, 111192 (2023)

34. Qiu, Y., Hua, C., Park, J.H., Wang, Y.: Interval approximation method for stability analysis of time-delay systems. J. Frankl. Inst. **360**, 3034–3046 (2023)
35. Sha, H., Park, J.H., Chen, J., Zhu, M., Nan, C.: Stability analysis of discrete-time systems with a time-varying delay via improved methods. IET Control Theory Appl. **18**, 951–959 (2024)
36. He, J., Han, Y., Yang, F.: Relaxed mixed convex combination lemmas: Application to stability analysis of time-varying delay systems. J. Frankl. Inst. **361**, 107125 (2024)
37. Chen, J., Ji, F., Zhuang, G., Sha, H.: Stability analysis of discrete-time delayed neural networks via delay-product-type Lyapunov-Krasovskii functionals. IET Control Theory Appl. **19**, e12778 (2025)
38. Knorn, F., Mason, O., Shorten, R.: On linear co-positive Lyapunov functions for sets of linear positive systems. Automatica **45**, 1943–1947 (2009)
39. Rami, M.A., Schönlein, M., Jordan, J.: Estimation of linear positive systems with unknown time-varying delays. European J. Control **19**, 179–187 (2013)
40. Zhang, J., Zhao, X., Zhang, R.: An improved approach to controller design of positive systems using controller gain decomposition. J. Frankl. Inst. **354**, 1356–1373 (2017)
41. Chen, X., Chen, M., Wang, L., Shen, J., Hu, J.: Static output-feedback controller synthesis for positive systems under ℓ_∞ performance. Int. J. Control Autom. Syst. **17**, 2871–2880 (2019)
42. Hong, M.T., Hien, L.V., Hang, T.T.M.: Static output-feedback peak-to-peak gain control of discrete-time positive linear systems with diverse interval delays. J. Syst. Control Eng. **236**, 1578–1589 (2022)
43. Ye, S., Li, J., Yao, J.: Robust H_∞ control for a class of 2-D discrete delayed systems. ISA Trans. **53**, 1456–1462 (2014)
44. Hien, L.V., Trinh, H.: Stability of two-dimensional Roesser systems with time-varying delays via novel 2D finite sum inequalities. IET Control Theory Appl. **10**, 1665–1674 (2016)
45. Huang, Y.B., An, J., He, Y., Wu, M.: Quasi-convex combination method and its application to the stability analysis of 2D discrete-time Roesser systems with time-varying delays. IET Control Theory Appl. **12**, 718–727 (2018)
46. Hien, L.V., Trinh, T.: Exponential stability of two-dimensional homogeneous monotone systems with bounded directional delays. IEEE Trans. Autom. Control **63**, 2694–2700 (2018)
47. Hien, L.V., Vu, L.H., Trinh, H.: Stability of two-dimensional descriptor systems with generalized directional delays. Syst. Control Lett. **112**, 42–50 (2018)
48. Hien, L.V., Trinh, H.: Stability analysis and control of two-dimensional fuzzy systems with directional time-varying delays. IEEE Trans. Fuzzy Syst. **26**, 1550–1564 (2018)
49. Peng, D., Su, H.S.: A novel approach to delay-variation-dependent stability analysis of 2-D discrete-time systems with mixed delays. IEEE Access **7**, 99817–99829 (2019)
50. Hien, L.V., Trinh, H.M., Pathirana, P.N.: On l_1-gain control of 2-D positive Roesser systems with directional delays: Necessary and sufficient conditions. Automatica **112**, 108720 (2020)
51. Yang, R., Zheng, W.X., Yu, Y.: Event-triggered sliding mode control of discrete-time two-dimensional systems in Roesser model. Automatica **114**, 108813 (2020)
52. Peng, D., Nie, H.: Abel lemma-based finite-sum inequality approach to stabilization for 2-D time-varying delay systems. Asian J. Control **23**, 1394–1406 (2021)
53. Li, L., Yang, R., Feng, Z., Wu, L.: Event-triggered dissipative control for 2-D switched systems. Info. Sci. **589**, 802–812 (2022)
54. Lv, Xinyu, Niu, Y., Park, J.H., Song, J.: Sliding mode control for 2D FMII systems: A bidirectional dynamic event-triggered strategy. Automatica **147**, 110727 (2023)
55. Ma, X., Dong, J., Tai, W., Zhou, J., Paszke, W.: Asynchronous event-triggered H_∞ control for 2D Markov jump systems subject to networked random packet losses. Commun. Nonlinear Sci. Numer. Simul. **126**, 107453 (2023)
56. Wang, M., Lam, H.K., Qiu, J., Yan, H., Li, Z.: Fuzzy-affine-model-based filtering design for continuous-time Roesser-type 2-D nonlinear systems. IEEE Trans. Cybern. **53**, 3220–3230 (2023)
57. Chelliq, E.H., Alfidi, M., Chalh, Z.: Admissibility and robust H_∞ controller design for uncertain 2D singular discrete systems with interval time-varying delays. Optim. Control Appl. Meth. **44**, 2476–2503 (2023)

58. Chen, W.H., Zhang, C.K., Liu, Z.Z., Wang, L., He, Y.: Stability and filtering for delayed discrete-time T-S fuzzy systems via membership-dependent approaches. IEEE Trans. Fuzzy Syst. **32**, 6101–6111 (2024)
59. Zhang, Y., Ji, Y., Jiang, T., Zhou, J.: Event-triggered control for Roesser model-based 2D Markov jump systems under stochastic communication protocol. Circuit Syst. Sig. Process **43**, 6953–6976 (2024)
60. Hien, L.V., Vu, L.H.: Finite-region boundedness and (Q, S, R)-dissipativity of 2-D singular Roesser systems with mixed time-varying delays. J. Frankl. Inst. **361**, 1416–1435 (2024)
61. Vu, L.H.: Exponential stability and H_∞ controller design for 2-D discrete-state systems in Roesser model with directional delays. Optim. Control Appl. Method **46**, 726–736 (2025)
62. Zhu, K., Wang, Z., Ding, D., Dong, H., Wei, G.: Encryption-decryption-based set-membership filtering for two-dimensional systems: on security and boundedness. Automatica **173**, 112091 (2025)

Chapter 2
Stability and Stabilization of Two-Dimensional Systems with State Delays

Abstract In this chapter, the problems of stability analysis and controller design are presented for some classes of discrete-time 2-D systems with directional state delays. Inspired by the conventional Lyapunov–Krasovskii functional method, an analysis scheme is first developed. The proposed scheme is then utilized in combination with extended 2-D summation inequalities to derive delay-dependent stability conditions in terms of linear matrix inequalities for 2-D Roesser systems with interval delays. Analogous results for 2-D in Fornasini–Marchesini model with delays are also discussed. On the basis of stability conditions, the stabilization problems via conventional state-feedback controllers and event-triggered state-feedback controllers are presented. The effectiveness of the obtained results is illustrated by numerical examples and simulations.

Keywords 2-D systems · Roesser model · 2-D summation inequalities · Directional delays

2.1 Introduction

Time-delay systems have a very strong background in practical and engineering systems according to technical limitations like the constrained speed of signal processing and data transmission between system components [1, 2]. The existence of delays usually leads to complicated behavior, degrades the system performance and even causes system instability [1]. Thus, the systems and control theory of dynamical systems with delays is one of the most active research topics which has been extensively studied during the past few decades. To mention a few, with up-to-date development, we refer the reader to recent works [3–9] and the cited references therein.

Looking at the existing literature so far, it can be realized that the use of Lyapunov–Krasovskii functional (LKF) method is one of the most efficient approaches to derive tractable stability conditions for time-delay systems [5]. In particular, for LTI systems with delays, stability conditions are typically derived in the form of linear matrix inequalities (LMIs) [9], which can be categorized into two kinds according to the involvement of the magnitude of delays. It has been well-recognized that most

L. V. Hien, *Two-Dimensional Time-Delay Systems*, Studies in Systems, Decision and Control 605, https://doi.org/10.1007/978-3-031-95678-2_2

practical systems are only stable with some finite threshold of time-delay. Thus, delay-dependent stability conditions, which explicitly depend on the size of delays, are more reasonable and less conservative than delay-independent ones. Therefore, much effort from researchers has been devoted to finding ways that can be utilized to derive less conservative delay-dependent stability conditions. In other words, how to enlarge the allowable delay range with as less as possible computational burden is the most attractive research topic.

The theory of systems and control has been extensively studied and developed for various types of one-dimensional (1-D) time-delay systems. Despite interesting properties and a wide range of practical applications, this theory for 2-D systems has attracted increasing attention recently (see, e.g., [10–16] and the references therein). Although the basic theory of 2-D systems can often be viewed as a generalization of the conventional 1-D systems theory, there exist deep and substantial differences of 1-D and 2-D systems. For example, since the information propagation in 2-D systems occurs in each of the two independent directions, existing methodologies developed for traditional 1-D systems generally fail when used to analyze these processes because such an approach ignores its inherent to 2-D systems structure. In addition, a 2-D framework has a more complex structure and the available preliminaries associated to 2-D systems are insufficient. This makes the analysis and synthesis of 2-D systems much more complicated than those of 1-D systems [17].

In this chapter, we consider the problems of stability analysis and controller design for some classes of discrete-time 2-D systems with directional state delays. An analysis scheme extended from the conventional LKF method is first developed. By utilizing the proposed scheme and some 2-D summation inequalities, delay-dependent stability conditions are derived. On the basis of stability conditions, stabilization problems via conventional state-feedback controllers (SFCs) and event-triggered state-feedback controllers (ETCs) are also presented. The effectiveness of the obtained results is illustrated by numerical examples.

2.2 Preliminaries

2.2.1 Model Description

Consider the following 2-D time-delay Roesser system

$$\begin{bmatrix} x^h(i+1,j) \\ x^v(i,j+1) \end{bmatrix} = A \begin{bmatrix} x^h(i,j) \\ x^v(i,j) \end{bmatrix} + A_d \begin{bmatrix} x^h(i-d_h(i),j) \\ x^v(i,j-d_v(j)) \end{bmatrix}, \tag{2.1}$$

where $x^h \in \mathbb{R}^{n_h}$ and $x^v \in \mathbb{R}^{n_v}$ are the horizontal state vector and the vertical state vector, respectively, A and $A_d \in \mathbb{R}^{n \times n}$ with $n = n_h + n_v$ are given real matrices, $d_h(i)$ and $d_v(j)$ respectively represent the time-delay along horizontal and vertical directions. It is assumed that the directional delays are time-varying and being confined in prescribed ranges

$$d_1^h \le d_h(i) \le d_2^h, \quad d_1^v \le d_v(j) \le d_2^v, \quad \text{for all } i, j \in \mathbb{N}_0, \tag{2.2}$$

where d_1^h, d_1^v and d_2^h, d_2^v are known positive integers involving the lower bounds and upper bounds of delays. The overall state vector of system (2.1) is defined as $x(i, j) = \text{col}\{x^h(i, j), x^v(i, j)\} \in \mathbb{R}^n$.

The initial condition of system (2.1) is specified as

$$\begin{aligned} x^h(i, j) &= \phi_h(i, j), \ i \in \mathbb{Z}[-d_2^h, 0], \ j \in \mathbb{N}_0, \\ x^v(i, j) &= \phi_v(i, j), \ i \in \mathbb{N}_0, \ j \in \mathbb{Z}[-d_2^v, 0]. \end{aligned} \tag{2.3}$$

A vector-valued sequence $\phi : \mathbb{N}_0 \to \mathbb{R}^r$ is said to be energy-bounded (i.e. belong to the space $l_2(\mathbb{R}^r)$) if the series $\sum_{k=0}^{\infty} \phi^\top(k)\phi(k)$ is finite. The sequence ϕ is said to have finite support if there exists a positive integer T such that $\phi(k) = 0$ for all $k \ge T$. In this chapter, we consider system (2.1) with initial sequences (2.3) satisfy $\phi_h(i, .) \in l_2(\mathbb{R}^{n_h})$ for all $i \in \mathbb{Z}[-d_2^h, 0]$ and $\phi_v(., j) \in l_2(\mathbb{R}^{n_v})$ for all $j \in \mathbb{Z}[-d_2^v, 0]$. This assumption is widely adopted in the area of analysis and control of 2-D systems [14, 16]. We define $\hat{\phi}_h(j) = \sup_{i \in \mathbb{Z}[-d_2^h, 0]} \|\phi_h(i, j)\|$ and $\hat{\psi}_v(i) = \sup_{j \in \mathbb{Z}[-d_2^v, 0]} \|\phi_v(i, j)\|$. Then, it can be verified that

$$\lim_{r \to \infty} \sum_{k=0}^{r} \left(\hat{\psi}_h^2(k) + \hat{\phi}_v^2(k) \right) < \infty. \tag{2.4}$$

Condition (2.4) is trivial if the initial sequences $\phi_h(i, .)$, $\phi_v(., j)$ have finite support [15, 18], that is, there exist positive integers T_h and T_v such that

$$\phi_h(i, j) = 0 \text{ for } j \ge T_h \text{ and } \phi_v(i, j) = 0 \text{ for } i \ge T_v. \tag{2.5}$$

2.2.2 *Problem Formulation*

Consider 2-D systems with directional time-varying delays as given in (2.1). By extending the concept of asymptotic stability proposed in [18], we introduce the following definition.

Definition 2.1 System (2.1) is said to be asymptotically stable if the following two conditions hold

(i) (ϵ-δ stability)
For any $\epsilon > 0$, there exists a $\delta(\epsilon) > 0$ such that if $\hat{\phi}_h(j) < \delta(\epsilon)$ and $\hat{\phi}_v(i) < \delta(\epsilon)$ then $\|x(i, j)\| < \epsilon$ for all $i, j \in \mathbb{N}_0$, where $x(i, j) = \text{col}\{x^h(i, j), x^v(i, j)\} \in \mathbb{R}^n$ is the overall state vector of system (2.1);

(ii) (2-D attractivity)

$$\lim_{i+j\to\infty} \|x(i, j)\| = 0$$

whenever $\lim_{j\to\infty} \hat{\phi}_h(j) = 0$ and $\lim_{i\to\infty} \hat{\phi}_v(i) = 0$.

Remark 2.1 In Definition 2.1, the assumption that initial conditions must go to zero at infinity is required. To verify asymptotic stability of system (2.1), in general, one needs to check both the ϵ-δ stability and 2-D attractivity conditions. This is because that, unlike the case of 1-D systems where asymptotic stability implies all trajectories go to zero at infinity whatever the initial conditions are, in 2-D systems, when taking initial conditions in the first quadrant, one cannot expect to have every trajectory approaching zero simply because it may not be the case for the initial conditions [18].

Remark 2.2 Condition (2.4) is obviously satisfied for finite support initial sequences. In addition, (2.4) also implies that $\lim_{j\to\infty} \hat{\phi}_h(j) = 0$ and $\lim_{i\to\infty} \hat{\phi}_v(i) = 0$. Therefore, as discussed in [20], to verify asymptotic stability of 2-D linear systems as given in (2.1), what we should do is to demonstrate that $\lim_{i+j\to\infty} \|x(i, j)\| = 0$.

One of the main objectives of this chapter is to derive delay-dependent conditions ensuring asymptotic stability of system (2.1). To tackle this problem, we first develop an analysis scheme by extending the LKF method. We then utilize the proposed scheme and efficient tools to derive less conservative delay-dependent stability conditions for system (2.1). In the second part of this chapter, we address the stabilization problem for 2-D systems in the form of (2.1) via some feedback control schemes.

2.3 Auxiliary Results

2.3.1 Extended Lyapunov–Krasovskii Functional Method

In this section, we formulate a technical lemma based on an extension of the 2-D LKF method. For a two indices sequence $\psi : \mathbb{Z} \times \mathbb{Z} \to \mathbb{R}^n$, we denote by $\Delta_h\psi$ and $\Delta_v\psi$ the partial differences of ψ with respect to the first and the second arguments (also referred to horizontal and vertical arguments), that is,

$$\Delta_h\psi(i, j) = \psi(i+1, j) - \psi(i, j) \text{ and } \Delta_v\psi(i, j) = \psi(i, j+1) - \psi(i, j).$$

Let $V : \mathbb{R}^n \to \mathbb{R}$, $x \mapsto V(x)$, be a functional defined by $V(x) = V^h(x^h) + V^v(x^v)$, where $x = (x^h, x^v)$, $x^h \in \mathbb{R}^{n_h}$, $x^v \in \mathbb{R}^{n_v}$, and $V^h : \mathbb{R}^{n_h} \to \mathbb{R}$ and $V^v : \mathbb{R}^{n_v} \to \mathbb{R}$ are some functionals. The difference of $V(x)$ along state trajectories of system (2.1) is defined as

$$\Delta V(x(i, j)) = \Delta_h(V^h(x^h(i, j))) + \Delta_v(V^v(x^v(i, j))). \tag{2.6}$$

We have the following technical lemma which is essential for our next derivation.

Lemma 2.1 (2-D Functional-based scheme) *Assume that there exist a functional $V(x) = V^h(x^h) + V^v(x^v)$ and positive scalars c_0, c_1, c_2 that satisfy the following two conditions*

(i) *For any $x \in \mathbb{R}^n$, it holds that*

$$c_1\|x\|^2 \le V(x) \le c_2\left(\|x^h\|^2 + \|x^v\|^2\right); \tag{2.7}$$

(ii) *The difference of $V(x)$ along any state trajectory $x(i, j)$ of system (2.1) satisfies the following condition*

$$\Delta V(x(i, j)) \le -c_0\|x(i, j)\|^2 \tag{2.8}$$

for all $(i, j) \in \mathbb{N}_0^2$.

Then, system (2.1) is asymptotically stable. Moreover, along any state trajectory $x(i, j)$ of system (2.1), we also have

$$\lim_{k\to\infty} \sum_{i+j=k} V(x(i, j)) = 0. \tag{2.9}$$

Proof We define the function

$$E(k) = \sum_{i+j=k} \left[V^h(x^h(i, j)) + V^v(x^v(i, j))\right], \ k \in \mathbb{N}_0. \tag{2.10}$$

In classical stability theory, the functional $V(x)$ is used as an energy indicator of a dynamical system. To this regard, $E(k)$ defines the energy of system (2.1) stored in the diagonal line $\Gamma_k = \{(i, j) \in \mathbb{N}_0^2 : i + j = k\}$, which is also called an energy function of system (2.1). It follows from (2.8) that

$$\begin{aligned} V^h(x^h(i+1, j)) + V^v(x^v(i, j+1)) \le\ & V^h(x^h(i, j)) + V^v(x^v(i, j)) \\ & - c_0 x^\top(i, j)x(i, j), \ (i, j) \in \mathbb{N}_0^2. \end{aligned} \tag{2.11}$$

Therefore, together with (2.11), we have

$$\begin{aligned}
E(k+1) &= \sum_{i+j=k+1} \left[V^h(x^h(i,j)) + V^v(x^v(i,j))\right] \\
&= \sum_{i+j=k} \left[V^h(x^h(i+1,j)) + V^v(x^v(i,j+1))\right] \\
&\quad + V^h(x^h(0,k+1)) + V^v(x^v(k+1,0)) \\
&\le \sum_{i+j=k} \left[V^h(x^h(i,j)) + V^v(x^v(i,j))\right] - c_0 \sum_{i+j=k} \|x(i,j)\|^2 \\
&\quad + V^h(x^h(0,k+1)) + V^v(x^v(k+1,0)) \\
&\le E(k) + F(k+1) - c_0 \sum_{i+j=k} \|x(i,j)\|^2,
\end{aligned}$$

where $F(k) = V^h(x^h(0,k)) + V^v(x^v(k,0))$. Thus, by iteratively estimate, it follows from the inequality

$$E(k+1) \le E(k) + F(k+1) - c_0 \sum_{i+j=k} \|x(i,j)\|^2, \ k \in \mathbb{N}_0, \tag{2.12}$$

that

$$E(q) \le E(0) + \sum_{k=1}^{q} F(k) - c_0 \sum_{k=0}^{q-1} \sum_{i+j=k} \|x(i,j)\|^2 \tag{2.13}$$

for all positive integer q. Taking (2.4), (2.7) and (2.13) into account we then obtain

$$\begin{aligned}
c_0 \sum_{k=0}^{q-1} \sum_{i+j=k} \|x(i,j)\|^2 + E(q) &\le c_2 \sum_{k=0}^{q} \left(\hat{\phi}_h^2(k) + \hat{\phi}_v^2(k)\right) \\
&\le c_2 \sum_{k=0}^{\infty} \left(\hat{\phi}_h^2(k) + \hat{\phi}_v^2(k)\right) < \infty.
\end{aligned} \tag{2.14}$$

The result of (2.14) ensures that

$$\lim_{k\to\infty} \sum_{i+j=k} V(x(i,j)) = 0$$

and system (2.1) is asymptotically stable. The proof is completed. □

Remark 2.3 Lemma 2.1 provides a useful tool to derive stability conditions for 2-D systems with delays in the form of (2.1). It is important to note that, to derive less conservative stability conditions using LMI settings, not only the construction of LKF candidates $V(x)$ but also the bounding techniques used to manipulate the difference of $V(x)$ along state trajectories of system (2.1) to meet the requirement in (2.8) play a crucial role. To deal with this issue, various methods have been proposed in the literature. Among those, the use of improved Jensen-type summation inequalities has been recognized as one of the most effective approaches. In the next section, we will present some refined summation inequalities, which are useful for the derivation of tractable analysis and design conditions of 2-D systems with delays.

2.3.2 Summation Inequalities

Let $R \in \mathbb{S}_n^+$ be a symmetric positive definite matrix. It is clear that the function $q_R : \mathbb{R}^n \to \mathbb{R}$ defined by $q_R(v) = v^\top R v$ is a convex function and the inequality

$$q_R\left(\sum_{k=1}^{m} \lambda_k v_k\right) \leq \sum_{k=1}^{m} \lambda_k q_R(v_k) \tag{2.15}$$

holds for any vectors $v_k \in \mathbb{R}^n$ and scalars $\lambda_k \geq 0$ with $\sum_{k=1}^{m} \lambda_k = 1$, where m is some positive integer. From (2.15) we readily obtain the following 2-D Jensen-type inequalities.

Lemma 2.2 (Jensen summation inequalities) *Given integers $a \leq b$, $c \leq d$ and a vector-valued function $\omega : \mathbb{Z}[a, b] \times [c, d] \to \mathbb{R}^n$. Then, the following inequalities hold for any matrix $R \in \mathbb{S}_n^+$,*

$$\sum_{i=a}^{b} \omega^\top(i, j) R \omega(i, j) \geq \frac{1}{l_{\mathbb{Z}}(a, b)} \left(\sum_{i=a}^{b} \omega(i, j)\right)^\top R \left(\sum_{i=a}^{b} \omega(i, j)\right) \tag{2.16}$$

and

$$\sum_{j=c}^{d} \omega^\top(i, j) R \omega(i, j) \geq \frac{1}{l_{\mathbb{Z}}(c, d)} \left(\sum_{j=c}^{d} \omega(i, j)\right)^\top R \left(\sum_{j=a}^{b} \omega(i, j)\right). \tag{2.17}$$

For the brevity of notations, let us denote by $J_h^g(\omega)$ and $J_v^g(\omega)$ the gap of (2.16) and (2.17), respectively. Note that, when $a = b$, the estimate in (2.16) is reduced to an equality and thus $J_h^g(\omega) = 0$. We now derive a tighter lower bound for $J_h^g(\omega)$ in the case $a < b$.

Lemma 2.3 (Refined Jensen-based inequality) *Given integers $a < b$, $c \leq d$ and a vector-valued function $\omega : \mathbb{Z}[a, b] \times [c, d] \to \mathbb{R}^n$. Then, for any matrix $R \in \mathbb{S}_n^+$, the following inequality holds*

$$J_h^g(\omega) \geq \frac{3(l_{\mathbb{Z}}(a, b) + 1)}{l_{\mathbb{Z}}(a, b) - 1} \chi^\top R \chi, \tag{2.18}$$

where

$$\chi = \upsilon_{11} - \frac{2}{l_{\mathbb{Z}}(a,b)+1}\upsilon_{12}$$

and

$$\upsilon_{11} = \sum_{i=a}^{b} \omega(i,j), \quad \upsilon_{12} = \sum_{i=a}^{b}\sum_{k=a}^{i} \omega(k,j).$$

Proof We define an approximation function $\hat{\omega} : \mathbb{Z}[a,b] \times [c,d] \to \mathbb{R}^n$ as

$$\hat{\omega}(i,j) = \omega(i,j) - \frac{1}{l_{\mathbb{Z}}(a,b)}\upsilon_{11} + \rho_i \hat{\chi}, \tag{2.19}$$

where $\rho_i = (i-a) + \frac{1-l_{\mathbb{Z}}(a,b)}{2}$, $i \in \mathbb{Z}[a,b]$, and $\hat{\chi} \in \mathbb{R}^n$ is a constant vector which will be determined later. It is easy to verify that

$$\sum_{i=a}^{b} \rho_i = 0 \text{ and } \sum_{i=a}^{b} \rho_i^2 = \frac{l(l^2-1)}{12},$$

where, for short, we write $l = l_{\mathbb{Z}}(a,b)$ if it does not make any confusion. Then, it follows from (2.19) that

$$\sum_{i=a}^{b} \hat{\omega}^\top(i,j)R\hat{\omega}(i,j) = \frac{1}{l}J_h^g(\omega) + 2\hat{\chi}^\top R\left(\sum_{i=a}^{b} \rho_i\omega(i,j)\right) + \frac{l(l^2-1)}{12}\hat{\chi}^\top R\hat{\chi}. \tag{2.20}$$

We now define

$$\omega_s(i,j) = \sum_{k=a}^{i-1} \omega(k,j)$$

for $i > a$ and $\omega_s(a,j) = 0$ for $j \in \mathbb{Z}[c,d]$. Then, we have $\omega(i,j) = \Delta_h(\omega_s(i,j))$. Thus,

$$\rho_i\omega(i,j) = \Delta_h(\rho_i\omega_s(i,j)) - \omega_s(i+1,j).$$

Taking summation in i from a to b we then obtain

$$\sum_{i=a}^{b} \rho_i\omega(i,j) = \frac{l+1}{2}\upsilon_{11} - \upsilon_{12} = \frac{l+1}{2}\chi. \tag{2.21}$$

From (2.20), (2.21) and the fact that

$$\sum_{i=a}^{b} \hat{\omega}^\top(i,j)R\hat{\omega}(i,j) \geq 0,$$

we have

$$J_h^g(\omega) + (l^2 + l)\hat{\chi}^\top R\chi + \frac{l^2(l^2-1)}{12}\hat{\chi}^\top R\hat{\chi} \geq 0. \tag{2.22}$$

In (2.22) we specify $\hat{\chi} = -\lambda\chi$, where λ is a scalar, then we have

$$J_h^g(\omega) \geq (l^2 + l)\left(\lambda - \frac{l(l-1)}{12}\lambda^2\right)\chi^\top R\chi. \tag{2.23}$$

Clearly, the function $\lambda - \frac{l(l-1)}{12}\lambda^2$ attains its maximum $\frac{3}{l(l-1)}$ at $\lambda_* = \frac{6}{l(l-1)}$. With the critical value λ_*, we obtain $\hat{\chi} = \frac{-6}{l(l-1)}\chi$ and from (2.22), we finally reach

$$J_g^g(\omega) \geq \frac{3(l+1)}{l-1}\chi^\top R\chi.$$

The proof is now completed. □

Remark 2.4 It is clear that the summation inequality (2.18) provides a tighter lower bound than 2-D Jensen-type inequality (2.16) since a positive term is added into the right-hand side of (2.16).

Remark 2.5 An alternative approach to improve the Jensen-type inequalities (2.16), (2.17) can be derived using Abel lemma on summation by parts [16, 21]. Specifically, by similar lines utilizing Abel lemma presented in [21], the derived inequality in (2.18) is now reduced to

$$J_h^g(\omega) \geq \frac{3(l_{\mathbb{Z}}(a,b)-1)}{l_{\mathbb{Z}}(a,b)+1}\tilde{\chi}^\top R\tilde{\chi}, \tag{2.24}$$

where

$$\tilde{\chi} = \upsilon_{11} - \frac{2}{l_{\mathbb{Z}}(a,b)-1}\upsilon_{12}.$$

Basically, the use of Refined Jensen-based inequality (2.18) can deliver better results in comparison to that of (2.24) since $\frac{l+1}{l-1} > \frac{l-1}{l+1}$ for all integer $l > 1$.

Remark 2.6 Given a matrix $R \in \mathbb{S}_n^+$, integers $a \leq b$, $c < d$, and consider a vector-valued function $\omega : \mathbb{Z}[a,b] \times [c,d] \to \mathbb{R}^n$. By similar lines used in the proof of Lemma 2.3 we obtain the following estimate

$$J_v^g(\omega) \geq \frac{3(l_{\mathbb{Z}}(c,d)+1)}{l_{\mathbb{Z}}(c,d)-1}\xi^\top R\xi, \tag{2.25}$$

where

$$\xi = \upsilon_{21} - \frac{2}{l_{\mathbb{Z}}(c,d)+1}\upsilon_{22}$$

and

$$\upsilon_{21} = \sum_{j=c}^{d} \omega(i, j), \quad \upsilon_{22} = \sum_{j=c}^{d} \sum_{k=c}^{j} \eta(i, k).$$

By simplifying (2.18) and (2.25), we obtain the following corollary, which makes the results of (2.18) and (2.25) easier to utilize for 2-D systems with delays.

Corollary 2.1 *Given positive integers p, q and a function $\omega : \mathbb{Z}[-p, 0] \times [-q, 0] \to \mathbb{R}^n$. Then, for any matrix $R \in \mathbb{S}_n^+$, the following inequalities hold*

$$\sum_{i=-p}^{-1} \Delta_h(\omega(i, j))^\top R \Delta_h(\omega(i, j)) \geq \frac{1}{p} \omega_h^\top(j) \hat{R} \omega_h(j), \tag{2.26}$$

$$\sum_{j=-q}^{-1} \Delta_v(\omega(i, j))^\top R \Delta_v(\omega(i, j)) \geq \frac{1}{q} \omega_v^\top(i) \hat{R} \omega_v(i), \tag{2.27}$$

where $\hat{R} = \mathrm{diag}\{R, 3R\}$ *and*

$$\omega_h(j) = \begin{bmatrix} \omega(0, j) - \omega(-p, j) \\ \omega(0, j) + \omega(-p, j) - \frac{2}{p+1} \sum_{i=-p}^{0} \omega(i, j) \end{bmatrix},$$

$$\omega_v(i) = \begin{bmatrix} \omega(i, 0) - \omega(i, -q) \\ \omega(i, 0) + \omega(i, -q) - \frac{2}{q+1} \sum_{j=-q}^{0} \omega(i, j) \end{bmatrix}.$$

Proof It is clearly that $\frac{p+1}{p-1} > 1$ for all $p > 1$. Thus, it follows from (2.18) that

$$p \sum_{i=-p}^{-1} \Delta_h(\omega(i, j))^\top R \Delta_h(\omega(i, j)) \geq \tilde{\upsilon}_{11}^\top R \tilde{\upsilon}_{11} + 3 \left(\tilde{\upsilon}_{11} - \frac{2}{p+1} \tilde{\upsilon}_{12} \right)^\top R \left(\tilde{\upsilon}_{11} - \frac{2}{p+1} \tilde{\upsilon}_{12} \right), \tag{2.28}$$

where

$$\tilde{\upsilon}_{11} = \sum_{i=-p}^{-1} \Delta_h(\omega(i, j)) = \omega(0, j) - \omega(-p, j),$$

$$\tilde{\upsilon}_{12} = \sum_{i=-p}^{-1} \sum_{k=-p}^{i} \Delta_h(\omega(k, j)) = \sum_{i=-p}^{0} \omega(i, j) - (p+1)\omega(-q, j).$$

From (2.28) we readily obtain the estimate (2.26). The proof of (2.27) is similar to that of (2.26) and thus is omitted here. □

2.3.3 Reciprocally Convex Inequalities

Dealing with time-varying delays to produce less conservative stability conditions, one usually needs to bound summations, called reciprocally convex combinations [22], of the form

$$\mathscr{C} = \sum_{k=1}^{N} \frac{1}{\alpha_k} \chi_k^\top R_k \chi_k$$

for vectors $\chi_k \in \mathbb{R}^{n_k}$, matrices $R_k \in \mathbb{S}_{n_k}^{+}$ and scalars $\alpha_k > 0$ with $\sum_{k=1}^{N} \alpha_k = 1$. We illustrate the main idea with $N = 2$ and $a_1 = \chi_1^\top R_1 \chi_1 \geq 0$, $a_2 = \chi_2^\top R_2 \chi_2 \geq 0$. For any $\alpha \in (0, 1)$, it holds that

$$\begin{aligned}
\frac{a_1}{\alpha} + \frac{a_2}{1-\alpha} &= \left(\frac{a_1}{\alpha} + \frac{a_2}{1-\alpha}\right)(\alpha + 1 - \alpha) \\
&\geq \left(\sqrt{\frac{a_1}{\alpha}}\sqrt{\alpha} + \sqrt{\frac{a_2}{1-\alpha}}\sqrt{1-\alpha}\right)^2 \\
&= (\sqrt{a_1} + \sqrt{a_2})^2.
\end{aligned} \tag{2.29}$$

To make the result utilizable in LMIzation processes, one needs to linearize the term $2\sqrt{a_1 a_2}$. Without lost of generality, we can assume $a_1 > 0$. Then, for any real number b, if $\begin{bmatrix} a_1 & b \\ b & a_2 \end{bmatrix} \geq 0$ then

$$q(t) = a_1 t^2 + 2bt + a_2 = \begin{bmatrix} t \\ 1 \end{bmatrix}^\top \begin{bmatrix} a_1 & b \\ b & a_2 \end{bmatrix} \begin{bmatrix} t \\ 1 \end{bmatrix} \geq 0 \tag{2.30}$$

for all $t \in \mathbb{R}$. It follows from (2.30) that $b^2 - a_1 a_2 \leq 0$ and hence $2\sqrt{a_1 a_2} \geq 2b$. The results of (2.29) and (2.30) can be summarized as

$$\min_{\alpha \in (0,1)} \left(\frac{a_1}{\alpha} + \frac{a_2}{1-\alpha}\right) \geq a_1 + 2b + a_2 \text{ subject to } \begin{bmatrix} a_1 & b \\ b & a_2 \end{bmatrix} \geq 0. \tag{2.31}$$

As a consequence of (2.31), we have the following lemma.

Lemma 2.4 *For given matrices $R_1 = R_1^\top > 0$, $R_2 = R_2^\top > 0$ and X_{12} of appropriate dimensions, if*

$$\begin{bmatrix} R_1 & X_{12} \\ X_{12}^\top & R_2 \end{bmatrix} \geq 0$$

then the inequality

$$\frac{1}{\alpha}\chi_1^\top R_1 \chi_1 + \frac{1}{1-\alpha}\chi_2^\top R_2 \chi_2 \geq \begin{bmatrix} \chi_1 \\ \chi_2 \end{bmatrix}^\top \begin{bmatrix} R_1 & X_{12} \\ X_{12}^\top & R_2 \end{bmatrix} \begin{bmatrix} \chi_1 \\ \chi_2 \end{bmatrix} \tag{2.32}$$

holds for all $\alpha \in (0, 1)$ and vectors χ_1, χ_2 of appropriate dimensions.

An extended result of Lemma 2.4 can be obtained as follows.

Lemma 2.5 *Given matrices $R_k \in \mathbb{S}_{n_k}^+$, $k = 1, 2, \ldots, N$. For any matrices X_{jk}, $j, k = 1, 2, \ldots, N$, $j \neq k$, of appropriate dimensions, if*

$$\begin{bmatrix} R_j & X_{jk} \\ X_{jk}^\top & R_k \end{bmatrix} \geq 0,$$

then the inequality

$$\sum_{k=1}^{N} \frac{1}{\alpha_k}\chi_k^\top R_k \chi_k \geq \sum_{k=1}^{N} \chi_k^\top R_k \chi_k + \sum_{j \neq k} \begin{bmatrix} \chi_j \\ \chi_k \end{bmatrix}^\top \begin{bmatrix} R_j & X_{jk} \\ X_{jk}^\top & R_k \end{bmatrix} \begin{bmatrix} \chi_j \\ \chi_k \end{bmatrix} \tag{2.33}$$

holds for all $\alpha_k \in (0, 1)$ with $\sum_{k=1}^{N} \alpha_k = 1$ and $\chi_k \in \mathbb{R}^{n_k}$.

2.3.4 Matrix Inequalities

Let $Q \in \mathbb{S}_n^+$ be a given symmetric positive definite matrix. For any matrices V, W of appropriate dimensions, we have

$$\begin{aligned} &V^\top Q^{-1} V + V^\top W + W^\top V + W^\top Q W \\ &\quad = \left(Q^{-1/2} V + Q^{1/2} W\right)^\top \left(Q^{-1/2} V + Q^{1/2} W\right) \geq 0. \end{aligned}$$

Based on this formulation, we have the following matrix inequality.

Lemma 2.6 (Cauchy matrix inequality) *For any matrices V, W of appropriate dimensions and a matrix $Q \in \mathbb{S}_n^+$, the following inequality holds*

$$- W^\top Q W \leq V^\top W + W^\top V + V^\top Q^{-1} V.$$

Lemma 2.7 (Schur complement lemma) *For given matrices $X = X^\top$, $Z = Z^\top$ and a matrix Y of appropriate dimensions, we have*

$$\begin{bmatrix} X & Y \\ Y^\top & Z \end{bmatrix} > 0 \Longleftrightarrow \begin{cases} Z > 0 \\ X - YZ^{-1}Y^\top > 0 \end{cases}$$
$$\Longleftrightarrow \begin{cases} X > 0 \\ Z - Y^\top X^{-1} Y > 0. \end{cases}$$

2.4 Delay-Dependent Stability Conditions

2.4.1 2-D Roesser Systems with Interval Time-Varying Delays

In this section, based on the scheme presented in Lemma 2.1 and by utilizing the Refined Jensen-based inequalities formulated in the preceding section, we derive delay-dependent stability conditions for 2-D Roesser systems with interval time-varying delays in the form of (2.1).

In light of the functional-based scheme proposed in Lemma 2.1, an augmented LKF candidate is constructed as follows

$$V(x(i,j)) = \underbrace{\begin{bmatrix} \tilde{x}^h(i,j) \\ \tilde{x}^v(i,j) \end{bmatrix}^\top P \begin{bmatrix} \tilde{x}^h(i,j) \\ \tilde{x}^v(i,j) \end{bmatrix}}_{V_0(x(i,j))} + \sum_{k=1}^{4} V_k(x(i,j)) \tag{2.34}$$

where, for the brevity, we write $V_k = V_k(x(i,j))(k = 1, 2, 3, 4)$ and

$$\tilde{x}^h(i,j) = \begin{bmatrix} x^h(i,j) \\ \sum_{k=i-d_1^h}^{i-1} x^h(k,j) \\ \sum_{k=i-d_2^h}^{i-d_1^h-1} x^h(k,j) \end{bmatrix}, \quad \tilde{x}^v(i,j) = \begin{bmatrix} x^v(i,j) \\ \sum_{l=j-d_1^v}^{j-1} x^v(i,l) \\ \sum_{l=j-d_2^v}^{j-d_1^v-1} x^v(i,l) \end{bmatrix},$$

$$V_1 = \sum_{k=i-d_1^h}^{i-1} x^{h\top}(k,j) Q_1^h x^h(k,j) + \sum_{l=j-d_1^v}^{j-1} x^{v\top}(i,l) Q_1^v x^v(i,l),$$

$$V_2 = \sum_{k=i-d_2^h}^{i-d_1^h-1} x^{h\top}(k,j) Q_2^h x^h(k,j) + \sum_{l=j-d_2^v}^{j-d_1^v-1} x^{v\top}(i,l) Q_2^v x^v(i,l),$$

$$V_3 = d_1^h \sum_{k=-d_1^h}^{-1} \sum_{l=i+k}^{i-1} \eta_h^\top(l,j) R_1^h \eta_h(l,j) + d_1^v \sum_{k=-d_1^v}^{-1} \sum_{l=j+k}^{j-1} \eta_v^\top(i,l) R_1^v \eta_v(i,l),$$

$$V_4 = d_{12}^h \sum_{k=-d_2^h}^{-d_1^h-1} \sum_{l=i+k}^{i-1} \eta_h^\top(l,j) R_2^h \eta_h(l,j) + d_{12}^v \sum_{k=-d_2^v}^{-d_1^v-1} \sum_{l=j+k}^{j-1} \eta_v^\top(i,l) R_2^v \eta_v(i,l),$$

$$\eta_h(i,j) = \Delta_h(x^h(i,j)), \quad \eta_v(i,j) = \Delta_v(x^v(i,j)),$$
$$d_{12}^h = d_2^h - d_1^h, \quad d_{12}^v = d_2^v - d_1^v.$$

Remark 2.7 In the existing literature, the functional $V_0(x(i,j))$ is usually constructed in the form [10, 11, 13]

$$\begin{bmatrix} x^h(i,j) \\ x^v(i,j) \end{bmatrix}^\top \operatorname{diag}(P^h, P^v) \begin{bmatrix} x^h(i,j) \\ x^v(i,j) \end{bmatrix}$$

which only uses the state vector. The functional (2.34) contains the state vector and its summations, which allows one to incorporate extra information on time-delay into its augmented vectors. Based using this functional, and by utilizing suitable 2-D Refined Jensen-based inequalities, less conservative delay-range-dependent stability conditions for system (2.1) can be derived.

For the brevity, we denote the block-row matrices $e_k = \begin{bmatrix} 0_{n\times(k-1)n} & I_n & 0_{n\times(7-k)n} \end{bmatrix}$ $(k = 1, 2, \ldots, 7)$, $\mathscr{A} = (A - I_n)e_1 + A_d e_3$ and augmented vector $\chi(i,j) = \begin{bmatrix} \chi_x(i,j) \\ \chi_s(i,j) \end{bmatrix}$, where

$$\chi_x(i,j) = \begin{bmatrix} x(i,j) \\ x_l(i,j) \\ x_d(i,j) \\ x_u(i,j) \end{bmatrix}, \quad \chi_s(i,j = \begin{bmatrix} x_l^s(i,j) \\ x_d^s(i,j) \\ x_u^s(i,j) \end{bmatrix},$$

$$x_l(i,j) = \begin{bmatrix} x^h(i-d_1^h, j) \\ x^v(i, j-d_1^v) \end{bmatrix}, \quad x_d(i,j) = \begin{bmatrix} x^h(i-d_h(i), j) \\ x^v(i, j-d_v(j)) \end{bmatrix},$$

$$x_u(i,j) = \begin{bmatrix} x^h(i-d_2^h, j) \\ x^v(i, j-d_2^v) \end{bmatrix}, \quad x_l^s(i,j) = \begin{bmatrix} \xi_{11}(i,j) \\ \xi_{21}(i,j) \end{bmatrix},$$

$$x_d^s(i,j) = \begin{bmatrix} \xi_{12}(i,j) \\ \xi_{22}(i,j) \end{bmatrix}, \quad x_u^s(i,j) = \begin{bmatrix} \xi_{13}(i,j) \\ \xi_{23}(i,j) \end{bmatrix},$$

$$\xi_{11}(i,j) = \frac{1}{1+d_1^h} \sum_{k=i-d_1^h}^{i} x^h(k,j),$$

$$\xi_{12}(i,j) = \frac{1}{1+d_h(i)-d_1^h} \sum_{k=i-d_h(i)}^{i-d_1^h} x^h(k,j),$$

$$\xi_{13}(i,j) = \frac{1}{1+d_2^h - d_h(i)} \sum_{k=i-d_2^h}^{i-d_h(i)} x^h(k,j),$$

$$\xi_{21}(i,j) = \frac{1}{1+d_1^v} \sum_{l=j-d_1^v}^{j} x^v(i,l)$$

$$\xi_{22}(i,j) = \frac{1}{1+d_v(j) - d_1^v} \sum_{l=j-d_v(j)}^{j-d_1^v} x^v(i,l),$$

$$\xi_{23}(i,j) = \frac{1}{1+d_2^v - d_v(j)} \sum_{l=j-d_2^v}^{j-d_v(j)} x^v(i,l).$$

We also denote the following matrices

$$\begin{aligned}
E(d_h) &= \left[e_1^\top \ (1+d_1^h)e_5^\top \ l_{\mathbb{Z}}(d_1^h, d_h)e_6^\top + l_{\mathbb{Z}}(d_h, d_2^h)e_7^\top\right]^\top, \\
G(d_v) &= \left[e_1^\top \ (1+d_1^v)e_5^\top \ l_{\mathbb{Z}}(d_1^v, d_v)e_6^\top + l_{\mathbb{Z}}(d_v, d_2^v)e_7^\top\right]^\top, \\
F_0 &= \left[\mathscr{A}^\top \ (e_1-e_2)^\top \ (e_2-e_4)^\top\right]^\top, \\
F_1 &= \left[0 \ e_1^\top \ (e_2+e_3)^\top\right]^\top, \quad F_2 = \left[-\mathscr{A}^\top \ e_2^\top \ (e_3+e_4)^\top\right]^\top, \\
\Gamma_1 &= \left[(e_1-e_2)^\top \ (e_1+e_2-2e_5)^\top\right]^\top, \\
\Gamma_2 &= \left[(e_2-e_3)^\top \ (e_2+e_3-2e_6)^\top\right]^\top, \\
\Gamma_3 &= \left[(e_3-e_4)^\top \ (e_3+e_4-2e_7)^\top\right]^\top, \\
\Pi_0(d_h, d_v) &= \mathsf{Sym}\left(E(d_h)^\top \overline{P^h} F_0 + G(d_v)^\top \overline{P^v} F_0\right) \\
&\quad + F_2^\top\left(\overline{P^h} + \overline{P^v}\right) F_2 - F_1^\top \left(\overline{P^h} + \overline{P^v}\right) F_1, \\
\overline{P^h} &= \mathbb{J}_{n_h}^{3\top} P^h \mathbb{J}_{n_h}^3, \quad \overline{P^v} = \mathbb{J}_{n_v}^{3\top} P^v \mathbb{J}_{n_v}^3, \\
\Pi_1 &= e_1^\top Q_1 e_1 - e_2^\top Q_1 e_2 + e_2^\top Q_2 e_2 - e_4^\top Q_2 e_4, \\
\Pi_2 &= \mathscr{A}^\top \left(\hat{R}_1 + \hat{R}_2\right)\mathscr{A}, \quad \Pi_3 = \Gamma_1^\top \Phi \Gamma_1, \\
\Pi_4 &= \begin{bmatrix}\Gamma_2 \\ \Gamma_3\end{bmatrix}^\top \left(\mathbb{J}_{n_h}^{4\top}\Psi_1\mathbb{J}_{n_h}^4 + \mathbb{J}_{n_v}^{4\top}\Psi_2\mathbb{J}_{n_v}^4\right)\begin{bmatrix}\Gamma_2 \\ \Gamma_3\end{bmatrix}, \\
\hat{R}_1 &= \mathrm{diag}\left((d_1^h)^2 R_1^h, (d_1^v)^2 R_1^v\right), \quad \hat{R}_2 = \mathrm{diag}\left((d_{12}^h)^2 R_2^h, (d_{12}^v)^2 R_2^v\right), \\
\Psi_1 &= \begin{bmatrix}\widetilde{R}_2^h & X \\ X^\top & \widetilde{R}_2^h\end{bmatrix}, \quad \Psi_2 = \begin{bmatrix}\widetilde{R}_2^v & Y \\ Y^\top & \widetilde{R}_2^v\end{bmatrix}, \\
\widetilde{R}_2^h &= \mathrm{diag}\left(R_2^h, 3R_2^h\right), \quad \widetilde{R}_2^v = \mathrm{diag}\left(R_2^v, 3R_2^v\right), \\
\mathbb{J}_{n_h}^k &= \underbrace{\mathrm{diag}\left([I_{n_h} \ \ 0_{n_h\times n_v}], \ldots, [I_{n_h} \ \ 0_{n_h\times n_v}]\right)}_{k \text{ blocks}},
\end{aligned}$$

$$\mathbb{J}^k_{n_v} = \underbrace{\text{diag}\left([0_{n_v \times n_h} \quad I_{n_v}], \ldots, [0_{n_v \times n_h} \quad I_{n_v}]\right)}_{k \text{ blocks}},$$

$$\Phi = \text{diag}\left(R_1, \frac{3(d_1^h + 1)}{d_1^h - 1} R_1^h, \frac{3(d_1^v + 1)}{d_1^v - 1} R_1^v\right).$$

Delay-range-dependent stability for 2-D time-varying delay systems in the form of (2.1) is presented as the following theorem.

Theorem 2.1 *Consider system (2.1) with time-varying delays $d_h(i) \in [d_1^h, d_2^h]$ and $d_v(j) \in [d_1^v, d_2^v]$. Assume that there exist symmetric positive definite matrices $P^h \in \mathbb{S}^+_{3n_h}$, $P^v \in \mathbb{S}^+_{3n_v}$, $Q_k = \text{diag}(Q_k^h, Q_k^v) \in \mathbb{S}^+_n$, $R_k = \text{diag}(R_k^h, R_k^v) \in \mathbb{S}^+_n$ $(k = 1, 2)$, and real matrices $X \in \mathbb{R}^{2n_h \times 2n_h}$, $Y \in \mathbb{R}^{2n_v \times 2n_v}$ such that*

$$\Psi_1 = \begin{bmatrix} \widetilde{R}_2^h & X \\ X^\top & \widetilde{R}_2^h \end{bmatrix} \geq 0, \quad \Psi_2 = \begin{bmatrix} \widetilde{R}_2^v & Y \\ Y^\top & \widetilde{R}_2^v \end{bmatrix} \geq 0, \tag{2.35}$$

and the following LMI-based condition holds for $d_h \in \{d_1^h, d_2^h\}$, $d_v \in \{d_1^v, d_2^v\}$,

$$\Pi(d_h, d_v) = \Pi_0(d_h, d_v) + \sum_{k=1}^{2} \Pi_k - \sum_{k=3}^{4} \Pi_k < 0. \tag{2.36}$$

Then, system (2.1) is asymptotically stable.

Proof We write the functional (2.34) in the form $V(x(i, j)) = V_h(x^h(i, j)) + V_v(x^v(i, j))$, where $P = \text{diag}(P^h, P^v)$. We recall here that the difference $\Delta V(x(i, j))$ is defined as

$$\begin{aligned} \Delta V(x(i, j)) &= \Delta_h(V_h(x^h(i, j))) + \Delta_v(V_v(x^v(i, j))) \\ &= V_h(x^h(i + 1, j)) - V_h(x^h(i, j)) \\ &\quad + V_v(x^v(i, j + 1)) - V^v(x^v(i, j)). \end{aligned} \tag{2.37}$$

Note at first that the functional $V_0(x(i, j))$ is given by

$$V_0(x(i, j)) = \underbrace{\tilde{x}^{h\top}(i, j) P^h \tilde{x}^h(i, j)}_{V_h^0(x^h(i,j))} + \underbrace{\tilde{x}^{v\top}(i, j) P^v \tilde{x}^v(i, j)}_{V_v^0(x^v(i,j))}, \tag{2.38}$$

where the augmented vectors $\tilde{x}^h(i, j)$ and $\tilde{x}^v(i, j)$ satisfy the identities

$$\begin{aligned} \tilde{x}^h(i, j) &= \mathbb{J}^3_{n_h}(E(d_h) - F_1)\chi(i, j), \\ \tilde{x}^v(i, j) &= \mathbb{J}^3_{n_v}(G(d_v) - F_1)\chi(i, j). \end{aligned}$$

In addition,

$$\tilde{x}^h(i+1, j) = \mathbb{J}^3_{n_h}(E(d_h) - F_2)\chi(i, j),$$
$$\tilde{x}^v(i, j+1) = \mathbb{J}^3_{n_v}(G(d_v) - F_2)\chi(i, j).$$

Therefore,

$$\begin{aligned}\Delta_h(V_h^0(x^h(i, j))) &= \chi^\top(i, j)\Big[(E(d_h) - F_2)^\top \overline{P^h}(E(d_h) - F_2) \\ &\quad - (E(d_h) - F_1)^\top \overline{P^h}(E(d_h) - F_1)\Big]\chi(i, j) \\ &= \chi^\top(i, j)\Big[\mathrm{Sym}\Big(E(d_h)^\top \overline{P^h} F_0\Big) + F_2^\top \overline{P^h} F_2 - F_1^\top \overline{P^h} F_1\Big]\chi(i, j) \quad (2.39)\end{aligned}$$

and

$$\begin{aligned}\Delta_v(V_v^0(x^v(i, j))) &= \chi^\top(i, j)\Big[(G(d_v) - F_2)^\top \overline{P^v}(G(d_v) - F_2) \\ &\quad - (G(d_v) - F_1)^\top \overline{P^v}(G(d_v) - F_1)\Big]\chi(i, j) \\ &= \chi^\top(i, j)\Big[\mathrm{Sym}\Big(G(d_v)^\top \overline{P^v} F_0\Big) + F_2^\top \overline{P^v} F_2 - F_1^\top \overline{P^v} F_1\Big]\chi(i, j). \quad (2.40)\end{aligned}$$

From (2.38)–(2.40), we have

$$\Delta V_0(x(i, j)) = \chi^\top(i, j)\Pi_0(d_h, d_v)\chi(i, j). \quad (2.41)$$

Next, the difference $\Delta V_1(x(i, j))$ is given by

$$\begin{aligned}\Delta V_1(x(i, j)) &= x^{h\top}(i, j)Q_1^h x^h(i, j) - x^{h\top}(i - d_1^h, j)Q_1^h x^h(i - d_1^h, j) \\ &\quad + x^{v\top}(i, j)Q_1^v x^v(i, j) - x^{v\top}(i, j - d_1^v)Q_1^v x^v(i, j - d_1^v) \\ &= \chi^\top(i, j)\left(e_1^\top Q_1 e_1 - e_2^\top Q_1 e_2\right)\chi(i, j). \quad (2.42)\end{aligned}$$

Similar to (2.42), we also have

$$\begin{aligned}\Delta V_2(x(i, j)) &= x^{h\top}(i - d_1^h, j)Q_2^h x^h(i - d_1^h, j) - x^{h\top}(i - d_2^h, j)Q_2^h x^h(i - d_2^h, j) \\ &\quad + x^{v\top}(i, j - d_1^v)Q_2^v x^v(i, j - d_1^v) - x^{v\top}(i, j - d_2^v)Q_2^v x^v(i, j - d_2^v) \\ &= \chi^\top(i, j)\left(e_2^\top Q_2 e_2 - e_4^\top Q_2 e_4\right)\chi(i, j). \quad (2.43)\end{aligned}$$

A combination of (2.42) and (2.43) gives

$$\Delta V_1(x(i, j)) + \Delta V_2(x(i, j)) = \chi^\top(i, j)\Pi_1\chi(i, j). \quad (2.44)$$

We now take the difference $\Delta V_3(x(i, j))$ and utilize the inequalities given in (2.18) and (2.24) to manipulate $\Delta V_3(x(i, j))$. Note that

$$
\begin{aligned}
\Delta V_3(x(i,j)) &= (d_1^h)^2 \eta_h^\top(i,j) R_1^h \eta_h(i,j) - d_1^h \sum_{k=i-d_1^h}^{i-1} \eta_h^\top(k,j) R_1^h \eta_h(k,j) \\
&\quad + (d_1^v)^2 \eta_v^\top(i,j) R_1^v \eta_v(i,j) - d_1^v \sum_{l=j-d_1^v}^{j-1} \eta_v^\top(i,l) R_1^v \eta_v(i,l)
\end{aligned} \tag{2.45}
$$

and

$$
\begin{bmatrix} \eta_h(i,j) \\ \eta_v(i,j) \end{bmatrix} = [(A - I_n)e_1 + A_d e_3]\chi(i,j) = \mathscr{A}\chi(i,j). \tag{2.46}
$$

Therefore, from (2.45) and (2.46), we get

$$
\begin{aligned}
\Delta V_3(x(i,j)) &= \chi^\top(i,j)\mathscr{A}^\top \hat{R}_1 \mathscr{A}\chi(i,j) - d_1^h \sum_{k=i-d_1^h}^{i-1} \eta_h^\top(k,j) R_1^h \eta_h(k,j) \\
&\quad - d_1^v \sum_{l=j-d_1^v}^{j-1} \eta_v^\top(i,l) R_1^v \eta_v(i,l).
\end{aligned} \tag{2.47}
$$

By utilizing the inequalities (2.18) and (2.24), we then obtain

$$
\begin{aligned}
&-d_1^h \sum_{k=i-d_1^h}^{i-1} \eta_h^\top(k,j) R_1^h \eta_h(k,j) \\
&\quad \le -\left[x^h(i,j) - x^h(i-d_1^h,j)\right]^\top R_1^h \left[x^h(i,j) - x^h(i-d_1^h,j)\right] \\
&\quad - \frac{3(d_1^h+1)}{d_1^h-1}\left[x^h(i,j) + x^h(i-d_1^h,j) - 2\xi_{11}(i,j)\right]^\top R_1^h \\
&\quad \times \left[x^h(i,j) + x^h(i-d_1^h,j) - 2\xi_{11}(i,j)\right]
\end{aligned} \tag{2.48}
$$

and

$$
\begin{aligned}
&-d_1^v \sum_{l=j-d_1^v}^{j-1} \eta_v^\top(i,l) R_1^v \eta_v(i,l) \\
&\quad \le -\left[x^v(i,j) - x^v(i,j-d_1^v)\right]^\top R_1^v \left[x^v(i,j) - x^v(i,j-d_1^v)\right] \\
&\quad - \frac{3(d_1^v+1)}{d_1^v-1}\left[x^v(i,j) + x^v(i,j-d_1^v) - 2\xi_{21}(i,j)\right]^\top R_1^v \\
&\quad \times \left[x^v(i,j) + x^v(i,j-d_1^v) - 2\xi_{21}(i,j)\right].
\end{aligned} \tag{2.49}
$$

The result of (2.47)–(2.49) yields

$$
\Delta V_3(x(i,j)) \le \chi^\top(i,j)\left(\mathscr{A}^\top \hat{R}_1 \mathscr{A} - \Pi_3\right)\chi(i,j). \tag{2.50}
$$

Similar to (2.47), the difference $\Delta V_4(i, j)$ is given by

$$\Delta V_4(x(i,j)) = \chi^T(i,j)\mathscr{A}^\top \hat{R}_2 \mathscr{A}\chi(i,j) - d_{12}^h \sum_{k=i-d_2^h}^{i-d_1^h-1} \eta_h^\top(k,j) R_2^h \eta_h(k,j)$$
$$- d_{12}^v \sum_{l=j-d_2^v}^{j-d_1^v-1} \eta_v^\top(i,l) R_2^v \eta_v(i,l). \tag{2.51}$$

By utilizing (2.26) and the reciprocally convex inequality formulated in Lemma 2.4, we have

$$\begin{aligned} -d_{12}^h \sum_{k=i-d_2^h}^{i-d_1^h-1} \eta_h^\top(k,j) R_2^h \eta_h(k,j) &= -d_{12}^h \sum_{k=i-d_2^h}^{i-d_h(i)-1} \eta_h^\top(k,j) R_2^h \eta_h(k,j) \\ &\quad - d_{12}^h \sum_{k=i-d_h(i)}^{i-d_1^h-1} \eta_h^\top(k,j) R_2^h \eta_h(k,j) \\ &\le -\frac{d_{12}^h}{d_h(i) - d_1^h} \tilde{\Omega}_1^\top \widetilde{R}_2^h \tilde{\Omega}_1 - \frac{d_{12}^h}{d_2^h - d_h(i)} \tilde{\Omega}_2^\top \widetilde{R}_2^h \tilde{\Omega}_2 \\ &\le -\begin{bmatrix} \tilde{\Omega}_1 \\ \tilde{\Omega}_2 \end{bmatrix}^\top \begin{bmatrix} \widetilde{R}_2^h & X \\ X^\top & \widetilde{R}_2^h \end{bmatrix} \begin{bmatrix} \tilde{\Omega}_1 \\ \tilde{\Omega}_2 \end{bmatrix}, \end{aligned} \tag{2.52}$$

where

$$\tilde{\Omega}_1 = \begin{bmatrix} x^h(i-d_1^h, j) - x^h(i-d_h(i), j) \\ x^h(i-d_1^h, j) + x^h(i-d_h(i), j) - 2\xi_{12} \end{bmatrix},$$
$$\tilde{\Omega}_2 = \begin{bmatrix} x^h(i-d_h(i), j) - x^h(i-d_2^h, j) \\ x^h(i-d_h(i), j) + x^h(i-d_2^h, j) - 2\xi_{13} \end{bmatrix}.$$

Note in additional that, when $d_h(i) = d_1^h$ or $d_h(i) = d_2^h$, then $\tilde{\Omega}_1 = 0$ and $\tilde{\Omega}_2 = 0$, respectively, and the last inequality in (2.52) is still valid. Similar to (2.52), we also have

$$- d_{12}^v \sum_{l=j-d_2^v}^{j-d_1^v-1} \eta_v^\top(i,l) R_2^v \eta_v(i,l) \le -\begin{bmatrix} \Theta_1 \\ \Theta_2 \end{bmatrix}^\top \begin{bmatrix} \widetilde{R}_2^v & Y \\ Y^\top & \widetilde{R}_2^v \end{bmatrix} \begin{bmatrix} \Theta_1 \\ \Theta_2 \end{bmatrix}, \tag{2.53}$$

where

$$\Theta_1 = \begin{bmatrix} x^v(i, j - d_1^v) - x^v(i, j - d_v(j)) \\ x^v(i, j - d_1^v) + x^v(i, j - d_v(j)) - 2\xi_{22} \end{bmatrix},$$

$$\Theta_2 = \begin{bmatrix} x^v(i, j - d_v(j)) - x^v(i, j - d_2^v) \\ x^v(i, j - d_v(j)) + x^v(i, j - d_2^v) - 2\xi_{23} \end{bmatrix}.$$

Since $\tilde{\Omega}_1 = \mathbb{J}_{n_h}^2 \Gamma_2 \chi(i, j), \tilde{\Omega}_2 = \mathbb{J}_{n_h}^2 \Gamma_3 \chi(i, j), \Theta_1 = \mathbb{J}_{n_v}^2 \Gamma_2 \chi(i, j)$ and $\Theta_2 = \mathbb{J}_{n_v}^2 \Gamma_3(i, j)$, from (2.51)–(2.53), we have

$$\Delta V_4(x(i, j)) \leq \chi^\top(i, j) \left(\mathscr{A}^\top \hat{R}_2 \mathscr{A} - \Pi_4 \right) \chi(i, j). \tag{2.54}$$

Finally, from (2.44), (2.50), and (2.54), we obtain

$$\Delta V(x(i, j)) \leq \chi^\top(i, j) \left(\Pi_0(d_h, d_v) + \sum_{k=1}^{2} \Pi_k - \sum_{k=3}^{4} \Pi_k \right) \chi(i, j). \tag{2.55}$$

It can be verified that the matrix

$$\Pi(d_h, d_v) = \Pi_0(d_h, d_v) + \sum_{k=1}^{2} \Pi_k - \sum_{k=3}^{4} \Pi_k$$

is linearly dependent (an affine function) to the delay parameters d_h and d_v. Thus,

$$\Pi(d_h, d_v) < 0$$

for all $(d_h, d_v) \in [d_1^h, d_2^h] \times [d_1^v, d_2^v]$ if and only if $\Pi(d_h, d_v) < 0$ for $d_h \in \{d_1^h, d_2^h\}$ and $d_v \in \{d_1^v, d_2^v\}$. Based on this fact, it can be deduced from (2.36) that there exists a positive scalar λ_0 such that

$$\begin{aligned} \Delta V(x(i, j)) &\leq -\lambda_0 \|\chi(i, j)\|^2 \\ &\leq -\lambda_0 \|x(i, j)\|^2. \end{aligned}$$

Thus, by Lemma 2.1, system (2.1) is asymptotically stable. The proof is completed. □

Remark 2.8 Theorem 2.1 provides a delay-dependent stability criterion for 2-D systems in the form of (2.1) with directional time-varying delays $d_h(i), d_v(j)$ whose lower bounds d_1^h and d_1^v are positive integers. When $d_1^h = 0$ and $d_1^v = 0$, the terms associated with the matrices Q_1 and R_1 in the expression of functional $V(x(i, j))$ given in (2.34) will be omitted and an alternative result of Theorem 2.1 can be obtained.

Remark 2.9 The result of Theorem is applicable for 2-D time-delay systems with arbitrary rate of varying. This means that the time delays can be fast varying in each of directions. For a more specific type of delays with prescribed ranges on the rate of change

$$\mu_1^h \le \Delta d_h(i) \le \mu_2^h, \quad \mu_1^v \le \Delta d_v(j) \le \mu_2^v, \tag{2.56}$$

where $\Delta d_h(i) = d_h(i+1) - d_h(i)$, $\Delta d_v(j) = d_v(j+1) - d_v(j)$, and μ_1^h, μ_2^h, μ_1^v, μ_2^v are known scalars, the matrix-separation methods [6, 7] can be utilized in combination with the techniques presented in the proof of Theorem 2.1 to derive delay-variation-dependent stability conditions. Based on this observation, extended results of Theorem 2.1 can be obtained.

Remark 2.10 For the case of constant delays (i.e. $d_1^h = d_2^h = d_h$ and $d_1^v = d_2^v = d_v$), the proposed functional (2.34) is reduced to the following simpler form

$$\begin{aligned}\hat{V}(x(i,j)) = {} & \begin{bmatrix}\hat{x}^h(i,j)\\ \hat{x}^v(i,j)\end{bmatrix}^\top P \begin{bmatrix}\hat{x}^h(i,j)\\ \hat{x}^v(i,j)\end{bmatrix} + \sum_{k=i-d_h}^{i-1} x^{h\top}(k,j)Q^h x^h(k,j) \\ & + \sum_{l=j-d_v}^{j-1} x^{v\top}(i,l)Q^v x^v(i,l) + d_h \sum_{k=-d_h}^{-1}\sum_{l=i+k}^{i-1} \eta_h^\top(l,j)R^h \eta_h(l,j) \\ & + d_v \sum_{k=-d_v}^{-1}\sum_{l=j+k}^{j-1} \eta_v^\top(i,l)R^v \eta_v(i,l),\end{aligned} \tag{2.57}$$

where

$$\hat{x}^h(i,j) = \begin{bmatrix} x^h(i,j) \\ \sum_{k=i-d_h}^{i-1} x^h(k,j)\end{bmatrix}, \quad \hat{x}^v(i,j) = \begin{bmatrix} x^v(i,j) \\ \sum_{l=j-d_1^v}^{j-1} x^v(i,l)\end{bmatrix}.$$

By some similar lines used in the proof of Theorem 2.1, we have the following result.

Corollary 2.2 *Consider a class of 2-D systems with constant delays d_h, d_v as given in (2.1). Assume that there exist matrices $P^h \in \mathbb{S}_{2n_h}^+$, $P^v \in \mathbb{S}_{2n_v}^+$, Q^h, $R^h \in \mathbb{S}_{n_h}^+$, Q^v, $R^v \in \mathbb{S}_{n_v}^+$, such that the following LMI condition holds*

$$\Xi_1 + \Xi_2 + e_1^\top Q e_1 - e_2^\top Q e_2 + \hat{\mathscr{A}}^\top \hat{R} \hat{\mathscr{A}} - \hat{\Gamma}^\top \hat{\Phi} \hat{\Gamma} < 0, \tag{2.58}$$

where $e_k \in \mathbb{R}^{n\times 3n}$ $(k = 1, 2, 3)$ denote the block-rows of the identity matrix I_{3n}, $\hat{\mathscr{A}} = (A - I_n)e_1 + A_d e_2$, $Q = \operatorname{diag}(Q^h, Q^v)$, $R = \operatorname{diag}(R^h, R^v)$, $\hat{R} = \operatorname{diag}\left(d_h^2 R^h, d_v^2 R^v\right)$ and

$$
\begin{aligned}
\Xi_1 &= \begin{bmatrix} \hat{\mathscr{A}} + e_1 \\ (1+d_h)e_3 - e_2 \end{bmatrix}^\top \widehat{P}^h \begin{bmatrix} \hat{\mathscr{A}} + e_1 \\ (1+d_h)e_3 - e_2 \end{bmatrix} \\
&\quad - \begin{bmatrix} e_1 \\ (1+d_h)e_3 - e_1 \end{bmatrix}^\top \widehat{P}^h \begin{bmatrix} e_1 \\ (1+d_h)e_3 - e_1 \end{bmatrix}, \\
\Xi_2 &= \begin{bmatrix} \hat{\mathscr{A}} + e_1 \\ (1+d_v)e_3 - e_2 \end{bmatrix}^\top \widehat{P}^v \begin{bmatrix} \hat{\mathscr{A}} + e_1 \\ (1+d_v)e_3 - e_2 \end{bmatrix} \\
&\quad - \begin{bmatrix} e_1 \\ (1+d_v)e_3 - e_1 \end{bmatrix}^\top \widehat{P}^v \begin{bmatrix} e_1 \\ (1+d_v)e_3 - e_1 \end{bmatrix}, \\
\hat{\Gamma} &= \left[(e_1 - e_2)^\top \ (e_1 + e_2 - 2e_3)^\top \right]^\top, \\
\hat{\Phi} &= \operatorname{diag}\left(R, \frac{3(d_h+1)}{d_h - 1} R^h, \frac{3(d_v+1)}{d_v - 1} R^v \right), \\
\widehat{P}^h &= \mathbb{J}_{n_h}^{2\top} P^h \mathbb{J}_{n_h}^2, \quad \widehat{P}^v = \mathbb{J}_{n_v}^{2\top} P^v \mathbb{J}_{n_v}^2.
\end{aligned}
$$

The, system (2.1) is asymptotically stable.

2.4.2 2-D Systems with Delays in FMLSS Model

Consider the following 2-D system described by Fornasini–Marchesini local state-space model (FMLSS)

$$
\begin{aligned}
x(i+1, j+1) &= A_1 x(i+1, j) + A_2 x(i, j+1) \\
&\quad + A_{d1} x(i+1, j - d_v(j)) + A_{d2} x(i - d_h(i), j+1),
\end{aligned} \tag{2.59}
$$

where $x(i, j) \in \mathbb{R}^n$ is the state vector, A_1, A_2, and A_{d1}, A_{d2} are given matrices of appropriate dimensions. The time-varying delays $d_h(i)$ and $d_v(j)$ satisfy

$$
d_1^h \le d_h(i) \le d_2^h, \quad d_1^v \le d_v(j) \le d_2^v.
$$

Note at first that, in some special cases, the FMLSS model (2.59) can be transformed to the Roesser model (2.1). For example, if either the delays are ignored in system (2.59), that is, $A_{d1} = 0$ and $A_{d2} = 0$, or $A_{d1} = 0$ and $A_{d2}A_1 = 0$, then by using the state transform

$$
x^h(i, j) = x(i, j+1) - A_1 x(i, j) \text{ and } x^v(i, j) = x(i, j),
$$

system (2.59) can be recast into system (2.1) with the system matrices

$$
A = \begin{bmatrix} A_2 & A_2 A_1 \\ I_n & A_1 \end{bmatrix}, \quad A_d = \begin{bmatrix} A_{d2} & 0 \\ 0 & 0 \end{bmatrix}.
$$

In such a case, stability conditions for system (2.59) can be reduced from those of Theorem 2.1. However, in general, 2-D systems with time-varying delays described by the Roesser model (2.1) and FMLSS model (2.59) are often studied separately in the literature due to their special structures and wide applications. By some suitable modifications, it can be verified that the results presented in the preceding section in this chapter can be extended to 2-D systems in FMLSS model as given in (2.59). Specifically, since the model (2.59) uses only one state vector to describe the information propagation in two directions, the functional $V(x(i, j))$ in (2.34) is now defined as

$$V(x(i, j)) = V^h(x(i, j)) + V^v(x(i, j)).$$

Similar to Lemma 2.1, if there exist positive scalars $\tilde{c}_0$ and c_1, c_2 that satisfy the following conditions

$$c_1\|x\|^2 \leq V(x) \leq c_2\|x\|^2 \tag{2.60}$$

for any $x \in \mathbb{R}^n$ and along any state trajectory $x(i, j)$ of system (2.59),

$$\Delta V \triangleq V(x(i+1, j+1)) - V^h(x(i, j+1)) - V^v(x(i+1, j)) \leq -\tilde{c}_0\|x(i, j)\|^2 \tag{2.61}$$

then, we have

$$\limsup_{k\to\infty} \sum_{i+j=k} V(x(i, j)) = 0.$$

Let $x_{\theta,\lambda} = x(i+\theta, j+\lambda)$. To illustrate the analysis scheme presented above, we select an LKF candidate as

$$V(x(i, j)) = V^h(x(i, j)) + V^v(x(i, j)), \tag{2.62}$$

where

$$\begin{aligned}
V^h(x(i, j)) &= x_{0,0}^{\top}(P - Q)x_{0,0} + \sum_{k=-d_2^h}^{-1} x_{k,0}^{\top} R^h x_{k,0} + \sum_{k=-d_2^h}^{-d_1^h}\sum_{l=k}^{-1} x_{l,0}^{\top} Q^h x_{l,0} \\
&\quad + d_1^h \sum_{k=-d_1^h}^{-1}\sum_{l=k}^{-1} \Delta_h x_{l,0}^{\top} S^h \Delta_h x_{l,0} + d_{12}^h \sum_{k=-d_2^h}^{-d_1^h-1}\sum_{l=k}^{-1} \Delta_h x_{l,0}^{\top} Z^h \Delta_h x_{l,0}, \\
V^v(x(i, j)) &= x_{0,0}^{\top} Q x_{0,0} + \sum_{k=-d_2^v}^{-1} x_{0,k}^{\top} R^v x_{0,k} + \sum_{k=-d_2^v}^{-d_1^v}\sum_{l=k}^{-1} x_{0,l}^{\top} Q^v x_{0,l} \\
&\quad + d_1^v \sum_{k=-d_1^v}^{-1}\sum_{l=k}^{-1} \Delta_v x_{0,l}^{\top} S^v \Delta_v x_{0,l} + d_{12}^v \sum_{k=-d_2^v}^{-d_1^v-1}\sum_{l=k}^{-1} \Delta_v x_{0,l}^{\top} Z^v \Delta_v x_{0,l},
\end{aligned}$$

and directional shifted operators $\Delta_h x_{\theta,\lambda} = x(i+1+\theta, j+\lambda) - x(i+\theta, j+\lambda)$, $\Delta_v x_{\theta,\lambda} = x(i+\theta, j+1+\lambda) - x(i+\theta, j+\lambda)$. In addition, for convenience, we denote the augmented vectors

$$\mathscr{X} = \begin{bmatrix} \mathscr{X}_{d1} \\ \mathscr{X}_{d2} \\ \mathscr{X}_{s1} \\ \mathscr{X}_{s2} \end{bmatrix}, \quad \mathscr{X}_{d1} = \begin{bmatrix} x_{1,0} \\ x_{0,1} \\ x_{1,-d_v(j)} \\ x_{-d_h(i),1} \end{bmatrix}, \quad \mathscr{X}_{d2} = \begin{bmatrix} x_{1,-d_1^v} \\ x_{-d_1^h,1} \\ x_{1,-d_2^v} \\ x_{-d_2^h,1} \end{bmatrix},$$

$$\mathscr{X}_{s1} = \begin{bmatrix} \frac{1}{d_{12}^v - 1} \sum\limits_{k=-d_2^v+1}^{-d_1^v-1} x_{1,k} \\ \frac{1}{d_{12}^h - 1} \sum\limits_{k=-d_2^h+1}^{-d_1^h-1} x_{k,1} \end{bmatrix}, \quad \mathscr{X}_{s2} = \begin{bmatrix} \frac{1}{d_1^v - 1} \sum\limits_{k=-d_1^v+1}^{-1} x_{1,k} \\ \frac{1}{d_1^h - 1} \sum\limits_{k=-d_1^h+1}^{-1} x_{k,1} \end{bmatrix},$$

and block-row matrices $e_k = \begin{bmatrix} 0_{n\times(k-1)n} & I_n & 0_{n\times(12-k)n} \end{bmatrix}$ $(k = 1, 2, \ldots, 12)$. Then, it is easy to verify from (2.59) that

$$x_{1,1} = \underbrace{(A_1 e_1 + A_2 e_2 + A_{d1} e_3 + A_{d2} e_4)}_{\mathscr{A}} \mathscr{X}.$$

According to (2.61), the difference of $V(x(i, j))$ along state trajectories of (2.59) is given as

$$\begin{aligned}
\Delta V = {} & \mathscr{X}^\top \mathscr{M} \mathscr{X} + (d_1^h)^2 \Delta_h x_{0,1}^\top S^h \Delta_h x_{0,1} + (d_1^v)^2 \Delta_v x_{1,0}^\top S^v \Delta_v x_{1,0} \\
& + (d_{12}^h)^2 \Delta_h x_{0,1}^\top Z^h \Delta_h x_{0,1} + (d_{12}^v)^2 \Delta_v x_{1,0}^\top Z^v \Delta_v x_{1,0} \\
& - d_1^h \sum_{k=-d_1^h}^{-1} \Delta_h x_{k,1}^\top S^h \Delta_h x_{k,1} - d_1^v \sum_{k=-d_1^v}^{-1} \Delta_v x_{1,k}^\top S^v \Delta_h x_{1,k} \\
& - d_{12}^h \sum_{k=-d_2^h}^{-d_1^h-1} \Delta_h x_{k,1}^\top Z^h \Delta_h x_{k,1} - d_{12}^v \sum_{k=-d_2^v}^{-d_1^v-1} \Delta_v x_{1,k}^\top Z^v \Delta_h x_{1,k},
\end{aligned}$$

where

$$\begin{aligned}
\mathscr{M} = {} & \mathscr{A}^\top P \mathscr{A} - e_1^\top Q e_1 - e_2^\top (P - Q) e_2 \\
& + e_1^\top R^v e_1 + e_2^\top R^h e_2 - e_7^\top R^v e_7 - e_8^\top R^h e_8 \\
& + (d_{12}^v + 1) e_1^\top Q^v e_1 + (d_{12}^h + 1) e_2^\top Q^h e_2 - e_3^\top Q^v e_3 - e_4^\top Q^h e_4. \qquad (2.63)
\end{aligned}$$

By utilizing the Abel-based summation inequality (2.24), we have

$$
\begin{aligned}
-d_1^h \sum_{k=-d_1^h}^{-1} \Delta_h x_{k,1}^\top S^h \Delta_h x_{k,1} \leq & -\mathscr{X}^\top (e_2 - e_6)^\top S^h (e_2 - e_6)\mathscr{X} \\
& - \frac{3(d_1^h - 1)}{d_1^h + 1} \mathscr{X}^\top (e_2 + e_6 - 2e_{12})^\top S^h (e_2 + e_6 - 2e_{12})\mathscr{X}.
\end{aligned}
$$

The summation terms $-d_1^v \sum_{k=-d_1^v}^{-1} \Delta_v x_{1,k}^\top S^v \Delta_h x_{1,k}$, $-d_{12}^h \sum_{k=-d_2^h}^{-d_1^h-1} \Delta_h x_{k,1}^\top Z^h \Delta_h x_{k,1}$, and $-d_{12}^v \sum_{k=-d_2^v}^{-d_1^v-1} \Delta_v x_{1,k}^\top Z^v \Delta_h x_{1,k}$ can be manipulated similarly. Note also that

$$
\begin{aligned}
\Delta_v x_{1,0} &= x_{1,1} - x_{1,0} = (\mathscr{A} - e_1)\mathscr{X}, \\
\Delta_h x_{0,1} &= x_{1,1} - x_{0,1} = (\mathscr{A} - e_2)\mathscr{X}.
\end{aligned}
$$

Therefore,

$$
\Delta V \leq \mathscr{X}^\top (\mathscr{M} - \mathscr{N} + \mathscr{O})\, \mathscr{X}, \tag{2.64}
$$

where

$$
\begin{aligned}
\mathscr{N} = & (e_1 - e_5)^\top S^v (e_1 - e_5) + (e_2 - e_6)^\top S^h (e_2 - e_6) \\
& + (e_5 - e_7)^\top Z^v (e_5 - e_7) + (e_6 - e_8)^\top Z^h (e_6 - e_8) \\
& + \frac{3(d_1^v - 1)}{d_1^v + 1} (e_1 + e_5 - 2e_{11})^\top S^v (e_1 + e_5 - 2e_{11}) \\
& + \frac{3(d_1^h - 1)}{d_1^h + 1} (e_2 + e_6 - 2e_{12})^\top S^h (e_2 + e_6 - 2e_{12}) \\
& + \frac{3(d_{12}^v - 1)}{d_{12}^v + 1} (e_5 + e_7 - 2e_9)^\top Z^v (e_5 + e_7 - 2e_9) \\
& + \frac{3(d_{12}^h - 1)}{d_{12}^h + 1} (e_6 + e_8 - 2e_{10})^\top Z^h (e_2 + e_6 - 2e_{10})
\end{aligned} \tag{2.65}
$$

and

$$
\begin{aligned}
\mathscr{O} = & (\mathscr{A} - e_1)^\top \left((d_1^v)^2 S^v + (d_{12}^v)^2 Z^v\right) (\mathscr{A} - e_1) \\
& + (\mathscr{A} - e_2)^\top \left((d_1^h)^2 S^h + (d_{12}^h)^2 Z^h\right) (\mathscr{A} - e_2).
\end{aligned} \tag{2.66}
$$

From (2.61) and (2.64), we have the following result.

Theorem 2.2 *Consider system (2.59) with directional delays $d_h(i) \in [d_1^h, d_2^h]$ and $d_v(j) \in [d_1^v, d_2^v]$. Assume that there exist symmetric positive definite matrices P, Q, Q^h, Q^v, R^h, R^v, S^h, S^v, and Z^h, Z^v of appropriate dimension that satisfy the following LMI-based condition*

$$
\mathscr{M} - \mathscr{N} + \mathscr{O} < 0, \tag{2.67}
$$

where the matrices $\mathscr{M}$, $\mathscr{N}$ and $\mathscr{O}$ are defined in (2.63), (2.65), and (2.66), respectively. Then, the 2-D system (2.59) is asymptotically stable.

Remark 2.11 Under the analysis scheme presented in this section, the result of Theorem 2.2 can be improved using more complex LKFs with augmented state vectors and some new 2-D polynomial-based summation inequalities [7, 24]. In addition, by slightly modifying the proof of Theorem 2.2 to incorporate the rate of change of time delays

$$\mu_1^h \le \Delta d_h(i) \le \mu_2^h, \quad \mu_1^v \le \Delta d_v(j) \le \mu_2^v,$$

where μ_1^h, μ_2^h, μ_1^v, μ_2^v are known scalars, analogous delay-variation-dependent stability conditions of Theorem 2.2 can be obtained.

2.4.3 Numerical Examples

In this section, we present some numerical examples to illustrate the efficacy of the obtained stability conditions.

Example 2.1 Figure 2.1 portrays a schematic control system of the thermal process in chemical reactors, heat exchanger or pipe furnaces which is described by the following Darboux-type equation [25]

$$\begin{aligned} \frac{\partial T(x,t)}{\partial x} + \frac{\partial T(x,t)}{\partial t} &= -a_0 T(x,t) - a_1 T(x-d_x,t) + bu(x,t), \\ T_{\text{out}} &= cT(x,t), \end{aligned} \tag{2.68}$$

where $T(x,t)$ is unknown variable (e.g. temperature) at space $x \in [0, x_f]$ and time t, $u_{\text{in}} = u(x,t)$ is the control input, T_{out} is the measurement output, d_x is a quantity involving space delay, and a_0, a_1, b, c are real constants.

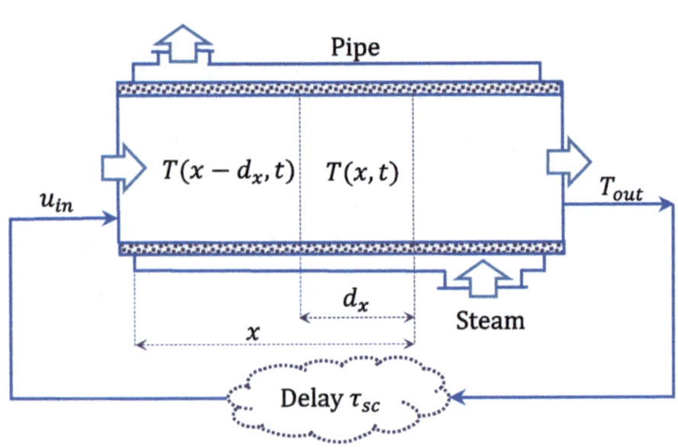

Fig. 2.1 A schematic heat exchanger control

For given increments Δx, Δt, we discretize equation (2.68) by simple forward Euler approximation

$$\frac{\partial T(x,t)}{\partial x} \approx \frac{T(i,j) - T(i-1,j)}{\Delta x}, \quad \frac{\partial T(x,t)}{\partial t} \approx \frac{T(i,j+1) - T(i,j)}{\Delta t},$$

and $T(i,j) = T(i\Delta x, j\Delta t)$, $u(i,j) = u(i\Delta x, j\Delta t)$. Then, equation (2.68) can be written in the form

$$\begin{aligned} T(i,j+1) = {} & \left(1 - \frac{\Delta t}{\Delta x} - a_0\Delta t\right) T(i,j) + \frac{\Delta t}{\Delta x} T(i-1,j) \\ & - a_1\Delta t T(i-1-d_s(i), j) + b\Delta t u(i,j), \end{aligned} \tag{2.69}$$

where $d_s(i) = \lfloor \frac{d_x}{\Delta x} \rfloor$ and $\lfloor . \rfloor$ denotes the floor function.

In control engineering, due to many practical reasons regarding the finite speed of data processing through a low-rate communication channel or sensor technology, the impact of time-delay cannot be ignored. We assume that the communication delay occurs in the signal processing. This time-delay is typically unknown or time-varying. Thus, a delayed static output-feedback for system (2.69) is taken in the form

$$u(i,j) = kT_{\text{out}}(i, j - d_c(j)) = kcT(i, j - d_c(j)).$$

We now define the state variables $x^h(i,j) = T(i-1,j)$ and $x^v(i,j) = T(i,j)$. Then, the closed-loop system of (2.69) can be recast into system (2.1), where

$$A = \begin{bmatrix} 0 & 1 \\ \frac{\Delta t}{\Delta x} & 1 - \frac{\Delta t}{\Delta x} - a_0\Delta t \end{bmatrix}, \quad A_d = \begin{bmatrix} 0 & 0 \\ -a_1\Delta t & bkc\Delta t \end{bmatrix}.$$

For illustrative purpose, we give $\Delta t = 0.1$, $\Delta x = 0.2$ and $a_0 = 1$, $a_1 = 1.25$. With the given parameters, simulation results show that the open-loop system of (2.69) is unstable for constant delay $d_s = 8$. The conducted simulations in Figs. 2.2 and 2.3 are taken with finite support initial conditions $x^h(i,j) = 2\cos(0.1j)$, $i \in \mathbb{Z}[-6,0]$, $j \le 100$, $x^v(i,0) = 2\sin(0.1i)$, $i \le 100$, and $x^h(0,j) = 0$, $x^v(i,0) = 0$ elsewhere. It can be seen that the state trajectories $x^h(i,j)$ and $x^v(i,j)$ are divergent as $i + j \to \infty$.

Now, we apply the analysis result presented in Theorem 2.1. Let $b = 1$, $c = 0.8$, $k = -0.5$ and time-varying delay $d_c(j) \in [2,4]$. By using the Matlab LMI toolbox, it is found that the derived stability conditions in Theorem 2.1 are feasible with

$$P^h = \begin{bmatrix} 13.1863 & -0.0956 & 0.0891 \\ -0.0956 & 0.0014 & -0.0149 \\ 0.0891 & -0.0149 & 4.7744 \end{bmatrix}, \quad P^v = \begin{bmatrix} 22.2257 & -0.6020 & 0.0161 \\ -0.6020 & 0.1801 & -0.0654 \\ 0.0161 & -0.0654 & 0.0383 \end{bmatrix},$$

$$Q_1 = \begin{bmatrix} 2.4925 & 0 \\ 0 & 0.0379 \end{bmatrix}, Q_2 = \begin{bmatrix} 1.4253 & 0 \\ 0 & 0.0086 \end{bmatrix}, R_1 = \begin{bmatrix} 7.0 & 0 \\ 0 & 9.0 \end{bmatrix},$$

$$R_2 = \begin{bmatrix} 8.0 & 0 \\ 0 & 10.0 \end{bmatrix}, X = \begin{bmatrix} -20.9594 & 0.0877 \\ -6.0660 & -13.9732 \end{bmatrix}, Y = \begin{bmatrix} -0.1090 & 0.0167 \\ -0.0260 & 0.0318 \end{bmatrix}.$$

Fig. 2.2 Open-loop state trajectory x^h

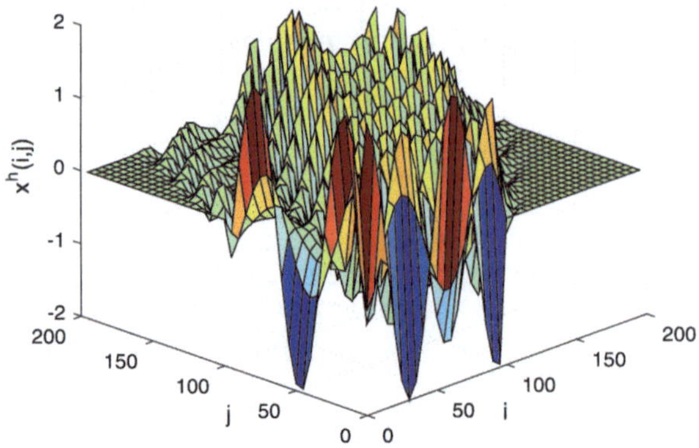

Fig. 2.3 Open-loop state trajectory x^v

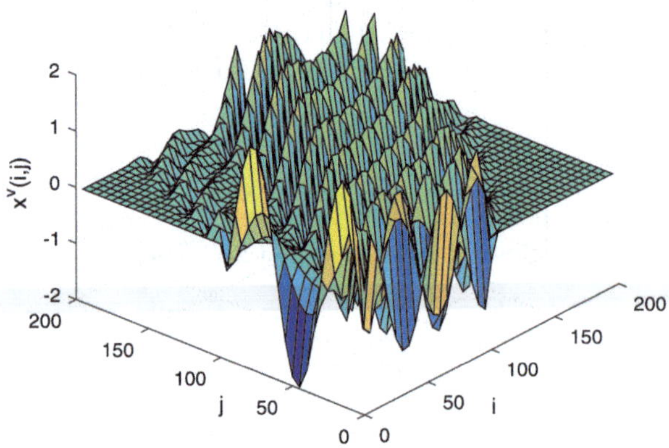

By Theorem 2.1, the closed-loop system of (2.69) is asymptotically stable. State trajectories x^h, x^v of the closed-loop system with $d_c(j) = 2 + 2|\sin(j\pi/4)|$ are presented in Figs. 2.4 and 2.5. It i clear that the closed-loop state trajectories $x^h(i, j)$ and $x^v(i, j)$ are convergent as $i + j \to \infty$.

For constant delay d_c, by the result of Theorem 1 in [10] (with $\alpha = 1$) and that of Corollary 2.2 in this chapter, the closed system of (2.69) is asymptotically stable with delays d_s, d_c in the domains $\mathscr{R}_1 = \{(1, l), (2, l), (3, 3) : l \in \mathbb{N}\}$ and $\mathscr{R}_2 = \{(k, l), (l, k) : k, l \in \mathbb{N}, k \leq 5\}$, respectively. Clearly, our conditions give better results than the ones proposed in [10]. In addition, the numbers of LMI decision variables (NoDv) of Corollary 2.2 and of Theorem 1 in [10] are $2(n_h^2 + n_v^2) + 3(n_h + n_v)$ and $5.5(n_h^2 + n_v^2) + 2.5(n_h + n_v)$, respectively. This shows the effectiveness of our method in reducing the conservativeness of the derived stability conditions.

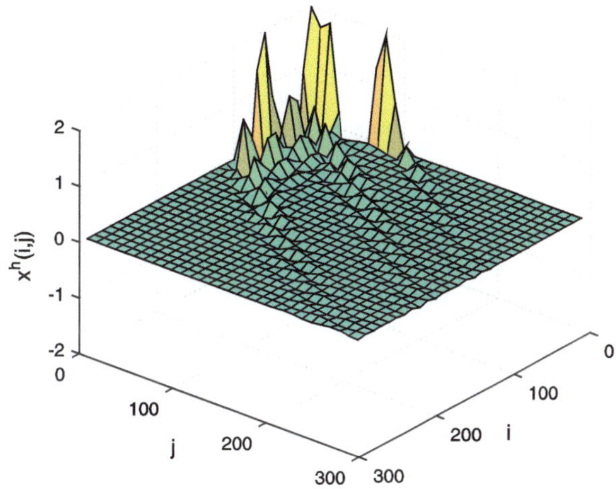

Fig. 2.4 Closed-loop state trajectory x^h

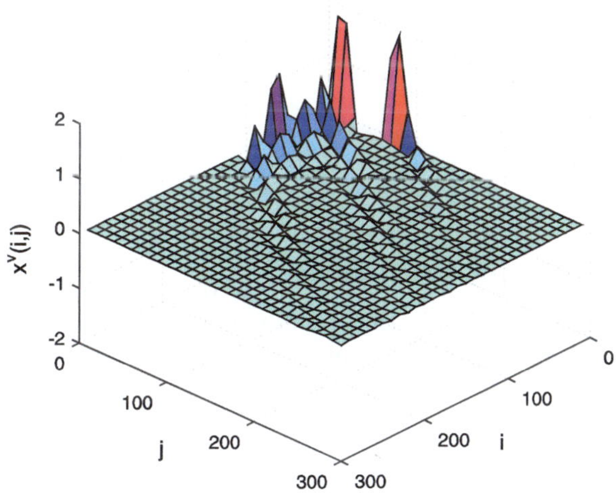

Fig. 2.5 Closed-loop state trajectory x^v

Example 2.2 Consider a 2-D system in the form of (2.1) with the system matrices

$$A = \left[\begin{array}{cc:c} 0.8 & -0.6 & 0.12 \\ 0.15 & 0.1 & 0 \\ \hdashline 0.15 & 0.1 & 0.25 \end{array}\right], \quad A_d = \left[\begin{array}{cc:c} 0.15 & -0.1 & 0 \\ 0.12 & -0.2 & -0.1 \\ \hdashline 0.05 & -0.06 & 0.11 \end{array}\right].$$

For time-varying delays $d_h(i) \in [d_1^h, d_2^h]$ and $d_v(j) \in [d_1^v, d_2^v]$ with $d_1^h < d_2^h$ and $d_1^v < d_2^v$, the results of [10] are not applicable. Alternatively, Theorem 1 in [26] can be applied to this case. To illustrate, let $d_1^h = 2$ and $d_1^v = 2$. The upper bound of d_h for various d_v obtained by Theorem 2.1 and Theorem 1 in [26] is presented in Table 2.1. Clearly, our method delivers better results than the one based on Jensen inequalities [26].

Table 2.1 Upper bound of d_h for various d_v

d_2^v	3	5	7	9	10
[26]	7	6	5	4	3
Theorem 2.1	10	10	9	8	6

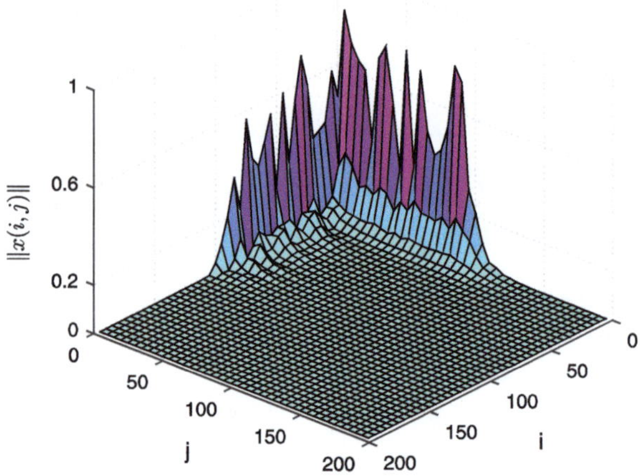

Fig. 2.6 Norm-state trajectory $\left\| \begin{bmatrix} x^h \\ x^v \end{bmatrix} \right\|$

The norm-state trajectory $\left\| \begin{bmatrix} x^h(i,j) \\ x^v(i,j) \end{bmatrix} \right\|$ of the system with time-varying delays $d_h(i) = 2 + 4|\sin(i\pi/2)$ and $d^v(j) = 2 + 8|\cos(j\pi/2)|$ is presented in Fig. 2.6. This validates the theoretical results.

2.5 Stabilization of 2-D Roesser Systems Time-Varying Delays

2.5.1 State-Feedback Controller Design

Consider the following control system

$$\begin{bmatrix} x^h(i+1,j) \\ x^v(i,j+1) \end{bmatrix} = A \begin{bmatrix} x^h(i,j) \\ x^v(i,j) \end{bmatrix} + A_d \begin{bmatrix} x^h(i-d_h(i),j) \\ x^v(i,j-d_v(j)) \end{bmatrix} + Bu(i,j), \tag{2.70}$$

where $x^h(i,j) \in \mathbb{R}^{n_h}$ and $x^v(i,j) \in \mathbb{R}^{n_v}$ are state vectors and $u(i,j) \in \mathbb{R}^m$ represents the control input. The system matrices $A, A_d \in \mathbb{R}^{n\times n}$ $(n = n_h + n_v)$ and $B \in \mathbb{R}^{n\times m}$ are given real matrices. Time-varying state delays d_h, d_v satisfy

$$d_m^h \leq d_h(i) \leq d_M^h, \quad d_m^v \leq d_v(j) \leq d_M^v,$$

where $d_m^h, d_m^v, d_M^h, d_M^v$ are known positive integers involving the upper and the lower bounds of delays. In this section, we demonstrate the design problem of a stabilizing state-feedback controller (SFC) for system (2.70) in the form

$$u(i, j) = K \begin{bmatrix} x^h(i, j) \\ x^v(i, j) \end{bmatrix}. \tag{2.71}$$

By incorporating the SFC (2.71), the closed-loop system of (2.70) is given by

$$\begin{bmatrix} x^h(i+1, j) \\ x^v(i, j+1) \end{bmatrix} = A_c \begin{bmatrix} x^h(i, j) \\ x^v(i, j) \end{bmatrix} + A_d \begin{bmatrix} x^h(i - d_h(i), j) \\ x^v(i, j - d_v(j)) \end{bmatrix} \tag{2.72}$$

where $A_c = A + BK$.

We first formulate stability conditions for the closed-loop system (2.72). For this, we construct a 2-D LKF candidate as

$$V \triangleq V(x(i, j)) = \sum_{k=1}^{7} V_k(x(i, j)),$$

where

$$V_1 = \begin{bmatrix} x^h(i, j) \\ x^v(i, j) \end{bmatrix}^\top \underbrace{\mathrm{diag}(P^h, P^v)}_{P} \begin{bmatrix} x^h(i, j) \\ x^v(i, j) \end{bmatrix},$$

$$V_2 = \sum_{k=i-d_m^h}^{i-1} x^{h\top}(k, j) Q_1^h x^h(k, j) + \sum_{l=j-d_m^v}^{j-1} x^{v\top}(i, l) Q_1^v x^v(i, l),$$

$$V_3 = \sum_{k=i-d_M^h}^{i-1} x^{h\top}(k, j) Q_2^h x^h(k, j) + \sum_{l=j-d_M^v}^{j-1} x^{v\top}(i, l) Q_2^v x^v(i, l),$$

$$V_4 = \sum_{k=i-d_h(i)}^{i-1} x^{h\top}(k, j) Q^h x^h(k, j) + \sum_{l=j-d_v(j)}^{j-1} x^{v\top}(i, l) Q^v x^v(i, l),$$

$$V_5 = \sum_{k=-d_M^h+1}^{-d_m^v} \sum_{l=i+k}^{i-1} x^{h\top}(l, j) Q^h x^h(l, j) + \sum_{k=-d_M^h+1}^{-d_m^v} \sum_{l=j+k}^{j-1} x^{v\top}(i, l) Q^v x^v(i, l),$$

$$V_6 = d_m^h \sum_{k=-d_m^h}^{-1} \sum_{l=i+k}^{i-1} \eta_h^\top(l, j) R_1^h \eta_h(l, j) + d_m^v \sum_{k=-d_m^v}^{-1} \sum_{l=j+k}^{j-1} \eta_v^\top(i, l) R_1^v \eta^v(i, l),$$

$$V_7 = d_M^h \sum_{k=-d_M^h}^{-1} \sum_{l=i+k}^{i-1} \eta_h^\top(l, j) R_2^h \eta^h(l, j) + d_M^v \sum_{k=-d_M^v}^{-1} \sum_{l=j+k}^{j-1} \eta_v^\top(i, t) R_2^v \eta_v(i, l),$$

and $\eta_h(i, j) = \Delta_h(x^h(i, j)), \Delta_v(x^v(i, j))$.

For convenience, we denote the block matrix $\mathbb{I}(\alpha, \beta) = \mathrm{diag}(\alpha I_{n_h}, \beta I_{n_v})$ for given scalars α, β, and augmented vectors

$$\begin{aligned} x_{dm}(i, j) &= \begin{bmatrix} x^h(i - d_m^h, j) \\ x^v(i, j - d_m^v) \end{bmatrix}, \ x_{dM}(i, j = \begin{bmatrix} x^h(i - d_M^h, j) \\ x^v(i, j - d_M^v) \end{bmatrix}, \\ x_d(i, j) &= \begin{bmatrix} x^h(i - d_h(i), j) \\ x^v(i, j - d_v(j)) \end{bmatrix}, \ \Upsilon(i, j) = \begin{bmatrix} \Upsilon_1(i, j) \\ \Upsilon_2(i, j) \end{bmatrix}, \\ \Upsilon_1(i, j) &= \begin{bmatrix} x(i, j) \\ x_d(i, j)) \end{bmatrix}, \ \Upsilon_2(i, j) = \begin{bmatrix} x_{dm}(i, j)) \\ x_{dM}(i, j) \end{bmatrix}. \end{aligned}$$

By splitting the functional $V(x(i, j))$ as $V(x(i, j)) = V^h(x^h(i, j)) + V^v(x^v(i, j))$, we compute the difference $\Delta V = \Delta V_k$ along state trajectory of system (2.71). It can be verified that

$$\begin{aligned} \Delta V_1 &= \Upsilon^\top(i, j) \left(\mathscr{A}_c^\top P \mathscr{A}_c - e_1^\top P e_1 \right) \Upsilon(i, j), \\ \Delta V_2 &= x^{h\top}(i, j) Q_1^h x^h(i, j) + x^{v\top}(i, j) Q_1^v x^v(i, j) \\ &\quad - x^{h\top}(i - d_m^h, j) Q_1^h x^h(i - d_m^h, j) - x^{v\top}(i, j - d_m^v) Q_1^v x^v(i, j - d_m^v) \\ &= \Upsilon^\top(i, j) \left(e_1^\top Q_1 e_1 - e_3^\top Q_1 e_3 \right) \Upsilon(i, j), \\ \Delta V_3 &= \Upsilon^\top(i, j) \left(e_1^\top Q_2 e_1 - e_4^\top Q_2 e_4 \right) \Upsilon(i, j), \end{aligned} \tag{2.73}$$

where $Q_l = \mathrm{diag}(Q_l^h, Q_l^v)$ $(l = 1, 2)$, $e_k = \left[0_{n \times (k-1)n} \ I_n \ 0_{n \times (4-k)n} \right]$ $(k = 1, \ldots, 4)$, and $\mathscr{A}_c = A_c e_1 + A_d e_2$. The difference ΔV_4 is computed and manipulated as follows

$$\begin{aligned} \Delta V_4 &\le x^\top(i, j) Q x(i, j) - x_d^\top(i, j) Q x_d(i, j) \\ &\quad + \sum_{k=i+1-d_M^h}^{i-1} x^{h\top}(k, j) Q^h x^h(k, j) - \sum_{k=i+1-d_m^h}^{i-1} x^{h\top}(k, j) Q^h x^h(k, j) \\ &\quad + \sum_{l=j+1-d_M^v}^{j-1} x^{v\top}(l, j) Q^v x^v(i, l) - \sum_{l=j+1-d_m^v}^{j-1} x^{v\top}(i, j) Q^v x^v(i, l) \\ &\le x^\top(i, j) Q x(i, j) - x_d^\top(i, j) Q x_d(i, j) \\ &\quad + \sum_{k=i+1-d_M^h}^{i-d_m^h} x^{h\top}(k, j) Q^h x^h(k, j) + \sum_{l=j+1-d_M^v}^{j-d_m^v} x^{v\top}(l, j) Q^v x^v(i, l), \end{aligned}$$

where $Q = \mathrm{diag}(Q^h, Q^v)$. In addition, since

$$\begin{aligned}\Delta V_5 = r_h x^{h\top}(i,j)Q^h x^h(i,j) - \sum_{k=i+1-d_M^h}^{i-d_m^h} x^{h\top}(k,j)Q^h x^h(k,j)\\ + r_v x^{v\top}(i,j)Q^v x^v(i,j) - \sum_{l=j+1-d_M^v}^{j-d_m^v} x^{v\top}(l,j)Q^v x^v(i,l),\end{aligned}$$

where $r_h = d_M^h - d_m^h$ and $r_v = d_M^v - d_m^v$, we have

$$\Delta V_4 + \Delta V_5 \le \Upsilon^\top(i,j)\left(e_1^\top \mathbb{I}(1+r_h, 1+r_v)Qe_1 - e_2^\top Qe_2\right)\Upsilon(i,j). \tag{2.74}$$

Next, we manipulate the differences of V_6 and V_7. Observe that

$$\begin{aligned}\Delta V_6 = (d_m^h)^2\eta_h^\top(i,j)R_1^h\eta_h(i,j) - d_m^h\sum_{k-i-d_m^h}^{i-1}\eta_h^\top(k,j)R_1^h\eta_h(k,j)\\ + (d_m^v)^2\eta_v^\top(i,j)R_1^v\eta_v(i,j) - d_m^v\sum_{l-j-d_m^v}^{j-1}\eta_v^\top(i,l)R_1^v\eta_v(i,l).\end{aligned}$$

By Lemma 2.2, we have

$$\begin{aligned}d_m^h\sum_{k=i-d_m^h}^{i-1}\eta_h^\top(k,j)R_1^h\eta_h(k,j) \ge \left(\sum_{k=i-d_m^h}^{i-1}\eta_h(k,j)\right)^\top R_1^h\left(\sum_{k=i-d_m^h}^{i-1}\eta_h(k,j)\right)\\ d_m^v\sum_{l=j-d_m^v}^{j-1}\eta_v^\top(i,l)R_1^v\eta_v(i,l) \ge \left(\sum_{l=j-d_m^v}^{j-1}\eta_v(i,l)\right)^\top R_1^v\left(\sum_{l=j-d_m^v}^{j-1}\eta_v(i,l)\right).\end{aligned}$$

Therefore,

$$\begin{aligned}\Delta V_6 \le \Upsilon^\top(i,j)\Big[(\mathscr{A}-e_1)^\top\mathbb{I}\left((d_m^h)^2,(d_m^v)^2\right)R_1(\mathscr{A}-e_1)\\ -(e_1-e_3)^\top R_1(e_1-e_3)\Big]\Upsilon(i,j),\end{aligned} \tag{2.75}$$

where $R_k = \mathrm{diag}(R_k^h, R_k^v)$ $(k=1,2)$. Similarly, we have

$$\begin{aligned}\Delta V_7 \le \Upsilon^\top(i,j)\Big[(\mathscr{A}_c-e_1)^\top\mathbb{I}\left((d_M^h)^2,(d_M^v)^2\right)R_2(\mathscr{A}_c-e_1)\\ -(e_1-e_4)^\top R_2(e_1-e_4)\Big]\Upsilon(i,j).\end{aligned} \tag{2.76}$$

A combination of (2.73) to (2.76) gives

$$\Delta V \le \Upsilon^\top(i,j)\,(\Phi + \Psi)\,\Upsilon(i,j), \tag{2.77}$$

where

$$\Phi = \begin{bmatrix} \Phi_{11} & 0 & R_1 & R_2 \\ * & -Q & 0 & 0 \\ * & * & -Q_1 - R_1 & 0 \\ * & * & * & -Q_2 - R_2 \end{bmatrix}$$

and

$$\begin{aligned} \Phi_{11} &= -P + Q_1 + Q_2 - R_1 - R_2 + \mathbb{I}(1 + r_h, 1 + r_v)Q, \\ \Psi &= \mathscr{A}_c^\top P \mathscr{A}_c + (\mathscr{A}_c - e_1)^\top \mathbb{I}\left((d_m^h)^2, (d_m^v)^2\right) R_1(\mathscr{A}_c - e_1) \\ &\quad + (\mathscr{A}_c - e_1)^\top \mathbb{I}\left((d_M^h)^2, (d_M^v)^2\right) R_2(\mathscr{A}_c - e_1) \end{aligned}$$

By the Schur complement (Lemma 2.7), we have the following result.

Theorem 2.3 *The closed-loop system (2.72) is asymptotically stable if there exist positive definite matrices* $P = \mathrm{diag}(P^h, P^v)$, $Q = \mathrm{diag}(Q^h, Q^v)$, $Q_k = \mathrm{diag}(Q_k^h, Q_k^v)$, $R_k = \mathrm{diag}(R_k^h, R_k^v)$ *($k = 1, 2$) that satisfy the following condition*

$$\begin{bmatrix} \Phi & \mathscr{A}_c^\top P & (\mathscr{A}_c - e_1)^\top \widehat{R}_1 & (\mathscr{A}_c - e_1)^\top \widehat{R}_2 \\ * & -P & 0 & 0 \\ * & * & -R_1 & 0 \\ * & * & * & -R_2 \end{bmatrix} < 0, \tag{2.78}$$

where $\widehat{R}_1 = \mathbb{I}(d_m^h, d_m^v)R_1$ *and* $\widehat{R}_2 = \mathbb{I}(d_M^h, d_M^v)R_2$

On the basis of the stability conditions given in (2.78), we now formulate design conditions of a stabilizing SFC in the form of (2.71). Let $\mathscr{D}_P = \mathrm{diag}(P, P, P, P)$. By pre- and post-multiplying with $\mathrm{diag}(\mathscr{D}_P^{-1}, P^{-1}, R_1^{-1}, R_2^{-1})$, condition (2.78) holds if and only if the following condition does

$$\begin{bmatrix} \mathscr{D}_P^{-1}\Phi\mathscr{D}_P^{-1} & \mathscr{D}_P^{-1}\mathscr{A}_c^\top & \mathscr{D}_P^{-1}(\mathscr{A}_c - e_1)^\top \mathbb{I}(d_m^h, d_m^v) & \mathscr{D}_P^{-1}(\mathscr{A}_c - e_1)^\top \mathbb{I}(d_M^h, d_M^v) \\ * & -P^{-1} & 0 & 0 \\ * & * & -R_1^{-1} & 0 \\ * & * & * & -R_2^{-1} \end{bmatrix} < 0. \tag{2.79}$$

We define the matrix $X = P^{-1}$. Then, it can see that

$$\mathscr{A}_c \mathscr{D}_P^{-1} = (A + BK)Xe_1 + A_d X e_2.$$

By changing the variable

$$KX = Y \tag{2.80}$$

we have

$$\mathscr{D}_P^{-1}\mathscr{A}_c^\top = \Psi_{X,Y}^\top, \quad \mathscr{D}_P^{-1}(\mathscr{A}_c - e_1)^\top = \hat{\Psi}_{X,Y}^\top,$$

where

$$\Psi_{X,Y} = (AX + BY)e_1 + A_d X e_2, \quad \hat{\Psi}_{X,Y} = (AX - X + BY)e_1 + A_d X e_2.$$

On the other hand, by utilizing the Cauchy matrix inequality (Lemma 2.6), we have

$$-R_k^{-1} \leq X R_k X - 2X, \quad k = 1, 2. \tag{2.81}$$

Finally, by defining the matrices

$$\widetilde{Q} = XQX, \quad \widetilde{Q}_k = XQ_kX, \quad \widetilde{R}_k = XR_kX, \quad k = 1, 2,$$

it can be verified that

$$\mathscr{D}_P^{-1}\Phi\mathscr{D}_P^{-1} \triangleq \widetilde{\Phi} = \begin{bmatrix} \widetilde{\Phi}_{11} & 0 & \widetilde{R}_1 & \widetilde{R}_2 \\ * & -\widetilde{Q} & 0 & 0 \\ * & * & -\widetilde{Q}_1 - \widetilde{R}_1 & 0 \\ * & * & * & -\widetilde{Q}_2 - \widetilde{R}_2 \end{bmatrix},$$

where

$$\widetilde{\Phi}_{11} = -X + \widetilde{Q}_1 + \widetilde{Q}_2 - \widetilde{R}_1 - \widetilde{R}_2 + \mathbb{I}(1 + r_h, 1 + r_v)\widetilde{Q}.$$

Summarizing the derivation in (2.79) and (2.81), we have the following result.

Theorem 2.4 *The 2-D system (2.70) is stabilizable via SFC (2.71) if there exist positive definite matrices* $X = \text{diag}(X^h, X^v)$, $\widetilde{Q} = \text{diag}(\widetilde{Q}^h, \widetilde{Q}^v)$, $\widetilde{Q}_k = \text{diag}(\widetilde{Q}_k^h, \widetilde{Q}_k^v)$, $\widetilde{R}_k = \text{diag}(\widetilde{R}_k^h, \widetilde{R}_k^v)$ $(k = 1, 2)$*, and a matrix* Y *of appropriate dimension that satisfy the following condition*

$$\begin{bmatrix} \widetilde{\Phi} & \Psi_{X,Y}^\top & \hat{\Psi}_{X,Y}^\top \mathbb{I}(d_m^h, d_m^v) & \hat{\Psi}_{X,Y}^\top \mathbb{I}(d_M^h, d_M^v) \\ * & -X & 0 & 0 \\ * & * & \widetilde{R}_1 - 2X & 0 \\ * & * & * & \widetilde{R}_2 - 2X \end{bmatrix} < 0. \tag{2.82}$$

The controller gain matrix is obtained as

$$K = YX^{-1}.$$

Remark 2.12 Theorem 2.4 provides a basic stabilization criterion via state-feedback control scheme for 2-D systems with time-varying delays in the form of (2.69). By similar derivation procedure of Theorem 2.4, design conditions via other conventional control schemes such as static/dynamic output-feedback or observer-based control can be formulated. In addition, the derived design conditions in Theorem 2.4 can be improved based on suitable augmented LKFs and extended manipulating progresses as presented in Theorem 2.1.

2.5.2 Event-Triggered State-Feedback Controller Design

In this section, we consider the design problem of an event-triggered state-feedback controller (ETC) for 2-D systems (2.70). It is apparent that how to operate a system with minimum used resource becomes more and more important in control engineering. In particular, to use limited networks resources more effectively, it is not necessary to transmit the data at every 2-D time instant (i, j) to enforce the stability with desired performance of the closed-loop system. In the following, we demonstrate a design scheme of an ETC which stabilizes system (2.70). Assume that the vertical and horizontal state vectors are updated at instants i_p, j_q, and the latest states will be denoted as $x^h(i_p, j)$ and $x^v(i, j_q)$. The errors $e^h(i, j)$, $e^v(i, j)$ between states $x^h(i, j)$, $x^v(i, j)$ and the latest transmitted states $x^h(i_p, j)$, $x^v(i, j_q)$ are defined by

$$\begin{bmatrix} e^h(i, j) \\ e^v(i, j) \end{bmatrix} = \begin{bmatrix} x^h(i_p, j) \\ x^v(i, j_q) \end{bmatrix} - \begin{bmatrix} x^h(i, j) \\ x^v(i, j) \end{bmatrix}.$$

From the latest transmitted instants i_p and j_q, the next horizontal and vertical time instants i_{p+1}, j_{q+1} are determined respectively by

$$i_{p+1} = i_p + \min_{s>0} \left\{ s = i - i_p \middle| e^{h\top}(i, j)\Pi_h e^h(i, j) > \alpha_h x^{h\top}(i, j)\Pi_h x^h(i, j) \right\} \tag{2.83}$$

and

$$j_{q+1} = j_q + \min_{t>0} \left\{ t = j - j_q \middle| e^{v\top}(i, j)\Pi_v e^v(i, j) > \alpha_v x^{v\top}(i, j)\Pi_v x^v(i, j) \right\}, \tag{2.84}$$

where α_h and α_v are positive scalars representing turnable triggered parameters and Π_h, Π_v are positive definite matrices, which will be determined. This means that the states $x^h(i, j)$ and $x^v(i, j)$ will not be transmitted within the rectangle region $[i_p, i_{p+1}) \times [j_q, j_{q+1})$. In this region (between two consecutive transmission instants), the following condition holds

$$\begin{cases} e^{h\top}(i, j)\Pi_h e^h(i, j) \le \alpha_h x^{h\top}(i, j)\Pi_h x^h(i, j), \\ e^{v\top}(i, j)\Pi_v e^v(i, j) \le \alpha_v x^{v\top}(i, j)\Pi_v x^v(i, j. \end{cases} \tag{2.85}$$

Fig. 2.7 Event-triggered control

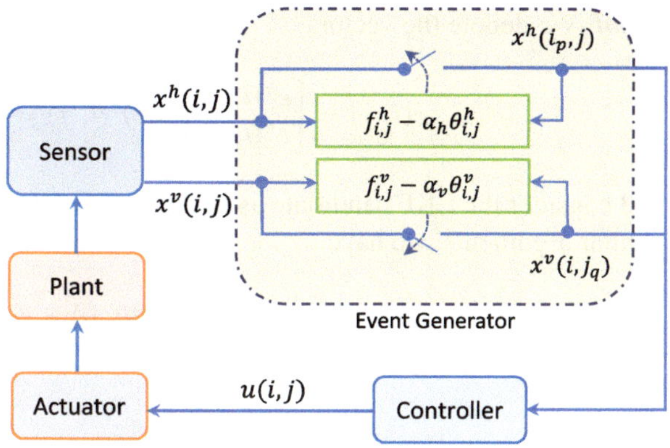

An ETC for system (2.70) will be designed in the form

$$u(i, j) = K \begin{bmatrix} x^h(i_p, j) \\ x^v(i, j_q) \end{bmatrix}, \tag{2.86}$$

where $x^h(i_p, j), x^v(i, j_q)$ are the latest transmitted states and triggering time instants i_p, j_q will be determined by an event generator. Figure 2.7 depicts a diagram of the proposed event-triggered scheme system (2.70), where $f_{i,j}^h = e^{h\top}(i, j)\Pi_h e^h(i, j)$, $\theta_{i,j}^h = x^{h\top}(i, j)\Pi_h x^h(i, j)$ and $f_{i,j}^v = e^{v\top}(i, j)\Pi_v e^v(i, j), \theta_{i,j}^v = x^{v\top}(i, j)\Pi_v x^v(i, j)$.

The closed-loop system of (2.70) subject to ETC (2.86) is given by

$$\begin{bmatrix} x^h(i+1, j) \\ x^v(i, j+1) \end{bmatrix} = A_c \begin{bmatrix} x^h(i, j) \\ x^v(i, j) \end{bmatrix} + A_d \begin{bmatrix} x^h(i - d_h(i), j) \\ x^v(i, j - d_v(j)) \end{bmatrix} + A_e \begin{bmatrix} e^h(i, j) \\ e^v(i, j) \end{bmatrix}, \tag{2.87}$$

where $A_c = A + BK$ and $A_e = BK$.

By similar arguments used in the proof of Theorem 2.3, we first derive stability conditions for the closed-loop system (2.87) as given in the following theorem.

Theorem 2.5 *For given scalars $\alpha_h > 0$, $\alpha_v > 0$, the closed-loop system (2.87) is asymptotically stable if there exist positive definite matrices $P = \mathrm{diag}(P^h, P^v)$, $Q = \mathrm{diag}(Q^h, Q^v)$, $Q_k = \mathrm{diag}(Q_k^h, Q_k^v)$, $R_k = \mathrm{diag}(R_k^h, R_k^v)$ $(k = 1, 2)$, and $\Pi = \mathrm{diag}(\Pi_h, \Pi_v)$ that satisfy the following condition*

$$\begin{bmatrix} \widehat{\Phi} & \hat{\mathscr{A}}_c^\top P & (\hat{\mathscr{A}}_c - \hat{e}_1)^\top \widehat{R}_1 & (\hat{\mathscr{A}}_c - \hat{e}_1)^\top \widehat{R}_2 \\ * & -P & 0 & 0 \\ * & * & -R_1 & 0 \\ * & * & * & -R_2 \end{bmatrix} < 0, \tag{2.88}$$

where $\widehat{\Phi} = \mathrm{diag}(\Phi, -\Pi) + \hat{e}_1^\top \mathbb{I}(\alpha_h, \alpha_v)\Pi\hat{e}_1$, $\hat{e}_k = \begin{bmatrix} 0_{n\times(k-1)n} & I_n & 0_{n\times(5-k)n} \end{bmatrix}$ $(k = 1, 2, \ldots, 5)$, $\hat{\mathscr{A}}_c = A_c\hat{e}_1 + A_d\hat{e}_2 + A_e\hat{e}_5$, and other matrix notations are defined as in Theorem 2.3.

Proof We denote the vectors

$$e(i,j)=\begin{bmatrix}e^h(i,j)\\ e^v(i,j)\end{bmatrix},\quad \hat{\Upsilon}(i,j)=\begin{bmatrix}\Upsilon(i,j)\\ e(i,j)\end{bmatrix}$$

and consider the LKF candidate as in the proof of Theorem 2.3. Then, by following similar arguments, we have

$$\Delta V \le \hat{\Upsilon}^\top(i,j)\left(\mathrm{diag}(\Phi,0)+\widehat{\Psi}\right)\hat{\Upsilon}(i,j), \tag{2.89}$$

where

$$\begin{aligned}\widehat{\Psi} &= \hat{\mathscr{A}}_c^\top P\hat{\mathscr{A}}_c + (\hat{\mathscr{A}}_c-\hat{e}_1)^\top \mathbb{I}\left((d_m^h)^2,(d_m^v)^2\right)R_1(\hat{\mathscr{A}}_c-\hat{e}_1)\\ &\quad + (\hat{\mathscr{A}}_c-\hat{e}_1)^\top \mathbb{I}\left((d_M^h)^2,(d_M^v)^2\right)R_2(\hat{\mathscr{A}}_c-\hat{e}_1).\end{aligned}$$

In addition, according to triggering condition (2.85), we have

$$x^\top(i,j)\mathbb{I}(\alpha_h,\alpha_v)\Pi x(i,j) - e^\top(i,j)\Pi e(i,j) \ge 0.$$

Therefore,

$$\hat{\Upsilon}^\top(i,j)\left(\hat{e}_1^\top \mathbb{I}(\alpha_h,\alpha_v)\Pi\hat{e}_1 - e_5^\top \Pi e_5\right)\hat{\Upsilon}(i,j)\ge 0.$$

This, in combination with (2.89), yields

$$\Delta V \le \hat{\Upsilon}^\top(i,j)\left(\widehat{\Phi}+\widehat{\Psi}\right)\hat{\Upsilon}(i,j).$$

By Lemma 2.7, condition (2.88) holds if and only if $\widehat{\Phi}+\widehat{\Psi}<0$. Thus, under condition (2.88), there exists a positive scalar λ_0 such that $\Delta V \le -\lambda_0\|x(i,j)\|^2$. Similar to Lemma 2.1, we can conclude that the closed-loop system (2.87) is asymptotically stable. The proof is completed. □

On the basis of stability conditions proposed in Theorem 2.5, we derive design conditions for an ETC (2.86) as given in the following theorem.

Theorem 2.6 *System (2.70) is stabilizable via event-triggered control (2.86) if, for given positive scalars α_h, α_v, there exist positive definite matrices $X=\mathrm{diag}(X^h,X^v)$, $\widetilde{Q}=\mathrm{diag}(\widetilde{Q}^h,\widetilde{Q}^v)$, $\widetilde{Q}_k=\mathrm{diag}(\widetilde{Q}_k^h,\widetilde{Q}_k^v)$, $\widetilde{R}_k=\mathrm{diag}(\widetilde{R}_k^h,\widetilde{R}_k^v)$ $(k=1,2)$, $\widetilde{\Pi}=\mathrm{diag}(\widetilde{\Pi}_h,\widetilde{\Pi}_v)$, and a matrix Y of appropriate dimension that satisfy the following condition*

$$\begin{bmatrix}\Omega & \Xi_{X,Y}^\top & \hat{\Xi}_{X,Y}^\top\mathbb{I}(d_m^h,d_m^v) & \hat{\Xi}_{X,Y}^\top\mathbb{I}(d_M^h,d_M^v)\\ * & -X & 0 & 0\\ * & * & \widetilde{R}_1-2X & 0\\ * & * & * & \widetilde{R}_2-2X\end{bmatrix}<0, \tag{2.90}$$

where $\Omega = \mathrm{diag}(\widetilde{\Phi}, -\widetilde{\Pi}) + \hat{e}_1^\top \mathbb{I}(\alpha_h, \alpha_v)\widetilde{\Pi}\hat{e}_1$, $\Xi_{X,Y} = (AX + BY)\hat{e}_1 + A_d X\hat{e}_2 + BY\hat{e}_5$, $\hat{\Xi}_{X,Y} = (AX - X + BY)\hat{e}_1 + A_d X\hat{e}_2 + BY\hat{e}_5$, *and the matrix* $\widetilde{\Phi}$ *is defined as in Theorem 2.4. The controller gain matrix is obtained as*

$$K = YX^{-1}.$$

Proof Let $\mathscr{D}_X = \mathrm{diag}(X, X, X, X, X)$. Then, we have

$$\mathscr{D}_X \hat{\mathscr{A}}_c^\top = \begin{bmatrix} X(A^\top + K^\top B^\top) \\ XA_d^\top \\ 0 \\ 0 \\ XK^\top B^\top \end{bmatrix} = \begin{bmatrix} (AX + BY)^\top \\ (A_d X)^\top \\ 0 \\ 0 \\ (BY)^\top \end{bmatrix} = \Xi_{X,Y}^\top.$$

The remaining of the proof can follow similar lines in the proof of Theorem 2.4. We omit it here. □

Remark 2.13 For 2-D systems (2.70) with constant delays d_h and d_v, the 2-D LKF candidate considered in Theorems 2.3, 2.5 will be simplified to as

$$\begin{aligned} V(x(i,j)) &= x^\top(i,j)\,\mathrm{diag}(P^h, P^v)x(i,j) \\ &+ \sum_{k=i-d_h}^{i-1} x^{h\top}(k,j)Q^h x^h(k,j) + \sum_{l=j-d_v}^{j-1} x^{v\top}(i,l)Q^v x^v(i,l) \\ &+ d_h \sum_{k=-d_h}^{-1}\sum_{l=i+k}^{i-1} \eta_h^\top(l,j)R^h\eta_h(l,j) + d_v \sum_{k=-d_v}^{-1}\sum_{l=j+k}^{j-1} \eta_v^\top(i,l)R^v\eta_v(i,l). \end{aligned}$$

Then, the design conditions of an ETC in the form of (2.86) are given in the following corollary.

Corollary 2.3 *System (2.70) with constant delays* d_h, d_v *is stabilizable via ETC (2.86) if, for given positive scalars* α_h, α_v, *there exist positive definite matrices* $X = \mathrm{diag}(X^h, X^v)$, $\check{Q} = \mathrm{diag}(\check{Q}^h, \check{Q}^v)$, $\check{R} = \mathrm{diag}(\check{R}^h, \check{R}^v)$, $\check{\Pi} = \mathrm{diag}(\check{\Pi}_h, \check{\Pi}_v)$, *and a matrix* Y *of appropriate dimension that satisfy the following condition*

$$\begin{bmatrix} \check{\Omega} & \check{R} & 0 & (AX+BY)^\top & (AX - X + BY)^\top \mathbb{I}(d_h, d_v) \\ * & -\check{Q} - \check{R} & 0 & (A_d X)^\top & (A_d X)^\top \mathbb{I}(d_h, d_v) \\ * & * & -\check{\Pi} & (BY)^\top & (BY)^\top \mathbb{I}(d_h, d_v) \\ * & * & * & -X & 0 \\ * & * & * & * & \check{R} - 2X \end{bmatrix} < 0, \quad (2.91)$$

where $\check{\Omega} = -X + \check{Q} - \check{R} + \mathbb{I}(\alpha_h, \alpha_v)\check{\Pi}$. *The controller gain matrix is obtained as*

$$K = YX^{-1}.$$

Remark 2.14 For delay-free systems

$$\begin{bmatrix} x^h(i+1,j) \\ x^v(i,j+1) \end{bmatrix} = A \begin{bmatrix} x^h(i,j) \\ x^v(i,j) \end{bmatrix} + Bu(i,j),$$

condition (2.91) is reduced to the following LMI

$$\begin{bmatrix} -X + \mathbb{I}(\alpha_h, \alpha_v)\check{\Pi} & 0 & XA^\top + Y^\top B^\top \\ * & -\check{\Pi} & Y^\top B^\top \\ * & * & -X \end{bmatrix} < 0. \tag{2.92}$$

According to triggering conditions (2.83)–(2.84), an event-triggered controller is obtained as

$$u(i,j) = YX^{-1} \begin{bmatrix} x^h(i_p, j) \\ x^v(i, j_q) \end{bmatrix}.$$

2.5.3 *Numerical Examples*

In this section, we give some examples with simulations to illustrate the efficacy of the control design methods presented in this chapter.

Example 2.3 Consider the following model of control system described by a Darboux-type equation

$$\begin{aligned} \frac{\partial^2 w(x,t)}{\partial x \partial t} &= a_0 w(x,t) + a_1 \frac{\partial w(x,t)}{\partial t} + a_2 \frac{\partial w(x,t)}{\partial x} \\ &\quad + a_d w(x, t-\tau(t)) + bu(x,t), \end{aligned} \tag{2.93}$$

where $w(x,t)$ is an unknown function of variable x and time t, $\tau(t)$ represents a time-dependent delay, $u(x,t)$ is the control input, a_0, a_1, a_2, a_d, and b are real coefficients. We denote

$$x^h(x,t) = \frac{\partial w(x,t)}{\partial t} - a_2 w(x,t), \quad x^v(x,t) = w(x,t),$$

then equation (2.93) can be written in the form

$$\begin{bmatrix} \frac{\partial x^h}{\partial x} \\ \frac{\partial x^v}{\partial t} \end{bmatrix} = \begin{bmatrix} a_1 & a_0 + a_1 a_2 \\ 1 & a_2 \end{bmatrix} \begin{bmatrix} x^h \\ x^v \end{bmatrix} + \begin{bmatrix} 0 & a_d \\ 0 & 0 \end{bmatrix} \begin{bmatrix} 0 \\ x_d^v \end{bmatrix} + \begin{bmatrix} b \\ 0 \end{bmatrix} u, \tag{2.94}$$

where $x_d^v(x,t) = w(x, t-\tau(t))$. For given increments Δx, Δt, we define state variables $x^h(i,j) = x^h(i\Delta x, j\Delta t)$, $x^v(i,j) = x^v(i\Delta x, j\Delta t)$, $u(i,j) = u(i\Delta x, j\Delta t)$ and $d_v(j) = \lfloor \tau(j\Delta t) \rfloor$. By using an Euler approximation scheme to discretize equation (2.94) we can obtain

$$\begin{bmatrix} x^h(i+1,j) \\ x^v(i,j+1) \end{bmatrix} = A \begin{bmatrix} x^h(i,j) \\ x^v(i,j) \end{bmatrix} + A_d \begin{bmatrix} x^h(i-d_h(i),j) \\ x^v(i,j-d_v(j)) \end{bmatrix} + Bu(i,j), \quad (2.95)$$

where the system matrices are given by

$$A = \begin{bmatrix} 1+a_1\Delta x & (a_0+a_1a_2)\Delta x \\ \Delta t & 1+a_2\Delta t \end{bmatrix}, \quad A_d = \begin{bmatrix} 0 & a_d\Delta x \\ 0 & 0 \end{bmatrix}, \quad B = \begin{bmatrix} b\Delta x \\ 0 \end{bmatrix}.$$

For illustrative purpose, we specify the parameters as

$$a_0 = 0.2, \quad a_1 = -1, \quad a_2 = -1, \quad a_d = 0.8,$$
$$b = 1, \quad \Delta x = 0.1, \quad \Delta t = 0.05.$$

With the above given values, simulation results show that the open system of (2.95) is unstable. State trajectories of the open system with $d_v = 6$ are given in Figs. 2.8 and 2.9. It can be seen that the conducted trajectories are divergent as $i + j \to \infty$, which illustrate the unstable behavior of the open system.

We now apply Theorem 2.4 to design a stabilizing SFC for system (2.95). For constant delays d_h and d_v, by using the Matlab LMI toolbox, we find that condition (2.82) is feasible with

$$X = \text{diag}(20.4056, 43.1229), \quad Y = \begin{bmatrix} 11.8714 & -80.3904 \end{bmatrix},$$

and therefore, the controller gain is obtained as

$$K = \begin{bmatrix} 0.5818 & -1.8642 \end{bmatrix}.$$

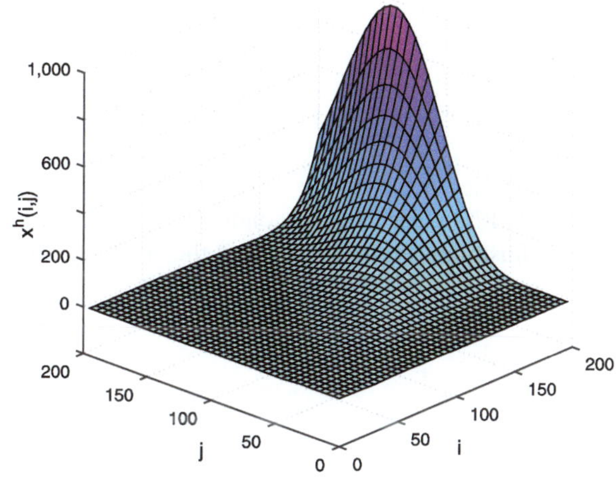

Fig. 2.8 Open-loop state trajectory x^h

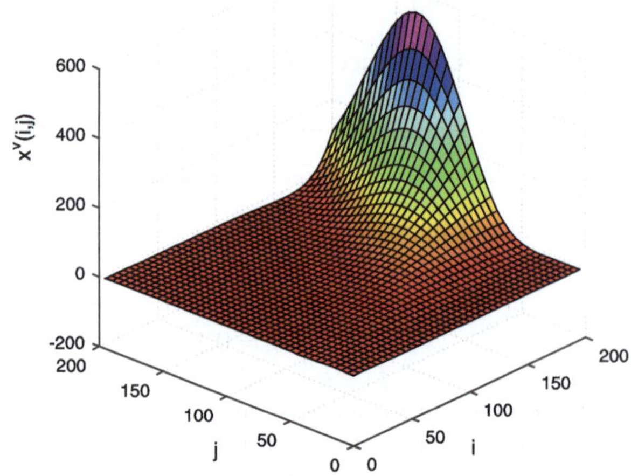

Fig. 2.9 Open-loop state trajectory x^v

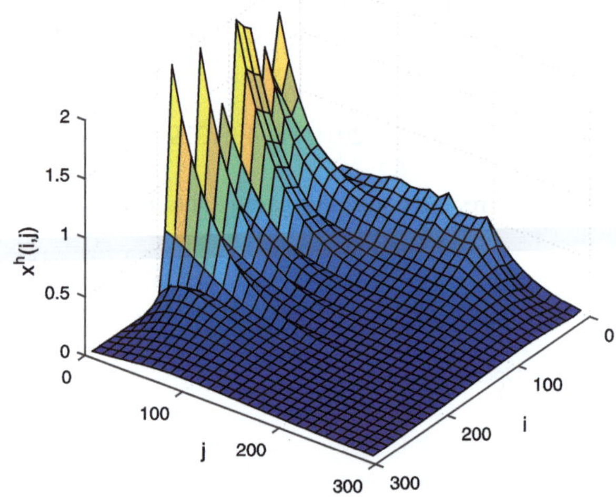

Fig. 2.10 Closed state trajectory x^h

State trajectories of the closed system with $d_v = 6$ are given in Figs. 2.10 and 2.11. It is clear that the conducted trajectories are convergent as $i + j \to \infty$, which illustrate the effectiveness of the design method given in Theorem 2.4.

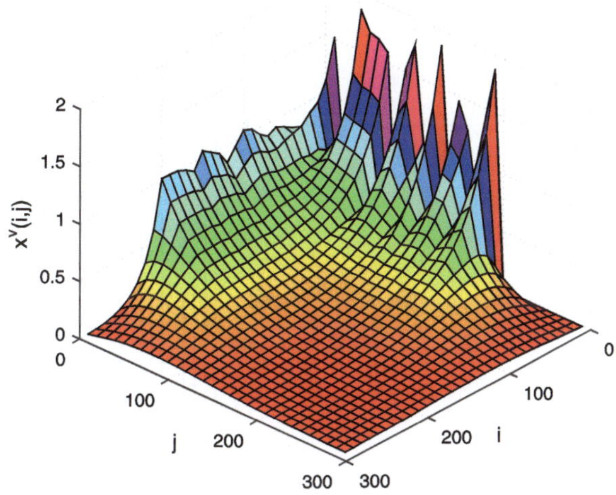

Fig. 2.11 Closed state trajectory x^v

Example 2.4 Consider system (2.95) with the parameters given in Example 2.5.1. For delays $d_h = d_v = 6$ and scalars $\alpha_h = \alpha_v = 0.5$, it is found that condition (2.91) is feasible with the matrices

$$X = \text{diag}(125.1773, 43.8572), \quad \check{Q} = \text{diag}(4.0017, 1.0470),$$
$$\check{R} = \text{diag}(21.6205, 6.7296), \quad \check{\Pi} = \text{diag}(36.0826, 3.4584),$$
$$Y = \begin{bmatrix} -246.5528 & -77.1567 \end{bmatrix}.$$

By Corollary 2.3, system (2.95) is stabilizable via the ETC (2.86), where the controller gain is given as

$$K = YX^{-1} = \begin{bmatrix} -1.9696 & -1.7593 \end{bmatrix}.$$

Simulation results with the ETC (2.86) are presented in Figs. 2.12, 2.13, 2.14, 2.15, 2.16 and 2.17. First, directional ranges of triggering times are given in Figs. 2.12 and 2.13, respectively, whereas 2-D triggering instants are plotted in Figs. 2.14 and 2.15. State trajectories of the closed system with $d_v = 6$ are given in Figs. 2.16 and 2.17. It is clear that the conducted trajectories are convergent as $i + j \to \infty$, which illustrate the effectiveness of the design method of Theorem 2.6 and Corollary 2.3.

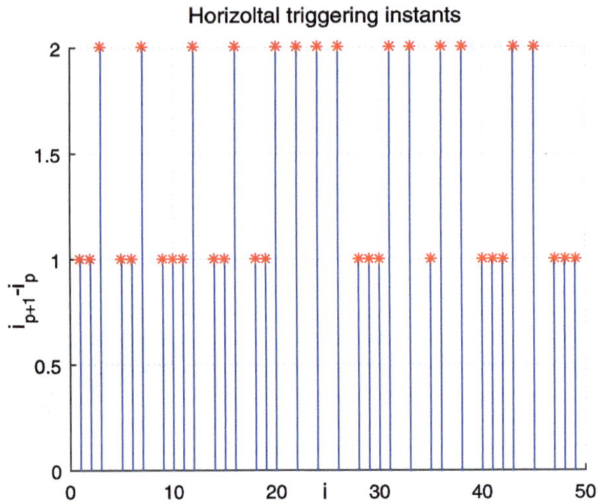

Fig. 2.12 Horizontal triggering instants $i_{p+1} - i_p$

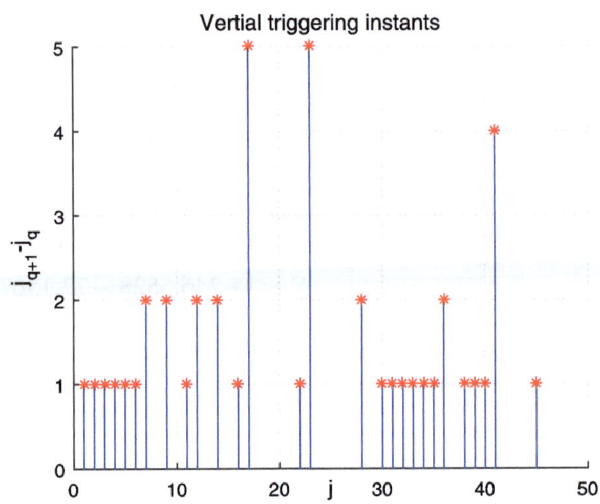

Fig. 2.13 Vertical triggering instants $j_{q+1} - j_q$

2.6 Conclusion

In this chapter, the problems of stability analysis and controller design have been addressed for some classes of 2-D systems with directional state delays. Analysis schemes have been developed and utilized in combination with 2-D summation inequalities to derive delay-dependent LMI-based stability conditions. The design problems of state-feedback controllers and event-triggered state-feedback controllers have also been presented to derive tractable stabilizing conditions. Numerical examples and simulations have been given to validate the effectiveness of the proposed methods.

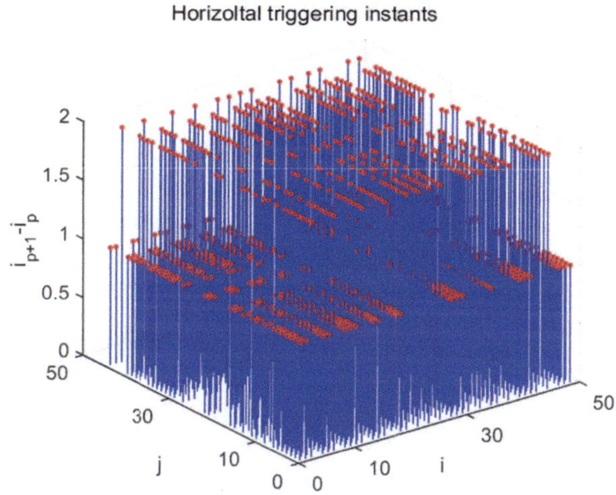

Fig. 2.14 2-D triggering instants $i_{p+1} - i_p$

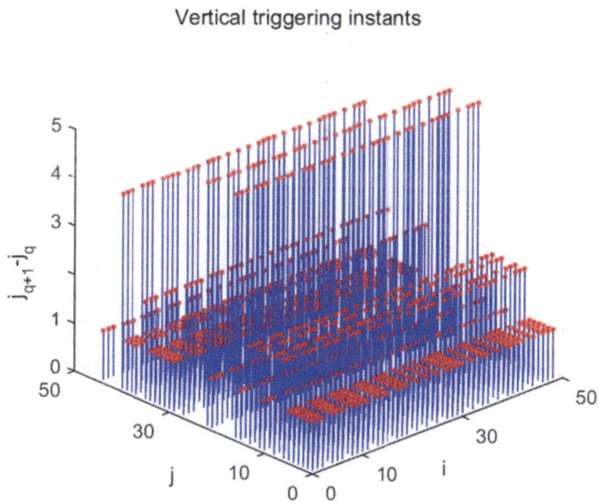

Fig. 2.15 2-D triggering instants $j_{q+1} - j_q$

As an additional concluding remark, it can be noticed that the analysis and design methods presented in this chapter can be effectively utilized for various problems in the systems and control theory of 2-D systems with directional time-varying delays such as H_∞ performance analysis and control, filtering or state estimation. The proposed methods in this chapter can also be extended to complex 2-D systems with or without jumping parameters in deterministic or stochastic models. However, specific applications to such problems especially for useful classes of realistic 2-D systems require certain modifications. This will be addressed in the future works.

Fig. 2.16 Closed state trajectory x^h

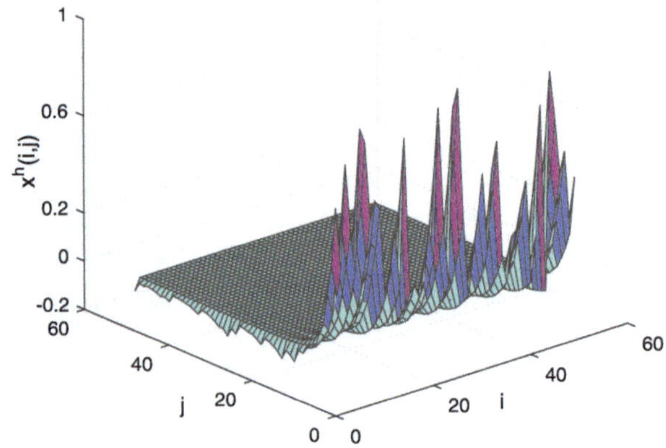

Fig. 2.17 Closed state trajectory x^v

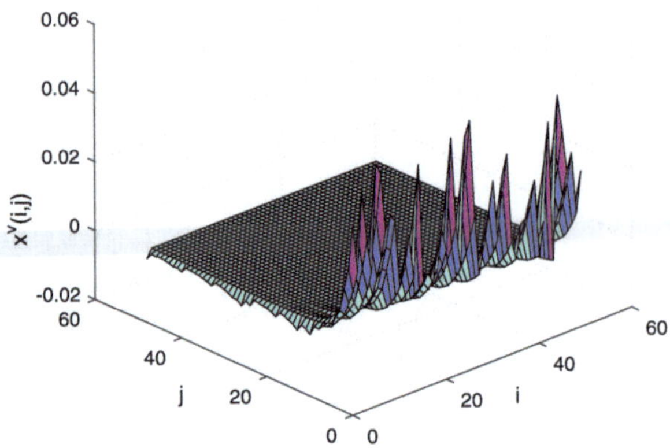

References

1. Sipahi, R., Niculescu, S.-I., Abdallah, C.T., Michiels, W., Gu, K.: Stability and stabilization of systems with time delay. IEEE Control Syst. Magaz. **31**, 38–65 (2011)
2. Witrant, E., Fridman, E., Sename, O., Dugard, L.: Recent Results on Time-Delay Systems: Analysis and Control. Springer, Basel (2016)
3. Hien, L.V., Trinh, H.: New finite-sum inequalities with applications to stability of discrete time-delay systems. Automatica **71**, 197–201 (2016)
4. Xiao, S., Xu, L., Zeng, H.B., Teo, K.L.: Improved stability criteria for discrete-time delay systems via novel summation inequalities. Int. J. Control Autom. Syst. **16**, 1592–1602 (2018)
5. Zhang, X.M., Han, Q.L., Seuret, A., Gouaisbaut, F., He, Y.: Overview of recent advances in stability of linear systems with time-varying delays. IET Control Theory Appl. **13**, 1–16 (2019)
6. Nam, P.T., Hiep, L.T.: A new delay-variation-dependent stability criterion for delayed discrete-time systems. J. Frankl. Inst. **357**, 6951–6967 (2020)

7. Zhang, C.K., Chen, W.H., Zhu, C., He, Y., Wu, M.: Stability analysis of discrete-time systems with time-varying delay via a delay-dependent matrix-separation-based inequality. Automatica **156**, 111192 (2023)
8. Qiu, Y., Hua, C., Park, J.H., Wang, Y.: Interval approximation method for stability analysis of time-delay systems. J. Frankl. Inst. **360**, 3034–3046 (2023)
9. Sha, H., Park, J.H., Chen, J., Zhu, M., Nan, C.: Stability analysis of discrete-time systems with a time-varying delay via improved methods. IET Control Theory Appl. **18**, 951–959 (2024)
10. Huang, S., Xiang, Z.: Delay-dependent stability for discrete 2D switched systems with state delays in the Roesser model. Circuit Syst. Sign. Process **32**, 2821–2837 (2013)
11. Yao, J., Wang, W., Zou, Y.: The delay-range-dependent robust stability analysis for 2-D state-delayed systems with uncertainty. Multidim. Syst. Sign. Process. **24**, 87–103 (2013)
12. Dey, A., Kar, H.: LMI-based criterion for robust stability of 2-D discrete systems with interval time-varying delays employing quantisation/overflow nonlinearities. Multidim. Syst. Sign. Process. **25**, 473–492 (2014)
13. Ye, S., Li, J., Yao, J.: Robust H_∞ control for a class of 2-D discrete delayed systems. ISA Trans. **53**, 1456–1462 (2014)
14. Hien, L.V., Trinh, H.: Stability of two-dimensional Roesser systems with time-varying delays via novel 2D finite sum inequalities. IET Control Theory Appl. **10**, 1665–1674 (2016)
15. Huang, Y.B., An, J., He, Y., Wu, M.: Quasi-convex combination method and its application to the stability analysis of 2D discrete-time Roesser systems with time-varying delays. IET Control Theory Appl. **12**, 718–727 (2018)
16. Peng, D., Nie, H.: Abel lemma-based finite-sum inequality approach to stabilization for 2-D time-varying delay systems. Asian J. Control **23**, 1394–1406 (2021)
17. Li, X., Lam, J., Gao, H., Gu, Y.: A frequency-partitioning approach to stability analysis of two-dimensional discrete systems. Multidim. Syst. Sign. Process. **26**, 67–93 (2015)
18. Yeganefar, N., Yeganefar, N., Ghamgui, M., Moulay, E.: Lyapunov theory for 2-D nonlinear Roesser models: Application to asymptotic and exponential stability. IEEE Trans. Autom. Control **58**, 1299–1304 (2013)
19. Hien, L.V., Trinh, H.: Exponential stability of two-dimensional homogeneous monotone systems with bounded directional delays. IEEE Trans. Autom. Control **63**, 2694–2700 (2018)
20. Wu, L., Yang, R., Shi, P., Su, X.: Stability analysis and stabilization of 2-D switched systems under arbitrary and restricted switchings. Automatica **59**, 206–215 (2015)
21. Zhang, X.M., Han, Q.L.: Abel lemma-based finite-sum inequality and its application to stability analysis for linear discrete time-delay systems. Automatica **57**, 199–202 (2015)
22. Park, P.G., Ko, J.W., Jeong, C.: Reciprocally convex approach to stability of systems with time-varying delays. Automatica **47**, 235–238 (2011)
23. Zhang, X.M., Han, Q.L., Seuret, A., Gouaisbaut, F.: An improved reciprocally convex inequality and an augmented Lyapunov-Krasovskii functional for stability of linear systems with time-varying delay. Automatica **84**, 221–226 (2017)
24. Peng, D., Su, H.S.: A novel approach to delay-variation-dependent stability analysis of 2-D discrete-time systems with mixed delays. IEEE Access **7**, 99817–99829 (2019)
25. Kaczorek, T.: Two-Dimensional Linear Systems. Springer, Berlin (1985)
26. Huang, S., Xiang, Z.: Delay-dependent robust H_∞ control for 2-D discrete nonlinear systems with state delays. Multidim. Syst. Sign. Process. **25**, 775–794 (2014)

Chapter 3
Asymptotic Estimation of Reachable Set for Two-Dimensional Time-Delay Systems

Abstract This chapter deals with the problem of asymptotic estimation of reachable set for 2-D time-delay systems with exogenous disturbances. The directional time-varying delays are confined within certain ranges and the input disturbances are assumed to belong to energy-type norm bounded spaces. Based on an analysis scheme extended from the Lyapunov–Krasovskii functional method, and by utilizing some 2-D weighted summation inequalities presented in this chapter combining with advance techniques to handle interval time-varying delays, delay-dependent conditions in terms of tractable linear matrix inequalities are derived to ensure the existence of an ellipsoid that attracts exponentially the system states in the presence of bounded disturbances. An application to exponential stability of 2-D systems with delays is also presented. The effectiveness of the obtained results are illustrated by numerical examples.

Keywords 2-D systems · State bounding · Reachable set estimation · Time-varying delay · Jensen-based inequalities

3.1 Introduction

Exogenous disturbances, encountered from the data processing, linearization approximations, noisy environment or measurement errors, are unavoidable in modeling practical systems in control engineering. In the presence of disturbances, system state trajectories are typically unpredictable and do not converge to steady states. Thus, for dynamical systems with certain type of disturbances, it is relevant and important to determine or estimate the set of reachable states starting from the original [1]. This is known as the reachable set estimation (RSE) or reachable set bounding problem. Practical applications of RSE problem are found in numerous areas such as state and parameter estimations, safety region realization, optimization or control problems with saturating actuators [1–4]. Dating back to the late 1960s, this problem has been extensively studied for various phenomena of 1-D systems with time delays and bounded peak inputs (see, e.g., [5–10] and the references therein).

L. V. Hien, *Two-Dimensional Time-Delay Systems*, Studies in Systems, Decision and Control 605, https://doi.org/10.1007/978-3-031-95678-2_3

Adaptive schemes have been developed based on the LKF method and LMIs techniques to derive conditions by which the boundedness of employed LKF candidates gives an ellipsoid that bounds the reachable set of the system [7]. For 2-D systems, despite potential applications, for example, in safety monitoring or saturated control, considerably less attention, except [11], has been devoted to developing the study of reachable set estimation for 2-D systems with disturbances. It is noticed also that relevant existing works of 1-D systems are no longer utilizable for 2-D systems due to the nature of their dynamics propagation [12].

Besides the effects of external disturbances, the behavior of state trajectories and performance of a dynamical system also depend heavily on admissible initial states. Thus, it is essential to study the asymptotic behavior of the set of reachable states subject to disturbances coupled with non trial initial states. This gives rise to the problem of determining an ellipsoid, called ellipsoidal ultimate bound (EUB), that attracts exponentially system state trajectories [13, 14]. This problem, on one hand, can be regarded as an extension of the RSE since the attracting set EUB turns to be a reachable set of system states starting from the origin. On the other hand, the methodologies developed for such a problem can be utilized to address various problems in analysis and control of time-delay systems involving dissipativity and stability [15, 16]. There have been a few results concerning this problem for 1-D systems with delays. We refer the reader to [17–20] for a few references.

In this chapter, we address the asymptotic reachable set estimation problem for 2-D systems described by the Roesser model with interval time-varying delays and energy-bounded input disturbances. An analysis scheme is presented and delay-dependent LMI-based conditions are derived to obtain an exponential EUB attracting set. As an application, exponential stability (when external disturbances are ignored) of 2-D systems with delays is also discussed. Numerical examples and simulations are provided to demonstrate the applicability and effectiveness of the derived theoretical results.

3.2 Preliminaries

3.2.1 Model Description

Consider a class of discrete-time 2-D systems described by the following Roesser model with time-varying delays

$$\begin{bmatrix} x^h(i+1,j) \\ x^v(i,j+1) \end{bmatrix} = A \begin{bmatrix} x^h(i,j) \\ x^v(i,j) \end{bmatrix} + A_d \begin{bmatrix} x^h(i-d_h(i),j) \\ x^v(i,j-d_v(j)) \end{bmatrix} + Bw(i,j), \tag{3.1}$$

where $x^h(i,j) \in \mathbb{R}^{n_h}$, $x^v(i,j) \in \mathbb{R}^{n_v}$ and $w(i,j) \in \mathbb{R}^{n_d}$ are the horizontal state, vertical state and disturbance input vectors, respectively; A, $A_d \in \mathbb{R}^{n \times n}$, where

$n = n_h + n_v$, and $B \in \mathbb{R}^{n \times n_d}$ are given real matrices; $d_h(i)$ and $d_v(j)$ represent directional time-varying delays which are assumed to satisfy the following conditions

$$d_1^h \le d_h(i) \le d_2^h, \quad d_1^v \le d_v(j) \le d_2^v, \quad \text{for all } i, j \in \mathbb{N}_0, \tag{3.2}$$

where d_1^h, d_1^v and d_2^h, d_2^v are known positive integers involving the lower bounds and upper bounds of delays. The overall state vector of system (3.1) is defined as $x(i, j) = [x^{h\top}(i, j) \;\; x^{v\top}(i, j)]^\top$.

Initial states of system (3.1) are specified as

$$\begin{aligned} x^h(i, j) &= \phi_h(i, j), \; i \in \mathbb{Z}[-d_2^h, 0], \; j \in \mathbb{N}_0, \\ x^v(i, j) &= \phi_v(i, j), \; i \in \mathbb{N}_0, \; j \in \mathbb{Z}[-d_2^v, 0]. \end{aligned} \tag{3.3}$$

In the area of analysis and control of 2-D systems, it is widely adopted that initial sequences ϕ_h, ϕ_v in (3.3) have finite support [21, 22], that is, there exist positive integer T_h, T_v such that

$$\phi_h(i, j) = 0 \text{ for } j \ge T_h \text{ and } \phi_v(i, j) = 0 \text{ for } i \ge T_v. \tag{3.4}$$

In system (3.1), the disturbance $w(i, j)$ is assumed to be bounded in the sense that there exists a scalar $w_\infty \ge 0$ such that

$$\sup_{k \ge 0} \sum_{i+j=k} w^\top(i, j) w(i, j) \le w_\infty^2. \tag{3.5}$$

For a given $w_\infty \ge 0$, let $l_{w_\infty}^\infty$ denote the set of admissible disturbances that satisfy (3.5) and $\mathscr{D}_\infty(\mathbb{N}_0^2) = \cup_{w_\infty \ge 0} l_{w_\infty}^\infty$. Obviously, $\mathscr{D}_\infty(\mathbb{N}_0^2)$ contains the set of energy-norm bounded sequences $l_2(\mathbb{N}_0^2)$, which is widely considered in the context of performance analysis of 2-D systems.

3.2.2 *Problem Formulation*

Consider 2-D systems with directional time-varying delays as given in (3.1). Let us first introduce the following definitions.

Definition 3.1 [22] System (3.1) without disturbance (i.e. $w(i, j) = 0$ for all $i, j \in \mathbb{N}_0$) is said to be exponentially stable if there exist positive constants $\alpha \in (0, 1)$ and $\beta > 0$ such that any solution $x(i, j)$ of system (3.1) with initial condition (3.3)–(3.4) satisfies the inequality

$$\|x(i, j)\|^2 \le \beta \left(\sum_{k=0}^{T_h} \frac{\hat{\phi}_h^2(k)}{\alpha^k} + \sum_{l=0}^{T_v} \frac{\hat{\phi}_v^2(l)}{\alpha^l} \right) \alpha^{i+j} \text{ for all } i, j \in \mathbb{N}_0, \tag{3.6}$$

where $\hat{\phi}_h(k) = \max_{-d_2^h \le i \le 0} \|\phi_h(i, k)\|$ and $\hat{\phi}_v(l) = \max_{-d_2^v \le j \le 0} \|\phi_v(l, j)\|$.

In the effect of non-vanishing disturbance $w(i, j)$, the property of exponential stability of system (3.1) is usually not preserved. Thus, one may expect state trajectories of the system can be asymptotically bounded within a certain range, normally a compact set. This is referred to an ultimate bound of system (3.1) as given in the following definition.

Definition 3.2 A compact set $\Omega \subset \mathbb{R}^n$ is said to be an ultimate bound of system (3.1) if it holds that

$$\limsup_{i+j \to \infty} \rho(x(i, j), \Omega) = 0, \tag{3.7}$$

where, for a vector $x \in \mathbb{R}^n$, $\rho(x, \Omega)$ denotes the distance from x to Ω defined as

$$\rho(x, \Omega) = \inf_{y \in \Omega} \|x - y\|.$$

In particular, if the aforementioned compact set Ω is a ball $\Omega = \{x \in \mathbb{R}^n : \|x\| \le r_b\}$ then r_b is called an upper ultimate bound of system (3.1).

Consider system (3.1) with zero initial condition (i.e. $\phi_h = 0$ and $\phi_v = 0$). Then, the reachable set of (3.1) is defined as

$$\mathscr{R}_x = \left\{ x(i, j) \in \mathbb{R}^n \;\middle|\; x(i, j), w(i, j) \text{ satisfy (3.1) and (3.5) for all } i, j \in \mathbb{N}_0 \right\}. \tag{3.8}$$

The main problem addressed in this chapter is to determine a smallest possible ellipsoid

$$\Omega_e(P_0, r_e) = \left\{ x \in \mathbb{R}^n : x^\top P_0 x \le r_e \right\}$$

which covers the reachable set $\mathscr{R}_x$ defined by (3.8) and is an EUB of system (3.1), where $P_0 \in \mathbb{S}_n^+$ is a symmetric positive definite matrix and r_e is some positive scalar. As an application, exponential stability of system (3.1) without disturbance is also presented.

3.3 Auxiliary Results

3.3.1 Functional-Based Technical Lemmas

In this section, we formulate technical lemmas based on an extension of the 2-D LKF method. First, for a vector function $\eta : \mathbb{Z} \times \mathbb{Z} \to \mathbb{R}^n$, we denote by $\Delta_h \eta$ and $\Delta_v \eta$ the partial differences of η with respect to the horizontal and vertical arguments, that is

$$\Delta_h \eta(i, j) = \eta(i + 1, j) - \eta(i, j) \text{ and } \Delta_v \eta(i, j) = \eta(i, j + 1) - \eta(i, j).$$

Let $V : \mathbb{R}^n \to \mathbb{R}$ be a functional defined by $V(x) = V^h(x^h) + V^v(x^v)$, where $V^h : \mathbb{R}^{n_h} \to \mathbb{R}$ and $V^v : \mathbb{R}^{n_v} \to \mathbb{R}$ are some functionals. The difference of $V(x)$ along state trajectories of system (3.1) is defined as

$$\Delta_p V(x(i, j)) = \Delta_h(V^h(x^h(i, j))) + \Delta_v(V^v(x^v(i, j))).$$

We have the following technical lemma which is essential for our next derivation.

Lemma 3.1 *Consider system* (3.1) *with bounded disturbances satisfying* (3.5). *Assume that there exist a functional* $V(x) = V^h(x^h) + V^v(x^v)$ *with* $V(0) = 0$, $V^h(x^h) \geq 0$, $V^v(x^v) \geq 0$, *and scalars* $\alpha \in (0, 1)$, $\mu \geq 0$, *such that the difference of* $V(x)$ *along any state trajectory* $x(i, j)$ *of system* (3.1) *satisfies the following condition*

$$\Delta_p V(x(i, j)) + (1 - \alpha)V(x(i, j)) - \mu w^\top(i, j)w(i, j) \leq 0 \tag{3.9}$$

for all $(i, j) \in \mathbb{N}_0^2$. *Then, it holds that*

$$\limsup_{r\to\infty} \sum_{i+j=r} V(x(i, j)) \leq \frac{\mu w_\infty^?}{1 - \alpha}. \tag{3.10}$$

Proof For the brevity of notation, we define the function

$$E(s) = \sum_{i+j=s} \left[V^h(x^h(i, j)) + V^v(x^v(i, j))\right], \quad s \in \mathbb{N}_0.$$

In Lyapunov theory, the functionals $V^h(x^h)$ and $V^v(x^v)$ are often involving energy of the system. In this regard, $E(s)$ defines the energy of system (3.1) stored in the diagonal line $\Gamma_s = \{(i, j) \in \mathbb{N}_0^2 : i + j = s\}$ and $E(s)$ is called an energy function of system (3.1).

From (3.9), we have

$$\begin{aligned} V^h(x^h(i + 1, j)) + V^v(x^v(i, j + 1)) \leq \alpha \left[V^h(x^h(i, j)) + V^v(x^v(i, j))\right] \\ + \mu w^\top(i, j)w(i, j), \quad (i, j) \in \mathbb{N}_0^2. \end{aligned} \tag{3.11}$$

Therefore, together with (3.11), we have

$$\begin{aligned} E(s + 1) &= \sum_{i+j=s+1} \left[V^h(x^h(i, j)) + V^v(x^v(i, j))\right] \\ &= \sum_{i+j=s} \left[V^h(x^h(i + 1, j)) + V^v(x^v(i, j + 1))\right] \\ &\quad + V^h(x^h(0, s + 1)) + V^v(x^v(s + 1, 0)) \\ &\leq \alpha \sum_{i+j=s} \left[V^h(x^h(i, j)) + V^v(x^v(i, j))\right] + \mu \sum_{i+j=s} w^\top(i, j)w(i, j) \end{aligned}$$

$$
\begin{aligned}
&+ V^h(x^h(0, s+1)) + V^v(x^v(s+1, 0)) \\
&\leq \alpha E(s) + F(s+1) + \mu \hat{w}(s),
\end{aligned}
$$

where $F(s) = V^h(x^h(0, s)) + V^v(x^v(s, 0))$ and $\hat{w}(s) = \sum_{i+j=s} w^\top(i, j) w(i, j)$. By iteratively estimate, from the following linear inequality

$$
E(s+1) \leq \alpha E(s) + F(s+1) + \mu \hat{w}(s), \ s \in \mathbb{N}_0, \tag{3.12}
$$

we obtain

$$
E(r) \leq \alpha^r E(0) + \sum_{s=1}^{r} \alpha^{r-s} F(s) + \frac{\mu}{\alpha} \sum_{s=0}^{r-1} \alpha^{r-s} \hat{w}(s), \ r \geq 1. \tag{3.13}
$$

Note also that $V^h(x^h(0, j)) = 0$ for $j \geq T_h$ and $V^v(x^v(i, 0)) = 0$ for $i \geq T_v$. This gives

$$
\sum_{s=0}^{r} \alpha^{r-s} F(s) \leq \left[\sum_{j=0}^{T_h} \alpha^{-j} V^h(x^h(0, j)) + \sum_{i=0}^{T_v} \alpha^{-i} V^v(x^v(i, 0)) \right] \alpha^r, \ \forall r \geq 1. \tag{3.14}
$$

Taking (3.5), (3.13) and (3.14) into account we finally obtain

$$
E(r) \leq \frac{\mu w_\infty^2}{1-\alpha} \left(1 - \alpha^r\right) + \hat{V}_\alpha(\phi_h, \phi_v) \alpha^r, \ r \geq 1, \tag{3.15}
$$

where

$$
\hat{V}_\alpha(\phi_h, \phi_v) = \sum_{j=0}^{T_h} \alpha^{-j} V^h(x^h(0, j)) + \sum_{i=0}^{T_v} \alpha^{-i} V^v(x^v(i, 0)).
$$

Since $\alpha^r \to 0$ as $r \to \infty$, it is clear from (3.15) that

$$
\limsup_{r \to \infty} \sum_{i+j=r} V(x(i, j)) = \limsup_{r \to \infty} E(r) \leq \frac{\mu w_\infty^2}{1-\alpha}.
$$

This completes the proof of Lemma 3.1. □

Lemma 3.2 *Let $V(x)$ be a functional satisfying the assumptions of Lemma 3.1. Assume that there exists a positive definite matrix $P_0 \in \mathbb{S}_n^+$ such that*

$$
V(x) \geq x^\top P_0 x, \quad x \in \mathbb{R}^n.
$$

Then, the ellipsoid $\Omega_e\left(P_0, \frac{\mu w_\infty^2}{1-\alpha}\right)$ is an EUB of system (3.1)–(3.5).

Proof Let $x = x(i, j)$ be an arbitrary state trajectory of system (3.1). By Lemma 3.1, for any $\epsilon > 0$, there exists an $N \in \mathbb{N}$ such that

$$V(x(i, j)) < \bar{V} + \epsilon$$

for all (i, j) subject to $i + j = r > N$, where $\bar{V} = \frac{\mu w_\infty^2}{1-\alpha}$.
If $x(i, j) \in \Omega_e(P_0, \bar{V})$ then $\rho\big(x(i, j), \Omega_e(P_0, \bar{V})\big) = 0$.
If $x(i, j) \notin \Omega_e(P_0, \bar{V})$ then,

$$\bar{V} < x^\top(i, j) P_0 x(i, j) \leq V(x(i, j)) < \bar{V} + \epsilon.$$

By defining

$$\delta_{ij} = \sqrt{x^\top(i, j) P_0 x(i, j)} \text{ and } \hat{x}(i, j) = \frac{\sqrt{\bar{V}}}{\delta_{ij}} x(i, j),$$

we have $\hat{x}^\top(i, j) P_0 \hat{x}(i, j) = \bar{V}$, and hence $\hat{x}(i, j) \in \Omega_e(P_0, \bar{V})$. Note also that

$$\delta_{i,j}^2 = x^\top(i, j) P_0 x(i, j) \geq \lambda_{\min}(P_0) \|x(i, j)\|^2$$

and

$$\begin{aligned} 0 < \delta_{i,j} - \sqrt{\bar{V}} &< \sqrt{\bar{V} + \epsilon} - \sqrt{\bar{V}} \\ &= \frac{\epsilon}{\sqrt{\bar{V} + \epsilon} + \sqrt{\bar{V}}} < \frac{\epsilon}{2\sqrt{\bar{V}}}. \end{aligned}$$

Therefore,

$$\begin{aligned} \rho^2\big(x(i, j), \Omega_e(P_0, \bar{V})\big) &\leq \|x(i, j) - \hat{x}(i, j)\|^2 \\ &= \left(\delta_{i,j} - \sqrt{\bar{V}}\right)^2 \frac{\|x(i, j)\|^2}{\delta_{ij}^2} \\ &< \frac{\epsilon^2}{4\bar{V}\lambda_{\min}(P_0)}. \end{aligned} \tag{3.16}$$

The result of (3.16) ensures that

$$\lim_{i+j\to\infty} \rho\big(x(i, j), \Omega_e(P_0, \bar{V})\big) = 0,$$

by which $\Omega_e(P_0, \bar{V})$ is an EUB of system (3.1)–(3.5). The proof is completed. □

Remark 3.1 With zero initial condition (i.e. $\phi_h = 0$ and $\phi_v = 0$), the estimate

$$V(x(i, j)) \leq \frac{\mu w_\infty^2}{1 - \alpha}$$

holds for all state trajectories $x(i, j)$ subject to disturbances satisfying (3.5). Thus, the reachable set $\mathscr{R}_x$ of system (3.1) is confined within $\Omega_V = \{x \in \mathbb{R}^n : V(x) \leq \bar{V}\}$, where $\bar{V}$ defined in Lemma 3.2. In addition, the functional $V(x)$ is usually chosen as an LKF candidate for 2-D systems, which consists a quadratic term $V_0(x) = x^\top P_0 x$ and nonnegative functionals of delayed state segments. Then, the ellipsoid

$$\Omega_e\Big(P_0, \frac{\mu w_\infty^2}{1-\alpha}\Big) = \Big\{x \in \mathbb{R}^n : x^\top P_0 x \leq r_e \triangleq \frac{\mu w_\infty^2}{1-\alpha}\Big\}$$

simultaneously contains the reachable set $\mathscr{R}_x$ (in the case of zero initial condition) and is an EUB of the system subject to bounded disturbances. To this regard, Lemma 3.1 provides a basic tool to address the state bounding and RSE problems for 2-D systems with delays and bounded disturbances as presented in (3.1).

Remark 3.2 Lemma 3.1 presented in this chapter extends many existing results, for example, [17, Lemma 4], [23, Lemma 2.1], [24, Lemma 4] or [25, Lemma 2.1], which were specifically established to deal with the problems of state bounding and RSE for 1-D discrete-time systems.

Remark 3.3 When the disturbance $w(i, j)$ is ignored in system (3.1), by specifying $w_\infty = 0$, it follows from (3.15) and Lemma 3.2 that

$$\|x(i, j)\|^2 \leq \frac{\hat{V}_\alpha(\phi_h, \phi_v)}{\lambda_{\min}(P_0)}\alpha^{i+j}, \ i, j \in \mathbb{N}_0.$$

Thus, Lemmas 3.1 and 3.2 are utilizable to derive exponential stability conditions for 2-D systems with delays.

3.3.2 Weighted Summation Inequalities

In this section, we gives some weighted summation inequalities (WSIs), which will be used to manipulate the difference of suitable functional candidates as mentioned in the preceding section. First, the following Jensen-type inequality can be obtained.

Lemma 3.3 *Given a matrix $R \in \mathbb{S}_n^+$, scalar $\alpha \in (0, 1)$ and integers k, h_1, h_2, r_1, r_2. For any function $\chi : \mathbb{Z}[k-h_2, k-h_1] \times [k-r_2, k-r_1] \to \mathbb{R}^n$, it holds that*

$$\sum_{i=k-h_2}^{k-h_1} \alpha^{k-i}\chi^\top(i, j)R\chi(i, j) \geq \hat{\alpha}_{h_1,h_2}\Bigg(\sum_{i=k-h_2}^{k-h_1}\chi(i, j)\Bigg)^\top R\Bigg(\sum_{i=k-h_2}^{k-h_1}\chi(i, j)\Bigg) \tag{3.17}$$

and

$$\sum_{j=k-r_2}^{k-r_1} \alpha^{k-j} \chi^\top(i,j) R \chi(i,j) \ge \hat{\alpha}_{r_1,r_2} \left(\sum_{j=k-r_2}^{k-r_1} \chi(i,j) \right)^\top R \left(\sum_{j=k-r_2}^{k-r_1} \chi(i,j) \right) \tag{3.18}$$

where $\hat{\alpha}_{h_1,h_2} = \frac{(1-\alpha)\alpha^{h_2}}{1-\alpha^{h_2-h_1+1}}$ *and* $\hat{\alpha}_{r_1,r_2} = \frac{(1-\alpha)\alpha^{r_2}}{1-\alpha^{r_2-r_1+1}}$.

Proof The proof is straightforward by using Schur complement lemma and is omitted here. See also [17, Lemma 2] for more details. □

Remark 3.4 Observe that

$$\lim_{\alpha\to 1} \hat{\alpha}_{h_1,h_2} = \frac{1}{h_2 - h_1 + 1} \text{ and } \lim_{\alpha\to 1} \hat{\alpha}_{r_1,r_2} = \frac{1}{r_2 - r_1 + 1}.$$

Thus, when α approaches 1, inequalities (3.17) and (3.18) are reduced to 2-D Jensen-type inequalities, which are widely used in the area of analysis and control of 2-D time-delay systems [26–28].

In the remaining of this section, we introduce new 2-D WSIs by refining Jensen-type inequalities given in (3.17) and (3.18). Let $G_h^\alpha(\chi)$ and $G_v^\alpha(\chi)$ denote the gap of inequalities (3.17) and (3.18), respectively

$$G_h^\alpha(\chi) = \sum_{i=k-h_2}^{k-h_1} \alpha^{k-i} \chi^\top(i,j) R \chi(i,j) - \hat{\alpha}_{h_1,h_2} \left(\sum_{i=k-h_2}^{k-h_1} \chi(i,j) \right)^\top R \left(\sum_{i=k-h_2}^{k-h_1} \chi(i,j) \right),$$

$$G_v^\alpha(\chi) = \sum_{j=k-r_2}^{k-r_1} \alpha^{k-j} \chi^\top(i,j) R \chi(i,j) - \hat{\alpha}_{r_1,r_2} \left(\sum_{j=k-r_2}^{k-r_1} \chi(i,j) \right)^\top R \left(\sum_{j=k-r_2}^{k-r_1} \chi(i,j) \right).$$

If $h_1 = h_2$ (or $r_1 = r_2$) then (3.17) (or (3.18)) becomes an equality and there is no gap. In the following, we derive tighter bounds for $G_h^\alpha(\chi)$ and $G_v^\alpha(\chi)$. For this, we first introduce some useful notations.
For any integers $h_1 \le h_2$ and scalar $\alpha \in (0, 1)$, we denote the summations

$$s_\alpha(h_1,h_2) = \sum_{i=-h_2}^{-h_1} (i + h_2 + 1)\alpha^i = \sum_{k=1}^{h_2-h_1+1} k\alpha^{k-1-h_2},$$

$$d_\alpha(h_1,h_2) = \sum_{i=-h_2}^{-h_1} (i + h_2 + 1)^2\alpha^i = \sum_{k=1}^{h_2-h_1+1} k^2\alpha^{k-1-h_2}.$$

By direct computation, we have

$$s_\alpha(h_1,h_2) = \frac{\alpha^{-h_2} - \sigma\alpha^{-h_1+1}}{(1-\alpha)^2},$$

$$d_\alpha(h_1,h_2) = \frac{(1+\alpha)\alpha^{-h_2} - \alpha^{-h_1+1}(\sigma^2+\alpha)}{(1-\alpha)^3},$$

where $\sigma = 1 + (1-\alpha)h_{12}$ and $h_{12} = h_2 - h_1 + 1$.
Next, let us define

$$\beta_\alpha(h_1, h_2) = \frac{h_{12}}{1-\alpha^{h_{12}}} - \frac{\alpha}{1-\alpha},$$
$$\gamma_\alpha(h_1, h_2) = \frac{\alpha}{(1-\alpha)^2} - \frac{h_{12}^2\alpha^{h_{12}}}{(1-\alpha^{h_{12}})^2}.$$

It can be verified using the Bernoulli's inequality that $\beta_\alpha(h_1, h_2) > 0$. In addition, since

$$\gamma_\alpha(h_1, h_2) = \hat{\alpha}_{h_1,h_2}\left[d_\alpha(h_1, h_2) - \hat{\alpha}_{h_1,h_2}s_\alpha^2(h_1, h_2)\right],$$

by Lemma 3.3 (inequality (3.17)), we can conclude that $\gamma_\alpha(h_1, h_2) > 0$ for all integers $h_1 < h_2$ and $\alpha \in (0, 1)$.

New 2-D WSIs, which improve the results of Lemma 3.3, will be presented in the following lemma.

Lemma 3.4 *Given a matrix $R \in \mathbb{S}_n^+$, integers k, $h_1 < h_2$, $r_1 < r_2$ and an $\alpha \in (0, 1)$. For any function $\chi : \mathbb{Z}[k-h_2, k-h_1] \times [k-r_2, k-r_1] \to \mathbb{R}^n$, the following inequalities hold*

$$\sum_{i=k-h_2}^{k-h_1} \alpha^{k-i}\chi^\top(i,j)R\chi(i,j) \geq \hat{\alpha}_{h_1,h_2}\begin{bmatrix}\chi_{11}\\ \chi_{12}\end{bmatrix}^\top \begin{bmatrix}R & 0\\ 0 & \tilde{\alpha}_{h_1,h_2}R\end{bmatrix}\begin{bmatrix}\chi_{11}\\ \chi_{12}\end{bmatrix}, \tag{3.19}$$

$$\sum_{j=k-r_2}^{k-r_1} \alpha^{k-j}\chi^\top(i,j)R\chi(i,j) \geq \hat{\alpha}_{r_1,r_2}\begin{bmatrix}\chi_{21}\\ \chi_{22}\end{bmatrix}^\top \begin{bmatrix}R & 0\\ 0 & \tilde{\alpha}_{r_1,r_2}R\end{bmatrix}\begin{bmatrix}\chi_{21}\\ \chi_{22}\end{bmatrix}, \tag{3.20}$$

where the associated coefficients and augmented vectors are defined as

$$\tilde{\alpha}_{h_1,h_2} = \frac{\beta_\alpha^2(h_1, h_2)}{\gamma_\alpha(h_1, h_2)}, \quad \tilde{\alpha}_{r_1,r_2} = \frac{\beta_\alpha^2(r_1, r_2)}{\gamma_\alpha(r_1, r_2)},$$
$$\chi_{11} = \sum_{i=k-h_2}^{k-h_1}\chi(i,j), \quad \chi_{21} = \sum_{j=k-r_2}^{k-r_1}\chi(i,j),$$
$$\chi_{12} = \sum_{i=k-h_2}^{k-h_1}\chi(i,j) - \frac{1}{\beta_\alpha(h_1,h_2)}\sum_{i=k-h_2}^{k-h_1}\sum_{s=k-h_2}^{i}\chi(s,j),$$
$$\chi_{22} = \sum_{j=k-r_2}^{k-r_1}\chi(i,j) - \frac{1}{\beta_\alpha(r_1,r_2)}\sum_{j=k-r_2}^{k-r_1}\sum_{t=k-r_2}^{j}\chi(i,t).$$

Proof We define a modified function $\hat{\chi} : \mathbb{Z}[k-h_2, k-h_1] \times [k-r_2, k-r_1] \to \mathbb{R}^n$ as

$$\hat{\chi}(i,j) = \chi(i,j) - \hat{\alpha}_{h_1,h_2}\alpha^{i-k}\chi_{11} + \theta_i F, \tag{3.21}$$

where $\theta_i = \alpha^{i-k}(i-k+1+h_2)$, $i \in \mathbb{Z}[k-h_2, k-h_1]$, and $\mathcal{F} \in \mathbb{R}^n$ is a constant vector which will be defined later. By Lemma 3.3, we have $G_h^\alpha(\hat{\chi}) \geq 0$. Taking (3.21) into account we then obtain

$$G_h^\alpha(\chi) \geq \left[\hat{\alpha}_{h_1,h_2}\Big(\sum_{i=k-h_2}^{k-h_1} \theta_i\Big)^2 - \sum_{i=k-h_2}^{k-h_1} \alpha^{k-i}\theta_i^2\right]\mathcal{F}^\top R\mathcal{F}$$
$$+ 2\hat{\alpha}_{h_1,h_2} \sum_{i=k-h_2}^{k-h_1} \theta_i \mathcal{F}^\top R\chi_{11} - 2\mathcal{F}^\top R \sum_{i=k-h_2}^{k-h_1} \alpha^{k-i}\theta_i \chi(i,j). \tag{3.22}$$

We also define an auxiliary function $\chi_s(i,j)$, $j \in \mathbb{Z}[k-r_2, k-r_1]$, as

$$\chi_s(i,j) = \begin{cases} \sum_{l=k-h_2}^{i-1} \eta(l,j) & \text{if } i > k-h_2 \\ 0 & \text{if } i = k-h_2. \end{cases}$$

Then, $\chi(i,j) = \Delta_h(\chi_s(i,j))$ and, consequently,

$$\alpha^{k-i}\theta_i\chi(i,j) = \Delta_h(\alpha^{k-i}\theta_i\chi_s(i,j)) - \chi_s(i+1,j).$$

Taking summation in the first argument from $k-h_2$ to $k-h_1$, we obtain

$$\sum_{i=k-h_2}^{k-h_1} \alpha^{k-i}\theta_i\chi(i,j) = (1+h_{12})\chi_{11} - \sum_{i=k-h_2}^{k-h_1}\sum_{l=k-h_2}^{i} \chi(l,j). \tag{3.23}$$

In addition, observe that

$$\sum_{i=k-h_2}^{k-h_1} \theta_i = s_\alpha(h_1,h_2), \quad \sum_{i=k-h_2}^{k-h_1} \alpha^{k-i}\theta_i^2 = d_\alpha(h_1,h_2). \tag{3.24}$$

This, in combination of (3.22)–(3.24), gives

$$W_h^\alpha(\chi) \geq -2\beta_\alpha(h_1,h_2)\mathcal{F}^\top R\chi_{11} - \frac{\gamma_\alpha(h_1,h_2)}{\hat{\alpha}_{h_1,h_2}}\mathcal{F}^\top R\mathcal{F}. \tag{3.25}$$

Now, we define the vector $\mathcal{F}$ in the form $\mathcal{F} = -\lambda\chi_{11}$, where $\lambda \in \mathbb{R}$ is a scalar variable. From (3.25), we have

$$W_h^\alpha(\chi) \geq \left[2\beta_\alpha(h_1,h_2)\lambda - \frac{\gamma_\alpha(h_1,h_2)}{\hat{\alpha}_{h_1,h_2}}\lambda^2\right]\chi_{12}^\top R\chi_{12}. \tag{3.26}$$

The function $\psi(\lambda) = 2\beta_\alpha(h_1, h_2)\lambda - \frac{\gamma_\alpha(h_1,h_2)}{\hat{\alpha}_{h_1,h_2}}\lambda^2$ attains its maximum as $\hat{\alpha}_{h_1,h_2}\tilde{\alpha}_{h_1,h_2}$. Thus, by maximizing the function $\psi(\lambda)$ we obtain inequality (3.19). The proof of (3.20) is similar to that of (3.19) and is omitted. □

Remark 3.5 Inequalities (3.19) and (3.20) are clearly tighter than the ones derived in Lemma 3.3 since positive terms are added into the right-hand sides of (3.19) and (3.20). Thus, the use of Lemma 3.4 will give less conservative conditions than that of Lemma 3.3.

The following simpler version of Lemma 3.4 is provided to make it easier to use in the framework of functional-based method.

Corollary 3.1 *Given a matrix $R \in \mathbb{S}_n^+$ and positive integers h, r. For any function $\chi : \mathbb{Z}[-h, 0] \times [-r, 0] \to \mathbb{R}^n$, the following inequalities hold*

$$\sum_{i=-h}^{-1} \alpha^{-i}\chi_h^\top(i,j)R\chi_h(i,j) \geq \hat{\alpha}_h\left(\Omega_{11}^\top R\Omega_{11} + \frac{\beta_h^2}{\gamma_h}\Omega_{12}^\top R\Omega_{12}\right), \tag{3.27}$$

$$\sum_{j=-r}^{-1} \alpha^{-j}\chi_v^\top(i,j)R\chi_v(i,j) \geq \hat{\alpha}_r\left(\Omega_{21}^\top R\Omega_{21} + \frac{\beta_r^2}{\gamma_r}\Omega_{22}^\top R\Omega_{22}\right), \tag{3.28}$$

where $\chi_h(i,j) = \Delta_h(\chi(i,j))$, $\chi_v(i,j) = \Delta_v(\chi(i,j))$ and other notations are defined as

$$\Omega_{11} = \chi(0,j) - \chi(-h,j), \quad \Omega_{21} = \chi(i,0) - \chi(i,-r),$$

$$\Omega_{12} = \chi(0,j) + \left(\frac{h+1}{\beta_h} - 1\right)\chi(-h,j) - \frac{1}{\beta_h}\sum_{i=-h}^{0}\chi(i,j),$$

$$\Omega_{22} = \chi(i,0) + \left(\frac{r+1}{\beta_r} - 1\right)\chi(i,-r) - \frac{1}{\beta_r}\sum_{j=-r}^{0}\chi(i,j),$$

$$\hat{\alpha}_h = \frac{(1-\alpha)\alpha^h}{1-\alpha^h}, \quad \beta_h = \frac{h}{1-\alpha^h} - \frac{\alpha}{1-\alpha}, \quad \gamma_h = \frac{\alpha}{(1-\alpha)^2} - \frac{h^2\alpha^h}{(1-\alpha^h)^2},$$

$$\hat{\alpha}_r = \frac{(1-\alpha)\alpha^r}{1-\alpha^r}, \quad \beta_r = \frac{r}{1-\alpha^r} - \frac{\alpha}{1-\alpha}, \quad \gamma_r = \frac{\alpha}{(1-\alpha)^2} - \frac{r^2\alpha^r}{(1-\alpha^r)^2}.$$

Remark 3.6 In critical case, since

$$\lim_{\alpha\to 1}\hat{\alpha}_h = \frac{1}{h}, \quad \lim_{\alpha\to 1}\beta_h = \frac{h+1}{2}, \text{ and } \lim_{\alpha\to 1}\gamma_h = \frac{(h+1)^2}{12},$$

when α approaches 1, inequalities (3.19) and (3.20) are reduced to the following ones

$$\sum_{i=-h}^{-1} \chi_h^\top(i,j)R\chi_h(i,j) \geq \frac{1}{h}\left(\Omega_{11}^\top R\Omega_{11} + 3\tilde{\Omega}_{12}^\top R\tilde{\Omega}_{12}\right), \tag{3.29}$$

$$\sum_{j=-r}^{-1} \chi_v^\top(i,j) R \chi(i,j) \geq \frac{1}{r}\left(\Omega_{21}^\top R \Omega_{21} + 3\tilde{\Omega}_{22}^\top R \tilde{\Omega}_{22}\right), \tag{3.30}$$

where

$$\tilde{\Omega}_{12} = \chi(0,j) + \chi(-h,j) - \frac{2}{h+1}\sum_{i=-h}^{0} \chi(i,j),$$

$$\tilde{\Omega}_{22} = \chi(i,0) + \chi(i,-r) - \frac{2}{r+1}\sum_{j=-r}^{0} \chi(i,j).$$

3.4 Reachable Set Estimation

3.4.1 Ellipsoidal Ultimate Bound

In this section, based on the scheme presented in Lemma 3.1 and by utilizing the 2-D WSIs formulated in the preceding section, we derive tractable conditions for the existence of an EUB of system (3.1).

By virtue of the functional-based scheme proposed in Lemma 3.1, an augmented functional is constructed as follows

$$V(x(i,j)) = \underbrace{\begin{bmatrix} \tilde{x}^h(i,j) \\ \tilde{x}^v(i,j) \end{bmatrix}^\top P \begin{bmatrix} \tilde{x}^h(i,j) \\ \tilde{x}^v(i,j) \end{bmatrix}}_{V_0(x(i,j))} + \sum_{k=1}^{4} V_k(x(i,j)), \tag{3.31}$$

where, for the brevity, we write $V_k = V_k(x(i,j))$ $(k = 1, 2, 3, 4)$ and

$$\tilde{x}^h(i,j) = \begin{bmatrix} x^h(i,j) \\ \sum_{k=i-d_1^h}^{i-1} x^h(k,j) \end{bmatrix}, \quad \tilde{x}^v(i,j) = \begin{bmatrix} x^v(i,j) \\ \sum_{l=j-d_1^v}^{j-1} x^v(i,l) \end{bmatrix},$$

$$V_1 = \sum_{k=i-d_1^h}^{i-1} \alpha^{i-k-1} x^{h\top}(k,j) Q_1^h x^h(k,j) + \sum_{l=j-d_1^v}^{j-1} \alpha^{j-l-1} x^{v\top}(i,l) Q_1^v x^v(i,l),$$

$$V_2 = \sum_{k=i-d_2^h}^{i-d_1^h-1} \alpha^{i-k-1} x^{h\top}(k,j) Q_2^h x^h(k,j) + \sum_{l=j-d_2^v}^{j-d_1^v-1} \alpha^{j-l-1} x^{v\top}(i,l) Q_2^v x^v(i,l),$$

$$V_3 = \sum_{k=-d_1^h}^{-1} \sum_{l=i+k}^{i-1} \alpha^{i-l-1} \chi_h^\top(l,j) R_1^h \chi_h(l,j) + \sum_{k=-d_1^v}^{-1} \sum_{l=j+k}^{j-1} \alpha^{j-l-1} \chi_v^\top(i,l) R_1^v \chi_v(i,l),$$

$$V_4 = \sum_{k=-d_2^h}^{-d_1^h-1} \sum_{l=i+k}^{i-1} \alpha^{i-l-1} \chi_h^\top(l,j) R_2^h \chi_h(l,j) + \sum_{k=-d_2^v}^{-d_1^v-1} \sum_{l=j+k}^{j-1} \alpha^{j-l-1} \chi_v^\top(i,l) R_2^v \chi_v(i,l),$$

$$\chi_h(i,j) = \Delta_h x^h(i,j), \quad \chi_v(i,j) = \Delta_v x^v(i,j).$$

Remark 3.7 The functional (3.31) is composed of state vector and its summations, which incorporates extra information of delays into its augmented vectors. This improved functional, together with new 2-D WSIs, can give less conservative conditions for RSE and state bounding of system (3.1).

To simplify the notations, we denote block matrices

$$\begin{aligned}
e_i &= [0_{n\times(i-1)n} \quad I_n \quad 0_{n\times[(5-i)n+n_d]}],\ i = 1, 2, \ldots, 5, \\
e_6 &= [0_{s\times 5n} \quad I_s], \quad \mathscr{A} = Ae_1 + A_d e_3 + Be_6, \\
\hat{\mathscr{A}} &= (A - I_n)e_1 + A_d e_3 + Be_6,
\end{aligned}$$

and delayed state vectors

$$\begin{aligned}
x_l(i,j) &= \begin{bmatrix} x^h(i-d_1^h, j) \\ x^v(i, j-d_1^v) \end{bmatrix}, \quad x_d(i,j) = \begin{bmatrix} x^h(i-d_h(i), j) \\ x^v(i, j-d_v(j)) \end{bmatrix}, \\
x_u(i,j) &= \begin{bmatrix} x^h(i-d_2^h, j) \\ x^v(i, j-d_2^v) \end{bmatrix}, \quad x_s(i,j) = \begin{bmatrix} \xi_h(i,j) \\ \xi_v(i,j) \end{bmatrix}, \\
\xi_h(i,j) &= \sum_{k=i-d_1^h}^{i} x^h(k,j), \quad \xi_v(i,j) = \sum_{l=j-d_1^v}^{j} x^v(i,l).
\end{aligned}$$

We also define the following matrices

$$\begin{aligned}
F_1 &= \left[\mathscr{A}^\top \ (e_5 - e_2)^\top\right]^\top, \quad F_2 = \left[e_1^\top \ (e_5 - e_1)^\top\right]^\top, \\
\Gamma_1 &= \begin{bmatrix} \mathbb{J}_{n_h}\left(\beta_{d_1^h} e_1 + (d_1^h + 1 - \beta_{d_1^h})e_2 - e_5\right) \\ \mathbb{J}_{n_v}\left(\beta_{d_1^v} e_1 + (d_1^v + 1 - \beta_{d_1^v})e_2 - e_5\right) \end{bmatrix}, \\
\Gamma_2 &= \left[(e_2 - e_3)^\top \ (e_3 - e_4)^\top\right]^\top, \\
\Pi_0 &= F_1^\top\left(\overline{P^h} + \overline{P^v}\right)F_1 - \alpha F_2^\top\left(\overline{P^h} + \overline{P^v}\right)F_2, \\
\overline{P^h} &= \mathbb{J}_{n_h}^{2\top} P^h \mathbb{J}_{n_h}^2, \quad \overline{P^v} = \mathbb{J}_{n_v}^{2\top} P^v \mathbb{J}_{n_v}^2, \\
\Pi_1 &= e_1^\top Q_1 e_1 - e_2^\top \widehat{Q}_1 e_2 + e_2^\top \widehat{Q}_2 e_2 - e_4^\top \widetilde{Q}_2 e_4, \\
\Pi_2 &= \hat{\mathscr{A}}^\top\left(\overline{R_1} + \overline{R_2}\right)\hat{\mathscr{A}}, \quad \Pi_3 = (e_1 - e_2)^\top \widehat{R}_1 (e_1 - e_2) + \Gamma_1^\top \widetilde{R}_1 \Gamma_1, \\
\Pi_4 &= \Gamma_2^\top\left(\tilde{\alpha}_{d_h} \mathbb{J}_{n_h}^{2\top} \Psi_1 \mathbb{J}_{n_h}^2 + \tilde{\alpha}_{d_v} \mathbb{J}_{n_v}^{2\top} \Psi_2 \mathbb{J}_{n_v}^2\right)\Gamma_2, \quad \Pi_5 = \mu e_6^\top e_6, \\
\Psi_1 &= \begin{bmatrix} R_2^h & X \\ X^\top & R_2^h \end{bmatrix}, \quad \Psi_2 = \begin{bmatrix} R_2^v & Y \\ Y^\top & R_2^v \end{bmatrix},
\end{aligned}$$

$$\mathbb{J}^k_{n_h} = \underbrace{\mathrm{diag}([I_{n_h} \quad 0_{n_h\times n_v}], \ldots, [I_{n_h} \quad 0_{n_h\times n_v}])}_{k \text{ blocks}},$$
$$\mathbb{J}^k_{n_v} = \underbrace{\mathrm{diag}([0_{n_v\times n_h} \quad I_{n_v}], \ldots, [0_{n_v\times n_h} \quad I_{n_v}])}_{k \text{ blocks}},$$
$$\tilde{\alpha}_{d_h} = \frac{(1-\alpha)\alpha^{d_2^h}}{1-\alpha^{d_2^h-d_1^h}}, \quad \tilde{\alpha}_{d_v} = \frac{(1-\alpha)\alpha^{d_2^v}}{1-\alpha^{d_2^v-d_1^v}},$$
$$\widehat{Q}_\nu = \mathrm{diag}\left(\alpha^{d_1^h} Q^h_\nu, \alpha^{d_1^v} Q^v_\nu\right), \quad \widetilde{Q}_\nu = \mathrm{diag}\left(\alpha^{d_2^h} Q^h_\nu, \alpha^{d_2^v} Q^v_\nu\right), \nu = 1, 2,$$
$$\overline{R_1} = \mathrm{diag}\left(d_1^h R_1^h, d_1^v R_1^v\right), \quad \overline{R_2} = \mathrm{diag}\left((d_2^h - d_1^h) R_2^h, (d_2^v - d_1^v) R_2^v\right),$$
$$\widehat{R}_1 = \mathrm{diag}\left(\alpha^{d_1^h} R_1^h, \alpha^{d_1^v} R_1^v\right), \quad \widetilde{R}_1 = \mathrm{diag}\left(\frac{\alpha^{d_1^h}}{\gamma_{d_1^h}} R_1^h, \frac{\alpha^{d_1^v}}{\gamma_{d_1^v}} R_1^v\right).$$

Main result on ellipsoidal state bounding of system (3.1) is presented as in the following theorem.

Theorem 3.1 *Consider system* (3.1) *with interval directional delays* $d_h(i) \in [d_1^h, d_2^h]$, $d_v(j) \in [d_1^v, d_2^v]$. *Assume that there exist scalars* $\alpha \in (0, 1)$, $\mu \geq 0$, *positive definite matrices* $P_0 \in \mathbb{S}^+_n$, $P^h \in \mathbb{S}^+_{2n_h}$, $P^v \in \mathbb{S}^+_{2n_v}$, $Q_\nu = \mathrm{diag}(Q^h_\nu, Q^v_\nu)$, $R_\nu = \mathrm{diag}(R^h_\nu, R^v_\nu) \in \mathbb{S}^+_n$ $(\nu = 1, 2)$ *and matrices* $X \in \mathbb{R}^{n_h\times n_h}$, $Y \in \mathbb{R}^{n_v\times n_v}$ *such that the following LMIs hold*

$$\Psi_1 = \begin{bmatrix} R_2^h & X \\ X^\top & R_2^h \end{bmatrix} \geq 0, \quad \Psi_2 = \begin{bmatrix} R_2^v & Y \\ Y^\top & R_2^v \end{bmatrix} \geq 0, \tag{3.32}$$

$$\Phi = \Pi_0 + \sum_{k=1}^{2} \Pi_k - \sum_{k=3}^{5} \Pi_k < 0, \tag{3.33}$$

$$\mathrm{diag}(P^h, P^v) \geq J^\top P_0 J, \tag{3.34}$$

where $J = \left[\mathbb{J}^\top_{n_h} \; 0_{n\times n_h} \; \mathbb{J}^\top_{n_v} \; 0_{n\times n_v}\right]$. *Then, the ellipsoid* $\Omega_e\left(P_0, \frac{\mu w_\infty^2}{1-\alpha}\right)$ *is an EUB of system* (3.1)–(3.5). *Furthermore, the reachable set of system* (3.1) *subject to disturbances* (3.5) *is also confined within* $\Omega_e\left(P_0, \frac{\mu w_\infty^2}{1-\alpha}\right)$.

Proof Let $P = \mathrm{diag}(P^h, P^v)$ be matrix determined by conditions (3.33)–(3.34). It can be seen that the functional (3.31) can be written in the form

$$V(x) = V_h(x^h) + V_v(x^v)$$

and the difference of $V(x)$ along state trajectory $x(i, j)$ is then defined by

$$\begin{aligned} \Delta_p V(x(i, j)) &= \Delta_h V_h(x^h(i, j)) + \Delta_v V_v(x^v(i, j)) \\ &= V_h(x^h(i+1, j)) - V_h(x^h(i, j)) \\ &\quad + V_v(x^v(i, j+1)) - V^v(x^v(i, j)). \end{aligned} \tag{3.35}$$

First, by (3.31), we have

$$V_0 = V_0(x(i, j)) = V_h^0(x^h(i, j)) + V_v^0(x^v(i, j)) \tag{3.36}$$

where $V_h^0(x^h(i, j)) = \tilde{x}^{h\top}(i, j)P^h\tilde{x}^h(i, j)$ and $V_v^0(x^v(i, j)) = \tilde{x}^{v\top}(i, j)P^v\tilde{x}^v(i, j)$. Note also that

$$\begin{aligned}
&\tilde{x}^h(i+1, j) = \mathbb{J}_{n_h}^2 F_1\mathscr{X}(i, j), \quad \tilde{x}^v(i, j+1) = \mathbb{J}_{n_v}^2 F_1\mathscr{X}(i, j),\\
&\tilde{x}^h(i, j) = \mathbb{J}_{n_h}^2 F_2\mathscr{X}(i, j), \quad \tilde{x}^v(i, j) = \mathbb{J}_{n_v}^2 F_2\mathscr{X}(i, j),
\end{aligned}$$

where

$$\mathscr{X}(i, j) = \left[x^\top(i, j)\; x_l^\top(i, j)\; x_d^\top(i, j)\; x_u^\top(i, j)\; x_s^\top(i, j)\; w^\top(i, j)\right]^\top.$$

Thus, the differences $\Delta_h V_h^0(x^h(i, j))$ and $\Delta_v V_v^0(x^v(i, j))$ can be obtained as

$$\begin{aligned}
\Delta_h V_h^0(x^h(i, j)) &= \mathscr{X}^\top(i, j)\left(F_1^\top\overline{P^h}F_1 - F_2^\top\overline{P^h}F_2\right)\mathscr{X}(i, j),\\
\Delta_v V_v^0(x^v(i, j)) &= \mathscr{X}^\top(i, j)\left(F_1^\top\overline{P^v}F_1 - F_2^\top\overline{P^v}F_2\right)\mathscr{X}(i, j).
\end{aligned}$$

As a consequence, we get

$$\Delta_p V_0 + (1-\alpha)V_0 = \mathscr{X}^\top(i, j)\Pi_0\mathscr{X}(i, j). \tag{3.37}$$

Next, the difference $\Delta_p V_1$ is given by

$$\begin{aligned}
\Delta_p V_1 = &\sum_{k=i-d_1^h+1}^{i}\alpha^{i-k}x^{h\top}(k, j)Q_1^h x^h(k, j) - \sum_{k=i-d_1^h}^{i-1}\alpha^{i-k}x^{h\top}(k, j)Q_1^h x^h(k, j)\\
&+\sum_{l=j-d_1^v+1}^{j}\alpha^{j-l}x^{v\top}(i, l)Q_1^v x^v(i, l)\\
&-\sum_{l=j-d_1^v}^{j-1}\alpha^{j-l}x^{v\top}(i, l)Q_1^v x^v(i, l) - (1-\alpha)V_1\\
= &\,\mathscr{X}^\top(i, j)\left(e_1^\top Q_1 e_1 - e_2^\top\widehat{Q}_1 e_2\right)\mathscr{X}(i, j) - (1-\alpha)V_1.
\end{aligned} \tag{3.38}$$

Similarly, the difference $\Delta_p V_2$ can be obtained

$$
\begin{aligned}
\Delta_p V_2 &= x^{h\top}(i-d_1^h, j)\alpha^{d_1^h} Q_2^h x^h(i-d_1^h, j) \\
&\quad - x^{h\top}(i-d_2^h, j)\alpha^{d_2^h} Q_2^h x^h(i-d_2^h, j) \\
&\quad + x^{v\top}(i, j-d_1^v)\alpha^{d_1^v} Q_2^v x^v(i, j-d_1^v) \\
&\quad - x^{v\top}(i, j-d_2^v)\alpha^{d_2^v} Q_2^v x^v(i, j-d_2^v) - (1-\alpha)V_2 \\
&= \mathscr{X}^\top(i,j)\left(e_2^\top \widehat{Q}_2 e_2 - e_4^\top \widetilde{Q}_2 e_4\right)\mathscr{X}(i,j) - (1-\alpha)V_2. \qquad (3.39)
\end{aligned}
$$

From (3.38) and (3.39) we readily obtain

$$
\Delta_p(V_1+V_2) = \mathscr{X}^\top(i,j)\Pi_1\mathscr{X}(i,j) - (1-\alpha)(V_1+V_2). \qquad (3.40)
$$

Now, we compute and manipulate the difference $\Delta_p V_3$ by utilizing the inequalities given in Corollary 3.1. Observe that

$$
\begin{aligned}
\Delta_p V_3 = &\begin{bmatrix}\chi_h(i,j)\\ \chi_v(i,j)\end{bmatrix}^\top \begin{bmatrix} d_1^h R_1^h & 0 \\ 0 & d_1^v R_1^v\end{bmatrix}\begin{bmatrix}\chi_h(i,j)\\ \chi_v(i,j)\end{bmatrix} \\
&- \sum_{k=i-d_1^h}^{i-1} \alpha^{i-k}\chi_h^\top(k,j)R_1^h\chi_h(k,j) \\
&- \sum_{l=j-d_1^v}^{j-1} \alpha^{j-l}\chi_v^\top(i,l)R_1^v\chi_v(i,l) - (1-\alpha)V_3 \qquad (3.41)
\end{aligned}
$$

and $[\chi_h^\top(i,j) \quad \chi_v^\top(i,j)]^\top = \hat{\mathscr{A}}\mathscr{X}(i,j)$. Therefore,

$$
\begin{aligned}
\Delta_p V_3 &= \mathscr{X}^\top(i,j)\hat{\mathscr{A}}^\top\overline{R_1}\hat{\mathscr{A}}\mathscr{X}(i,j) - (1-\alpha)V_3 \\
&\quad - \sum_{k=i-d_1^h}^{i-1}\alpha^{i-k}\chi_h^\top(k,j)R_1^h\chi_h(k,j) - \sum_{l=j-d_1^v}^{j-1}\alpha^{j-l}\chi_v^\top(i,l)R_1^v\chi_v(i,l). \qquad (3.42)
\end{aligned}
$$

By utilizing the inequalities (3.27) and (3.28), we get

$$
-\sum_{k=i-d_1^h}^{i-1}\alpha^{i-k}\chi_h^\top(k,j)R_1^h\chi_h(k,j) \le -\hat{\alpha}_{d_1^h}\left(\psi_{11}^\top R_1^h\psi_{11} + \frac{\beta_{d_1^h}^2}{\gamma_{d_1^h}}\psi_{12}^\top R_1^h\psi_{12}\right) \qquad (3.43)
$$

and

$$
-\sum_{l=j-d_1^v}^{j-1}\alpha^{j-l}\chi_v^\top(i,l)R_1^v\chi_v(i,l) \le -\hat{\alpha}_{d_1^v}\left(\psi_{21}^\top R_1^v\psi_{21} + \frac{\beta_{d_1^v}^2}{\gamma_{d_1^v}}\psi_{22}^\top R_1^v\psi_{22}\right), \qquad (3.44)
$$

where

$$\psi_{11} = x^h(i,j) - x^h(i-d_1^h,j), \quad \psi_{21} = x^v(i,j) - x^v(i,j-d_1^v),$$
$$\psi_{12} = x^h(i,j) + \left(\frac{d_1^h+1}{\beta_{d_1^h}} - 1\right)x^h(i-d_1^h,j) - \frac{1}{\beta_{d_1^h}}\xi_h(i,j),$$
$$\psi_{22} = x^v(i,j) + \left(\frac{d_1^v+1}{\beta_{d_1^v}} - 1\right)x^v(i,j-d_1^v) - \frac{1}{\beta_{d_1^v}}\xi_v(i,j).$$

From (3.41)–(3.44), we have

$$\Delta_p V_3 \le \mathscr{X}^\top(i,j)\left(\hat{\mathscr{A}}^\top \overline{R_1}\hat{\mathscr{A}} - \Pi_3\right)\mathscr{X}(i,j) - (1-\alpha)V_3. \tag{3.45}$$

Similar to (3.41), the difference $\Delta_p V_4$ is given by

$$\begin{aligned}\Delta_p V_4 &= \mathscr{X}^\top(i,j)\hat{\mathscr{A}}^\top \overline{R_2}\hat{\mathscr{A}}\mathscr{X}(i,j) - (1-\alpha)V_4\\ &\quad - \sum_{k=i-d_2^h}^{i-d_1^h-1} \alpha^{i-k}\chi_h^\top(k,j)R_2^h\chi_h(k,j)\\ &\quad - \sum_{l=j-d_2^v}^{j-d_1^v-1} \alpha^{j-l}\chi_v^\top(i,l)R_2^v\chi_v(i,l).\end{aligned} \tag{3.46}$$

To deal with interval delays encountered in bounding $\Delta_p V_4$, we need to suitably use the reciprocally convex combination [17, 29]. For more details, we refer the reader to Sect. 2.3.3 in Chap. 2. By Lemma 3.3, we have

$$\begin{aligned}&- \sum_{k=i-d_2^h}^{i-d_1^h-1} \alpha^{i-k}\chi_h^\top(k,j)R_2^h\chi_h(k,j)\\ &\le -\frac{(1-\alpha)\alpha^{d_h(i)}}{1-\alpha^{d_h(i)-d_1^h}}\phi_{11}^\top R_2^h\phi_{11} - \frac{(1-\alpha)\alpha^{d_2^h}}{1-\alpha^{d_2^h-d_h(i)}}\phi_{12}^\top R_2^h\phi_{12}\\ &\le -\frac{(1-\alpha)\alpha^{d_2^h}}{1-\alpha^{d_2^h-d_1^h}}\begin{bmatrix}\phi_{11}\\ \phi_{12}\end{bmatrix}^\top \begin{bmatrix}R_2^h & X\\ X^\top & R_2^h\end{bmatrix}\begin{bmatrix}\phi_{11}\\ \phi_{12}\end{bmatrix},\end{aligned} \tag{3.47}$$

where

$$\phi_{11} = x^h(i-d_1^h,j) - x^h(i-d_h(i),j)$$

and

$$\phi_{12} = x^h(i-d_h(i),j) - x^h(i-d_2^h,j).$$

Note that when $d_h(i) = d_1^h$ and $d_h(i) = d_2^h$ then $\phi_{11} = 0$ and $\phi_{12} = 0$, respectively, and the last inequality in (3.47) is still valid.

By similar lines used to derive (3.47), we also have

$$
\begin{aligned}
&-\sum_{l=j-d_2^v}^{j-d_1^v-1} \alpha^{j-l} \chi_v^\top(i,l) R_2^v \chi_v(i,l) \\
&\le -\frac{(1-\alpha)\alpha^{d_v(j)}}{1-\alpha^{d_v(j)-d_1^v}} \phi_{21}^\top R_2^v \phi_{21} - \frac{(1-\alpha)\alpha^{d_2^v}}{1-\alpha^{d_2^v-d_v(j)}} \phi_{22}^\top R_2^v \phi_{22} \\
&\le -\frac{(1-\alpha)\alpha^{d_2^v}}{1-\alpha^{d_2^v-d_1^v}} \begin{bmatrix} \phi_{21} \\ \phi_{22} \end{bmatrix}^\top \begin{bmatrix} R_2^v & Y \\ Y^\top & R_2^v \end{bmatrix} \begin{bmatrix} \phi_{21} \\ \phi_{22} \end{bmatrix},
\end{aligned} \tag{3.48}
$$

where

$$
\phi_{21} = x^v(i, j - d_1^v) - x^v(i, j - d_v(j))
$$

and

$$
\phi_{22} = x^v(i, j - d_v(j)) - x^v(i, j - d_2^v).
$$

Note also that $[\phi_{11}^\top \;\; \phi_{12}^\top]^\top = \mathbb{J}_{n_h}^2 \Gamma_2 \mathscr{X}(i,j)$ and $[\phi_{21}^\top \;\; \phi_{22}^\top]^\top = \mathbb{J}_{n_v}^2 \Gamma_2 \mathscr{X}(i,j)$. Therefore, from (3.46)–(3.48), we readily obtain

$$
\Delta_p V_4 \le \mathscr{X}^\top(i,j) \left(\mathscr{A}^\top R_2 \mathscr{A} - \Pi_4 \right) \mathscr{X}(i,j) - (1-\alpha) V_4. \tag{3.49}
$$

Finally, from (3.37), (3.40), (3.45) and (3.49), we obtain

$$
\Delta_p V(x(i,j)) + (1-\alpha) V(x(i,j)) - \mu w^\top(i,j) w(i,j) \le \mathscr{X}^\top(i,j) \Phi \mathscr{X}(i,j). \tag{3.50}
$$

If condition (3.33) holds then, by Lemma 3.1, from (3.50) we get

$$
\limsup_{i+j\to\infty} V(x(i,j)) \le \frac{\mu w_\infty^2}{1-\alpha}.
$$

In addition, under condition (3.34) and the positivity of functionals V_k ($k = 1, 2, 3, 4$), we have

$$
V(x(i,j)) \ge x^\top(i,j) P_0 x(i,j), \quad \forall (i,j) \in \mathbb{N}_0^2.
$$

According to Lemma 3.2, the ellipsoid $\Omega_e\left(P_0, \frac{\mu w_\infty^2}{1-\alpha}\right)$ is an EUB of system (3.1)–(3.5). The proof is completed. □

Remark 3.8 In the proof of Theorem 3.1, 2-D WSIs are utilized to manipulate the difference of LKF candidates and the reciprocally convex combination technique is used to tackle the relationship between time-varying delays and their intervals, which is effective to reduce conservatism of the resulting conditions. Moreover, unlike the existing methods where free-weighting matrices are incorporated to improve feasibility of the derived LMI conditions, the approach presented in this chapter can also reduce computational burden because no extra slack variable is introduced.

Remark 3.9 Although the derived conditions in Theorem 3.1 involve a non-convex scalar α, by fixing the scalar α, condition (3.33) is reduced to an LMI which can be effectively solved by various computational tools. In addition, as discussed in the literature, e.g., [7, 24], if the conditions in (3.32)–(3.34) are feasible for some $\alpha \in (0, 1)$, it is not difficult to search for the desired value of α.

Remark 3.10 For a prescribed scalar $\alpha \in (0, 1)$, the ellipsoid $\Omega_e\big(P_0, \frac{\mu w_\infty^2}{1-\alpha}\big)$, which contains the reachable set of system (3.1), can be determined by solving the derived LMIs (3.32)–(3.34). To make this reachable set ellipsoidal bounding as small as possible, some generic search procedures, for example, developed in [17, 24] can be employed.

3.4.2 *Exponential Stability Conditions*

As an application, in this section we derive exponential stability conditions for system (3.1) without disturbance (i.e. $w = 0$). Specifically, with $w = 0$, from (3.15), we have

$$x^\top(i, j) P_0 x(i, j) \le V(x(i, j)) \le \hat{V}_\alpha(\phi_h, \phi_v)\alpha^{i+j}, \tag{3.51}$$

where

$$\hat{V}_\alpha(\phi_h, \phi_v) = \sum_{j=0}^{T_h} \alpha^{-j} V^h(x^h(0, j)) + \sum_{i=0}^{T_v} \alpha^{-i} V^v(x^v(i, 0)).$$

Taking (3.31) into account we then obtain

$$\hat{V}_\alpha(\phi_h, \phi_v) \le \lambda_1 \sum_{k=0}^{T_h} \frac{\hat{\phi}_h^2(k)}{\alpha^k} + \lambda_2 \sum_{l=0}^{T_v} \frac{\hat{\phi}_v^2(l)}{\alpha^l}, \tag{3.52}$$

where $\hat{\phi}_h(k)$ and $\hat{\phi}_v(l)$ are defined as in (3.6) and

$$\lambda_1 = (1 + d_1^h)\lambda_{\max}(P^h) + d_2^h \left(\lambda_{\max}(Q_1^h) \vee \lambda_{\max}(Q_2^h) + \lambda_{\max}(R_1^h) \vee \lambda_{\max}(R_2^h)\right),$$
$$\lambda_2 = (1 + d_1^v)\lambda_{\max}(P^v) + d_2^v \left(\lambda_{\max}(Q_1^v) \vee \lambda_{\max}(Q_2^v) + \lambda_{\max}(R_1^v) \vee \lambda_{\max}(R_2^v)\right).$$

Therefore, from (3.51) and (3.52) we obtain

$$\|x(i, j)\|^2 \le \frac{\lambda_1 + \lambda_2}{\lambda_0} \left(\sum_{k=0}^{T_h} \frac{\hat{\phi}_h^2(k)}{\alpha^k} + \sum_{l=0}^{T_v} \frac{\hat{\phi}_v^2(l)}{\alpha^l} \right) \alpha^{i+j}, \ i, j \in \mathbb{N}_0, \tag{3.53}$$

where $\lambda_0 = \lambda_{\min}(P^h) \wedge \lambda_{\min}(P^v)$. This shows that system (3.1) without disturbance is exponentially stable.

We summarize this result in the following corollary.

Corollary 3.2 *Consider system* (3.1) *with* $w = 0$. *Assume that there exist an* $\alpha \in (0, 1)$, *positive definite matrices* $P^h \in \mathbb{S}^+_{2n_h}$, $P^v \in \mathbb{S}^+_{2n_v}$, $Q_\nu = \mathrm{diag}(Q^h_\nu, Q^v_\nu)$, $R_\nu = \mathrm{diag}(R^h_\nu, R^v_\nu) \in \mathbb{S}^+_n$ $(\nu = 1, 2)$ *and matrices* $X \in \mathbb{R}^{n_h \times n_h}$, $Y \in \mathbb{R}^{n_v \times n_v}$ *such that condition* (3.32) *and the following LMI hold*

$$\tilde{\Phi} = \tilde{\Pi}_0 + \sum_{k=1}^{2} \tilde{\Pi}_k - \sum_{k=3}^{4} \tilde{\Pi}_k < 0, \tag{3.54}$$

where $\tilde{e}_k = [0_{n\times(k-1)n}\ I_n\ 0_{n\times(5-k)n}]$ $(k = 1, 2, \ldots, 5)$ *denote the block-row matrices of* I_{5n} *and*

$$\begin{aligned}
\tilde{\Pi}_0 &= \tilde{F}_1^\top \left(\overline{P^h} + \overline{P^v}\right) \tilde{F}_1 - \alpha \tilde{F}_2^\top \left(\overline{P^h} + \overline{P^v}\right) \tilde{F}_2, \\
\tilde{\Pi}_1 &= \tilde{e}_1^\top Q_1 \tilde{e}_1 - \tilde{e}_2^\top \widehat{Q}_1 \tilde{e}_2 + \tilde{e}_2^\top \widehat{Q}_2 \tilde{e}_2 - \tilde{e}_4^\top \widetilde{Q}_2 \tilde{e}_4, \\
\tilde{\Pi}_2 &= \tilde{\mathscr{A}}_2^\top \left(\overline{R_1} + \overline{R_2}\right) \tilde{\mathscr{A}}_2, \quad \tilde{\Pi}_3 = (\tilde{e}_1 - \tilde{e}_2)^\top \widehat{R}_1 (\tilde{e}_1 - \tilde{e}_2) + \tilde{\Gamma}_1^\top \widetilde{R}_1 \tilde{\Gamma}_1, \\
\tilde{\Pi}_4 &= \tilde{\Gamma}_2^\top \left\{\tilde{\alpha}_{d_h} \mathbb{J}^{2\top}_{n_h} \Psi_1 \mathbb{J}^2_{n_h} + \tilde{\alpha}_{d_v} \mathbb{J}^{2\top}_{n_v} \Psi_2 \mathbb{J}^2_{n_v}\right\} \tilde{\Gamma}_2, \\
\tilde{F}_1 &= [\tilde{\mathscr{A}}_1^\top \ \ (\tilde{e}_5 - \tilde{e}_2)^\top]^\top, \quad \tilde{F}_2 = [\tilde{e}_1^\top \ \ (\tilde{e}_5 - \tilde{e}_1)^\top]^\top, \\
\tilde{\Gamma}_1 &= \begin{bmatrix} \mathbb{J}_{n_h}\left(\beta_{d_1^h} \tilde{e}_1 + (d_1^h + 1 - \beta_{d_1^h}) \tilde{e}_2 - \tilde{e}_5\right) \\ \mathbb{J}_{n_v}\left(\beta_{d_1^v} \tilde{e}_1 + (d_1^v + 1 - \beta_{d_1^v}) \tilde{e}_2 - \tilde{e}_5\right) \end{bmatrix}, \\
\tilde{\Gamma}_2 &= [(\tilde{e}_2 - \tilde{e}_3)^\top \ \ (\tilde{e}_3 - \tilde{e}_4)^\top]^\top, \\
\tilde{\mathscr{A}}_1 &= A\tilde{e}_1 + A_d \tilde{e}_3, \quad \tilde{\mathscr{A}}_2 = (A - I_n)\tilde{e}_1 + A_d \tilde{e}_3.
\end{aligned}$$

Then, system (3.1) *is exponentially stable.*

Remark 3.11 For the case of constant delays (i.e. $d_1^h = d_2^h$ and $d_1^v = d_2^v$), condition (3.54) of Corollary 3.2 is not directly applicable. In that case, by slightly modifying the functional (3.31) where the terms associated with Q_2 and R_2 are omitted, we obtain exponential stability conditions for system (3.1) as stated below.

Corollary 3.3 *With constant delays* d_h *and* d_v, *system* (3.1) *without disturbance is exponentially stable if there exist an* $\alpha \in (0, 1)$, *positive definite matrices* $P^h \in \mathbb{S}^+_{2n_h}$, $P^v \in \mathbb{S}^+_{2n_v}$, $Q = \mathrm{diag}(Q^h, Q^v)$ *and* $R = \mathrm{diag}(R^h, R^v) \in \mathbb{S}^+_n$ *such that the following LMI holds*

$$\begin{aligned}
&\begin{bmatrix} \hat{\mathscr{A}}_1 \\ \hat{e}_3 - \hat{e}_2 \end{bmatrix}^\top \left(\overline{P^h} + \overline{P^v}\right) \begin{bmatrix} \hat{\mathscr{A}}_1 \\ \hat{e}_3 - \hat{e}_2 \end{bmatrix} \\
&\qquad - \alpha \begin{bmatrix} \hat{e}_1 \\ \hat{e}_3 - \hat{e}_1 \end{bmatrix}^\top \left(\overline{P^h} + \overline{P^v}\right) \begin{bmatrix} \hat{e}_1 \\ \hat{e}_3 - \hat{e}_1 \end{bmatrix} \\
&\qquad + \hat{e}_1^\top Q \hat{e}_1 - \hat{e}_2^\top \widehat{Q} \hat{e}_2 + \hat{\mathscr{A}}_2^\top \overline{R} \hat{\mathscr{A}}_2 \\
&\qquad - (\hat{e}_1 - \hat{e}_2)^\top \widehat{R} (\hat{e}_1 - \hat{e}_2) - \hat{\Gamma}^\top \widetilde{R} \hat{\Gamma} < 0,
\end{aligned} \tag{3.55}$$

where $\hat{e}_k = [0_{n\times(k-1)n}\ I_n\ 0_{n\times(3-k)n}]$ $(k = 1, 2, 3)$ *and*

$$\hat{\mathscr{A}}_1 = A\hat{e}_1 + A_d\hat{e}_2, \quad \hat{\mathscr{A}}_2 = (A - I_n)\hat{e}_1 + A_d\hat{e}_2,$$
$$\widehat{Q} = \operatorname{diag}(\alpha^{d_h} Q^h, \alpha^{d_v} Q^v), \quad \overline{R} = \operatorname{diag}(d_h R^h, d_v R^v),$$
$$\widehat{R} = \operatorname{diag}(\alpha^{d_h} R^h, \alpha^{d_v} R^v), \quad \widetilde{R} = \operatorname{diag}\left(\frac{\alpha^{d_h}}{\gamma_{d_h}} R^h, \frac{\alpha^{d_v}}{\gamma_{d_v}} R^v\right),$$
$$\hat{\Gamma} = \begin{bmatrix} \mathbb{J}_{n_h}\left(\beta_{d_h}\hat{e}_1 + (d_h + 1 - \beta_{d_h})\hat{e}_2 - \hat{e}_3\right) \\ \mathbb{J}_{n_v}\left(\beta_{d_v}\hat{e}_1 + (d_v + 1 - \beta_{d_v})\hat{e}_2 - \hat{e}_3\right) \end{bmatrix}.$$

3.5 2-D Systems with Delays in FMLSS Model

Consider the following 2-D system described by the FMLSS model

$$\begin{aligned} x(i+1, j+1) = {} & A_1 x(i+1, j) + A_2 x(i, j+1) \\ & + A_{d1} x(i+1, j - d_v(j)) + A_{d2} x(i - d_h(i), j+1) \\ & + B_1 w(i, j+1) + B_2 w(i+1, j), \end{aligned} \tag{3.56}$$

where $x(i, j) \in \mathbb{R}^n$ and $w(i, j) \in \mathbb{R}^{n_d}$ are the state and exogenous disturbance vectors, respectively; A_k, A_{dk} and B_k $(k = 1, 2)$ are given matrices of appropriate dimensions.

For 2-D systems with time-varying delays and exogenous disturbances described by Roesser model and FMLSS model as given in (3.1) and (3.56), respectively, they are often studied separately in the literature due to their special structures and wide applications. With suitable modifications, it can be verified that the results presented in the preceding sections in this chapter can be extended to 2-D systems in FMLSS model as given in (3.56). More specifically, since the model (3.56) uses only one state vector to describe the information propagation in two directions, the functional $V(x(i, j))$ in (3.9) is now defined as

$$V(x(i, j)) = V^h(x(i, j)) + V^v(x(i, j)).$$

Similar to Lemma 3.1, if there exist scalars $\alpha \in (0, 1)$ and $\mu \geq 0$ that satisfy

$$\begin{aligned} & V(x(i+1, j+1)) - \alpha\left(V^h(x(i, j+1)) + V^v(x(i+1, j))\right) \\ & \quad - \mu \begin{bmatrix} w(i, j+1) \\ w(i+1, j) \end{bmatrix}^{\top} \begin{bmatrix} w(i, j+1) \\ w(i+1, j) \end{bmatrix} \leq 0, \quad (i, j) \in \mathbb{N}_0^2, \end{aligned} \tag{3.57}$$

then it holds that

$$\limsup_{r \to \infty} \sum_{i+j=r} V(x(i, j)) \leq \frac{\mu \widetilde{w}_\infty^2}{1 - \alpha},$$

where $\widetilde{w}_\infty$ is an upper bound of the admissible disturbances defined by

$$\sup_{k\geq 0}\sum_{i+j=k}\begin{bmatrix} w(i,j+1) \\ w(i+1,j)\end{bmatrix}^{\top}\begin{bmatrix} w(i,j+1) \\ w(i+1,j)\end{bmatrix} \leq \widetilde{w}_{\infty}^2.$$

By suitably modifying the functional candidates like (3.31) according to (3.57), similar results on RSE, state bounding and exponential stability analysis of system (3.1) can be obtained for system (3.56).

3.6 Numerical Examples

In this section, we use some numerical examples with simulations to illustrate the effectiveness of the results presented in this chapter.

Example 3.1 Consider a 2-D system in the form of (3.1), where

$$A = \left[\begin{array}{cc|c} 0.55 & -0.61 & 0.12 \\ 0.15 & 0.11 & 0 \\ 0.15 & 0.12 & 0.25 \end{array}\right], \quad A_d = \left[\begin{array}{cc|c} 0.15 & -0.11 & 0 \\ 0.12 & -0.21 & -0.11 \\ \hline 0.05 & -0.08 & 0.11 \end{array}\right], \quad B = \left[\begin{array}{c} 0.1 \\ -0.1 \\ \hline 0.05 \end{array}\right].$$

Time-varying delays $d_h(i)$ and $d_v(j)$ satisfy condition (3.2) with $d_1^h = d_1^v = 2, d_2^h = 4$ and $d_2^v = 5$. We fix $\alpha = 0.9$, $\mu = 0.1$. Using a search algorithm in Matlab LMI toolbox to maximize $\delta > 0$ subject to LMI conditions (3.33), (3.34) and $P_0 \geq \delta I_n$ we then obtain maximal possible $\delta_* = 1.16$ and

$$P_0 = \left[\begin{array}{cc|c} 1.1611 & -0.0012 & 0 \\ -0.0012 & 1.1620 & 0 \\ \hline 0 & 0 & 1.3650 \end{array}\right].$$

For admissible disturbances satisfying (3.5) with $w_{\infty}^2 = 0.1$, by Theorem 3.1, all state trajectories of system (3.1)–(3.5) converge exponentially within the ellipsoid $\Omega_e(P_0, 0.1)$ defined below and is shown in Fig. 3.1.

$$\Omega_e(P_0, 0.1) = \left\{x \in \mathbb{R}^3 : 1.1611x_1^2 - 0.0024x_1x_2 + 1.162x_2^2 + 1.365x_3^2 \leq 0.1\right\}.$$

For illustrative purpose, let the disturbance

$$w(i,j) = \frac{0.3162\cos(0.1i\pi/2 + 0.2j\pi/3)}{\sqrt{i+j}} \text{ for } i, j > 0,$$

with the trajectory presented in Fig. 3.2 and $d_h(i) = 2 + 2\left|\cos(\frac{i\pi}{2})\right|$, $d_v(j) = 2 + 3\left|\sin(\frac{j\pi}{2})\right|$. State trajectories $x_1^h(i,j), x_2^h(i,j)$ and $x^v(i,j)$ with zero initial condition are portrayed in Figs. 3.3, 3.4 and 3.5, respectively. It can be seen from Figs. 3.3, 3.4 and 3.5 that $x(i,j) = [x_1^h(i,j), x_2^h(i,j), x^v(i,j)]^{\top}$ belongs to $\Omega_e(P_0, 0.1)$ for

Fig. 3.1 The EUB $\Omega_e(P_0, 0.1)$

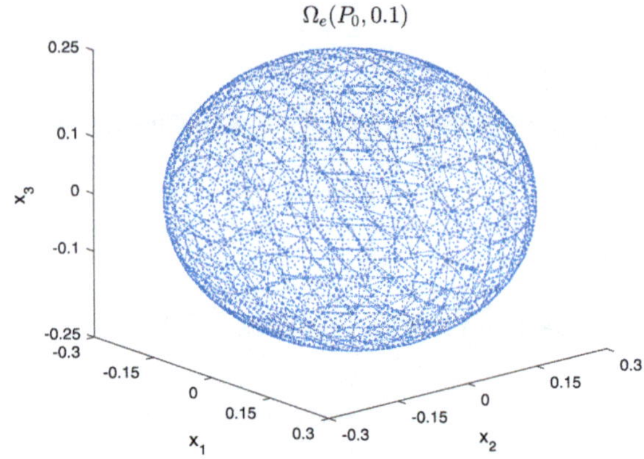

Fig. 3.2 A trajectory of disturbance $w(i, j)$

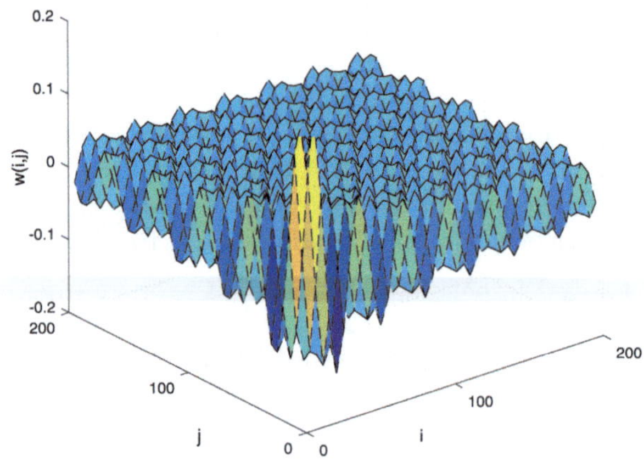

all (i, j). This shows that $\Omega_e(P_0, 0.1)$ is an EUB, which also contains the reachable set of the system.

Example 3.2 Consider a thermal process in chemical reactors, heat exchanger or pipe furnaces, which is described by the following partial differential equation [30]

$$\frac{\partial T(x,t)}{\partial x} + \frac{\partial T(x,t)}{\partial t} = -a_0 T(x,t) - a_1 T(x - x_d, t) - a_2 T(x, t - t_d), \quad (3.58)$$

where $T(x, t)$ is the unknown function represent the temperature at $x \in [0, x_f]$ (space) and $t \in [0, \infty)$ (time); x_d, t_d are space and time delays, respectively; a_0, a_1, a_2 are real constants. We discretize equation (3.58) using Euler approximation scheme by defining $T(i, j) = T(i\Delta x, j\Delta t)$ and

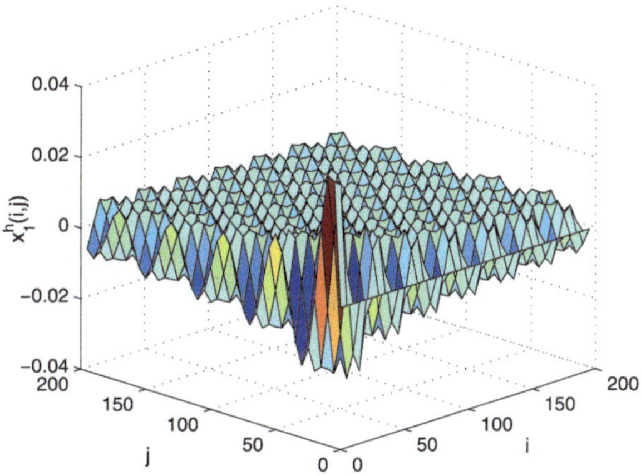

Fig. 3.3 State trajectory of $x_1^h(i, j)$

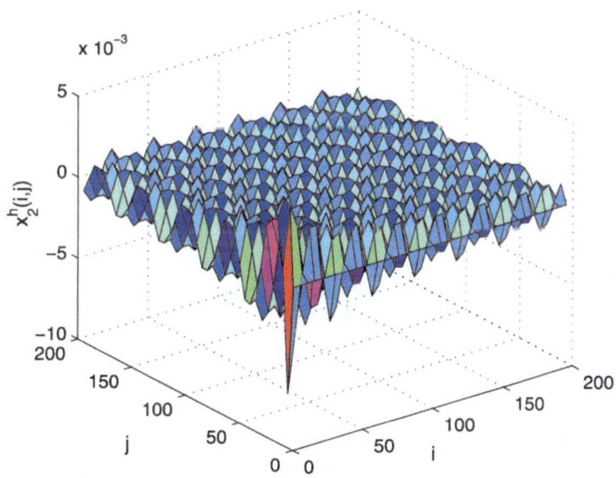

Fig. 3.4 State trajectory of $x_2^h(i, j)$

$$\frac{\partial T(x,t)}{\partial x} \approx \frac{T(i,j) - T(i-1,j)}{\Delta x}, \quad \frac{\partial T(x,t)}{\partial t} \approx \frac{T(i,j+1) - T(i,j)}{\Delta t}.$$

Then, Eq. (3.58) can be written in the form

$$\begin{aligned} T(i, j+1) = &\Big(1 - \frac{\Delta t}{\Delta x} - a_0 \Delta t\Big) T(i, j) + \frac{\Delta t}{\Delta x} T(i-1, j) \\ &- a_1 \Delta t T(i-1-d_1, j) - a_2 \Delta t T(i, j-d_2), \end{aligned} \tag{3.59}$$

where $d_1 = \lfloor \frac{x_d}{\Delta x} \rfloor$, $d_2 = \lceil \frac{t_d}{\Delta t} \rceil$ and $\lfloor . \rfloor$, $\lceil . \rceil$ denote the floor and the ceiling functions, respectively. We now define state variables $x^h(i, j) = T(i-1, j)$, $x^v(i, j) = T(i, j)$. Equation (3.59) is recast into Roesser system (3.1) with constant delays, where

$$A = \begin{bmatrix} 0 & 1 \\ \frac{\Delta t}{\Delta x} & 1 - \frac{\Delta t}{\Delta x} - a_0 \Delta t \end{bmatrix}, \quad A_d = \begin{bmatrix} 0 & 0 \\ -a_1 \Delta t & -a_2 \Delta t \end{bmatrix}.$$

Fig. 3.5 State trajectory of $x^v(i, j)$

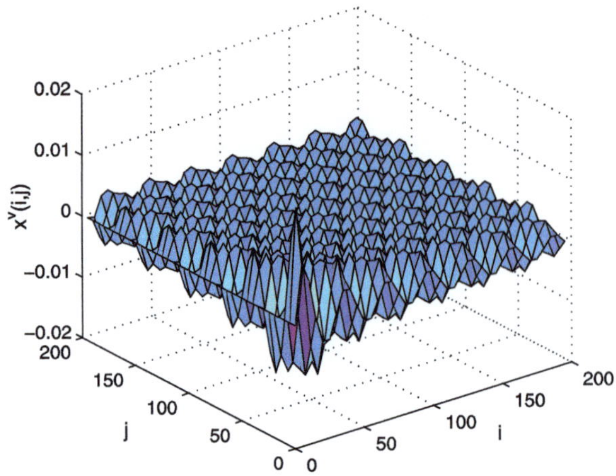

Table 3.1 Upper bound of d_v for various d_h and $\alpha = 0.99$

d_h	7	9	11	15	20	NoDv
Huang and Xiang [31] (Theorem 1)	4	3	3	3	3	$5.5(n_h^2 + n_v^2) + 2.5(n_h + n_v)$
Corollary 3.3	6	6	6	5	5	$3(n_h^2 + n_v^2) + 2(n_h + n_v)$

NoDv: Number of decision variables

For given increments $\Delta t = 0.1$, $\Delta x = 0.4$ and parameters $a_0 = 1$, $a_1 = 1.25$, $a_2 = 0.5$, the results delivered by Corollary 3.3 and Theorem 1 in [31] for $\alpha = 0.99$ are given in Table 3.1. It is clear that the derived conditions in Corollary 3.3 are less conservative than those of [31], not only for the feasibility but also the computational burden.

For constant delays $d_h = 11$, $d_v = 6$, condition (3.55) is feasible with $\alpha = 0.99$ and the following matrix variables

$$P^h = \begin{bmatrix} 280.8902 & -0.7581 \\ -0.7581 & 0.0928 \end{bmatrix}, \quad P^v = \begin{bmatrix} 444.6021 & -0.9794 \\ -0.9794 & 0.1204 \end{bmatrix},$$
$$Q = \begin{bmatrix} 51.1892 & 0 \\ 0 & 0.7121 \end{bmatrix}, \quad R = \begin{bmatrix} 2.8174 & 0 \\ 0 & 23.6252 \end{bmatrix}.$$

By Corollary 3.3, Eq. (3.59) is exponentially stable. Simulation results presented in Figs. 3.6 and 3.7 are taken with initial conditions in the ranges $0 \le i \le 100$, $0 \le j \le 100$ and delays $d_h = 11$, $d_v = 6$. Figures 3.6 and 3.7 indicate that the state trajectories $x^h(i, j)$ and $x^v(i, j)$ converge exponentially to zero which illustrates the obtained result. It is note also that, to this case, the obtained results are very closed to the analytical limit. Specifically, for $d_h = 11$ and $d_v > 6$, simulation results given in Figs. 3.8 and 3.9 exhibit unstable behavior. These results show the effectiveness and advantage of the method presented in this chapter.

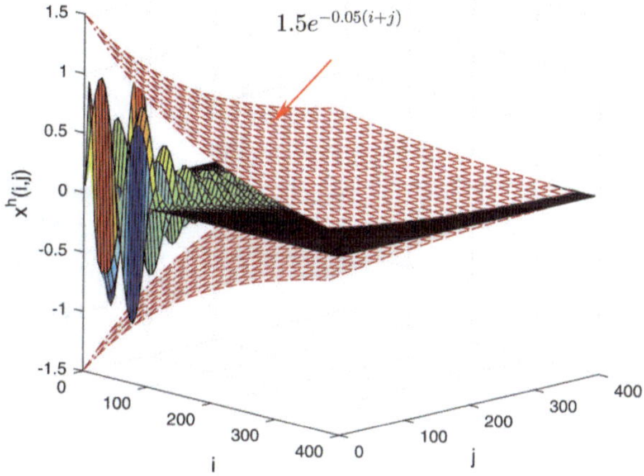

Fig. 3.6 State trajectory $x^h(i, j)$ with delays $d_h = 11$ and $d_v = 6$

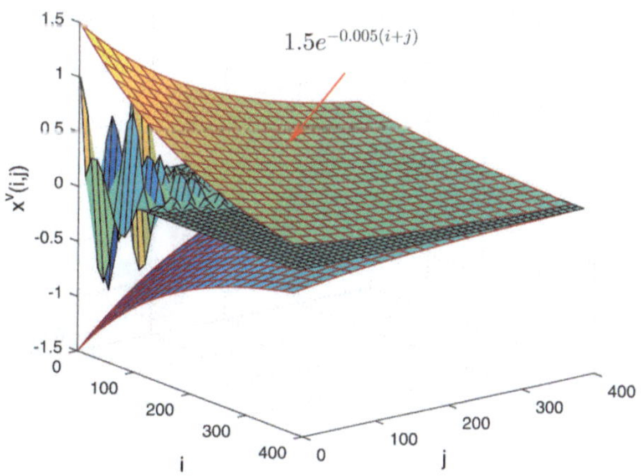

Fig. 3.7 State trajectory $x^v(i, j)$ with delays $d_h = 11$ and $d_v = 6$

Example 3.3 Consider system (3.1) with $w = 0$ and the matrices

$$A = \left[\begin{array}{c:c} A_{11} & A_{12} \\ \hdashline 0_{3\times 3} & A_{22} \end{array}\right], \quad A_d = \left[\begin{array}{c:c} A_{d11} & 0_{3\times 3} \\ \hdashline 0_{3\times 3} & A_{d22} \end{array}\right],$$

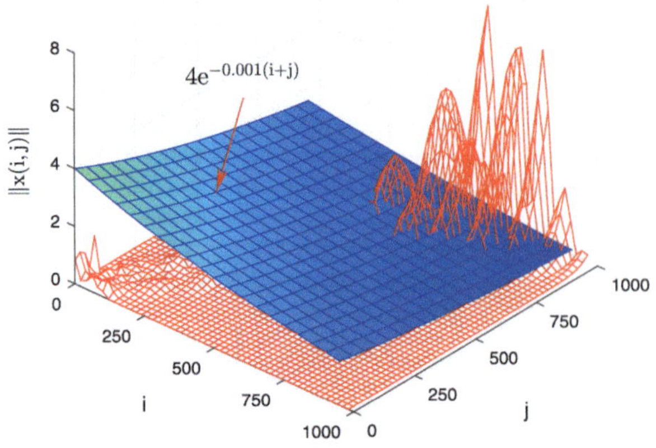

Fig. 3.8 Norm-state trajectory $\|x(i, j)\|$ of the system with $d_h = 11$ and $d_v = 7$

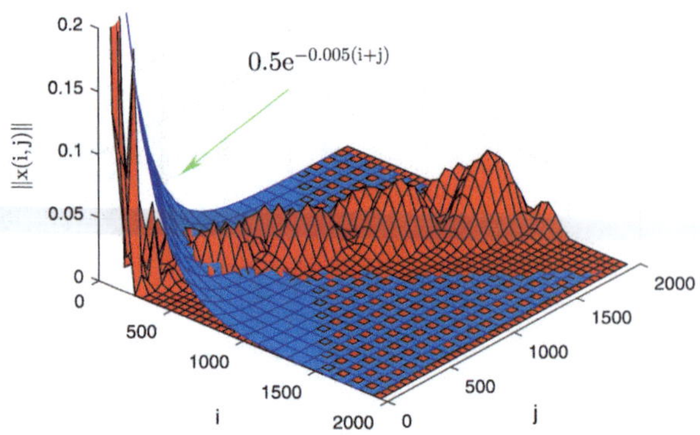

Fig. 3.9 Norm-state trajectory $\|x(i, j)\|$ of the system with $d_h = 11$ and $d_v = 8$

where

$$A_{11} = \begin{bmatrix} 0.8638 & 0.2719 & -0.039 \\ -0.2719 & -0.5951 & 0.3608 \\ -0.039 & -0.3608 & 0.3562 \end{bmatrix}, \quad A_{12} = \begin{bmatrix} 0.4294 & 0.3382 & 0.13 \\ 0.3382 & 0.2664 & 0.1024 \\ 0.13 & 0.1024 & 0.0394 \end{bmatrix}$$

$$A_{22} = \begin{bmatrix} 0.8638 & -0.2719 & -0.039 \\ 0.2719 & 0.5951 & -0.3608 \\ -0.039 & 0.3608 & 0.3562 \end{bmatrix}, \quad A_{d11} = \begin{bmatrix} -0.11 & 0 & 0.06 \\ -0.02 & -0.03 & 0 \\ 0.05 & 0.05 & -0.1 \end{bmatrix},$$

$$A_{d22} = \begin{bmatrix} 0.1 & 0 & 0.05 \\ -0.1 & 0.02 & 0 \\ 0.03 & 0.12 & -0.03 \end{bmatrix}.$$

Table 3.2 Upper bounds of d_2^h, d_2^v for various α

α	0.95	0.97	0.99	0.995	NoDv
Huang and Xiang [32] (Lemma 4)	–	9, 3	10, 3	10, 3	$12.5(n_h^2 + n_v^2) + 4.5(n_h + n_v)$
Corollary 3.2	9, 9	9, 10	10, 11	10, 11	$5(n_h^2 + n_v^2) + 3(n_h + n_v)$

Table 3.3 Lower bound of α for $d_2^v = 3$ and various d_2^h

d_2^h	3	6	7	9	10
Huang and Xiang [32] (Lemma 4)	0.9615	0.9615	0.9615	0.9661	0.99
Corollary 3.2	0.7269	0.7761	0.8447	0.9476	0.9856

To compare exponential stability conditions derived by the approach presented in this chapter with existing methods, for example, free-weighting matrices [32], let $d_1^h = 2$, $d_1^v = 2$. Table 3.2 gives the results obtained by Corollary 3.2 Lemma 4 in [32] for $d_1^h = 2$, $d_1^v = 2$ and various α. Table 3.3 gives the results of decaying rate α for various d_2^h and $d_2^v = 3$. Clearly, our approach gives significantly improved results than the method proposed in [32].

3.7 Conclusion

In this chapter, the problems of state bounding and reachable set estimation have been studied for 2-D systems with interval time-varying delays and bounded disturbances. A functional-based scheme for obtaining ultimate bounds has been presented. The proposed method is also utilizable to address the reachable set estimation and exponential stability problems. Based on some new 2-D WSIs, which provide tighter lower bound than the conventional Jensen-type inequality, and an improved LKF candidate, delay-dependent conditions have been derived for the existence of an ellipsoid that bounds the system states in the presence of bounded disturbances. Exponential stability conditions of 2-D time-delay systems have also been derived in this chapter. Numerical examples and simulations have been given to demonstrate the effectiveness of the obtained results.

References

1. Boyd, S., El Ghaoui, L., Feron, E., Balakrishnan, V.: Linear Matrix Inequalities in System and Control Theory. SIAM, Philadelphia (1994)
2. Fridman, E., Shaked, U.: On reachable sets for linear systems with delay and bounded peak inputs. Automatica **39**, 2005–2010 (2003)
3. Hu, T., Teel, A.R., Zaccarian, L.: Stability and performance for saturated systems via quadratic and nonquadratic Lyapunov functions. IEEE Trans. Autom. Control **51**, 1770–1786 (2006)
4. Liu, C., Jiang, B.: H_∞ fault-tolerant control for time-varied actuator fault of nonlinear system. Int. J. Syst. Sci. **45**, 1–11 (2014)
5. Zuo, Z., Fu, Y., Chen, Y., Wang, Y.: A new method of reachable set estimation for time delay systems with polytopic uncertainties. Appl. Math. Comput. **221**, 639–646 (2013)
6. Zhang, B., Lam, J., Xu, S.: Reachable set estimation and controller design for distributed delay systems with bounded disturbances. J. Frank. Inst. **351**, 3068–3088 (2014)
7. Zhang, B., Lam, J., Xu, S.: Relaxed results on reachable set estimation of time-delay systems with bounded peak inputs. Int. J. Robust. Nonlinear Control **26**, 1994–2007 (2016)
8. Chen, Y., Lam, J.: Estimation and synthesis of reachable set for discrete-time periodic systems. Opt. Control Appl. Meth. **37**, 885–901 (2016)
9. Fei, Z., Guan, C., Shi, P.: Reachable set estimation for discrete-time switched system with application to time-delay system. Int. J. Robust Nonlinear Control **28**, 2468–2483 (2018)
10. Zhang, Z., Wang, Z., Tang, W., Feng, Z., Shen, Y.: Zonotopic reachable set estimation for linear discrete-time systems with time delay. Int. J. Robust Nonlinear Control **30**, 5542–5558 (2020)
11. Trinh, H., Hien, L.V.: On reachable set estimation of two-dimensional systems described by the Roesser model with time-varying delays. Int. J. Robust Nonlinear Control **28**, 227–246 (2018)
12. Hien, L.V., Trinh, H.: Stability analysis of two-dimensional Markovian jump state-delayed systems in the Roesser model with uncertain transition probabilities. Inf. Sci. **368**, 403–417 (2016)
13. Oucheriah, S.: Robust exponential convergence of a class of linear delayed systems with bounded controllers and disturbances. Automatica **42**, 1863–1867 (2006)
14. Nam, P.T., Pathirana, P.N., Trinh, H.: Exponential convergence of time-delay systems in the presence of bounded disturbances. J. Optim. Theory Appl. **157**, 843–852 (2013)
15. Ding, Y., Liu, H., Xu, H., Zhong, S.: On uniform ultimate boundedness of linear systems with time-varying delays and peak-bounded disturbances. Appl. Math. Comput. **349**, 381–392 (2019)
16. Tian, Y., Sun, Y.: Practical stability and stabilisation of switched delay systems with non-vanishing perturbations. IET Control Theory Appl. **13**, 1329–1335 (2019)
17. Hien, L.V., An, N.T., Trinh, H.: New results on state bounding for discrete-time systems with interval time-varying delay and bounded disturbance inputs. IET Control Theory Appl. **8**, 1405–1414 (2014)
18. Hien, L.V., Trinh, H.: A new approach to state bounding for linear time-varying systems with delay and bounded disturbances. Automatica **50**, 1735–1738 (2014)
19. Nam, P.T., Trinh, H., Pathirana, P.N.: Minimization of state bounding for perturbed positive systems with delays. SIAM J. Control Optim. **56**,(2018). https://doi.org/10.1137/16M1104846
20. Shen, T., Petersen, I.R.: An ultimate state bound for a class of linear systems with delay. Automatica **87**, 447–449 (2018)
21. Yeganefar, N., Yeganefar, N., Ghamgui, M., Moulay, E.: Lyapunov theory for 2-D nonlinear Roesser models: Application to asymptotic and exponential stability. IEEE Trans. Autom. Control **58**, 1299–1304 (2013)
22. Hien, L.V., Trinh, T.: Exponential stability of two-dimensional homogeneous monotone systems with bounded directional delays. IEEE Trans. Autom. Control **63**, 2694–2700 (2018)
23. That, N.D., Nam, P.T., Ha, Q.P.: Reachable set bounding for linear discrete-time systems with delays and bounded disturbances. J. Optim. Theory Appl. **157**, 96–107 (2013)

24. Lam, J., Zhang, B., Chen, Y., Xu, S.: Reachable set estimation for discrete-time linear systems with time delays. Int. J. Robust Nonlinear Control **25**, 269–281 (2015)
25. Hien, L.V., Dzung, N.T., Minh, H.B.: A novel approach to state bounding for discrete-time Markovian jump systems with interval time-varying delay. IMA Math. Control Inf. **33**, 293–307 (2016)
26. Chen, S.F.: Delay-dependent stability for 2D systems with time-varying delay subject to state saturation in the Roesser model. Appl. Math. Comput. **216**, 2613–2622 (2010)
27. Dey, A., Kar, H.: An LMI based criterion for the global asymptotic stability of 2-D discrete state-delayed systems with saturation nonlinearities. Digital Sig. Process **22**, 633–639 (2012)
28. Dey, A., Kar, H.: LMI-based criterion for robust stability of 2-D discrete systems with interval time-varying delays employing quantisation/overflow nonlinearities. Multidim. Syst. Sig. Process **25**, 473–492 (2014)
29. Hien, L.V., Trinh, H.: New finite-sum inequalities with applications to stability of discrete time-delay systems. Automatica **71**, 197–201 (2016)
30. Kaczorek, T.: Two-Dimensional Linear Systems. Springer, Berlin (1985)
31. Huang, S., Xiang, Z.: Delay-dependent robust H_∞ control for 2-D discrete nonlinear systems with state delays. Multidim. Syst. Sig. Process **25**, 775–794 (2014)
32. Huang, S., Xiang, Z.: Robust reliable control of uncertain 2D discrete switched systems with state delays. Trans. Inst. Meas. Control **36**, 119–130 (2014)

Chapter 4
Stability Analysis and Observer Design for Two-Dimensional Positive Time-Delay Systems

Abstract In the first part of this chapter, based on the monotone dynamical systems theory, necessary and sufficient delay-independent exponential stability conditions are formulated for 2-D nonlinear systems in the Roesser model with time-varying delays. Applications to the stability analysis of positive 2-D systems with delays are also obtained. In the second part, the design problem of positive observers for 2-D positive linear systems with delays is investigated. Extended Luenberger-type observers that generate nonnegative asymptotic estimations of state vectors are constructed for 2-D positive time-delay systems described by the Roesser model and Fornasini–Marchesini local state-space model. A reduced-order functional state observer that is an asymptotic approximation of a given linear functional of the state vector is also taken into consideration for 2-D positive time-delay systems in the Roesser model. Necessary and sufficient delay-independent design conditions are derived using linear programming approach. The efficacy of the design method is demonstrated by numerical examples.

Keywords Positive systems · Observers · Directional delays · Linear programming

4.1 Introduction

Positive systems [1] are widely used to describe dynamics of practical models whose states and outputs are always nonnegative. Applications of positive systems can be found in various areas in economics, biology, population dynamics or communication [2–4]. In most studies involving the analysis and control of positive systems, positivity characterization is often exploited using a special feature called the monotonicity. Monotone systems [4, 5], which induce the monotonicity in their state trajectories, form an important class of dynamical systems, and are of significant interest not only for their practical applications but also for their interesting theoretical properties. In the past few decades, considerable research attention from the mathematics and control communities has been devoted to the systems and control

L. V. Hien, *Two-Dimensional Time-Delay Systems*, Studies in Systems, Decision and Control 605, https://doi.org/10.1007/978-3-031-95678-2_4

theory of positive 1-D systems and monotone systems with delays, and a rich literature has been established. We refer the reader to [6–16] for a few references. The concept of positive systems and some problems in the control theory have also been developed and studied for some classes of 2-D systems with delays, recently (see, e.g., [17–20] and the references therein).

In the area of state estimation [21], positivity constraints imposed upon state observers are essential in practice as they possess real physical meaning. This makes the positive observation problem become a relevant research topic. So far, a few results in the literature dealing with positive observers design for 1-D positive systems have been reported [22–24]. However, this theory has not been well developed for positive 2-D systems with delays. There are several technical challenges that the designer has to overcome in order to successfully design positive state observers for 2-D positive time-delay systems. On one hand, due to a deep and substantial difference between the structures of 1-D and 2-D systems, attempts to analyze 2-D processes using conventional 1-D systems theory generally fail because such an approach ignores their inherent 2-D systems structure. The available methodologies used to analyze 2-D positive systems with delays are insufficient. On the other hand, the constraint on positivity of observers coupled with the actual states is also troublesome as the information propagation occurs in each of the two independent directions in 2-D processes.

In this chapter, we first investigate the exponential stability of 2-D homogeneous monotone systems with directional time-varying delays of the form

$$\begin{bmatrix} x^h(i+1,j) \\ x^v(i,j+1) \end{bmatrix} = f\left(\begin{bmatrix} x^h(i,j) \\ x^v(i,j) \end{bmatrix}\right) + g\left(\begin{bmatrix} x^h(i-\tau_h(i),j) \\ x^v(i,j-\tau_v(j)) \end{bmatrix}\right).$$

Based on the monotone property and homogeneity of the associated vector fields, necessary and sufficient delay-independent exponential stability conditions are derived. Applications to delay-independent stability of 2-D positive linear systems with delays are also obtained. We then address the design problem of positive observers for 2-D positive time-delay systems both in the Roesser model

$$\begin{aligned}
\begin{bmatrix} x^h(i+1,j) \\ x^v(i,j+1) \end{bmatrix} &= A\begin{bmatrix} x^h(i,j) \\ x^v(i,j) \end{bmatrix} + A_d\begin{bmatrix} x^h(i-\tau_h(i),j) \\ x^v(i,j-\tau_v(j)) \end{bmatrix} + Bu(i,j), \\
y(i,j) &= C\begin{bmatrix} x^h(i,j) \\ x^v(i,j) \end{bmatrix} + C_d\begin{bmatrix} x^h(i-\tau_h(i),j) \\ x^v(i,j-\tau_v(j)) \end{bmatrix},
\end{aligned}$$

and Fornasini-Marchesini model

$$\begin{aligned}
x(i+1,j+1) &= A_{01}x(i+1,j) + A_{02}x(i,j+1) \\
&\quad + A_{d1}x(i-\tau_h(i),j+1) + A_{d2}x(i+1,j-\tau_v(j)) \\
&\quad + B_1u(i+1,j) + B_2u(i,j+1), \\
y(i,j) &= Cx(i,j).
\end{aligned}$$

Extended Luenberger-type observers and a 2-D reduced-order functional observer are constructed to generate nonnegative asymptotic estimations of the state vector and a given linear functional of the state vector of 2-D positive time-delay systems. Necessary and sufficient delay-independent conditions for the existence of positive observers and an effective design procedure are formulated using LP which can be solved by various convex optimization algorithms to obtain the observer parameters.

4.2 Exponential Stability of 2-D Monotone Systems with Delays

4.2.1 Preliminaries

A vector field $f : \mathbb{R}^n \to \mathbb{R}^n$ is said to be monotone on $\mathbb{R}^n$ if $f(x) \preceq f(y)$ for any vectors $x, y \in \mathbb{R}^n$ satisfying $x \preceq y$ [4]; f is nowhere increasing on $\mathbb{R}^n$, written as $f \not\succeq id_n$, if there does not exist an $x \in \mathbb{R}^n \setminus \{0\}$ such that $f(x) \succeq x$, where id_n is the identity vector field of $\mathbb{R}^n$. If there exists a positive scalar α such that $f(\lambda x) = \lambda^\alpha f(x)$ for all $\lambda > 0, x \in \mathbb{R}^n_+$, then f is said to be homogeneous of degree α. Clearly, a linear function $f(x) = Ax$, where $A \in \mathbb{R}^{n\times n}$, is a homogeneous vector field of degree one ($\alpha = 1$). In addition, if $A \succeq 0$, then $f(x) = Ax$ is monotone on $\mathbb{R}^n_+$.

The following version of KKM (Knaster-Kuratowski-Mazurkiewicz) lemma, reformulated in [8], will also be used in this chapter. Let $e_1, e_2, \ldots, e_n$ be basic vectors of $\mathbb{R}^n$ and Δ_n be the simplex

$$\left\{\lambda_1 e_1 + \ldots + \lambda_n e_n : \lambda_i \geq 0, \sum_{i=1}^{n} \lambda_i = 1\right\}$$

with vertices e_i, $i \in [n]$. Given a set of indices $1 \leq i_1 < i_2 < \ldots < i_p \leq n$, the simplex

$$F = \left\{\lambda_1 e_{i_1} + \ldots + \lambda_p e_{i_p} : \lambda_i \geq 0, \sum_{i=1}^{p} \lambda_i = 1\right\}$$

with vertices $e_{i_1}, e_{i_2}, \ldots, e_{i_p}$ is called a face of Δ_n. We denote by $v_F = \{i_1, i_2, \ldots, i_p\}$ the set of vertex indices of F and $\mathscr{F}(\Delta_n)$ the set of all faces of Δ_n. We have the following lemma.

Lemma 4.1 *[8] Let $\Omega_1, \Omega_2, \ldots, \Omega_n$ be relatively open subsets of Δ_n. If each face $F \in \mathscr{F}(\Delta_n)$ is contained in the corresponding union $\bigcup_{i\in v_F} \Omega_i$, then $\bigcap_{i=1}^{n} \Omega_i \neq \emptyset$.*

4.2.2 Two-Dimensional Monotone Systems with Delays

Consider the following 2-D nonlinear system

$$\begin{bmatrix} x^h(i+1,j) \\ x^v(i,j+1) \end{bmatrix} = f\left(\begin{bmatrix} x^h(i,j) \\ x^v(i,j) \end{bmatrix}\right) + g\left(\begin{bmatrix} x^h(i-\tau_h(i),j) \\ x^v(i,j-\tau_v(j)) \end{bmatrix}\right), \tag{4.1}$$

where $x^h(i,j) \in \mathbb{R}^{n_h}$ and $x^v(i,j) \in \mathbb{R}^{n_v}$ are the horizontal and vertical state vectors, respectively, $f, g : \mathbb{R}^n \to \mathbb{R}^n$ $(n = n_h + n_v)$ are some vector fields, $f(0) = 0$, $g(0) = 0$ and $\tau_h(i)$, $\tau_v(j)$ are time-varying directional delays which satisfy

$$0 \le \tau_h(i) \le \bar{\tau}_h, \quad 0 \le \tau_v(j) \le \bar{\tau}_v,$$

where $\bar{\tau}_h$ and $\bar{\tau}_v$ are known positive integers involving the upper bounds of delays.

Initial states of system (4.1) are specified as

$$\begin{aligned} x^h(i,j) &= \phi_h(i,j), \ i \in \mathbb{Z}[-\bar{\tau}_h, 0], \ j \in \mathbb{N}_0, \\ x^v(i,j) &= \phi_v(i,j), \ i \in \mathbb{N}_0, \ j \in \mathbb{Z}[-\bar{\tau}_v, 0]. \end{aligned} \tag{4.2}$$

In the area of analysis and control of 2-D systems, it is widely adopted that initial sequences ϕ_h, ϕ_v in (4.2) have finite support, that is, there exist positive integer T_h, T_v such that

$$\phi_h(i,j) = 0 \text{ for } j \ge T_h \text{ and } \phi_v(i,j) = 0 \text{ for } i \ge T_v. \tag{4.3}$$

For convenience, each initial function satisfying (4.3) will be denoted as $\phi = \begin{bmatrix} \phi_h \\ \phi_v \end{bmatrix}$. We also write $\phi \preceq \psi$ if $\phi_h(i,j) \preceq \psi_h(i,j)$ for $i \in \mathbb{Z}[-\bar{\tau}_h, 0]$, $0 \le j \le T_h$ and $\phi_v(i,j) \preceq \psi_v(i,j)$ for $0 \le i \le T_v$, $j \in \mathbb{Z}[-\bar{\tau}_v, 0]$. Clearly, for a given initial function ϕ, system (4.1)–(4.2) possesses a unique solution, which is denoted by $x_\phi(i,j) = \begin{bmatrix} x^h(i,j) \\ x^v(i,j) \end{bmatrix}$ or simply as $x(i,j)$.

Definition 4.1 System (4.1) is said to be monotone if for any solutions x_ϕ, x_ψ of (4.1) with respect to initial functions $\phi \preceq \psi$, it holds that $x_\phi(i,j) \preceq x_\psi(i,j)$ for all $(i,j) \in \mathbb{N}_0^2$.

For notational simplicity, the above definition can be expressed as

$$\phi \preceq \psi \Longrightarrow x_\phi \preceq x_\psi.$$

It can be verified that, for 2-D linear systems in the form of (4.1) with the vector fields $f(x) = Ax$ and $g(x) = Bx$, where $A, B \in \mathbb{R}^{n \times n}$, the system is monotone if and only if A and B are nonnegative matrices [17]. This property is extended to 2-D nonlinear system (4.1) using the monotone property of vector fields f, g as presented in the following.

Proposition 4.1 *System (4.1) is monotone for arbitrary delays $\tau_h(i)$, $\tau_v(j)$ if and only if the vector fields f and g of (4.1) are monotone on $\mathbb{R}^n$.*

Proof (Necessity) By assigning constant delays $\tau_h(i) = \bar{\tau}_h$, $\tau_v(j) = \bar{\tau}_v$, we can assume without loss of generality that $\tau_h(0) < 0$ and $\tau_v(0) < 0$. By virtue of the contrary arguments, suppose that the vector field f is not monotone. Then, there exist two vectors $\bar{x}_1, \bar{x}_2 \in \mathbb{R}^n$ such that $\bar{x}_1 \preceq \bar{x}_2$, but $f(\bar{x}_1) \npreceq f(\bar{x}_2)$. We now specify initial functions ϕ, ψ as

$$\phi(i, j) = \begin{cases} \bar{x}_1 \text{ if } i = j = 0, \\ 0 \text{ else where,} \end{cases} \qquad \psi(i, j) = \begin{cases} \bar{x}_2 \text{ if } i = j = 0, \\ 0 \text{ else where.} \end{cases}$$

It is clear that $\phi \preceq \psi$, and thus, by the monotonicity, $x_\phi(i, j) \preceq x_\psi(i, j)$ for all i, j. However, with the above particular initial functions, and note also that $f(0) = 0$, $g(0) = 0$, from (4.1), we have that

$$\begin{bmatrix} x_\phi^h(1, 0) \\ x_\phi^v(0, 1) \end{bmatrix} = f(\bar{x}_1) \npreceq f(\bar{x}_2) = \begin{bmatrix} x_\psi^h(1, 0) \\ x_\psi^v(0, 1) \end{bmatrix}.$$

This yields a contradiction. By the same arguments, we also have that the vector field g is monotone.

(Sufficiency) Let $x_\phi(i, j)$ and $x_\psi(i, j)$ be any solutions of (4.1) with respect to initial functions $\phi \preceq \psi$. We will show that $x_\phi(i, j) \preceq x_\psi(i, j)$ for all $(i, j) \in \mathbb{N}_0^2$ by the mathematical induction. For this, we define by

$$\Gamma_q = \{(i, j) \in \mathbb{N}_0^2 : i + j = q\}$$

the diagonal line in $\mathbb{N}_0^2$ for each $q \in \mathbb{N}_0$. It is clear that $x_\phi(0, 0) \preceq x_\psi(0, 0)$.

Assume that the inequality $x_\phi(i, j) \preceq x_\psi(i, j)$ holds for $(i, j) \in \Gamma_k$ with $k \le q$. It follows from (4.1) and the monotonicity of vector fields f and g that

$$f\left(x_\phi(i, j)\right) \preceq f\left(x_\psi(i, j)\right), \quad g\left(x_{d\phi}(i, j)\right) \preceq g\left(x_{d\psi}(i, j)\right), \quad \forall (i, j) \in \Gamma_q,$$

where, for simplicity, we denote the delayed states as

$$x_d(i, j) = \begin{bmatrix} x^h(i - \tau_h(i), j) \\ x^v(i, j - \tau_v(j)) \end{bmatrix}.$$

This leads to

$$x_\phi^h(i + 1, j) \preceq x_\psi^h(i + 1, j) \text{ and } x_\phi^v(i, j + 1) \preceq x_\psi^v(i, j + 1).$$

On the other hand, since $(i+1, j-1) \in \Gamma_q$ for any $(i, j) \in \Gamma_q$, from (4.1), we have

$$\begin{aligned}\begin{bmatrix} x_\phi^h(i+2, j-1) \\ x_\phi^v(i+1, j) \end{bmatrix} &= f\left(x_\phi(i+1, j-1)\right) + g\left(x_{d\phi}(i+1, j-1)\right) \\ &\preceq f\left(x_\psi(i+1, j-1)\right) + g\left(x_{d\psi}(i+1, j-1)\right) \\ &= \begin{bmatrix} x_\psi^h(i+2, j-1) \\ x_\psi^v(i+1, j) \end{bmatrix}.\end{aligned}$$

Therefore, $x_\phi^v(i+1, j) \preceq x_\psi^v(i+1, j)$ and, as a consequence, we obtain that

$$x_\phi(i+1, j) \preceq x_\psi(i+1, j).$$

The comparison $x_\phi(i, j+1) \preceq x_\psi(i, j+1)$ can be similarly established. This shows that the inequality $x_\phi(i, j) \preceq x_\psi(i, j)$ holds for all $(i, j) \in \Gamma_{q+1}$. By the induction method, we can conclude that $x_\phi(i, j) \preceq x_\psi(i, j)$ holds for all $q \in \mathbb{N}_0$ and $(i, j) \in \Gamma_q$. This, together with the fact $\mathbb{N}_0^2 = \cup_{q \geq 0} \Gamma_q$, completes the proof. □

System (4.1) is said to be positive for any initial function $\phi \succeq 0$, the corresponding state trajectory $x_\phi(i, j)$ is always nonnegative (i.e. $x_\phi(i, j) \succeq 0$ for all $(i, j) \in \mathbb{N}_0^2$). Clearly, if system (4.1) is monotone then it is also positive. However, unlike 2-D linear systems, these two concepts are not identical for 2-D nonlinear systems. For an example, let us consider system (4.1) with the vector fields

$$f(x) = \begin{bmatrix} x_1|\sin(x_2)| \\ x_2|\cos(x_1)| \end{bmatrix}, \quad g(x) = \begin{bmatrix} \sin^2(x_1) \\ \cos^2(x_2) \end{bmatrix}, \quad x = \begin{bmatrix} x_1 \\ x_2 \end{bmatrix} \in \mathbb{R}_+^2. \tag{4.4}$$

System (4.1) with the vector fields given in (4.4) is positive. However, it is easy to verify that f and g are not monotone on $\mathbb{R}_+^2$ and, by Proposition 4.1, system (4.1) is not monotone.

Definition 4.2 System (4.1) is said to be globally exponentially stable (GES) if there exist scalars $\rho \in (0, 1)$ and $\beta > 0$ such that any solution $x(i, j)$ of (4.1)–(4.2) satisfies the following inequality

$$\|x(i, j)\|_\infty \leq \beta \left(\sum_{r=0}^{i} \frac{\hat{\phi}_v(r)}{\rho^{r+1}} + \sum_{s=0}^{j} \frac{\hat{\phi}_h(s)}{\rho^{s+1}} \right) \rho^{i+j}, \quad i, j \in \mathbb{N}_0, \tag{4.5}$$

where $\hat{\phi}_v(r) = \sup_{-\bar{\tau}_v \leq j \leq 0} \|\phi_v(r, j)\|_\infty$ and $\hat{\phi}_h(s) = \sup_{-\bar{\tau}_h \leq i \leq 0} \|\phi_h(i, s)\|_\infty$.

Definition 4.2 is an extension of the conventional exponential stability concept developed for 1-D systems with delays. Specifically, by similar lines used in the proof of Proposition 2.4 in [25], it can be verified from (4.5) that for any $\epsilon > 0$,

$$\begin{cases} \hat{\phi}_h(j) < \epsilon(1-\rho)/2\beta \\ \hat{\phi}_v(i) < \epsilon(1-\rho)/2\beta \end{cases} \Longrightarrow \|x(i,j)\|_\infty < \epsilon.$$

This property is referred to (ϵ-δ) stability of 2-D system (4.1). Besides that (4.5) ensures $x(i, j) \to 0$ as $i + j \to \infty$. Thus, exponential stability of system (4.1) implies asymptotic stability.

4.2.3 Exponential Stability Conditions

For 1-D linear systems, the positivity can dominate the impact of certain classes of time delays [7]. A key factor is that, for linear systems, the positivity is equivalent to the monotone property of the associated vector fields. In addition, with linear functions, which are homogeneous of degree one, exponential comparisons can be preserved. In this section, we establish delay-independent necessary and sufficient conditions for the exponential stability of system (4.1) under the assumption that the vector fields f and g are monotone and homogeneous of degree one on $\mathbb{R}^n_+$.

Theorem 4.1 *Assume the vector fields f, g associated with system (4.1) are continuous, homogeneous of degree one and monotone on $\mathbb{R}^n_+$. Then, the 2-D monotone system (4.1) is GES for all bounded directional delays $\tau_h(.)$ and $\tau_v(.)$ if and only if there exists a positive vector $0 \prec \eta = \begin{bmatrix} \eta_h \\ \eta_v \end{bmatrix} \in \mathbb{R}^n$ such that*

$$f(\eta) + g(\eta) - \eta \prec 0. \tag{4.6}$$

Proof (Necessity) Let $\tau_h(.) = 0$ and $\tau_v(.) = 0$, the delay-free system

$$\begin{bmatrix} x^h(i+1, j) \\ x^v(i, j+1) \end{bmatrix} = f\left(\begin{bmatrix} x^h(i, j) \\ x^v(i, j) \end{bmatrix}\right) + g\left(\begin{bmatrix} x^h(i, j) \\ x^v(i, j) \end{bmatrix}\right) \tag{4.7}$$

is GES. Thus, there exist scalars $\rho_0 \in (0, 1)$ and $\beta_0 > 0$ satisfying (4.5). We first show that the vector field $\Gamma \triangleq f + g$ is nowhere increasing on $\mathbb{R}^n_+$.

Suppose in contrary that there exists a vector $x_* = \begin{bmatrix} x^h_* \\ x^v_* \end{bmatrix} \in \mathbb{R}^n_+$, $x_* \neq 0$, such that $\Gamma(x_*) \succeq x_*$. Let $x(i, j)$ be the solution of (4.1) corresponding to the initial condition

$$x^h(0, j) = \delta^j x^h_*, \; j \le T_h, \quad x^v(i, 0) = \delta^i x^v_*, \; i \le T_v,$$

where $\delta = \sqrt{\rho_0}$. By induction, it can be deduced from the monotone property of the vector field Γ that $x(i, j) \succeq \delta^{i+j} x_*$. Thus,

$$\begin{aligned}\delta^{i+j}\|x_*\|_\infty &\leq \|x(i,j)\|_\infty \\ &\leq \beta_0\left(\sum_{r=0}^{i}\frac{\delta^r}{\rho_0^{r+1}}\|x_*^v\|_\infty + \sum_{s=0}^{j}\frac{\delta^s}{\rho_0^{s+1}}\|x_*^h\|_\infty\right)\rho_0^{i+j}\end{aligned}$$

which yields

$$0 < \|x_*\|_\infty \leq \frac{\beta_0}{\rho_0(1-\delta)}\left(\delta^i\|x_*^h\|_\infty + \delta^j\|x_*^v\|_\infty\right) \to 0 \tag{4.8}$$

as $i, j \to \infty$. By choosing sufficiently large T_h, T_v, (4.8) raises a contradiction.

Therefore, there does not exist any $x \in \mathbb{R}_+^n \setminus \{0\}$ such that $\Gamma(x) \succeq x$. This shows that, for any $x \in \mathbb{R}_+^n \setminus \{0\}$, there exists an index $i \in [n]$ such that $\Gamma_i(x) - x_i < 0$. We now apply the KKM lemma to show the existence of a vector $\eta \succ 0$ satisfying (4.6). For this, we define

$$\Omega_i = \{x \in \Delta_n : \Gamma_i(x) - x_i < 0\}, \ i \in [n].$$

Then $\bigcup_{i=1}^n \Omega_i = \Delta_n$. By the continuity of Γ, $\Omega_1, \ldots, \Omega_n$ are open subsets of Δ_n.

Let $F \in \mathscr{F}(\Delta_n)$ be a face of Δ_n and $x \in F$. Then, $x_j = 0$ for all $j \notin v_F$. On the other hand, $\Gamma_j(x) \geq 0$. Thus, $x \notin \Omega_j$ for all $j \notin v_F$. Consequently, $x \in \bigcup_{i \in v_F} \Omega_i$. This shows that $F \subset \bigcup_{i \in v_F} \Omega_i$. By KKM lemma, $\bigcap_{i=1}^n \Omega_i \neq \emptyset$. Let $\eta \in \bigcap_{i=1}^n \Omega_i$, then $\eta \succ 0$ and $\Gamma_i(\eta) - \eta_i < 0$ for all $i \in [n]$. Thus, $\Gamma(\eta) - \eta \prec 0$.

(Sufficiency) Let $\tau_m = \max(\bar{\tau}_h, \bar{\tau}_v)$ and $\eta \succ 0$ satisfying condition (4.6). For each $i \in [n]$, there exists a unique solution $\rho_i \in (0, 1)$ of the scalar equation

$$f_i(\eta) + g_i(\eta)\rho_i^{-\tau_m} - \rho_i\eta_i = 0. \tag{4.9}$$

Denote $\rho = \max_{1 \leq i \leq n} \rho_i \in (0, 1)$. It follows from (4.9) that

$$f_i(\eta) + g_i(\eta)\rho^{-\tau_m} \leq \rho\eta_i, \ i \in [n].$$

Let $x(i, j)$ be a solution (4.1)–(4.2). We will prove by the induction method employing the monotone property of the vector fields f, g that

$$\|x(i,j)\|_\infty \leq C_\eta\left(\sum_{r=0}^{i}\frac{\hat{\phi}_v(r)}{\rho^{r+1}} + \sum_{s=0}^{j}\frac{\hat{\phi}_h(s)}{\rho^{s+1}}\right)\rho^{i+j}, \ i, j \in \mathbb{N}_0, \tag{4.10}$$

where $C_\eta = \eta^+/\eta_-$, $\eta^+ = \max_{1 \leq i \leq n} \eta_i$ and $\eta_- = \min_{1 \leq i \leq n} \eta_i$.
Let Γ_q be the diagonal line defined as

$$\Gamma_q = \{(i, j) \in \mathbb{N}_0^2 : i + j = q\}, \ q \in \mathbb{N}_0.$$

We have

$$\|x(0,0)\|_\infty \le \hat{\phi}_h(0) + \hat{\phi}_v(0) \le \frac{1}{\eta_-}\Big(\frac{\hat{\phi}_h(0)}{\rho} + \frac{\hat{\phi}_v(0)}{\rho}\Big)\eta_i.$$

Thus,

$$x(i,j) \preceq \underbrace{\frac{1}{\eta_-}\Big(\sum_{r=0}^{i}\frac{\hat{\phi}_v(r)}{\rho^{r+1}} + \sum_{s=0}^{j}\frac{\hat{\phi}_h(s)}{\rho^{s+1}}\Big)}_{\kappa_\phi}\rho^{i+j}\eta \tag{4.11}$$

for $(i,j) \in \Gamma_0$.

Suppose that (4.11) holds for $(i,j) \in \Gamma_k$, $k \le q$. Then, using the properties that f and g are monotone and homogeneous of degree one, we have

$$f(x(i,j)) \preceq \kappa_\phi \rho^{i+j} f(\eta) \tag{4.12}$$

and

$$\begin{aligned} x^h(i-\tau_h(i), j) &\preceq \kappa_\phi \rho^{i+j-\tau_h(i)}\eta_h \preceq \kappa_\phi \rho^{i+j-\tau_m}\eta_h, \\ x^v(i, j-\tau_v(j)) &\preceq \kappa_\phi \rho^{i+j-\tau_v(j)}\eta_v \preceq \kappa_\phi \rho^{i+j-\tau_m}\eta_v. \end{aligned}$$

Therefore,

$$g\left(\begin{bmatrix} x^h(i-\tau_h(i), j) \\ x^v(i, j-\tau_v(j)) \end{bmatrix}\right) \preceq \kappa_\phi \rho^{i+j} g(\eta)\rho^{-\tau_m}. \tag{4.13}$$

A combination of (4.12) and (4.13) gives

$$\begin{aligned} \begin{bmatrix} x^h(i+1, j) \\ x^v(i, j+1) \end{bmatrix} &\preceq \kappa_\phi \rho^{i+j}\left(f(\eta) + g(\eta)\rho^{-\tau_m}\right) \\ &\preceq \kappa_\phi \rho^{i+j+1}\eta. \end{aligned} \tag{4.14}$$

On the other hand, let

$$S_x(i,j) = f\left(\begin{bmatrix} x^h(i,j) \\ x^v(i,j) \end{bmatrix}\right) + g\left(\begin{bmatrix} x^h(i-\tau_h(i), j) \\ x^v(i, j-\tau_v(j)) \end{bmatrix}\right)$$

denote the solution map of (4.1). Then,

$$x^v(i+1, j) = J_v S_x(i+1, j-1),$$

where $J_v = [0_{n_v \times n_h} \quad I_{n_v}]$. Similar to (4.14), we have

$$x^v(i+1, j) \preceq \kappa_\phi \rho^{i+j+1}(J_v \eta). \tag{4.15}$$

Thus, according to (4.14) and (4.15),

$$x(i+1, j) \preceq \kappa_\phi \rho^{i+j+1}\eta \text{ for all } (i, j) \in \Gamma_q.$$

The fact $x(i, j+1) \preceq \kappa_\phi \rho^{i+j+1}\eta$ can be proved similarly. Note also that κ_ϕ is nondecreasing in i, j. Consequently, the inequality (4.11) holds for all $(i, j) \in \Gamma_{q+1}$. By induction, and the fact that $\mathbb{N}_0^2 = \cup_{q=0}^{\infty}\Gamma_q$, estimation (4.11) holds for all $i, j \in \mathbb{N}_0$. From (4.11) we readily get the exponential estimate (4.10). The proof is completed. □

Remark 4.1 Although the derived stability condition (4.6) is delay-independent, the magnitudes of delays $\bar{\tau}_h$, $\bar{\tau}_v$ have an impact on the exponential convergence rate. Specifically, let $\rho_i \in (0, 1)$ is the unique solution of the scalar equation

$$\frac{f_i(\eta)}{\eta_i} + \frac{g_i(\eta)}{\eta_i}\rho_i^{-\max(\bar{\tau}_h, \bar{\tau}_v)} - \rho_i = 0, \ i \in [n].$$

Then, the exponential decay rate of system (4.1) is determined by $\rho = \min_{1 \le i \le n} \rho_i$.

Remark 4.2 If the vector fields f and g are monotone and homogeneous of degree $\alpha > 1$, inequality (4.6) is only necessary and sufficient condition for the local exponential stability of system (4.1).

4.2.4 Applications to 2-D Systems with Multi-directional Delays

The result of Theorem 4.1 can be extended to 2-D homogeneous monotone systems with multiple time-varying delays of the form

$$\begin{bmatrix} x^h(i+1, j) \\ x^v(i, j+1) \end{bmatrix} = f\left(\begin{bmatrix} x^h(i, j) \\ x^v(i, j) \end{bmatrix}\right) + \sum_{k=1}^{N} g_k\left(\begin{bmatrix} x^h(i-\tau_{hk}(i), j) \\ x^v(i, j-\tau_{vk}(j)) \end{bmatrix}\right), \tag{4.16}$$

where the directional delays $\tau_{hk}(i)$ and $\tau_{vk}(j)$, $k \in [N]$, satisfy $0 \le \tau_{hk}(i) \le \bar{\tau}_{hk}$ and $0 \le \tau_{vk}(j) \le \bar{\tau}_{vk}$. By similar arguments used in the proof of Theorem 4.1, we have the following result.

Theorem 4.2 *Assume that the system vector fields f and g_s, $s \in [N]$, are continuous, monotone and homogeneous of degree one on $\mathbb{R}_+^n$. Then, system (4.16) is GES for all bounded directional delays $\tau_{hs}(i)$ and $\tau_{vs}(j)$, $s \in [N]$, if and only if there exists a positive vector $0 \prec \eta \in \mathbb{R}^n$ which satisfies the following condition*

$$f(\eta) + \sum_{k=1}^{N} g_k(\eta) - \eta \prec 0. \tag{4.17}$$

As a special case of (4.16), consider the following 2-D linear system with delays

$$\begin{bmatrix} x^h(i+1,j) \\ x^v(i,j+1) \end{bmatrix} = A_0 \begin{bmatrix} x^h(i,j) \\ x^v(i,j) \end{bmatrix} + \sum_{k=1}^{N} A_k \begin{bmatrix} x^h(i-\tau_{hk}(i),j) \\ x^v(i,j-\tau_{vk}(j)) \end{bmatrix}, \tag{4.18}$$

where $A_0 = \left(a_{ij}^{(0)}\right)$ and $A_k = \left(a_{ij}^{(k)}\right)$, $k \in [N]$, are given matrices in $\in \mathbb{R}^{n\times n}$.

Proposition 4.2 *System (4.18) is positive for any delays $\tau_{hk}(i)$, $\tau_{vk}(j)$, $k \in [N]$, if and only if the matrices A_0 and A_k, $k \in [N]$, are nonnegative.*

A delay-independent exponential stability criterion for 2-D linear positive systems in the form of (4.18) is presented in the following theorem.

Theorem 4.3 *The positive system (4.18) is GES for all bounded time-varying delays $\tau_{hk}(i)$, $\tau_{vk}(j)$, $k \in [N]$, if and only if there exists a vector $0 \prec \eta \in \mathbb{R}^n$ such that*

$$\left(A_0 + \sum_{k=1}^{N} A_k\right)\eta - \eta \prec 0. \tag{4.19}$$

Moreover, any solution $x(i,j)$ of system (4.18) satisfies the exponential estimate

$$\|x(i,j)\|_\infty \le C_\eta\Big(\sum_{r=0}^{i} \frac{\hat{\phi}_v(r)}{\rho^{r+1}} + \sum_{s=0}^{j} \frac{\hat{\phi}_h(s)}{\rho^{s+1}}\Big)\rho^{i+j},\ i,j \in \mathbb{N}_0,$$

where $\rho = \max_{1\le i\le n} \rho_i$ and ρ_i is the unique positive solution of the equation

$$\frac{1}{\eta_i}\sum_{j=1}^{n} a_{ij}^{(0)}\eta_j + \left(\frac{1}{\eta_i}\sum_{k=1}^{N}\sum_{j=1}^{n} a_{ij}^{(k)}\eta_j\right)\rho_i^{-\tau_m} - \rho_i = 0,\ i \in [n], \tag{4.20}$$

where $\tau_m = \max_{1\le k\le N}(\bar{\tau}_{hk}, \bar{\tau}_{vk})$.

Remark 4.3 Condition (4.19) is formulated as a linear programming (LP) problem in η, which can be solved numerically in polynomial time by various computational tools [7]. In addition, to find a smallest possible decay rate ρ, a similar generic convex optimization procedure as developed in [7] can be utilized.

Remark 4.4 By the spectral properties of nonnegative matrices [27], condition (4.19) is feasible for a vector $\eta \succ 0$ if and only if there exists a vector $\zeta \in \mathbb{R}^n$, $\zeta \succ 0$, such that

$$\zeta^\top\left(A_0 + \sum_{k=1}^{N} A_k\right) - \zeta^\top \prec 0. \tag{4.21}$$

Remark 4.5 For general 2-D linear systems in the form of (4.18), we have

$$\begin{bmatrix} |x^h(i+1,j)| \\ |x^v(i,j+1)| \end{bmatrix} \preceq |A_0| \begin{bmatrix} |x^h(i,j)| \\ |x^v(i,j)| \end{bmatrix} + \sum_{k=1}^{N} |A_k| \begin{bmatrix} |x^h(i-\tau_{hk}(i),j)| \\ |x^v(i,j-\tau_{vk}(j))| \end{bmatrix}.$$

Thus, for any solution $x(i,j)$ of (4.18) with initial condition (4.2), it holds that $|x(i,j)| \preceq \hat{x}(i,j)$ for all $i, j \in \mathbb{N}_0$, where $\hat{x}(i,j)$ is the corresponding solution of the following positive system

$$\begin{bmatrix} \hat{x}^h(i+1,j) \\ \hat{x}^v(i,j+1) \end{bmatrix} = |A_0| \begin{bmatrix} \hat{x}^h(i,j) \\ \hat{x}^v(i,j) \end{bmatrix} + \sum_{k=1}^{N} |A_k| \begin{bmatrix} \hat{x}^h(i-\tau_{hk}(i),j) \\ \hat{x}^v(i,j-\tau_{vk}(j)) \end{bmatrix} \tag{4.22}$$

with initial condition $\hat{x}^h(i,j) = |\phi_h(i,j)|$ and $\hat{x}^v(i,j) = |\phi_v(i,j)|$. Similar to (4.19), we have the following corollary.

Corollary 4.1 *The 2-D linear system (4.18) is GES for all bounded time-varying delays $\tau_{hk}(i)$ and $\tau_{vk}(j)$, $k \in [N]$, if there exists a vector $\eta \in \mathbb{R}^n$, $\eta > 0$, such that*

$$\left(|A_0| + \sum_{k=1}^{N} |A_k| \right) \eta - \eta \prec 0. \tag{4.23}$$

In the remaining of this section, let us consider the following 2-D system described by the Fornasini–Marchesini local state-space (FMLSS) model [26]

$$\begin{aligned} x(i+1,j+1) &= A_{01}x(i+1,j) + A_{02}x(i,j+1) \\ &\quad + \sum_{k=1}^{N} \left[A_{d1k}x\left(i-\tau_{hk}(i), j+1\right) + A_{d2k}x\left(i+1, j-\tau_{vk}(j)\right) \right], \end{aligned} \tag{4.24}$$

where A_{01}, A_{02} and A_{d1k}, A_{d2k}, $k \in [N]$, are given matrices in $\in \mathbb{R}^{n\times n}$. The initial condition of (4.24) is similarly specified as (4.2)–(4.3), that is,

$$x(i,j) = \begin{cases} \phi_h(i,j), & i \in \mathbb{Z}[-\bar{\tau}_h, 0],\ j \in \mathbb{Z}[0, T_h], \\ \phi_v(i,j), & i \in \mathbb{Z}[0, T_v],\ j \in \mathbb{Z}[-\bar{\tau}_v, 0], \\ 0, & i > T_h \text{ and } j > T_v, \end{cases} \tag{4.25}$$

where $0 \le \tau_{hk}(i) \le \bar{\tau}_h$ and $0 \le \tau_{vk}(j) \le \bar{\tau}_v$.

The vector field $f : \mathbb{R}^n \times \mathbb{R}^n \to \mathbb{R}^n$, $f(\mathscr{X}) = A_{01}x_1 + A_{02}x_2$, is monotone if and only if the matrices A_{01}, A_{02} are nonnegative, where $\mathscr{X} = (x_1, x_2)$. Similarly, the vector field $g : (\mathbb{R}^n)^N \times (\mathbb{R}^n)^N \to \mathbb{R}^n$, $g(\bar{\mathscr{X}}) = \sum_{k=1}^{N}[A_{d1k}x_{1k} + A_{d2k}x_{2k}]$, is monotone if and only if A_{d1k} and A_{d2k} are all nonnegative matrices, where $\bar{\mathscr{X}} = (x_{11}, \ldots, x_{1N}, x_{21}, \ldots, x_{2N})$. This shows that the 2-D system (4.24) is positive if and only if all the system matrices A_{01}, A_{02} and A_{d1k}, A_{d2k} are nonnegative. In addition, by similar comparison method proposed in the proof of Theorem 4.1, we obtain the following result.

Proposition 4.3 *Assume that the 2-D system in FMLSS model (4.24) is positive. Then, the following statements are equivalent.*

(i) *System (4.24) is exponentially stable for any bounded delays $\tau_{hk}(i)$, $\tau_{vk}(j)$.*
(ii) *There exists a vector $\eta \in \mathbb{R}^n$, $\eta \succ 0$, such that*

$$(A_{01} + A_{02} + A_{ds} - I_n)\,\eta \prec 0, \tag{4.26}$$

where $A_{ds} = \sum_{k=1}^{N} (A_{d1k} + A_{d2k})$.
(iii) *There exists a vector $\zeta \in \mathbb{R}^n$, $\zeta \succ 0$, such that*

$$(A_{01} + A_{02} + A_{ds} - I_n)^{\top}\,\zeta \prec 0. \tag{4.27}$$

(iv) *The matrix $A_{01} + A_{02} + A_{ds} - I_n$ is invertible and $(A_{01} + A_{02} + A_{ds} - I_n)^{-1} \preceq 0$.*

4.2.5 An Illustrative Example

Example 4.1 Consider system (4.1) with the vector fields

$$f(x) = \begin{bmatrix} 0.11 & 0.15 \\ 0.16 & 0.09 \end{bmatrix} \begin{bmatrix} x^h \\ x^v \end{bmatrix} + \sqrt{(x^h)^2 + (x^v)^2} \begin{bmatrix} 0.12 \\ 0.11 \end{bmatrix},$$
$$g(x) = \begin{bmatrix} \dfrac{0.8x^h x^v}{\sqrt{3(x^h)^2 + (x^v)^2}} \\ \dfrac{0.8x^h x^v}{\sqrt{(x^h)^2 + 3(x^v)^2}} \end{bmatrix} \quad \text{for } x = \begin{bmatrix} x^h \\ x^v \end{bmatrix} \in \mathbb{R}^2_+ \backslash \{0\}, \;\; g(0) = 0. \tag{4.28}$$

Clearly, the vector fields f and g given in (4.28) are monotone and homogeneous of degree one on $\mathbb{R}^2_+$. Moreover, let $\eta = [1 \quad 1]^{\top}$, we have

$$f(\eta) + g(\eta) = \begin{bmatrix} 0.8297 \\ 0.8056 \end{bmatrix} \prec \eta.$$

Thus, by Theorem 4.1, system (4.1) with the vector fields f, g given in (4.28) is GES for all bounded time-varying delays $\tau_h(i)$, $\tau_v(j)$.

Let $\tau_h(i) = 1 + 4|\sin(i\pi/2)|$ and $\tau_v(j) = 1 + 4|\cos(j\pi/2)|$. Then, $\bar{\tau}_h = \bar{\tau}_v = 5$. By solving equation (4.9) with $\tau_m = 5$ and $\eta = [1 \quad 1]^{\top}$ we obtain $\rho_1 \simeq 0.9491$ and $\rho_2 \simeq 0.9427$ and $\rho = \max(\rho_1, \rho_2) = 0.9491$. By the estimation (4.10), any solution $x(i, j)$ of (4.1)–(4.2) satisfies

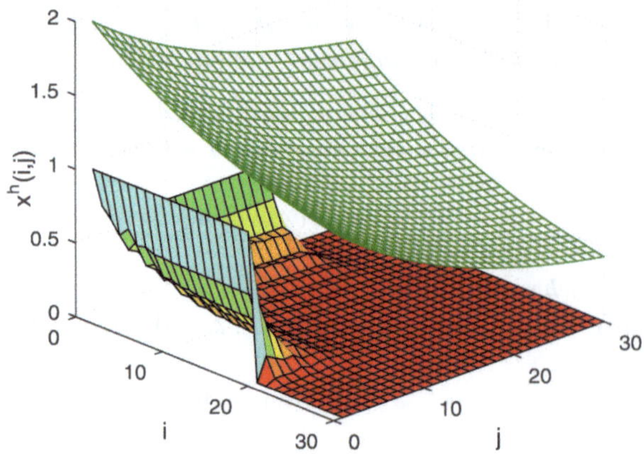

Fig. 4.1 State trajectory $x^h(i, j)$

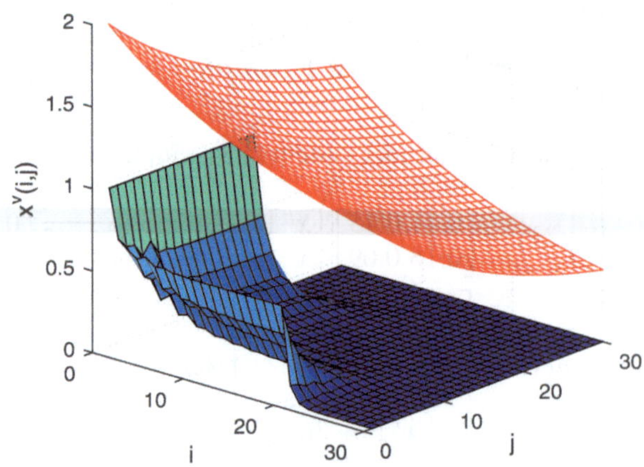

Fig. 4.2 State trajectory $x^v(i, j)$

$$\|x(i, j)\|_\infty \leq \left(\sum_{r=0}^{i} \frac{\hat{\phi}_v(r)}{0.9491^{r+1}} + \sum_{s=0}^{j} \frac{\hat{\phi}_h(s)}{0.9491^{s+1}} \right) 0.9491^{i+j}, \ i, j \in \mathbb{N}_0.$$

A simulation result with initial condition $\phi = [1 \quad 1]^\top$ and $\tau_h(i) = 1 + 4|\sin(i\pi/2)|$, $\tau_v(j) = 1 + 4|\cos(j\pi/2)|$ is presented in Figs. 4.1 and 4.2. It can be seen that the corresponding state trajectories $x^h(i, j)$ and $x^v(i, j)$ converge exponentially as revealed by the theoretical results.

4.3 Observers Design

4.3.1 Problem Statement

Consider a class of 2-D linear Roesser systems with delays

$$\begin{bmatrix} x^h(i+1, j) \\ x^v(i, j+1) \end{bmatrix} = A \begin{bmatrix} x^h(i, j) \\ x^v(i, j) \end{bmatrix} + A_d \begin{bmatrix} x^h(i-\tau_h(i), j) \\ x^v(i, j-\tau_v(j)) \end{bmatrix} + Bu(i, j) \tag{4.29}$$

$$y(i, j) = C \begin{bmatrix} x^h(i, j) \\ x^v(i, j) \end{bmatrix} + C_d \begin{bmatrix} x^h(i-\tau_h(i), j) \\ x^v(i, j-\tau_v(j)) \end{bmatrix} \tag{4.30}$$

where $x^h \in \mathbb{R}^{n_h}$, $x^v \in \mathbb{R}^{n_v}$ are the horizontal, vertical state vectors; $u \in \mathbb{R}^{n_u}$ and $y \in \mathbb{R}^{n_y}$ are the input and measurement output vectors; $A, A_d \in \mathbb{R}^{n\times n}$ $(n = n_h + n_v)$, $B \in \mathbb{R}^{n\times n_u}$ and $C, C_d \in \mathbb{R}^{n_y\times n}$ are given matrices, $\tau_h(i)$ and $\tau_v(j)$ are time-varying delays along the horizontal and vertical directions with upper bounds $0 \leq \tau_h(i) \leq \bar{\tau}_h$ and $0 \leq \tau_v(j) \leq \bar{\tau}_v$. The initial condition of (4.29) is specified similarly to (4.2) with finite support sequences ϕ_h, ϕ_v.

The positivity of 2-D systems with inputs and outputs in the form of (4.29)–(4.30) is presented in following definition.

Definition 4.3 System (4.29)–(4.30) is said to be positive if for any nonnegative initial condition $\phi_h(i, j) \succeq 0$, $\phi_v(i, j)$ defined by (4.2)–(4.3) and nonnegative input $u(i, j) \succeq 0$, the state vector $x(i, j) = \begin{bmatrix} x^h(i, j) \\ x^v(i, j) \end{bmatrix}$ and the output vector $y(i, j)$ are always nonnegative, that is, $x(i, j) \succeq 0$ and $y(i, j) \succeq 0$ for all $i, j \in \mathbb{N}_0$.

Similar to Proposition 4.1, the following result can be obtained directly using the properties of monotone vector fields.

Proposition 4.4 *System (4.29)–(4.30) is positive for any bounded delays $\tau_h(i)$, $\tau_v(j)$ if and only if the matrices A, A_d, B and C, C_d are nonnegative.*

The main objective of this section is to derive necessary and sufficient conditions to design nonnegative full-order state observers and reduced-order/functional state observers for positive 2-D systems in the form of (4.29). Analogous design issues for 2-D systems in FMLSS model are also discussed.

4.3.2 Full-Order State Observers

In this section, we derive necessary and sufficient conditions to determine a nonnegative state observer $\hat{x}(i, j)$ that converges to $x(i, j)$. In the theoretical derivation, the time delays $\tau_h(i)$, $\tau_v(j)$ are time-varying and possibly unknown but bounded. However, from practical meaning, $\tau_h(i)$ and $\tau_v(j)$ are assumed to be known.

Consider system (4.29). To determine an approximation $\hat{x}(i, j)$ of the state vector $x(i, j)$, an extended Luenberger-type observer will be constructed in the form

$$\begin{aligned}\begin{bmatrix}\hat{x}^h(i+1, j)\\ \hat{x}^v(i, j+1)\end{bmatrix} &= A\begin{bmatrix}\hat{x}^h(i, j)\\ \hat{x}^v(i, j)\end{bmatrix} + A_d\begin{bmatrix}\hat{x}^h(i-\tau_h(i), j)\\ \hat{x}^v(i, j-\tau_v(j))\end{bmatrix}\\ &\quad - L\big(\hat{y}(i, j) - y(i, j)\big) + Bu(i, j), \qquad (4.31)\\ \hat{y}(i, j) &= C\begin{bmatrix}\hat{x}^h(i, j)\\ \hat{x}^v(i, j)\end{bmatrix} + C_d\begin{bmatrix}\hat{x}^h(i-\tau_h(i), j)\\ \hat{x}^v(i, j-\tau_v(j))\end{bmatrix}, \qquad (4.32)\end{aligned}$$

where $L \in \mathbb{R}^{n\times n_y}$ is the observer gain which will be determined. From equations (4.29) and (4.31), the associated error vector $e(i, j) = \hat{x}(i, j) - x(i, j)$ is governed by the following 2-D system

$$\begin{bmatrix}e^h(i+1, j)\\ e^v(i, j+1)\end{bmatrix} = A_c\begin{bmatrix}e^h(i, j)\\ e^v(i, j)\end{bmatrix} + A_{dc}\begin{bmatrix}e^h(i-\tau_h(i), j)\\ e^v(i, j-\tau_v(j))\end{bmatrix}, \qquad (4.33)$$

where $A_c = A - LC$ and $A_{dc} = A_d - LC_d$.

Theorem 4.4 *Assume that the system (4.29)–(4.30) is positive. Then, there exists a positive observer in the form of (4.31)–(4.32) if and only if the following conditions hold*

$$\begin{bmatrix}LC\\ LC_d\end{bmatrix} \succeq 0, \quad \begin{bmatrix}A - LC\\ A_d - LC_d\end{bmatrix} \succeq 0, \qquad (4.34)$$

$$\exists \chi \in \mathbb{R}^n, \chi \succ 0 : (A_c + A_{dc} - I_n)^\top \chi \prec 0, \qquad (4.35)$$

where the matrices A_c and A_{dc} are defined in (4.33).

Proof We define the following augmented state vectors

$$\bar{x}^h(i, j) = \begin{bmatrix}x^h(i, j)\\ \hat{x}^h(i, j)\end{bmatrix}, \quad \bar{x}^v(i, j) = \begin{bmatrix}x^v(i, j)\\ \hat{x}^v(i, j)\end{bmatrix}.$$

From equation (4.29) and (4.31), we have

$$\begin{bmatrix}\bar{x}^h(i+1, j)\\ \bar{x}^v(i, j+1)\end{bmatrix} = \mathscr{A}_\Pi\begin{bmatrix}\bar{x}^h(i, j)\\ \bar{x}^v(i, j)\end{bmatrix} + \mathscr{A}_{d\Pi}\begin{bmatrix}\bar{x}^h(i-\tau_h(i), j)\\ \bar{x}^v(i, j-\tau_v(j))\end{bmatrix} + \mathscr{B}_\Pi u(i, j), \qquad (4.36)$$

where $\mathscr{A}_\Pi = \Pi^{-1}\mathscr{A}\Pi$, $\mathscr{A}_{d\Pi} = \Pi^{-1}\mathscr{A}_d\Pi$, $\mathscr{B}_\Pi = \Pi^{-1}\begin{bmatrix}B\\ B\end{bmatrix}$ and

$$\mathscr{A} = \begin{bmatrix}A & 0\\ LC & A - LC\end{bmatrix}, \quad \mathscr{A}_d = \begin{bmatrix}A_d & 0\\ LC_d & A_d - LC_d\end{bmatrix}, \quad \Pi = \begin{bmatrix}I_{n_h} & 0 & 0 & 0\\ 0 & 0 & I_{n_v} & 0\\ 0 & I_{n_h} & 0 & 0\\ 0 & 0 & 0 & I_{n_v}\end{bmatrix}.$$

By Proposition 4.4, it can be deduced from (4.36) that an observer in the form of (4.31)–(4.32) generates a positive trajectory $\hat{x}(i, j) \succeq 0$ if and only if

$$\mathscr{A}_{\Pi} \succeq 0, \ \mathscr{A}_{d\Pi} \succeq 0 \text{ and } \mathscr{B}_{\Pi} \succeq 0.$$

Since the matrices A, A_d and B are nonnegative, the above conditions are fulfilled if and only if condition (4.34) holds.

On the other hand, due to (4.34), the system (4.33) is positive. Thus, $\hat{x}(i, j)$ converges asymptotically to $x(i, j)$ as $i, j \to \infty$ if and only if system (4.33) is asymptotically stable. By Theorem 4.3, this condition is satisfied if and only if the LP condition (4.35) is feasible for a vector $\chi \in \mathbb{R}^n$, $\chi \succ 0$. The proof is completed. □

To implement the observer gain L, we need to convert conditions (4.34) and (4.35) to an LP problem. For this, let $D_\chi = \operatorname{diag}(\chi) \in \mathbb{R}^{n \times n}$ denote the diagonal matrix formulated by stacking components of the vector $\chi \in \mathbb{R}^n$. It can be directly verified that, for any vector $\chi \in \mathbb{R}^n$, one has $\chi = D_\chi \mathbf{1}_n$, where $\mathbf{1}_n$ is the vector in $\mathbb{R}^n$ with all entries equal one. Then, condition (4.35) can be rewritten as

$$\left(A^\top + A_d^\top\right) \chi - \left(C^\top + C_d^\top\right) L^\top D_\chi \mathbf{1}_n \prec \chi. \tag{4.37}$$

By using the transformation

$$L^\top D_\chi = Z \in \mathbb{R}^{n_y \times n},$$

condition (4.37) holds if and only if the following LP-based condition is feasible for a matrix $Z \in \mathbb{R}^{n_y \times n}$ and a vector $0 \prec \chi \in \mathbb{R}^n$

$$\left(A^\top + A_d^\top - I_n\right) \chi - \left(C^\top + C_d^\top\right) Z \mathbf{1}_n \prec 0. \tag{4.38}$$

On the other hand, $A - LC \succeq 0$ if and only if

$$A^\top - C^\top L^\top \succeq 0. \tag{4.39}$$

By post-multiplying with $D_\chi \succ 0$, condition (4.39) holds if and only if

$$A^\top D_\chi - C^\top Z \succeq 0. \tag{4.40}$$

Other terms in the condition (4.34) can be similarly converted. In summary, a full-order positive state observer in the form of (4.31)–(4.32) can be implemented by the LP problem given in the following theorem.

Theorem 4.5 *Assume that the system (4.29)–(4.30) is positive. Then, there exists a positive observer in the form of (4.31)–(4.32) if and only if the following LP problem is feasible for a matrix* $Z \in \mathbb{R}^{n_y \times n}$ *and a vector* $\chi \in \mathbb{R}^n$

$$\begin{cases} \left(A^\top + A_d^\top - I_n\right)\chi - \left(C^\top + C_d^\top\right) Z\mathbf{1}_n \prec 0, \\ A^\top D_\chi - C^\top Z \succeq 0, \\ A_d^\top D_\chi - C_d^\top Z \succeq 0, \\ [C \;\; C_d]^\top Z \succeq 0, \\ \chi \succ 0. \end{cases} \tag{4.41}$$

The observer gain L is obtained as

$$L = D_\chi^{-1} Z^\top, \tag{4.42}$$

where (χ, Z) is a feasible solution of (4.41).

4.3.3 Reduced-Order Functional Observers

Let $z(i, j) = Fx(i, j) \in \mathbb{R}_+^{n_z}$ be a linear functional of the state vector, where $F \in \mathbb{R}_+^{n_z \times n}$ is a given matrix. We now consider the design problem of a reduced-order observer $\hat{z}(i, j) \in \mathbb{R}_+^{n_z}$ that converges to the functional $z(i, j)$. To facilitate for the derivation process, we assume the kernel matrix of the estimated linear functional is of the form $F = \mathrm{diag}(F_h, F_v)$, where $F_h \in \mathbb{R}_+^{n_{z1} \times n_h}$, $F_v \in \mathbb{R}_+^{n_{z2} \times n_v}$ and $n_z = n_{z_1} + n_{z_2}$.

Consider the following reduced-order functional state observer

$$\begin{aligned} \begin{bmatrix} \hat{z}^h(i+1, j) \\ \hat{z}^v(i, j+1) \end{bmatrix} &= A^F \begin{bmatrix} \hat{z}^h(i, j) \\ \hat{z}^v(i, j) \end{bmatrix} + A_d^F \begin{bmatrix} \hat{z}^h(i - \tau_h(i), j) \\ \hat{z}^v(i, j - \tau_v(j)) \end{bmatrix} \\ &\quad + Hy(i, j) + FBu(i, j), \end{aligned} \tag{4.43}$$

where A^F, A_d^F and H are the observer gains which will be determined.

The existence conditions of a positive observer in the form of (4.43) are given in the following theorem.

Theorem 4.6 *The 2-D system (4.43) generates an asymptotic positive estimation of the functional $Fx(i, j)$ if and only if the following conditions are satisfied*

$$A^F \succeq 0, \quad A_d^F \succeq 0, \quad HC \succeq 0, \quad HC_d \succeq 0, \tag{4.44}$$

$$\exists \eta \in \mathbb{R}^{n_z}, \eta \succ 0 : \left(A^{F\top} + A_d^{F\top}\right)\eta - \eta \prec 0, \tag{4.45}$$

$$HC + A^F F - FA = 0, \tag{4.46}$$

$$HC_d + A_d^F F - FA_d = 0. \tag{4.47}$$

Proof We define the following augmented state vectors

$$\tilde{x}^h(i, j) = \begin{bmatrix} x^h(i, j) \\ \hat{z}^h(i, j) \end{bmatrix}, \quad \tilde{x}^v(i, j) = \begin{bmatrix} x^v(i, j) \\ \hat{z}^v(i, j) \end{bmatrix}.$$

Similar to equation (4.36), we have

$$\begin{bmatrix} \tilde{x}^h(i+1,j) \\ \tilde{x}^v(i,j+1) \end{bmatrix} = \tilde{\mathscr{A}}_\Pi \begin{bmatrix} \tilde{x}^h(i,j) \\ \tilde{x}^v(i,j) \end{bmatrix} + \tilde{\mathscr{A}}_{d\Pi} \begin{bmatrix} \tilde{x}^h(i-\tau_h(i),j) \\ \tilde{x}^v(i,j-\tau_v(j)) \end{bmatrix} + \tilde{\mathscr{B}}_\Pi u(i,j), \tag{4.48}$$

where $\tilde{\mathscr{A}}_\Pi = \tilde{\Pi}^{-1}\tilde{\mathscr{A}}\Pi$, $\tilde{\mathscr{A}}_{d\Pi} = \tilde{\Pi}^{-1}\tilde{\mathscr{A}}_d\Pi$, $\tilde{\mathscr{B}}_\Pi = \tilde{\Pi}^{-1}\begin{bmatrix} B \\ FB \end{bmatrix}$ and

$$\tilde{\mathscr{A}} = \begin{bmatrix} A & 0 \\ HC & A^F \end{bmatrix}, \quad \tilde{\mathscr{A}}_d = \begin{bmatrix} A_d & 0 \\ HC_d & A_d^F \end{bmatrix}, \quad \tilde{\Pi} = \begin{bmatrix} I_{n_h} & 0 & 0 & 0 \\ 0 & 0 & I_{n_v} & 0 \\ 0 & I_{n_{z_1}} & 0 & 0 \\ 0 & 0 & 0 & I_{n_{z_2}} \end{bmatrix}.$$

It is clear that, subject to the positivity of system (4.29)–(4.30), system (4.48) is positive if and only if condition (4.44) holds.

The error vector $e(i,j) = \hat{z}(i,j) - Fx(i,j)$ is represented by the following 2-D system

$$\begin{aligned} \begin{bmatrix} e^h(i+1,j) \\ e^v(i,j+1) \end{bmatrix} &= A^F \begin{bmatrix} e^h(i,j) \\ e^v(i,j) \end{bmatrix} + A_d^F \begin{bmatrix} e^h(i-\tau_h(i),j) \\ e^v(i,j-\tau_v(j)) \end{bmatrix} \\ &+ \left(HC - FA + A^F F\right) \begin{bmatrix} x^h(i,j) \\ x^v(i,j) \end{bmatrix} \\ &+ \left(HC_d - FA_d + A_d^F F\right) \begin{bmatrix} x^h(i-\tau_h(i),j) \\ x^v(i,j-\tau_v(j)) \end{bmatrix}. \end{aligned} \tag{4.49}$$

Thus, if the two conditions (4.46) and (4.47) hold, then system (4.49) is reduced to systems (4.18) or (4.33). By Theorem 4.3, the error $e(i,j) \to 0$ as $i, j \to \infty$ if and only if the LP-based condition (4.45) holds.

If one of the derived conditions in (4.46) and (4.47) is not satisfied then, by similar arguments in the proof of Theorem 1 in [28], we can find an input $u(i,j)$ by which the corresponding state trajectory $x(i,j)$ makes $e(i,j) \nrightarrow 0$ as $i, j \to \infty$. Thus, there exists a reduced-order observer in the form of (4.43) which is an asymptotic positive estimation of the functional $z(i,j) = Fx(i,j)$ if and only if the derived conditions (4.44)–(4.47) are fulfilled. The proof is completed. □

The following lemma provides a useful tool to formulate testable conditions for the existence of a reduced-order observer (4.43).

Lemma 4.2 *The system of two matrix equations (4.46) and (4.47) admits at least a solution if and only if the following rank condition holds*

$$\operatorname{rank} \begin{bmatrix} \Psi \\ \Omega \end{bmatrix} = \operatorname{rank} \Psi, \tag{4.50}$$

where $\Psi = \begin{bmatrix} C & C_d \\ F & 0 \\ 0 & F \end{bmatrix} \in \mathbb{R}^{(n_y+2n_z)\times 2n}$ *and* $\Omega = \begin{bmatrix} FA & FA_d \end{bmatrix} \in \mathbb{R}^{n_z \times 2n}$.

Proof By grouping condition (4.46) and (4.47) we obtain

$$\begin{bmatrix} HC + A^F F & HC_d + A_d^F F \end{bmatrix} = \Omega,$$

which can be written in the form

$$\begin{bmatrix} H & A^F & A_d^F \end{bmatrix} \Psi = \Omega. \tag{4.51}$$

It is clear that equation (4.51) is solvable (admits at least a solution) with respect to matrix variable $X = \begin{bmatrix} H & A^F & A_d^F \end{bmatrix}$ if and only if the rank condition (4.50) is satisfied. This completes the proof. □

By utilizing Lemma 4.2, we now implement design parameters of a positive functional observer in the form of (4.43). Note at first that, subject to condition (4.50), general solution of (4.51) is given by

$$X = \Omega\Psi^+ + L\left(I_{n_y+2n_z} - \Psi\Psi^+\right), \tag{4.52}$$

where $\Psi^+ \in \mathbb{R}^{2n\times(n_y+2n_z)}$ denotes the Moore-Penrose pseudoinverse of Ψ and $L \in \mathbb{R}^{n_z\times(n_y+2n_z)}$ is an arbitrary matrix. From (4.52), we have

$$\begin{aligned} H &= \Omega\Psi^+ X_1 + L\left(I_{n_y+2n_z} - \Psi\Psi^+\right) X_1, \\ A^F &= \Omega\Psi^+ X_2 + L\left(I_{n_y+2n_z} - \Psi\Psi^+\right) X_2, \\ A_d^F &= \Omega\Psi^+ X_3 + L\left(I_{n_y+2n_z} - \Psi\Psi^+\right) X_3, \end{aligned} \tag{4.53}$$

where

$$X_1 = \begin{bmatrix} I_{n_y} \\ 0 \\ 0 \end{bmatrix}, \quad X_2 = \begin{bmatrix} 0 \\ I_{n_z} \\ 0 \end{bmatrix}, \quad X_3 = \begin{bmatrix} 0 \\ 0 \\ I_{n_z} \end{bmatrix}.$$

Finally, we recast conditions (4.44)–(4.45) to formulate an LP-based problem for checking the design parameters of a positive functional observer in the form of (4.43). For convenience, we denote

$$\Phi = \Omega\Psi^+, \quad \Gamma = I_{n_y+2n_z} - \Psi\Psi^+.$$

Using the general form of solutions (4.53), condition (4.45) can be represented as

$$(\Phi e_z)^\top \eta + (\Gamma e_z)^\top L^\top D_\eta \mathbf{1}_n - \eta \prec 0, \tag{4.54}$$

where $e_z = X_2 + X_3$ and, similar to Theorem 4.5, D_η denotes the diagonal matrix formulated by stacking components of the vector η. By changing the variable

$$L^\top D_\eta = Z,$$

condition (4.54) is reduced to the following one

$$(\Phi e_z)^\top \eta + (\Gamma e_z)^\top Z \mathbf{1}_n - \eta \prec 0. \tag{4.55}$$

Let $\bar{C} = \begin{bmatrix} C^\top & 0_{n\times 2n_z} \end{bmatrix}$ and $\bar{C}_d = \begin{bmatrix} C_d^\top & 0_{n\times 2n_z} \end{bmatrix}$ then condition $HC \succeq 0$ holds if and only if

$$\bar{C}\Phi^\top D_\eta + \bar{C}\Gamma^\top Z \succeq 0. \tag{4.56}$$

By similar lines used to derive condition (4.56), it can be deduced that the four conditions in (4.44) are fulfilled if and only if the following condition is satisfied

$$\begin{cases} (\Phi X_2)^\top D_\eta + (\Gamma X_2)^\top Z \succeq 0, \\ (\Phi X_3)^\top D_\eta + (\Gamma X_3)^\top Z \succeq 0, \\ \bar{C}\Phi^\top D_\eta + \bar{C}\Gamma^\top Z \succeq 0, \\ \bar{C}_d\Phi^\top D_\eta + \bar{C}_d\Gamma^\top Z \succeq 0. \end{cases} \tag{4.57}$$

In summary, the existence conditions of a positive functional observer in the form of (4.43) are given in the following theorem.

Theorem 4.7 *Assume that the system (4.29)–(4.30) is positive. Then, there exists a positive functional observer in the form of (4.43) that is an asymptotic estimation of the functional $z(i, j) = Fx(i, j)$ if and only if the following two conditions are satisfied.*

(i) *Rank compatible*

$$\operatorname{rank}\begin{bmatrix} C & C_d \\ F & 0 \\ 0 & F \\ FA & FA_d \end{bmatrix} = \operatorname{rank}\begin{bmatrix} C & C_d \\ F & 0 \\ 0 & F \end{bmatrix}; \tag{4.58}$$

(ii) *The LP-based problem*

$$\begin{aligned} &(\Phi X_2)^\top D_\eta + (\Gamma X_2)^\top Z \succeq 0, \\ &(\Phi X_3)^\top D_\eta + (\Gamma X_3)^\top Z \succeq 0, \\ &\bar{C}\Phi^\top D_\eta + \bar{C}\Gamma^\top Z \succeq 0, \\ &\bar{C}_d\Phi^\top D_\eta + \bar{C}_d\Gamma^\top Z \succeq 0, \\ &(\Phi e_z)^\top \eta + (\Gamma e_z)^\top Z \mathbf{1}_n - \eta \prec 0, \\ &\eta \succ 0, \end{aligned} \tag{4.59}$$

is feasible for a vector $\eta \in \mathbb{R}^{n_z}$ and a matrix $Z \in \mathbb{R}^{(n_y+2n_z)\times n_z}$.

The observer gain $X = \begin{bmatrix} A^F & A_d^F & H \end{bmatrix}$ *is defined by (4.53) with* $L = D_\eta^{-1} Z^\top$.

Remark 4.6 The Luenberger-type observer (4.31)–(4.32) is a special case of the functional observer (4.43). Specifically, when F is the identity matrix, condition (4.58) (or (4.50)) is obviously satisfied since the matrix Ψ has full-column rank. By designating the matrix H as the observer gain and $A^F = A - HC$, $A_d^F = A_d - HC_d$, conditions (4.46) and (4.47) are also obvious. The design conditions of Theorem 4.7 are now reduced to those of Theorem 4.5.

4.3.4 Observer Design for 2-D Positive Systems in FMLSS Model

In this section, we consider the problem of observers design for 2-D positive time-delay systems described by the following Fornasini–Marchesini model

$$\begin{aligned} x(i+1, j+1) &= A_{01}x(i+1, j) + A_{02}x(i, j+1) \\ &\quad + A_{d1}x(i+1, j-\tau_v(j)) + A_{d2}x(i-\tau_h(i), j+1) \\ &\quad + B_1u(i+1, j) + B_2u(i, j+1), \\ y(i, j) &= Cx(i, j). \end{aligned} \tag{4.60}$$

Assume that system (4.60) is positive. A Luenberger-type observer that generates a nonnegative asymptotic estimation of the state vector of system (4.60) is constructed in the form

$$\begin{aligned} \hat{x}(i+1, j+1) &= A_{01}\hat{x}(i+1, j) + A_{02}\hat{x}(i, j+1) + A_{d1}\hat{x}(i+1, j-\tau_v(j)) \\ &\quad + A_{d2}\hat{x}(i-\tau_h(i), j+1) + B_1u(i+1, j) + B_2u(i, j+1) \\ &\quad - L_1\left(\hat{y}(i+1, j) - y(i+1, j)\right) \\ &\quad - L_2\left(\hat{y}(i, j+1) - y(i, j+1)\right), \\ \hat{y}(i, j) &= C\hat{x}(i, j), \end{aligned} \tag{4.61}$$

where the matrices L_1, L_2 define an observer gain, which will be determined. By using the augmented state vector $\widetilde{x}(i, j) = \begin{bmatrix} x(i, j) \\ \hat{x}(i, j) \end{bmatrix}$, it can be verified that system (4.61) generates nonnegative states $\hat{x}(i, j)$ if and only if the following conditions are satisfied

$$\begin{bmatrix} L_1 \\ L_2 \end{bmatrix} C \succeq 0, \quad \begin{bmatrix} A_{01} \\ A_{02} \end{bmatrix} - \begin{bmatrix} L_1 \\ L_2 \end{bmatrix} C \succeq 0. \tag{4.62}$$

On the other hand, the error vector $e(i, j) = \hat{x}(i, j) - x(i, j)$ is governed by the system

$$\begin{aligned} e(i+1, j+1) &= (A_{01} - L_1C)\,e(i+1, j) + (A_{02} - L_2C)\,x(i, j+1) \\ &\quad + A_{d1}e(i+1, j-\tau_v(j)) + A_{d2}e(i-\tau_h(i), j+1). \end{aligned} \tag{4.63}$$

In addtion, by (4.62), system (4.63) is positive. Thus, according to Proposition 4.3, $\hat{x}(i, j)$ is an asymptotic approximation of $x(i, j)$ (i.e. $e(i, j) \to 0$ as $i, j \to \infty$) if and only if the LP-based condition

$$(A_{0s} + A_{ds} - (L_1 + L_2)C - I_n)^\top \eta \prec 0 \tag{4.64}$$

is feasible for a vector $\eta \in \mathbb{R}^n, \eta \succ 0$, where $A_{0s} = A_{01} + A_{02}$ and $A_{ds} = A_{d1} + A_{d2}$.

We define the following transformation

$$L_1^\top D_\eta = Z_1, \quad L_2^\top D_\eta = Z_2. \tag{4.65}$$

Then, condition (4.64) becomes

$$\left(A_{0s}^\top + A_{ds}^\top\right)\eta - C^\top(Z_1 + Z_2)\mathbf{1}_n \prec \eta. \tag{4.66}$$

Moreover, condition (4.62) can be recast into the following condition

$$\begin{bmatrix} Z_1 & Z_2 \end{bmatrix}^\top C \succeq 0, \tag{4.67}$$

$$D_{\eta,\eta}\begin{bmatrix} A_{01} \\ A_{02} \end{bmatrix} - \begin{bmatrix} Z_1 & Z_2 \end{bmatrix}^\top C \succeq 0. \tag{4.68}$$

where $D_{\eta,\eta}$ is the diagonal matrix of blocks D_η, D_η.

From the above derivation, a positive state observer for system (4.60) in the form of (4.61) can be implemented by the following LP-based problem.

Theorem 4.8 *Assume that the system (4.60) is positive. Then, there exists a positive observer in the form of (4.61) if and only if the following LP problem is feasible for a vector $\eta \in \mathbb{R}^n$ and matrices $Z_1.Z_2 \in \mathbb{R}^{n \times n}$*

$$\begin{cases} \left(A_{0s}^\top + A_{ds}^\top\right)\eta - C^\top(Z_1 + Z_2)\mathbf{1}_n \prec \eta, \\ A^\top D_\chi - C^\top Z \succeq 0, \\ D_{\eta,\eta}\begin{bmatrix} A_{01} \\ A_{02} \end{bmatrix} - \begin{bmatrix} Z_1 & Z_2 \end{bmatrix}^\top C \succeq 0, \\ \begin{bmatrix} Z_1 & Z_2 \end{bmatrix}^\top C \succeq 0, \\ \eta \succ 0. \end{cases} \tag{4.69}$$

The observer gain $L = \begin{bmatrix} L_1 & L_2 \end{bmatrix}$ is obtained as

$$L = D_\eta^{-1}\begin{bmatrix} Z_1 \\ Z_2 \end{bmatrix}^\top, \tag{4.70}$$

where (η, Z_1, Z_2) is a feasible solution of (4.69).

4.3.5 Numerical Examples

In this section, we give some numerical examples and simulations to illustrate the effectiveness of the design method.

Example 4.2 Consider system (4.29)–(4.30) with the system matrices

$$A = \begin{bmatrix} 0.082 & 0.158 \\ 0.206 & 0.119 \end{bmatrix}, \quad A_d = \begin{bmatrix} 1.105 & 0.551 \\ 0.625 & 1.019 \end{bmatrix},$$
$$C = \begin{bmatrix} 0 & 0 \end{bmatrix}, \quad C_d = \begin{bmatrix} 1 & 1 \end{bmatrix}.$$

Let $0.01\mathbf{1}_2 \preceq \chi \preceq 2\mathbf{1}_2$. By solving the LP problem (4.41) with respect to χ and Z, a feasible solution is obtained as

$$\chi = \begin{bmatrix} 1.8899 \\ 1.8232 \end{bmatrix}, \quad Z = \begin{bmatrix} 1.0356 & 1.1189 \end{bmatrix}.$$

By Theorem 4.5, an asymptotic full-order positive observer in the form of (4.31)–(4.32) can be implemented. The observer gain is given by

$$L = D_\eta^{-1} Z^\top = \begin{bmatrix} 0.5480 \\ 0.6137 \end{bmatrix}. \tag{4.71}$$

With the observer gain (4.71), the error system (4.33) is asymptotically stable. A state trajectory $e(i, j)$ of system (4.33) with time-delays $\tau_h(i) = 2 + 3\,|\sin(i\pi/2)|$ and $\tau_v(j) = 1 + 5\,|\cos(j\pi/2)|$ is presented in Figs. 4.3 and 4.4. It can be seen from Figs. 4.3 and 4.4 that the error vector $e(i, j)$ converges to zero as revealed by the theoretical results.

Example 4.3 Consider system (4.29)–(4.30) with the system matrices

$$A = \begin{bmatrix} 0.152 & 0.267 & 0 & 0.215 \\ 0.348 & 0.233 & 0 & 0.089 \\ 0.097 & 0.107 & 0.128 & 0.152 \\ 0.118 & 0.118 & 0 & 0.358 \end{bmatrix}, \quad A_d = \begin{bmatrix} 0.318 & 0.105 & 0 & 0.165 \\ 0.125 & 0.338 & 0 & 0.236 \\ 0.432 & 0.705 & 0.457 & 0.686 \\ 0.217 & 0.217 & 0 & 0.528 \end{bmatrix},$$
$$C = \begin{bmatrix} 1 & 1 & 0 & 0 \end{bmatrix}, \quad C_d = \begin{bmatrix} 1 & 1 & 0 & 1 \end{bmatrix}.$$

The estimated linear functional is given by $z(i, j) = Fx(i, j)$, where

$$F = \begin{bmatrix} 1 & 1 & 0 & 0 \\ 0 & 0 & 0 & 1 \end{bmatrix}.$$

It can be verified that the compatible rank condition (4.58) is satisfied. More precisely, we have

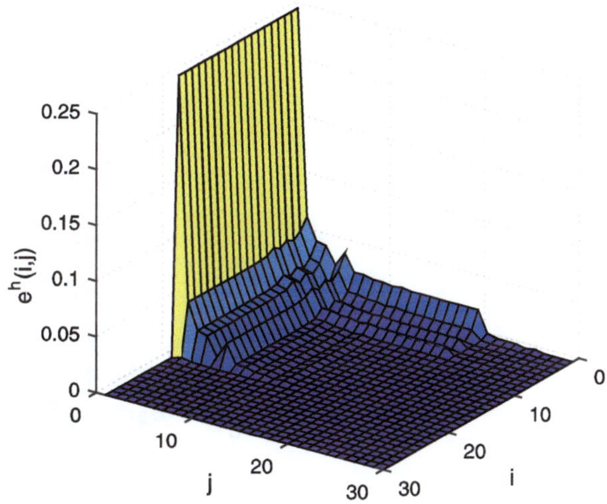

Fig. 4.3 A trajectory of error $e^h(i, j)$ with $\tau_h(i) = 2 + 3\,|\sin(i\pi/2)|$ and $\tau_v(j) = 1 + 5\,|\cos(j\pi/2)|$

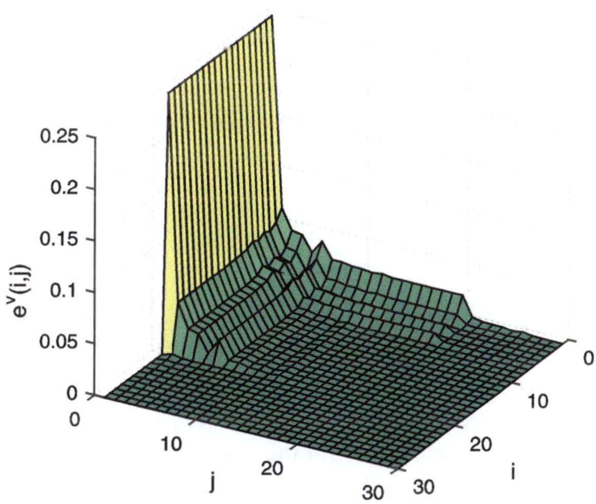

Fig. 4.4 A trajectory of error $e^v(i, j)$ with $\tau_h(i) = 2 + 3\,|\sin(i\pi/2)|$ and $\tau_v(j) = 1 + 5\,|\cos(j\pi/2)|$

$$\text{rank}\begin{bmatrix}\Psi \\ \Omega\end{bmatrix} = \text{rank}\,\Psi = 4.$$

By solving the LP-based problem (4.59) with $0.01\mathbf{1}_2 \preceq \eta \preceq 2\mathbf{1}_2$, a feasible solution is obtained as

$$\eta = \begin{bmatrix}0.8779 \\ 1.8534\end{bmatrix}, \quad Z^\top = \begin{bmatrix}2.2515 & 0.7485 & -2.0 & 0.7485 & 0.7485 \\ 2.0272 & 0.9728 & -2.0 & 0.9728 & 0.9728\end{bmatrix}.$$

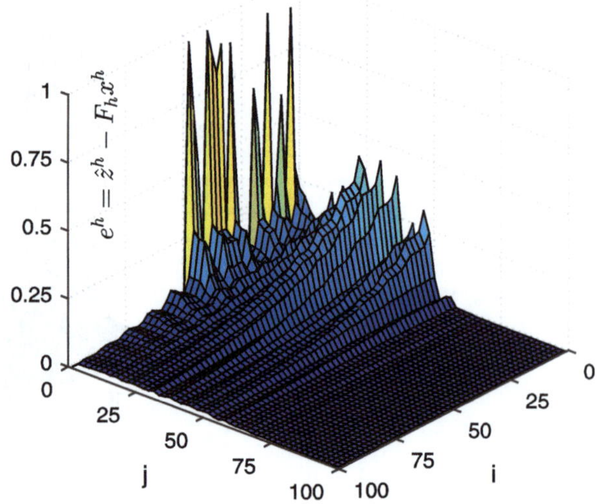

Fig. 4.5 A trajectory of error $e^h = \hat{z}^h - F_h x^h$ with $\tau_h = 6$, $\tau_v = 8$

According to Theorem 4.7, an asymptotic positive functional observer in the form of (4.43) can be implemented. The parameters of a desired observer gain $X = \begin{bmatrix} A^F & A_d^F & H \end{bmatrix}$ are given by

$$A^F = \begin{bmatrix} 0.1623 & 0.304 \\ 0.0224 & 0.358 \end{bmatrix}, \quad A_d^F = \begin{bmatrix} 0.1053 & 0.0633 \\ 0.1214 & 0.4324 \end{bmatrix}, \quad H = \begin{bmatrix} 0.3377 \\ 0.0956 \end{bmatrix}. \quad (4.72)$$

Let

$$F_h = [1 \quad 1], \quad F_v = [0 \quad 1].$$

A trajectory of error $e^h(i, j) = \hat{z}^h(i, j) - F_h x^h(i, j)$ and a trajectory of $e^v(i, j) = \hat{z}^v(i, j) - F_v x^v(i, j)$ with the obtained observer parameters in (4.72) and constant delays $\tau_h = 6$, $\tau_v = 8$ are given in Figs. 4.5 and 4.6. It can be seen from Figs. 4.5 and 4.6 that the error $\hat{z}(i, j)$ is an asymptotic approximation of the functional $Fx(i, j)$ as revealed by Theorem 4.7.

4.4 Conclusion

This chapter has dealt with the problems of stability analysis and observers design for positive 2-D systems with time-varying delays. Necessary and sufficient stability conditions have first been formulated for 2-D nonlinear systems in the Roesser model with time-varying delays based on monotone dynamical systems theory. Stability conditions of positive 2-D linear systems both in Roesser model and FMLSS model with delays have also been obtained as some particular applications. On the

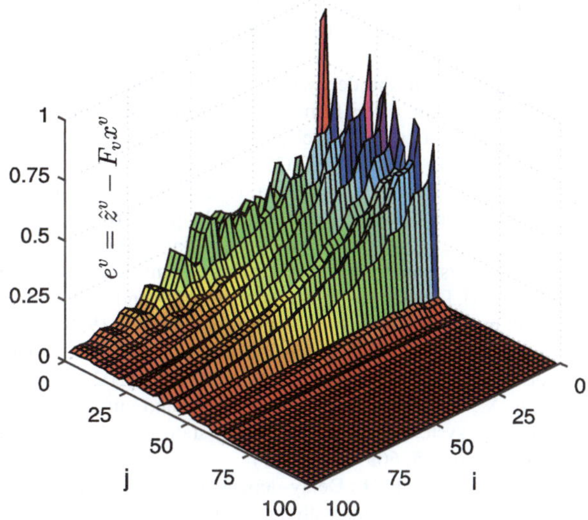

Fig. 4.6 A trajectory of error $e^v = \hat{z}^v - F_v x^v$ with $\tau_h = 6$, $\tau_v = 8$

basis of the obtained positivity and stability conditions, the design problem of positive observers for 2-D positive linear systems with delays has been addressed. Extended Luenberger-type observers and reduced-order functional state observers have been implemented using necessary and sufficient delay-independent conditions in the linear programming setting. Numerical examples and simulations have been given to demonstrate the effectiveness of the obtained results.

References

1. Farina, L., Rinaldi, S.: Positive Linear Systems: Theory and Applications. John Wiley & Sons, New York (2000)
2. Godfrey, K.: Compartmental models and their applications. Academic Press, London (1983)
3. Leenheer, P.D., Angeli, D., Sontag, E.D.: Monotone chemical reaction networks. J Math Chem **41**, 295–314 (2007)
4. Smith, H.: Monotone Dynamical Systems: An Introduction to the Theory of Competitive and Cooperative Systems. American Mathematical Society, Providence, RI (2008)
5. Hirsch MW, Smith H (2005) Monotone dynamical systems. Hand book of Differential Equations: Ordinary Differential Equations, vol. 2, Chapter 4, 239–357
6. Ngoc, P.H.A.: Stability of positive differential systems with delay. IEEE Trans Autom Control **58**, 203–209 (2013)
7. Feyzmahdavian, H.R., Charalambous, T., Johanson, M.: Exponential stability of homogeneous positive systems of degree one with time-varying delays. IEEE Trans Autom Control **59**, 1594–1599 (2014)
8. Bokharaie, V.S., Mason, O.: On delay-independent stability of a class of nonlinear positive time-delay systems. IEEE Trans Autom Control **59**, 1974–1977 (2014)
9. Feyzmahdavian, H.R., Charalambous, T., Johanson, M.: Asymptotic stability and decay rates of homogeneous positive systems with bounded and unbounded delays. SIAM J Control Optim **52**, 2623–2650 (2014)

10. Zaidi, I., Chaabane, M., Tadeo, F., Benzaouia, A.: Static state-feedback controller and observer design for interval positive systems with time delay. IEEE Trans Circuit Syst II **62**, 506–510 (2015)
11. Dong, J.G.: On the decay rates of homogeneous positive systems of any degree with time-varying delays. IEEE Trans Autom Control **60**, 2983–2988 (2015)
12. Devane, E., Lestas, I.: Delay-independent asymptotic stability of monotone systems. IEEE Trans Autom Control **61**, 2625–2631 (2016)
13. Nam, P.T., Trinh, H., Pathirana, P.N.: Minimization of state bounding for perturbed positive systems with delays. SIAM J Control Optim **56**,(2018). https://doi.org/10.1137/16M1104846
14. Hien, L.V., Hong, M.T.: An optimization approach to static output-feedback control of LTI positive systems with delayed measurements. J Frank Inst **356**, 5087–5103 (2019)
15. Hong, M.T., Hien, L.V.: Solvability of L_1-induced controller synthesis for positive systems with multiple delays. Int J Control Autom Syst **19**, 3569–3579 (2021)
16. Hien, L.V.: Delay-dependent positivity and stability analysis of discrete-time systems with delay. IEEE Control Syst Lett **6**, 2575–2580 (2022)
17. Liu, X., Yu, W., Wang, L.: Necessary and sufficient asymptotic stability criterion for 2-D positive systems with time-varying state delays described by Roesser model. IET Control Theory Appl **5**, 663–668 (2011)
18. Hien, L.V., Trinh, H.: Delay-dependent stability and stabilization of two-dimensional positive Markov jump systems with delays. IET Control Theory Appl **11**, 1603–1610 (2017)
19. Hien, L.V., Trinh, H., Pathirana, P.N.: On ℓ_1-gain control of 2-D positive Roesser systems with directional delays: Necessary and sufficient conditions. Automatica **112**, 108720 (2020)
20. Fu, J., Duan, Z., Ghous, I.: ℓ_1-induced norm and controller synthesis for positive 2D systems with multiple delays. J Frank Inst **357**, 7904–7920 (2020)
21. Trinh, H., Fernando, T.: Functional Observers for Dynamical Systems. Springer, Berlin (2012)
22. Rami, M.A., Tadeo, F., Helmke, U.: Positive observers for linear positive systems, and their implications. Int J Control **84**, 716–725 (2011)
23. Rami, M.A., Schönlein, M., Jordan, J.: Estimation of linear positive systems with unknown time-varying delays. Eur J Control **19**, 179–187 (2013)
24. Trinh, H., Huong, D.C., Hien, L.V., Nahavandi, S.: Design of reduced-order positive linear functional observers for positive time-delay systems. IEEE Trans Circuit Syst II **64**, 555–559 (2017)
25. Yeganefar, N., Yeganefar, N., Ghamgui, M., Moulay, E.: Lyapunov theory for 2-D nonlinear Roesser models: Aplication to asymptotic and exponential stability. IEEE Trans Autom Control **58**, 1299–1304 (2013)
26. Duan, Z., Han, Y., Xiang, Z., Ghous, I.: On l_1-gain control for 2D delayed positive systems in FM LSS models: necessary and sufficient conditions. Int J Syst Sci **53**, 3449–3464 (2022)
27. Berman, A., Plemmons, R.J.: Nonnegative Matrices in the Mathematical Sciences. SIAM, Philadelphia (1994)
28. Darouach, M.: Linear functional observers for systems with delays in state variables. IEEE Trans Autom Control **46**, 491–496 (2001)

Chapter 5
Performance Analysis and l_1-Gain Control of Two-Dimensional Positive Systems with Delays

Abstract This chapter addresses the problem of l_1-gain control for two-dimensional (2-D) positive systems in Roesser model and Fornasini–Marchesini local state-space (FMLSS) model with directional delays. Exact value characterizations of l_1-induced norm of the input-output operators are obtained using 2-D Z-transform approach. The obtained l_1-gain characterizations are then utilized to establish necessary and sufficient conditions in terms of linear programming (LP) settings for l_1-induced performance with a prescribed attenuation level. Finally, by utilizing a linear convex optimization approach, necessary and sufficient conditions for the existence of a state-feedback controller (SFC) that makes the closed-loop system positive, stable and has prescribed l_1-induced performance are formulated. Numerical examples and simulations are given to illustrate the effectiveness of the proposed method.

Keywords 2-D systems · Roesser model · FMLSS model · Positive systems · l_1-gain control · Feasibility · Linear programming

5.1 Introduction

Positive systems [1] have been intensively studied in the past few decades due to their elegant properties that have no counterpart in general dynamical systems [2, 3]. Practical applications of positive systems have been found in various areas, from economics, biology, ecology and epidemiology to chemistry, pharmacokinetics or population dynamics [4–6]. Many results concerning stability analysis and controller/observer design for 2-D positive systems with or without delays have also been published. We refer the reader to [7–12] for a few references.

In control theory of positive systems, the problem of l_1-gain (or L_1-gain) control is an essential topic. Roughly speaking, l_1-gain control gives an interpretation of mass balance with linear supply rates and linear co-positive storage functions, which naturally arise in dissipative theory of positive systems [13, 14]. During the past two decades, considerable research attention has been devoted to develop the problems of performance analysis and controller/filter/observer synthesis under l_1/L_1-gain schemes for various classes of one-dimensional (1-D) positive systems with and

L. V. Hien, *Two-Dimensional Time-Delay Systems*, Studies in Systems, Decision and Control 605, https://doi.org/10.1007/978-3-031-95678-2_5

without delays [15–17]. In particular, the problem of L_1-gain control via state-feedback or static-output feedback schemes was tackled using linear programming approach [18, 19] or iteration search method [20] to solve bilinear matrix inequality (BMI) conditions. By utilizing the Laplace transform combined with linear optimization techniques, L_1-induced controller synthesis was investigated for positive systems with multiple delays in [21]. For linear positive discrete-time systems, an explicit expression of l_1-induced norm was first formulated using the fundamental matrix representation [22]. Then, an l_1-induced stabilizing SFC was designed using a particular form of controller gains involving the so-called directional matrices. The theory of l_1-gain control for positive 2-D systems has not been well developed. By utilizing lifting techniques, l_1-induced norm characterizations were derived for 2-D systems described by FMLSS model in [23, 24]. The design problem of SFCs was then considered using matrix transformation approach which, in fact, only gives sufficient conditions according to matrix changing variable equations.

In this chapter, motivated by recent work [25], we address the problem of l_1-gain control for positive 2-D systems in Roesser and FMLSS models with state delays. Explicit expressions of l_1-induced norm of the input-output operators are obtained using 2-D Z-transform approach. The obtained l_1-gain characterizations are then utilized to establish necessary and sufficient conditions in terms of linear programming for l_1-induced performance with a prescribed attenuation level. Finally, by utilizing linear optimization procedures, necessary and sufficient conditions for the existence of an SFC that makes the closed-loop system positive, stable and has prescribed l_1-induced performance are formulated.

5.2 Preliminaries

5.2.1 Model Description

Consider a class of 2-D time-delay systems described by the Roesser model

$$\begin{bmatrix} x^h(i+1,j) \\ x^v(i,j+1) \end{bmatrix} = A \begin{bmatrix} x^h(i,j) \\ x^v(i,j) \end{bmatrix} + A_d \begin{bmatrix} x^h(i-\tau_h,j) \\ x^v(i,j-\tau_v) \end{bmatrix} + Bu(i,j) + B_w w(i,j), \tag{5.1}$$

$$z(i,j) = C \begin{bmatrix} x^h(i,j) \\ x^v(i,j) \end{bmatrix} + C_d \begin{bmatrix} x^h(i-\tau_h,j) \\ x^v(i,j-\tau_v) \end{bmatrix} + Du(i,j) + D_w w(i,j), \tag{5.2}$$

where $x^h(i,j) \in \mathbb{R}^{n_h}$ and $x^v(i,j) \in \mathbb{R}^{n_v}$ are the horizontal and vertical state vectors, $u(i,j) \in \mathbb{R}^{n_u}$ is the control input, $z(i,j) \in \mathbb{R}^{n_z}$ and $w(i,j) \in \mathbb{R}^{n_w}$ are the controlled output and exogenous disturbance input vectors, respectively. τ_h and τ_v are nonnegative integers representing the time delays along the horizontal and vertical directions. $A, A_d \in \mathbb{R}^{n\times n}$ $(n = n_h + n_v)$, $B \in \mathbb{R}^{n\times n_u}$, $B_w \in \mathbb{R}^{n\times n_w}$, $C, C_d \in \mathbb{R}^{n_z\times n}$,

$D \in \mathbb{R}^{n_z \times n_u}$ and $D_w \in \mathbb{R}^{n_z \times n_w}$ are given real matrices. The set of admissible exogenous disturbances $w(i, j)$ is defined as

$$l_1(\mathbb{R}^{n_w}) = \left\{ f : \mathbb{N}_0 \times \mathbb{N}_0 \to \mathbb{R}^{n_w} : \|f\|_{l_1} = \sum_{i,j=0}^{\infty} \|f(i, j)\|_1 < \infty \right\}.$$

Initial conditions of system (5.1) are specified as

$$\begin{aligned} x^h(i, j) &= \phi_h(i, j), \quad i \in \mathbb{Z}[-\tau_h, 0], \quad j \in \mathbb{N}_0, \\ x^v(i, j) &= \phi_v(i, j), \quad i \in \mathbb{N}_0, \quad j \in \mathbb{Z}[-\tau_v, 0], \end{aligned} \tag{5.3}$$

where $\phi_h : \mathbb{Z}[-\tau_h, 0] \times \mathbb{N}_0 \to \mathbb{R}^{n_h}$ and $\phi_v : \mathbb{N}_0 \times \mathbb{Z}[-\tau_v, 0] \to \mathbb{R}^{n_v}$ are given functions. It is assumed that $\phi_h(k, .) \in l_1(\mathbb{R}^{n_h})$ and $\phi_v(., l) \in l_1(\mathbb{R}^{n_v})$ for each $k \in \mathbb{Z}[-\tau_h, 0]$ and $l \in \mathbb{Z}[-\tau_v, 0]$ with $\|\phi_h\| = \max_k \|\phi_h(k, .)\|_{l_1}$ and $\|\phi_v\| = \max_l \|\phi_v(., l)\|_{l_1}$.

5.2.2 *Problem Formulation*

Similar to Chap. 4, the concept of positivity of 2-D systems in the form of (5.1)–(5.2) is presented in following definition.

Definition 5.1 System (5.1)–(5.2) is said to be (internally) positive if for any nonnegative initial functions $\phi_h(i, j) \succeq 0$, $\phi_v(i, j)$ defined by (5.3) and nonnegative inputs $u(i, j) \succeq 0$, $w(i, j) \succeq 0$, the state vector $x(i, j) = \begin{bmatrix} x^h(i, j) \\ x^v(i, j) \end{bmatrix}$ and the output vector $z(i, j)$ are always nonnegative, that is, $x(i, j) \succeq 0$ and $z(i, j) \succeq 0$ for all $i, j \in \mathbb{N}_0$.

Proposition 5.1 *System* (5.1)–(5.2) *is positive for any delays τ_h and τ_v if and only if the system matrices A, A_d, B, B_w, C, C_d, D and D_w are nonnegative.*

Proof The proof is similar to that of Proposition 4.1. We omit it here. □

Without control input (i.e. $u = 0$), the open-loop system of (5.1) is given by

$$\begin{bmatrix} x^h(i+1, j) \\ x^v(i, j+1) \end{bmatrix} = A \begin{bmatrix} x^h(i, j) \\ x^v(i, j) \end{bmatrix} + A_d \begin{bmatrix} x^h(i - \tau_h, j) \\ x^v(i, j - \tau_v) \end{bmatrix} + B_w w(i, j). \tag{5.4}$$

First, we give the following definitions for some notions of stability of system (5.4).

Definition 5.2 System (5.4) without exogenous disturbance (i.e. $w = 0$) is said to be

(i) *stable* (S) if for any $\varepsilon > 0$, there exists a $\delta = \delta(\varepsilon) > 0$ such that if $\|\phi_h(k, j)\|_1 < \delta$ and $\|\phi_v(i, l)\|_1 < \delta$ then $\|x(i, j)\|_1 < \varepsilon$ for all $i, j \in \mathbb{N}_0$;

(ii) *globally asymptotically stable* (GAS) if it is stable and attractive, that is, for any initial condition (5.3), the corresponding solution $x(i, j)$ of (5.4) satisfies $\lim_{i+j\to\infty} \|x(i, j)\|_1 = 0$.

Definition 5.3 System (5.4) with $w = 0$ is said to be *globally exponentially stable* (GES) if there exist scalars $\alpha \in (0, 1)$ and $\beta > 0$ such that any solution $x(i, j)$ of system (5.4) subject to initial condition (5.3) satisfies the following estimate

$$\|x(i, j)\|_1 \le \beta\Bigg(\sum_{r=0}^{i} \frac{\hat{\phi}_v(r)}{\alpha^{r+1}} + \sum_{s=0}^{j} \frac{\hat{\phi}_h(s)}{\alpha^{s+1}}\Bigg)\alpha^{i+j},\ i, j \in \mathbb{N}_0, \tag{5.5}$$

where $\hat{\phi}_v(r) = \max_l \|\phi_v(r, l)\|_1$ and $\hat{\phi}_h(s) = \max_k \|\phi_h(k, s)\|_1$.

Definition 5.4 System (5.4) is said to be l_1-*stable* (l_1-S) if the following two requirements are fulfilled

(1) With $w = 0$, system (5.4) is GAS;
(2) For any nonzero disturbance $w \in l_1(\mathbb{R}^{n_w})$, and initial functions $\phi_h(k, .) \in l_1(\mathbb{R}^{n_h})$, $\phi_v(., l) \in l_1(\mathbb{R}^{n_v})$ defined by (5.3), the corresponding solution $x(i, j)$ of (5.4) belongs to $l_1(\mathbb{R}^n)$, that is, $\sum_{i,j=0}^{\infty} \|x(i, j)\|_1 < \infty$.

As discussed in Chap. 4, Definition 5.3 is an extension of the conventional exponential stability concept developed for 1-D systems with delays. Moreover, it follows from Definitions 5.2-5.4 that the properties of (GES) and (l_1-S) imply (GAS) and (S), which is demonstrated by Fig. 5.1. Exact relationships between these stability concepts for positive systems in the form of Eq. (5.4) will be discussed in the next section.

Consider system (5.1)–(5.2) without control input and assume that the open-loop system is stable in the sense of (l_1-S). We define an input-output operator as

$$\mathscr{L} : l_1(\mathbb{R}^{n_w}) \longrightarrow l_1(\mathbb{R}^{n_z}), \quad w \mapsto z$$

and l_1-gain of system (5.1) under zero initial condition is defined by

$$\|\mathscr{L}\|_{l_1-l_1} = \sup_{\|w\|_{l_1}\neq 0} \frac{\|z\|_{l_1}}{\|w\|_{l_1}} = \sup_{\|w\|_{l_1}=1} \|z\|_{l_1}. \tag{5.6}$$

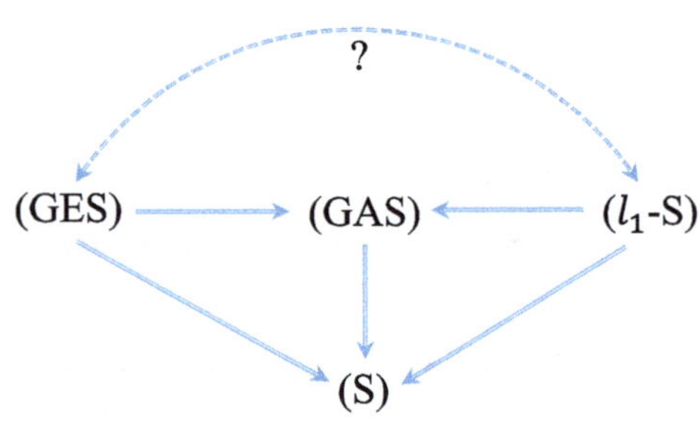

Fig. 5.1 Relations between stability concepts (S), (l_1-S), (GAS) and (GES) by the definition

Definition 5.5 For a given $\gamma > 0$, system (5.1) (with $u = 0$) is said to have l_1-gain performance of level γ if, under zero initial condition, it holds that $\|\mathscr{L}\|_{l_1-l_1} < \gamma$. In other words, for any nonzero disturbance $w \in l_1(\mathbb{R}^{n_w})$, and under zero initial condition, the output trajectory $z(i, j)$ of system (5.1) satisfies

$$\|z(i, j)\|_1 < \gamma \|w(i, j)\|_1 \text{ for all } i, j \geq 0.$$

Main objectives of this chapter are as follows

- Formulate the exact value of l_1-gain $\|\mathscr{L}\|_{l_1-l_1}$ for system (5.1).
- Establish necessary and sufficient conditions of l_1-gain performance index.
- Derive necessary and sufficient tractable conditions for the existence of a state-feedback controller which makes the resultant closed-loop system positive, stable and has prescribed l_1-gain performance level.

5.2.3 Stability Analysis

In this section, we reformulate necessary and sufficient conditions and show that, for positive 2-D linear systems with delays in the form of (5.4), the stability notions presented in the proceeding section are equivalent.

Theorem 5.1 *Consider a positive 2-D system as given in* (5.4) *with* $w = 0$. *The following statements are equivalent.*

(i) *System* (5.4) *is GES.*
(ii) *System* (5.4) *is GAS.*
(iii) *There exists a vector* $\eta \in \mathbb{R}^n$, $\eta \succ 0$, *such that*

$$(A + A_d - I_n)\,\eta \prec 0. \tag{5.7}$$

(iv) *There exists a vector* $\zeta \in \mathbb{R}^n$, $\zeta \succ 0$, *such that*

$$\zeta^\top (A + A_d - I_n) \prec 0. \tag{5.8}$$

(v) *The spectral radius of the matrix* $A + A_d$ *satisfies*

$$\rho(A + A_d) < 1. \tag{5.9}$$

Proof By the properties of nonnegative matrices [26], it is found that condition (5.7) is feasible for a positive vector $\eta \in \mathbb{R}^n$ if and only if $\rho(A + A_d) < 1$ [1, 26]. In addition, it is clear that $\rho(A + A_d) < 1$ if and only if $\rho\big(A^\top + A_d^\top\big) < 1$. This validates the equivalence of items (iii), (iv) and (v).

The proof of (ii) $\Longrightarrow$ (iii) and (iii) $\Longrightarrow$ (i) is similar to [Chap. 4, Theorem 4.1]. We omit some details here. □

Remark 5.1 The positivity of system (5.4) plays a key role for the formation of necessary and sufficient stability condition as (5.7). First, for general 2-D systems, condition (5.7) does not guarantee the matrix $A + A_d$ is Schur stable and therefore the delay-free system of (5.4) is not necessary to be stable. Second, without positivity, we cannot use co-positive Lyapunov functionals (linear functionals of state vectors) [10]. Instead, quadratic Lyapunov–Krasovskii functionals will be used to derive sufficient stability conditions [27–29]. Such conditions are typically delay-dependent and in the form of linear matrix inequalities, which are not tractable for the problem of l_1-gain control.

The following result shows that one of the equivalent conditions derived in Theorem 5.1 is also necessary and sufficient for l_1-stability of positive 2-D system (5.4).

Proposition 5.2 *The positive 2-D system* (5.4) *is l_1-stable for given delays τ_h, τ_v if and only if there exists a vector $\lambda \in \mathbb{R}^n$, $\lambda \succ 0$, that satisfies the following LP condition*

$$\lambda^\top (A + A_d) - \lambda^\top \prec 0. \tag{5.10}$$

Proof The *Necessity* follows from Theorem 5.1. We now prove the *Sufficiency*. Let $\lambda = [\lambda_h^\top \;\; \lambda_v^\top]^\top \succ 0$ be a positive vector defined by (5.10). We construct a co-positive 2-D Lyapunov functional as

$$V(x(i,j)) = V^h(x^h(i,j)) + V^v(x^v(i,j)), \tag{5.11}$$

where

$$V^h(x^h(i,j)) = \lambda_h^\top x^h(i,j) + \sum_{k=-\tau_h}^{-1} \lambda_h^\top([I_{n_h} \;\; 0]A_d)x^h(i+k,j),$$

$$V^v(x^v(i,j)) = \lambda_v^\top x^v(i,j) + \sum_{l=-\tau_v}^{-1} \lambda_v^\top([0 \;\; I_{n_v}]A_d)x^v(i,j+l).$$

The partial difference of the functional (5.11) along state trajectories of system (5.4) is given as

$$\begin{aligned}
\Delta V(x(i,j)) &= V^h(x^h(i+1,j)) - V^h(x^h(i,j)) \\
&\quad + V^v(x^v(i,j+1)) - V^v(x^v(i,j)) \\
&= \lambda^\top \left(\begin{bmatrix} x^h(i+1,j) \\ x^v(i,j+1) \end{bmatrix} - x(i,j) \right) \\
&\quad + \lambda^\top A_d \left(x(i,j) - \begin{bmatrix} x^h(i-\tau_h,j) \\ x^v(i,j-\tau_v) \end{bmatrix} \right).
\end{aligned}$$

Thus, it follows from (5.4) that

$$\Delta V(x(i,j)) = \lambda^\top (A + A_d - I_n)x(i,j) + \lambda^\top B_w w(i,j).$$

Taking (5.10) into account, we then obtain

$$\Delta V(x(i,j)) \le -\varepsilon\lambda^\top x(i,j) + \lambda^\top B_w w(i,j), \tag{5.12}$$

for sufficiently small $\varepsilon > 0$ satisfying $\lambda^\top(A + A_d) \prec (1-\varepsilon)\lambda^\top$.
For any positive integers T, K, by taking summation both sides of (5.12), we can get

$$\begin{aligned}\varepsilon\lambda_+ \sum_{i=0}^{T}\sum_{j=0}^{K} \mathbf{1}_n^\top x(i,j) &\le \sum_{j=0}^{K}[V^h(x^h(0,j)) - V^h(x^h(T+1,j))] \\ &\quad + \sum_{i-0}^{T}[V^v(x^v(i,0)) - V^v(x^v(i,K+1))] + \|\lambda^\top B_w\|_1 \|w\|_{l_1} \\ &\le \beta\,(\|\phi_h\| + \|\phi_v\|) + \|\lambda^\top B_w\|_1 \|w\|_{l_1},\end{aligned}$$

where $\beta = \lambda^+(1 + \max\{\|[I_{n_h}\ \ 0]A_d\|_1, \|[0\ \ I_{n_v}]A_d\|_1\})$, $\lambda_+ = \min \lambda_k$ and $\lambda^+ = \max \lambda_k$. Let $T, K \to \infty$, we get

$$\sum_{i,j=0}^{\infty} \|x(i,j)\|_1 \le \frac{\beta}{\varepsilon\lambda_+}(\|\phi_h\| + \|\phi_v\|) + \frac{\|\lambda^\top B_w\|_1}{\varepsilon\lambda_+}\|w\|_{l_1},$$

which indicates l_1-stability of system (5.4). The proof is completed. □

Remark 5.2 The equivalence of stability notions (GAS), (GES) and (l_1-S) of positive 2-D system (5.4) and LP-based condition (5.10) is illustrated in Fig. 5.2, where solid arrows indicate direct implications and dashed arrows refer to indirect implications. For example, the meaning of a solid arrow and a dashed arrow between (5.10) and (l_1-S) is that LP (5.10) directly implies l_1-S by the proof of Proposition 5.2, while we do not have a direct implication from (l_1-S) to (5.10). Since (l_1-S) and (GES) notations deal with the cases of l_1-disturbances and zero-disturbance, there exist no direct implications between them, which are represented by dashed arrows in Fig. 5.2. In addition, when $w = 0$, the equivalences of (GES) and (GAS) with LP (5.10) are shown directly by the method of [10] (see, also, Chap. 4, Theorem 4.1), respectively, which are also described by solid arrows in Fig. 5.2. For simplicity, hereafter in this chapter, we say that positive system (5.4) is stable if it is GAS, GES or l_1-S.

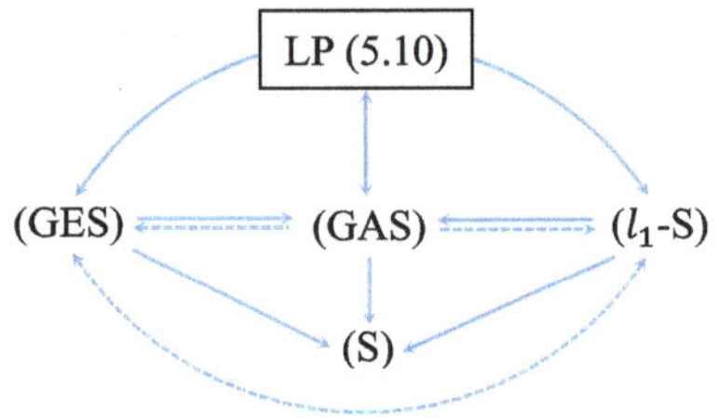

Fig. 5.2 The equivalence of stability notions (l_1-S), (GAS) and (GES) for positive system (5.4)

Remark 5.3 The result of Proposition 5.2 can be extended for positive 2-D systems with multiple delays

$$\begin{bmatrix} x^h(i+1,j) \\ x^v(i,j+1) \end{bmatrix} = A \begin{bmatrix} x^h(i,j) \\ x^v(i,j) \end{bmatrix} + \sum_{k=1}^{N} A_{dk} \begin{bmatrix} x^h(i-\tau_{hk},j) \\ x^v(i,j-\tau_{vk}) \end{bmatrix} + \sum_{k=1}^{M} B_{kw} w_k(i,j). \tag{5.13}$$

Specifically, the positive 2-D system (5.13) is l_1-stable if and only if there exists a vector $\lambda \in \mathbb{R}^n$, $\lambda \succ 0$, such that the following LP-based condition holds

$$\lambda^\top \left(A + \sum_{k=1}^{m} A_{dk} \right) - \lambda^\top \prec 0. \tag{5.14}$$

5.2.4 Stability of 2-D Positive Systems in FMLSS Model with Delays

Let us consider the following 2-D system described by the Fornasini-Marchesini local state-space model

$$\begin{aligned} x(i+1,j+1) = {} & A_1 x(i+1,j) + A_2 x(i,j+1) \\ & + [A_{d1} x(i-\tau_h, j+1) + A_{d2} x(i+1, j-\tau_v)] \\ & + B_{1w} w(i+1,j) + B_{2w} w(i,j+1), \end{aligned} \tag{5.15}$$

where A_k, A_{dk} and B_{kw} ($k = 1, 2$) are given real matrices of appropriate dimensions. For some details description of system (5.15) and its initial conditions, we refer the reader to Chap. 4 (Sect. 4.2.4). In addition, the concepts of (GAS), (GES) and (l_1-S) presented in Definitions 5.2-5.4 can be similarly developed for positive 2-D systems in the form of (5.15).

Proposition 5.3 *The positive 2-D system in FMLSS model* (5.15) *is GES, GAS or l_1-S if and only if one of the following conditions holds*

(i) *There exists a vector* $\eta \in \mathbb{R}^n$, $\eta \succ 0$, *such that*

$$\left(\sum_{k=1}^{2}(A_k + A_{dk}) - I_n\right)\eta \prec 0. \tag{5.16}$$

(ii) *There exists a vector* $\zeta \in \mathbb{R}^n$, $\zeta \succ 0$, *such that*

$$\left(\sum_{k=1}^{2}(A_k + A_{dk}) - I_n\right)^{\top} \zeta \prec 0. \tag{5.17}$$

(iii) *The matrix* $\sum_{k=1}^{2}(A_k + A_{dk}) - I_n$ *is invertible and* $\left(\sum_{k=1}^{2}(A_k + A_{dk}) - I_n\right)^{-1} \preceq 0$.

5.3 l_1-Gain and l_1-Induced Performance

In this section, using a 2-D Z-transform approach, we formulate an explicit expression of l_1-induced norm of input-output operators for 2-D positive time-delay systems in the form of (5.4) and (5.15). The obtained l_1-gain characterization is then utilized to derive necessary and sufficient conditions for l_1-induced performance of 2-D positive time-delay systems subject to a prescribed attenuation level.

5.3.1 l_1-Gain of 2-D Systems Described by the Roesser Model

For a bivariate sequence $\xi : \mathbb{N}_0 \times \mathbb{N}_0 \to \mathbb{R}^n$, we define a 2-D Z-transform

$$\widehat{\xi}(p, q) = Z[f; p, q] = \sum_{i,j=0}^{\infty} \xi(i, j) p^{-i} q^{-j} \tag{5.18}$$

for $(p, q) \in \mathbb{C}^2$ that the concerned series is well-defined.

Let $\xi_s^h(i, j) = \xi(i + 1, j)$ and $\xi_s^v(i, j) = \xi(i, j + 1)$ be one-step shift-forward sequences of $f(i, j)$, then

$$\widehat{\xi_s^h}(p, q) = p\left(\widehat{\xi}(p, q) - \widehat{\xi_0^h}(q)\right)$$

and

$$\widehat{\xi_s^v}(p, q) = q\left(\widehat{\xi}(p, q) - \widehat{\xi_0^v}(p)\right),$$

where $\widehat{\xi_0^h}(q)$ and $\widehat{\xi_0^v}(p)$ are, respectively, 1-D Z-transforms of $\xi_0^h(j) = \xi(0, j)$ and $\xi_0^v(i) = \xi(i, 0)$. In addition, for a given positive integer τ, let $\xi_{-\tau}^h(i, j) = \xi(i - \tau, j)$, $\phi_k(j) = \xi(-k, j)$ and $\xi_{-\tau}^v(i, j) = \xi(i, j - \tau)$, $\psi_k(i) = \xi(i, -k)$ be shift-backward sequences of $\xi(i, j)$, then we have

$$\widehat{\xi_{-\tau}^h}(p, q) = p^{-\tau}\left(\widehat{\xi}(p, q) + \sum_{k=1}^{\tau} \widehat{\phi}_k(q)\right)$$

and

$$\widehat{\xi_{-\tau}^v}(p, q) = q^{-\tau}\left(\widehat{\xi}(p, q) + \sum_{k=1}^{\tau} \widehat{\psi}_k(p)\right).$$

Under the derived condition (5.10), by Proposition 5.2, the positive 2-D system (5.4) is l_1-stable. Thus, for any $w \in l_1(\mathbb{R}^{n_w})$, the 2-D transforms $\widehat{x}(p, q)$ and $\widehat{z}(p, q)$ of $x(i, j)$ and $z(i, j)$ are well-defined for $(p, q) \in \mathbb{C}^2$ with $|p| \geq 1$ and $|q| \geq 1$. In addition, the feasibility of LP (5.10) also ensures that the matrix $I_n - (A + A_d)$ is invertible and

$$(I_n - A - A_d)^{-1} = \sum_{k=0}^{\infty} (A + A_d)^k . \tag{5.19}$$

On the other hand, as a key feature lay behind the positivity, system (5.4) is GAS for any delays τ_h, τ_v if and only if the corresponding delay-free system (i.e. $\tau_h = \tau_v = 0$) is GAS. Thus, by Theorem 1 [30], it follows from condition (5.10) that

$$\det\left[\mathbb{I}(p, q) - A^\top - \mathbb{I}\left(p^{-\tau_h}, q^{-\tau_v}\right) A_d^\top\right] \neq 0$$

for $z = (p, q) \in \mathbb{C}^2$, $|p| \geq 1$ and $|q| \geq 1$, at least in some neighborhood of $z = (1, 1)$, where $\mathbb{I}(z_1, z_2) = \mathrm{diag}(z_1 I_{n_h}, z_2 I_{n_v})$ with $z = (z_1, z_2) \in \mathbb{C}^2$. Thus, the matrix

$$H(p, q) = \mathbb{I}(p, q) - A - A_d\mathbb{I}\left(p^{-\tau_h}, q^{-\tau_v}\right)$$

is invertible.

The following auxiliary lemma is useful for the derivation of main results in this section.

Lemma 5.1 *Let x_w, z_w denote the state and output trajectories of system* (5.4) *with respect to input w and zero initial condition. Assume that system* (5.4) *is positive. Then, for any $w \in l_1(\mathbb{R}^{n_w})$, we have*

$$|x_w(i, j)| \preceq x_{|w|}(i, j) \text{ and } |z_w(i, j)| \preceq z_{|w|}(i, j) \text{ for all } i, j \geq 0.$$

Proof The proof follows the monotonicity induced by the positivity and the fact

$$-\,|w(i,j)| \preceq w(i,j) \preceq |w(i,j)| \text{ for all } i, j \geq 0.$$

The remaining will be left to reader. □

According to Lemma 5.1, we have

$$\|z_w\|_{l_1} = \sum_{i,j=0}^{\infty} \mathbf{1}_{n_z} |z_w(i,j)| \leq \sum_{i,j=0}^{\infty} \mathbf{1}_{n_z} z_{|w|}(i,j) = \|z_{|w|}\|_{l_1}.$$

Thus, to derive an upper bound for l_1-gain $\|\mathscr{L}\|_{l_1-l_1}$, it is only necessary to consider nonnegative inputs $w \in l_1^+(\mathbb{R}^{n_w}) = \{w \in l_1(\mathbb{R}^{n_w}) : w(i,j) \succeq 0\}$. Based on the above discussions, a characterization of l_1-induced norm of the input-output operator will be formulated in the following theorem.

Theorem 5.2 *Assume that system* (5.4) *is positive and one of the derived conditions in* (5.7)–(5.10) *holds. Then, the l_1-gain of system* (5.4) *is obtained as*

$$\|\mathscr{L}\|_{l_1-l_1} = \left\|(C + C_d)\,(I_n - A - A_d)^{-1}\, B_w + D_w\right\|_1. \tag{5.20}$$

Proof Let $w \in l_1^+(\mathbb{R}^{n_w})$ be an arbitrary nonnegative input sequence. Then, under zero initial condition, it follows from (5.4) that $z(i,j) \succeq 0$ for all $i, j \in \mathbb{N}_0$.

Let $\widehat{x}^h$, $\widehat{x}^v$ and $\widehat{w}$ denote the 2-D Z-transforms of x^h, x^v, and w, respectively. Taking 2-D Z-transform both sides of (5.4), with zero initial condition, we obtain

$$\begin{bmatrix} p\widehat{x}^h(p,q) \\ q\widehat{x}^v(p,q) \end{bmatrix} = A \begin{bmatrix} \widehat{x}^h(p,q) \\ \widehat{x}^v(p,q) \end{bmatrix} + A_d \begin{bmatrix} p^{-\tau_h}\widehat{x}^h(p,q) \\ q^{-\tau_v}\widehat{x}^v(p,q) \end{bmatrix} + B_w\widehat{w}(p,q) \tag{5.21}$$

for $(p,q) \in \mathscr{D}_{\widehat{x}} \cap \mathscr{D}_{\widehat{w}}$. Therefore, it follows from (5.21) that

$$H(p,q)\widehat{x}(p,q) = B_w\widehat{w}(p,q), \quad \widehat{x}(p,q) = \begin{bmatrix} \widehat{x}^h(p,q) \\ \widehat{x}^v(p,q) \end{bmatrix}, \tag{5.22}$$

where $H(p,q) = \mathbb{I}(p,q) - A - A_d\mathbb{I}\left(p^{-\tau_h}, q^{-\tau_v}\right)$. Similarly, from (5.2), we have

$$\widehat{z}(p,q) = \left(C + C_d\mathbb{I}\left(p^{-\tau_h}, q^{-\tau_v}\right)\right)\widehat{x}(p,q) + D_w\widehat{w}(p,q). \tag{5.23}$$

A combination of (5.22) and (5.23) gives

$$\widehat{z}(p,q) = \left(G(p,q)H^{-1}(p,q)B_w + D_w\right)\widehat{w}(p,q), \tag{5.24}$$

where $G(p,q) = C + C_d\mathbb{I}\left(p^{-\tau_h}, q^{-\tau_v}\right)$ and $|p| \geq 1$, $|q| \geq 1$.

Let $p = 1$ and $q = 1$, with regard to (5.18), from (5.24), we have

$$\begin{aligned}\|z\|_{l_1} &= \sum_{i,j=0}^{\infty} \|z(i,j)\|_1 \\ &= \sum_{i,j=0}^{\infty} \mathbf{1}_{n_z}^\top z(i,j) = \mathbf{1}_{n_z}^\top \widehat{z}(1,1) \\ &\le \left\| G(1,1)H^{-1}(1,1)B_w + D_w \right\|_1 \|\widehat{w}(1,1)\|_1 \\ &= \left\| (C + C_d)H^{-1}(1,1)B_w + D_w \right\|_1 \|w\|_{l_1}. \end{aligned} \tag{5.25}$$

Based on (5.25), we can conclude that

$$\begin{aligned}\|\mathscr{L}\|_{l_1-l_1} &= \sup_{w\neq 0} \frac{\|z\|_{l_1}}{\|w\|_{l_1}} \\ &\le \left\| (C + C_d)\,(I_n - A - A_d)^{-1}\, B_w + D_w \right\|_1 . \end{aligned} \tag{5.26}$$

On the other hand, for a special $w \in l_1^+(\mathbb{R}^{n_w})$, where $w_0 = w(0,0) \neq 0$ and $w(i,j) = 0$ for $i + j > 0$, we have

$$\|z\|_{l_1} = \mathbf{1}_{n_z}^\top \widetilde{G} w_0,$$

where $\widetilde{G} = (C + C_d)\,(I_n - A - A_d)^{-1}\, B_w + D_w$. Assume that $\|\widetilde{G}\|_1 = \max_{1\le l \le n_w} \|\widetilde{G}_l\|_1$ attains at l_*th column. By taking $w_0 = e_{l_*}$, the l_*th basis vector of $\mathbb{R}^{n_w}$, we have $\widetilde{G} w_0 = \widetilde{G}_{l_*}$ (l_*th column of $\widetilde{G}$). Therefore,

$$\|z\|_{l_1} = \|\widetilde{G}_{l_*}\|_1 = \left\| (C + C_d)\,(I_n - A - A_d)^{-1}\, B_w + D_w \right\|_1 . \tag{5.27}$$

It follows from (5.26) and (5.27) that

$$\|\mathscr{L}\|_{l_1-l_1} = \sup_{\|w\|_{l_1}\neq 0} \frac{\|z\|_{l_1}}{\|w\|_{l_1}} = \left\| (C + C_d)\,(I_n - A - A_d)^{-1}\, B_w + D_w \right\|_1 .$$

The proof is completed. □

5.3.2 l_1-Gain of 2-D Time-Delay Systems Described by FMLSS Model

Consider a 2-D system described by the following FMLSS model

$$\begin{aligned} x(i+1, j+1) &= A_1 x(i+1, j) + A_2 x(i, j+1) \\ &\quad + [A_{d1} x(i - \tau_h, j+1) + A_{d2} x(i+1, j - \tau_v)] \\ &\quad + B_{1w} w(i+1, j) + B_{2w} w(i, j+1), \\ z(i,j) &= C x(i,j) + D_w w(i,j), \end{aligned} \tag{5.28}$$

where $z(i, j) \in \mathbb{R}^{n_z}$ is the output vector, $C \in \mathbb{R}^{n_z \times n}$, $D_w \in \mathbb{R}^{n_z \times n_w}$ are given matrices and other notations are described as in system (5.15).

In this section, the derivation method presented in the preceding section will be extended to 2-D systems in FMLSS model as given in (5.28). Assume that the system (5.28) is positive and, with $w = 0$, is GES. By Proposition 5.3, the matrix $A_s = \sum_{k=1}^{2}(A_k + A_{dk})$ is Schur stable, that is, the spectral radius $\rho(A_s) < 1$. By taking 2-D Z-transformation, from the first equation in (5.28), we obtain

$$\begin{aligned} pq\widehat{x}(p,q) = &\left(pA_1 + qA_2 + p^{-\tau_h}A_{d1} + q^{-\tau_v}A_{d2}\right)\widehat{x}(p,q) \\ &+ (pB_{1w} + qB_{2w})\widehat{w}(p,q) \end{aligned} \tag{5.29}$$

for $(p, q) \in \mathbb{C}^2$ that the concerned transformations are well-defined. It follows from (5.29) that

$$\underbrace{\left(pqI_n - J(p,q)\mathscr{A} - J(p^{-\tau_h}, q^{-\tau_v})\mathscr{A}_d\right)}_{E(p,q)}\widehat{x}(p,q) = J(p,q)\mathscr{B}_w\widehat{w}(p,q), \tag{5.30}$$

where we denote $J(p, q) = \lfloor pI_n \;\; qI_n \rfloor$ and

$$\mathscr{A} = \begin{bmatrix} A_1 \\ A_2 \end{bmatrix}, \quad \mathscr{A}_d = \begin{bmatrix} A_{d1} \\ A_{d2} \end{bmatrix}, \quad \mathscr{B}_w = \begin{bmatrix} B_{1w} \\ B_{2w} \end{bmatrix}.$$

Thus, according to (5.30), we have

$$\widehat{x}(p,q) = E^{-1}(p,q)J(p,q)\mathscr{B}_w\widehat{w}(p,q). \tag{5.31}$$

In addition, from the second equation in (5.28), we also have

$$\widehat{z}(p,q) = C\widehat{x}(p,q) + D_w\widehat{w}(p,q),$$

which, together (5.31), yields the following expression for the 2-D transform of the input-output operator

$$\widehat{z}(p,q) = \left[CE^{-1}(p,q)J(p,q)\mathscr{B}_w + D_w\right]\widehat{w}(p,q). \tag{5.32}$$

Based on the representation obtained in (5.32), and by similar arguments in the proof of Theorem 5.2, we have the following result.

Theorem 5.3 *Assume that system* (5.28) *is positive and one of the derived conditions in Proposition 5.3 holds. Then, the l_1-gain of system* (5.28) *is obtained as*

$$\|\mathscr{L}\|_{l_1-l_1} = \left\|C\left(I_n - A_s\right)^{-1}B_{ws} + D_w\right\|_1, \tag{5.33}$$

where $A_s = \sum_{k=1}^{2}(A_k + A_{dk})$ and $B_{ws} = B_{1w} + B_{2w}$.

Proof The proof is similar to that of Theorem 5.2. We omit is here. □

Remark 5.4 The result of Theorem 5.3 can be extended to 2-D systems in FMLSS model with multiple delays of the form

$$
\begin{aligned}
x(i+1, j+1) = {} & A_1 x(i+1, j) + A_2 x(i, j+1) \\
& + \sum_{k=1}^{N} [A_{d1k} x(i - \tau_{hk}, j+1) + A_{d2k} x(i+1, j - \tau_{vk})] \\
& + \sum_{k=1}^{M} [B_{1kw} w_k(i+1, j) + B_{2kw} w_k(i, j+1)], \\
z(i, j) = {} & Cx(i, j) + \sum_{k=1}^{M} D_{kw} w_k(i, j),
\end{aligned}
\tag{5.34}
$$

where $w_k \in l_1(\mathbb{R}^{n_{w_k}})$. To this case, the l_1-gain derived in (5.33) becomes

$$
\|\mathscr{L}\|_{l_1 - l_1} = \left\| C \left(I_n - \widetilde{A}_s \right)^{-1} \widetilde{B}_{ws} + \widetilde{D}_w \right\|_1, \tag{5.35}
$$

where

$$
\begin{aligned}
\widetilde{A}_s &= A_1 + A_2 + \sum_{k=1}^{N} (A_{d1k} + A_{d2k}), \\
\widetilde{B}_{ws} &= \left[B_{11w} + B_{21w} \ \cdots \ B_{1Mw} + B_{2Mw} \right], \\
\widetilde{D}_w &= \left[D_{1w} \ D_{2w} \ \cdots \ D_{Mw} \right].
\end{aligned}
$$

Remark 5.5 In [23, 24], some l_1-gain results were derived for 2-D positive systems in FMLSS models with or without delays. By using lifting techniques to recast the considered systems into higher delay-free 2-D systems, expressions of exact value of l_1-gain were obtained. It is noted that the results proposed in [23, 24] can be attained from those given in (5.33) and (5.35) as particular cases. In addition, different from the method based on lifting techniques [23, 24], the method using 2-D Z-transform presented in this chapter can be easily adapted for 2-D systems with delays in both Roesser and FMLSS models, which yields explicit expressions of the transform of input-output operators. On the basis of such an representation, characterizations of l_1-gain can be obtained by simple manipulation processes.

5.3.3 l_1-Induced Performance

Theorems 5.2 and 5.3 provide a characterization of exact value of l_1-gain for 2-D positive systems with delays in Roesser model (5.4) and FMLSS model (5.28), respectively. Such a representation is still not tractable for the design problem as it is involved calculations of norms and inversions of parameter-dependent matrices, which are realized as NP-hard problems. In this section, the results of Theorems 5.2 and 5.3 will be employed to derive a necessary and sufficient conditions for l_1-induced performance with prescribed attenuation levels of 2-D positive systems with delays.

Theorem 5.4 *Assume that system* (5.4) *is positive. Then, for a given* $\gamma > 0$, *system* (5.4) *is stable and has an* l_1*-induced performance at level* γ *if and only if there exists a vector* $\mu \in \mathbb{R}^n$, $\mu \succ 0$, *satisfying the following conditions*

$$(A + A_d - I_n)^\top \mu + (C + C_d)^\top \mathbf{1}_{n_z} \prec 0, \tag{5.36}$$

$$B_w^\top \mu + D_w^\top \mathbf{1}_{n_z} - \gamma \mathbf{1}_{n_w} \prec 0. \tag{5.37}$$

Proof To prove the *Necessity*, we notice at first that the positive 2-D system (5.4) is stable if and only if one of the equivalent conditions derived in (5.7)–(5.10) is satisfied. Let $\lambda \in \mathbb{R}^n$ be a feasible vector of (5.10). For given $\varepsilon > 0$, we define the following vector

$$\mu = \varepsilon\lambda + \left(I_n - A_a^\top\right)^{-1} C_a^\top \mathbf{1}_{n_z},$$

where $A_a = A + A_d$ and $C_a = C + C_d$. Then, we have

$$(A_a - I_n)^\top \mu + C_a^\top \mathbf{1}_{n_z} = \varepsilon \left(A_a^\top \lambda - \lambda\right) \prec 0 \tag{5.38}$$

according to condition (5.10).

On the other hand, by Theorem 5.2, system (5.4) has l_1-induced performance at level γ, that is, $\|\mathscr{L}\|_{l_1-l_1} < \gamma$, if and only if

$$\left\| C_a(I_n - A_a)^{-1} B_w + D_w \right\|_1 < \gamma. \tag{5.39}$$

By the definition of 1-norm of a matrix, condition (5.39) holds if and only if

$$\eta^\top \triangleq \mathbf{1}_{n_z}^\top C_a(I_n - A_a)^{-1} B_w + \mathbf{1}_{n_z}^\top D_w - \gamma \mathbf{1}_{n_w}^\top \prec 0. \tag{5.40}$$

From (5.40) we readily obtain

$$B_w^\top \mu + D_w^\top \mathbf{1}_{n_z} - \gamma \mathbf{1}_{n_w} = \varepsilon B_w^\top \lambda + \eta \prec 0 \tag{5.41}$$

for sufficiently small $\varepsilon > 0$. Thus, the necessity of (5.36)–(5.37) clearly follows from (5.38) and (5.41).

We now prove the *Sufficiency*. Note from (5.36) that

$$\mu^\top (A + A_d - I_n) \prec 0.$$

Thus, condition (5.8) is satisfied. By Theorem 5.1, the positive system (5.4) is stable. To show the γ level of l_1-induced performance, observe from (5.25) that

$$\widehat{z}(1,1) = \left[C_a (I_n - A_a)^{-1} B_w + D_w\right] \widehat{w}(1,1).$$

If $B_w \widehat{w}(1,1) = 0$ then

$$\|z\|_{l_1} = \mathbf{1}_{n_z}^\top \widehat{z}(1,1) = \mathbf{1}_{n_z}^\top D_w \widehat{w}(1,1).$$

In addition, since $\widehat{w}(1,1) \succeq 0$ and, by (5.37), $\mathbf{1}_{n_z}^\top D_w \prec \gamma \mathbf{1}_{n_w}^\top - \mu^\top B_w$, we have

$$\begin{aligned}\|z\|_{l_1} &< \left(\gamma \mathbf{1}_{n_w}^\top - \mu^\top B_w\right) \widehat{w}(1,1) \\ &= \gamma \mathbf{1}_{n_w}^\top \widehat{w}(1,1) = \gamma \|w\|_{l_1}.\end{aligned}$$

We now assume that $B_w \widehat{w}(1,1) \neq 0$. Let

$$\chi^\top \triangleq \mathbf{1}_{n_z}^\top C_a - \mu^\top (I_n - A_a)$$

then $\chi \prec 0$ and we have

$$\mu^\top = \mathbf{1}_{n_z}^\top C_a (I_n - A_a)^{-1} - \chi^\top (I_n - A_a)^{-1}.$$

Therefore,

$$\mu^\top B_w + \mathbf{1}_{n_z}^\top D_w = \mathbf{1}_{n_z}^\top \left[C_a (I_n - A_a)^{-1} B_w + D_w\right] - \chi^\top (I_n - A_a)^{-1} B_w.$$

This gives

$$\begin{aligned}\|z\|_{l_1} &\leq \chi^\top (I_n - A_a)^{-1} B_w \widehat{w}(1,1) + \gamma \mathbf{1}_{n_w}^\top \widehat{w}(1,1) \\ &= \chi^\top (I_n - A_a)^{-1} B_w \widehat{w}(1,1) + \gamma \|w\|_{l_1}.\end{aligned}$$

Since $(I_n - A_a)^{-1} \succeq 0$, $\chi \prec 0$, $B_w \widehat{w}(1,1) \succeq 0$ and $B_w \widehat{w}(1,1) \neq 0$, from the above derivation, we have that

$$\|z\|_{l_1} < \gamma \|w\|_{l_1}.$$

The proof is completed. □

The l_1-gain performance characterization for positive 2-D systems in FMLSS model as (5.34) is presented as the following theorem.

Theorem 5.5 *For a given $\gamma > 0$, the positive 2-D system* (5.34) *is stable and has an l_1-induced performance at level γ if and only if there exists a vector $\eta \in \mathbb{R}^n$, $\eta \succ 0$, that satisfies the following conditions*

$$\mathbf{1}_{n_z}^\top C + \eta^\top \widetilde{A}_s - \eta^\top \prec 0, \tag{5.42}$$

$$\eta^\top \widetilde{B}_{ws} + \mathbf{1}_{n_z}^\top \widetilde{D}_w - \gamma \mathbf{1}_{n_w}^\top \prec 0, \tag{5.43}$$

where $n_w = \sum_{k=1}^{M} n_{w_k}$ and other matrices are as defined in Remark 5.4.

Proof The proof is similar to that of Theorem 5.4. We omit it here. □

5.4 State-Feedback l_1-Gain Control

In this section, we deal with the problem of l_1-gain control of positive 2-D systems with delays. l_1-gain performance characterizations derived in the preceding section will be employed to derive necessary and sufficient conditions for the existence of a state-feedback controller (SFC) that makes the resultant closed-loop systems positive, stable and have a prescribed l_1-induced performance level. The proposed results for positive 2-D systems described by Roesser and FMLSS models are established via linear optimization procedures and linear programming settings.

5.4.1 General Formulation

Consider a 2-D positive time-delay system in the form of (5.1)–(5.2). An SFC $u(i, j) = -Kx(i, j)$ will be designed to make the closed-loop system

$$\begin{bmatrix} x^h(i+1, j) \\ x^v(i, j+1) \end{bmatrix} = \underbrace{(A - BK)}_{A_c} \begin{bmatrix} x^h(i, j) \\ x^v(i, j) \end{bmatrix} + A_d \begin{bmatrix} x^h(i-\tau_h, j) \\ x^v(i, j-\tau_v) \end{bmatrix} + B_w w(i, j), \tag{5.44}$$

$$z(i, j) = \underbrace{(C - DK)}_{C_c} \begin{bmatrix} x^h(i, j) \\ x^v(i, j) \end{bmatrix} + C_d \begin{bmatrix} x^h(i-\tau_h, j) \\ x^v(i, j-\tau_v) \end{bmatrix} + D_w w(i, j), \tag{5.45}$$

positive, stable and has prescribed l_1-induced performance.

We say that the l_1-induced control problem of system (5.1)–(5.2) is *solvable* if there exists an SFC $u(i, j)$ that makes the closed-loop system (5.44)–(5.45) positive, stable (in sense of GAS, GES or l_1-S) and has a prescribed l_1-induced performance. By Proposition 5.1 and Theorem 5.4, the closed-loop system (5.44)–(5.45) is positive, stable and has l_1-induced performance at given level $\gamma > 0$ if and only if there exist a matrix $K \in \mathbb{R}^{n_u \times n}$ and a vector $\mu \in \mathbb{R}^n$, $\mu \succ 0$, such that the following conditions are satisfied

$$\begin{bmatrix} A \\ C \end{bmatrix} - \begin{bmatrix} B \\ D \end{bmatrix} K \succeq 0, \tag{5.46}$$

$$\left(\begin{bmatrix} A_a - I_n \\ C_a \end{bmatrix}^\top - K^\top \begin{bmatrix} B \\ D \end{bmatrix}^\top\right) \begin{bmatrix} \mu \\ \mathbf{1}_{n_w} \end{bmatrix} \prec 0, \tag{5.47}$$

$$\begin{bmatrix} B_w^\top & D_w^\top \end{bmatrix} \begin{bmatrix} \mu \\ \mathbf{1}_{n_w} \end{bmatrix} - \gamma \mathbf{1}_{n_w} \prec 0, \tag{5.48}$$

where $A_a = A + A_d$, $C_a = C + C_d$.

Condition (5.47) belongs to a type of bilinear matrix inequalities (BMIs). In general, the feasibility of BMIs has been recognized as an NP-hard problem. Thus, condition (5.47) is still not tractable for validation of the existence of a desired SFC. In the following, we give a complete description for the solution set of (5.46) and recast equivalently conditions (5.46)–(5.48) into LP-based conditions. For this purpose, we define the matrices

$$L = \begin{bmatrix} A \\ C \end{bmatrix}, \quad H = \begin{bmatrix} B \\ D \end{bmatrix}$$

and assume that the matrix H has full-column rank. This assumption encompasses the case of single-input multiple-output (SIMO) systems and is widely adopted in the area of controller design. We decompose the matrices K and L by columns as

$$K = \begin{bmatrix} k_1 & k_2 & \cdots & k_n \end{bmatrix}, \quad k_j \in \mathbb{R}^{n_u},$$
$$L = \begin{bmatrix} L_1 & L_2 & \cdots & L_n \end{bmatrix}, \quad L_j \in \mathbb{R}^{n+n_z}.$$

Then, condition (5.46) is equivalent to the following system

$$\begin{cases} Hk_j = z_j \\ z_j \preceq L_j, \quad j \in [n]. \end{cases} \tag{5.49}$$

The following lemma will be employed to describe feasible domain of (5.49).

Lemma 5.2 *[31] Let H^g be a generalized inverse of H (i.e. $HH^gH = H$). Then, for a given vector $z \in \mathbb{R}^{n+n_z}$, the linear system $Hk = z$ has a solution if and only if $HH^gz = z$. Moreover, the complete set of solutions to $Hk = z$ is given by*

$$k = H^g z + \left(I_{n_u} - H^g H\right) \hat{z},$$

where $\hat{z} \in \mathbb{R}^{n_u}$ is an arbitrary vector.

In view of Lemma 5.2, since H has full-column rank, the left inverse

$$H^\dagger = \left(H^\top H\right)^{-1} H^\top$$

of H exists and the system $Hk_j = z_j$ has a unique solution $k_j = H^\dagger z_j$ subject to $HH^\dagger z_j = z_j$. Therefore, system (5.49) can be written as

$$k_j = H^\dagger z_j, \quad z_j \in \mathscr{N}\left(I_{n+n_z} - HH^\dagger\right) \cap \mathbb{R}^{n+n_z}_{\preceq L_j}, \quad j \in [n], \tag{5.50}$$

where $\mathscr{N}(M)$ denotes the null-space of matrix M and

$$\mathbb{R}^{n+n_z}_{\preceq L_j} = \left\{x \in \mathbb{R}^{n+n_z} : x \preceq L_j\right\}, \quad j \in [n].$$

Note that $P = HH^\dagger$ is a projection of $\mathbb{R}^{n+n_z}$ onto the range-space $\mathscr{R}(H)$. Thus, $Q = I_{n+n_z} - HH^\dagger$ is also a projection of $\mathbb{R}^{n+n_z}$ and $\mathscr{N}(Q) = \mathscr{R}(H)$. By this, complete solution set of (5.46) can be described as

$$K = H^\dagger Z, \quad Z = \begin{bmatrix} z_1 & z_2 & \cdots & z_n \end{bmatrix}, \quad z_j \in \mathscr{R}(H) \cap \mathbb{R}^{n+n_z}_{\preceq L_j}, \quad j \in [n]. \tag{5.51}$$

According to (5.51), we have $HK = Z$. Thus, condition (5.47) is reduced to the following condition

$$\left(\begin{bmatrix} A_a - I_n \\ C_a \end{bmatrix}^\top - Z^\top\right)\begin{bmatrix} \mu \\ \mathbf{1}_{n_z} \end{bmatrix} \prec 0. \tag{5.52}$$

In addition, if $Z \preceq \widetilde{Z}$, where $z_j, \widetilde{z}_j \in \mathscr{R}(H) \cap \mathbb{R}^{n+n_z}_{\preceq L_j}$, $j \in [n]$, then

$$Z^\top \begin{bmatrix} \mu \\ \mathbf{1}_{n_z} \end{bmatrix} \preceq \widetilde{Z}^\top \begin{bmatrix} \mu \\ \mathbf{1}_{n_z} \end{bmatrix}$$

for any $\mu \in \mathbb{R}^n$, $\mu \succ 0$. In other words, if (5.52) is feasible with Z for some positive vector μ then it is also feasible with $\widetilde{Z}$. This allows us to derive necessary and sufficient tractable conditions for l_1-induced performance by maximizing Z in (5.51).

5.4.2 *The Case of SIMO 2-D Systems*

To illustrate the proposed ideas, in this section, we address the solvability control problem for SIMO 2-D systems in the form of (5.1)–(5.2). To this case, we have $n_u = 1$ and B, D are column-matrices. Specifically, let

$$b \triangleq B = (b_k) \in \mathbb{R}^n_+ \text{ and } d \triangleq D = (d_l) \in \mathbb{R}^{n_z}_+.$$

Then, we have

$$H^\dagger = \frac{1}{\delta}\begin{bmatrix} b^\top & d^\top \end{bmatrix}, \quad \delta = b^\top b + d^\top d > 0,$$

and

$$\mathscr{R}(H) \cap \mathbb{R}^{n+n_z}_{\preceq L_j} = \left\{\lambda \begin{bmatrix} b \\ d \end{bmatrix} : \lambda \in \mathbb{R}, \ \lambda \leq \delta_j^*\right\}, \tag{5.53}$$

where $\delta_j^* = \min_{i,l}\left\{\frac{a_{ij}}{b_i}, \frac{c_{lj}}{d_l} \;\middle|\; b_i d_l \neq 0\right\}$. We summarize this result in the following proposition.

Proposition 5.4 *The problem of l_1-induced control of SIMO positive 2-D systems given in* (5.1)–(5.2) *is solvable if and only if the LP-based problem governed by* (5.48) *and the following condition*

$$\left(\begin{bmatrix}(A_a - I_n)\\ C_a\end{bmatrix}^\top - Z_*^\top\right)\begin{bmatrix}\mu\\ \mathbf{1}_{n_z}\end{bmatrix} \prec 0, \tag{5.54}$$

where

$$Z_* = \begin{bmatrix}z_1^* & z_2^* & \cdots & z_n^*\end{bmatrix} \text{ and } z_j^* = \delta_j^*\begin{bmatrix}b\\ d\end{bmatrix},\ j \in [n],$$

is feasible for a vector $\mu \in \mathbb{R}^n$, $\mu \succ 0$. The optimal controller gain is obtained as

$$K_{op}^s = \begin{bmatrix}\delta_1^* & \delta_2^* & \cdots & \delta_n^*\end{bmatrix}. \tag{5.55}$$

Proof The proof can be deduced from (5.51) and (5.52). We omit some details here. □

Remark 5.6 Let $K = H^\dagger Z$, given in (5.51), be the gain of a desired SFC for SIMO positive 2-D systems (5.1). If there exists a column $0 \neq z_j \preceq 0$ then we can replace it by zero. Thus, the l_1-induced control problem of SIMO positive 2-D systems is solvable if and only if it is solvable with nonnegative SFCs. Regarding to this fact, the set of admissible controller gains can be characterized as $K^s = \begin{bmatrix}\delta_1 & \delta_2 & \cdots & \delta_n\end{bmatrix}$, where $0 \leq \delta_j \leq \delta_j^*$.

5.4.3 The Case of MIMO 2-D Systems

In applications to compartmental systems, there is typically only one controlling input per compartment. For such scenarios, the matrix H has *unitary rows*. This also includes SIMO systems as a special case. Extending from Proposition 5.4, we have the following result.

Theorem 5.6 *The l_1-induced control problem of system* (5.1)–(5.2)*, where B and D have unitary rows, is solvable if and only if the LP conditions* (5.48) *and* (5.54) *are feasible, where $z_j^* = Hx_j^*$, $j \in [n]$, and*

$$x_{ij}^* = \min_{k,l}\left\{\frac{a_{kj}}{b_{ki}}, \frac{c_{lj}}{d_{li}} \;\middle|\; b_{ki}d_{li} \neq 0\right\},\ i \in [m]. \tag{5.56}$$

The optimal controller gain is given by

$$K_{op}^{s} = \begin{bmatrix} x_1^* & x_2^* & \cdots & x_n^* \end{bmatrix}. \tag{5.57}$$

Proof By the assumption that the matrix H has unitary rows, it can be deduced that

$$\mathscr{R}(H) \cap \mathbb{R}^{n+n_z}_{\preceq L_j} = \left\{ Hx : x \in \mathbb{R}^{n_u},\ x \preceq x_j^* \right\},$$

and hence $Z \preceq Z_*$. Clearly, condition (5.52) is feasible if and only if condition (5.54) is feasible. In addition,

$$K_{op}^{s} = H^{\dagger} Z_* = \begin{bmatrix} x_1^* & x_2^* & \cdots & x_n^* \end{bmatrix}.$$

The proof is completed. □

Remark 5.7 To derive complete solutions of Proposition 5.4 and Theorem 5.6, the key feature is that the matrix H has nonnegative left inverse $H^{\dagger} \succeq 0$ if and only if, up to a permutation, it has unitary rows [26]. In general case, when $H^{\dagger} \not\succeq 0$, it is more challenging to equivalently recast simultaneously conditions (5.46) and (5.47) into a unique coupled LP condition.

As discussed earlier, to verify the existence of a stabilizing matrix Z satisfying (5.51) and (5.52), it is only necessary to consider the case that each column z_j of Z, which belongs to $\mathscr{R}(H)$, is not negative ($z_j \not\preceq 0$). Meanwhile, we would like to maximize Z in feasible solution set. This suggests an objective function as $\begin{bmatrix} \mu^{\top} & \mathbf{1}_{n_z}^{\top} \end{bmatrix} z_j$ for a fixed vector $\mu \in \mathbb{R}^n$, $\mu \succ 0$, an optimization problem

$$\texttt{maximize} \left\{ \begin{bmatrix} \mu^{\top} & \mathbf{1}_{n_z}^{\top} \end{bmatrix} (Hx) \ \texttt{s.t.}\ \ x \in \mathbb{R}^{n_u} :\ Hx \preceq L_j \right\}. \tag{5.58}$$

Lemma 5.3 *Assume that* $\text{rank}(H) = n_u$. *Then, there exists a scalar* $\lambda_0 > 0$ *such that* $\|Hx\|_1 \geq \lambda_0 \|x\|_1$ *for all* $x \in \mathbb{R}^{n_u}$.

Proof Let $\overline{B} = \{x \in \mathbb{R}^{n_u} : \|x\|_1 = 1\}$ be the unit sphere in $\mathbb{R}^{n_u}$. Since the function $\varphi : \overline{B} \to [0, \infty)$, $\varphi(x) = \|Hx\|_1$, is continuous on the compact $\overline{B}$, there exists $x_0 \in \overline{B}$ such that $\varphi(x_0) = \min_{\overline{B}} \varphi(x)$. If $\varphi(x_0) = 0$ then $Hx_0 = 0$, and thus $x_0 = 0$ as $\mathscr{N}(H) = 0$. This yields a contradiction since $x_0 \in \overline{B}$. Therefore, $\lambda_0 \triangleq \varphi(x_0) > 0$. For any $x \in \mathbb{R}^{n_u}$, $x \neq 0$, let $\tilde{x} = x/\|x\|_1 \in \overline{B}$, we have

$$\|Hx\|_1 = \varphi(\tilde{x})\|x\|_1 \geq \lambda_0 \|x\|_1.$$

The proof is completed. □

It is clear that $\mathscr{P}_j = \left\{ x \in \mathbb{R}^{n_u} :\ Hx \preceq L_j \right\}$ defines a nonempty convex polyhedron in $\mathbb{R}^{n_u}$ and $\begin{bmatrix} \mu^{\top} & \mathbf{1}_{n_z}^{\top} \end{bmatrix} (Hx) \geq 0$ for any efficient point of (5.58). By defining the index sets

$$J_+ = \{k : (Hx)_k \geq 0\} \text{ and } J_- = \{l : (Hx)_l < 0\},$$

we have

$$\sum_{l\in J_-}\begin{bmatrix}\mu\\ \mathbf{1}_{n_z}\end{bmatrix}_l (Hx)_l \geq -\sum_{k\in J_+}\begin{bmatrix}\mu\\ \mathbf{1}_{n_z}\end{bmatrix}_k (Hx)_k \geq -\sum_{k\in J_+}\begin{bmatrix}\mu\\ \mathbf{1}_{n_z}\end{bmatrix}_k L_{kj}.$$

The last inequality shows that

$$\mathbf{1}_{n+n_z}^\top Hx \geq -\mu^*\|L_j\|_1,$$

where $\mu^* = \max(\mu_i, 1)/\min(\mu_i, 1)$. Thus, the problem (5.58) is equivalent to that of maximizing the function $\begin{bmatrix}\mu^\top & \mathbf{1}_{n_z}^\top\end{bmatrix}(Hx)$ on the polyhedron

$$\widetilde{\mathscr{P}}_j = \left\{x \in \mathscr{P}_j : \mathbf{1}_{n+n_z}^\top Hx \geq -\mu^*\|L_j\|_1\right\}.$$

Since

$$\|Hx\|_1 = \sum_{k\in J_+}(Hx)_k - \sum_{l\in J_-}(Hx)_l \leq (1+\mu^*)\|L_j\|_1,$$

by Lemma 5.3, $\|x\|_1 \leq (1+\mu^*)\|L_j\|_1/\lambda_0$. This shows that $\widetilde{\mathscr{P}}_j$ is a convex compact subset of $\mathbb{R}^{n_u}$. Using the fact that the function $\begin{bmatrix}\mu^\top & \mathbf{1}_{n_z}^\top\end{bmatrix}(Hx)$ is continuous on the compact $\widetilde{\mathscr{P}}_j$, the solution set of (5.58) is nonempty. Let V_j denote the set of vertices of $\mathscr{P}_j$.

Lemma 5.4 *Let $x_* \in \mathscr{P}_j$ be a point that maximizes the function $h_\mu(x) = \begin{bmatrix}\mu^\top & \mathbf{1}_{n_z}^\top\end{bmatrix}Hx$ on $\mathscr{P}_j$. Then, there exists a vertex $x_j \in V_j$ such that $h_\mu(x_j) = h_\mu(x_*)$.*

Proof Let $h = H^\top \mathbf{1}_{n+n_z} \in \mathbb{R}^{n_u}$. If x_* is an interior point of $\mathscr{P}_j$ then, for sufficiently small $\varepsilon > 0$, $x_\varepsilon = x_* + \varepsilon h \in \mathscr{P}_j$, and we have

$$\begin{aligned} h_\mu(x_\varepsilon) &= \begin{bmatrix}\mu^\top & \mathbf{1}_{n_z}^\top\end{bmatrix} H\big(x_* + \varepsilon h\big) \\ &\geq h_\mu(x_*) + \varepsilon\underline{\mu}\|h\|_2^2 > h_\mu(x_*), \end{aligned}$$

where $\underline{\mu} = \min\{\mu_i, 1\}$. This contradicts with $h_\mu(x_*)$ is the maximum of $h_\mu(x)$ on $\mathscr{P}_j$. Thus, x_* must fall within the boundary of $\mathscr{P}_j$. By this fact we can represent $x_* = \sum_{l=1}^p \lambda_l x_l$ with $x_l \in V_j$ and $\lambda_l \geq 0$, $\sum_{l=1}^p \lambda_l = 1$. This leads to

$$0 = h_\mu(x_*) - \sum_{l=1}^p \lambda_l h_\mu(x_l) = \sum_{l=1}^p \lambda_l \underbrace{\big(h_\mu(x_*) - h_\mu(x_l)\big)}_{\geq 0}.$$

It is clear from the above presentation that $h_\mu(x_*) = h_\mu(x_l)$ corresponding to $\lambda_l > 0$. The proof is completed. □

Remark 5.8 Lemma 5.4 can be interpreted that the objective function $h_\mu(x)$ attains its maximum at one of the vertices of polyhedron $\mathscr{P}_j$. This result is known as the fundamental theorem of linear programming.

Based on the above derivation, a complete solution to l_1-gain control problem of MIMO 2-D systems is obtained as the following theorem.

Theorem 5.7 *Assume that the matrix* $H = \begin{bmatrix} B \\ D \end{bmatrix}$ *has full-column rank. The problem of* l_1*-induced control of positive system* (5.1)–(5.2) *is solvable if and only if there exists a set of vertices* $x_j^* \in V_j$, $j \in [n]$, *such that the LP-based conditions* (5.48) *and* (5.52) *are feasible for a vector* $0 \prec \mu \in \mathbb{R}^n$, *where* $Z = H \begin{bmatrix} x_1^* & x_2^* & \cdots & x_n^* \end{bmatrix}$. *A desired controller gain is obtained as*

$$K^s = \begin{bmatrix} x_1^* & x_2^* & \cdots & x_n^* \end{bmatrix}. \tag{5.59}$$

Proof Let K be the gain of a stabilizing controller which solves the l_1-induced control problem of system (5.1)–(5.2). According to (5.51), we have

$$\begin{aligned} K &= H^\dagger Z, \quad Z = \begin{bmatrix} z_1 & z_2 & \cdots & z_n \end{bmatrix}, \\ z_j &= H w_j \preceq L_j, \quad w_j \in \mathbb{R}^{n_u}, \ j \in [n]. \end{aligned}$$

In addition, condition (5.52) is feasible for some vector $\mu \in \mathbb{R}^n$, $\mu \succ 0$. If $w_j \notin V_j$ then we set the LP problem (5.58). By Lemma 5.4, there exists a vertex $x_j^* \in V_j$ such that

$$\begin{bmatrix} \mu^\top & \mathbf{1}_{n_z}^\top \end{bmatrix} H x_j^* = \max \left\{ \begin{bmatrix} \mu^\top & \mathbf{1}_{n_z}^\top \end{bmatrix} H x : x \in \mathscr{P}_j \right\}.$$

It is clear that

$$\begin{aligned} \begin{bmatrix} \mu^\top & \mathbf{1}_{n_z}^\top \end{bmatrix} z_j &= \begin{bmatrix} \mu^\top & \mathbf{1}_{n_z}^\top \end{bmatrix} H w_j \\ &\le \begin{bmatrix} \mu^\top & \mathbf{1}_{n_z}^\top \end{bmatrix} H x_j^*. \end{aligned}$$

By applying this process for other columns of Z we can conclude that there exists a set of vertices $x_j^* \in V_j$, $j \in [n]$, such that

$$\begin{bmatrix} \mu^\top & \mathbf{1}_{n_z}^\top \end{bmatrix} Z \preceq \begin{bmatrix} \mu^\top & \mathbf{1}_{n_z}^\top \end{bmatrix} Z^s,$$

where

$$Z^s = H \begin{bmatrix} x_1^* & x_2^* & \cdots & x_n^* \end{bmatrix}.$$

Moreover, the LP (5.52) is also feasible with respect to Z^s. This means that the gain matrix $K^s = H^\dagger Z^s$ also solves the l_1-induced control problem of system (5.1)–(5.2). This shows the *Necessity*. The *Sufficiency* is obvious. The proof is completed. □

Remark 5.9 Each vertex $x_j^* \in V_j$ in (5.59) is typically selected as an optimal solution to the problem $\max_{x \in \mathscr{P}_j} \left(\mathbf{1}_n^\top B + \mathbf{1}_{n_z}^\top D \right) x$. For a selected matrix $Z = H K^s$, the l_1-performance index γ can be minimized by the following LP problem:

$$\min \gamma > 0 \ \texttt{subject to}$$

$$\begin{cases} \left(A_a^\top - (BK^s)^\top - I_n\right)\mu + C_a^\top \mathbf{1}_{n_z} + \varepsilon_1 \mathbf{1}_n \preceq (DK^s)^\top \mathbf{1}_{n_z}, \\ B_w^\top \mu + D_w^\top \mathbf{1}_{n_z} + \varepsilon_2 \mathbf{1}_{n_w} \preceq \gamma \mathbf{1}_{n_w}, \\ \mu \succ 0, \end{cases} \tag{5.60}$$

where ε_1, ε_2 are prescribed modification parameters.

Remark 5.10 Since the controller gain K^s in (5.59) is sign-indefinite, for MIMO positive 2-D systems as (5.1)–(5.2), we cannot restrict to only consider nonnegative gain matrices as that for SIMO systems. This makes the problem of jointly minimizing all elements of $A - BK$ and $C - DK$ subject to l_1-induced performance is much more challenging.

Remark 5.11 The problem of l_1-induced control was studied for LTI positive 1-D systems without delay in [22]. Using a structured constraint, the controller gain K is taken in the form $K = \Pi\Delta$, where $\Pi = \begin{bmatrix} \pi_1 & \pi_2 & \cdots & \pi_n \end{bmatrix} \in \mathbb{R}^{n_u \times n}$, $\Pi \succ 0$, is referred to as a fixed *direction matrix* and $\Delta = \mathrm{diag}(\delta_1, \delta_2, \ldots, \delta_n)$ to be determined. With $B\pi_j \neq 0$, an optimal structured matrix K is obtained by $\Delta^* = \mathrm{diag}(\delta_1^*, \ldots, \delta_n^*)$, where $\delta_j^* = \min_{i,l}\left\{\frac{a_{ij}}{r_{bi}^\top \pi_j}, \frac{c_{lj}}{r_{dl}^\top \pi_j}\right\}$, $r_{bi}^\top$, $i \in [n]$, and $r_{dl}^\top$, $l \in [n_z]$, are rows of B and D, respectively. This method, on one hand, only provides sufficient conditions as $K = \Pi\Delta^* \succeq 0$ is not an optimal solution in general. On the other hand, when the closed-loop system is unstable with a fixed directional matrix Π (i.e. the spectral radius $\rho(A - B\Pi\Delta^*) \geq 1$), there does not exist any criterion to search for a desired matrix Π. Theorem 5.7 in this chapter provides a complete solution to l_1-induced control of 2-D positive systems with delays, which can also be adapted for positive 1-D systems with or without delay.

5.4.4 l_1-Gain Control of SIMO 2-D Systems in FMLSS Model

By a similar approach using linear optimization procedures, in this section, we address the l-induced control problem for a class of positive 2-D systems with delays described by the FMLSS model

$$\begin{aligned} x(i+1, j+1) &= A_1 x(i+1, j) + A_2 x(i, j+1) \\ &\quad + [A_{d1} x(i - \tau_h, j+1) + A_{d2} x(i+1, j - \tau_v)] \\ &\quad + B_1 u(i+1, j) + B_2 u(i, j+1), \\ &\quad + B_{1w} w(i+1, j) + B_{2w} w(i, j+1), \\ z(i, j) &= Cx(i, j) + Du(i, j) + D_w w(i, j), \end{aligned} \tag{5.61}$$

where $u(i, j)$ is the control input, $B_1 = (b_{1k})$, $B_2 = (b_{2k}) \in \mathbb{R}^n$ and $D = (d_k) \in \mathbb{R}^{n_z}$ are given matrices.

An SFC for system (5.61) will be designed in the form $u(i, j) = -Kx(i, j)$, where $K \in \mathbb{R}^{1\times n}$ is the controller gain. The closed-loop system of (5.61) can be written as

$$\begin{aligned} x(i+1, j+1) &= \underbrace{(A_1 - B_1 K)}_{A_{1c}} x(i+1, j) + \underbrace{(A_2 - B_2 K)}_{A_{c2}} x(i, j+1) \\ &\quad + [A_{d1} x(i - \tau_h, j+1) + A_{d2} x(i+1, j - \tau_v)] \\ &\quad + B_{1w} w(i+1, j) + B_{2w} w(i, j+1), \\ z(i, j) &= \underbrace{(C - DK)}_{C_c} x(i, j) + Du(i, j) + D_w w(i, j). \end{aligned} \tag{5.62}$$

By Theorem 5.3, and similar to (5.46)–(5.48), the l_1-induced control problem of system (5.61) is solvable if and only if there exist a matrix $K \in \mathbb{R}^{n_u \times n}$ and a vector $\mu \in \mathbb{R}^n$, $\mu \succ 0$, such that the following conditions are satisfied

$$\widetilde{L} - \widetilde{H} K \succeq 0, \tag{5.63}$$

$$\left(\begin{bmatrix} A_s - I_n \\ C \end{bmatrix}^{\top} - K^{\top} \begin{bmatrix} B_s \\ D \end{bmatrix}^{\top} \right) \begin{bmatrix} \mu \\ \mathbf{1}_{n_w} \end{bmatrix} \prec 0, \tag{5.64}$$

$$\begin{bmatrix} B_{ws}^{\top} & D_w^{\top} \end{bmatrix} \begin{bmatrix} \mu \\ \mathbf{1}_{n_w} \end{bmatrix} - \gamma \mathbf{1}_{n_w} \prec 0, \tag{5.65}$$

where $A_s = A_1 + A_2$, $B_s = B_1 + B_2$, $B_{ws} = B_{1w} + B_{2w}$ and

$$\widetilde{L} = \begin{bmatrix} A_1 \\ A_2 \\ C \end{bmatrix}, \quad \widetilde{H} = \begin{bmatrix} B_1 \\ B_2 \\ D \end{bmatrix}.$$

Let $K = \begin{bmatrix} k_1 & k_2 & \cdots & k_n \end{bmatrix} \in \mathbb{R}^{1\times n}$, where $k_j \in \mathbb{R}$. Similar to (5.53), it can be verified that condition (5.63) is satisfied if and only if

$$k_j \leq k_j^* = \min_{i,k,l} \left\{ \frac{a_{ij}^{(1)}}{b_{1i}}, \frac{a_{kj}^{(2)}}{b_{2k}}, \frac{c_{lj}}{d_l} \;\middle|\; b_{i1} b_{2k} d_l \neq 0 \right\}, \tag{5.66}$$

where $A_1 = (a_{ij}^{(1)})$, $A_2 = (a_{kj}^{(2)})$ and $C = (c_{lj})$. In regard to (5.66) and the fact

$$K^{\top} \begin{bmatrix} B_s \\ D \end{bmatrix}^{\top} \begin{bmatrix} \mu \\ \mathbf{1}_{n_w} \end{bmatrix} \preceq K_s^{*\top} \begin{bmatrix} B_s \\ D \end{bmatrix}^{\top} \begin{bmatrix} \mu \\ \mathbf{1}_{n_w} \end{bmatrix}$$

for any $K \preceq K_{op}^*$, where $K_{op}^* = \begin{bmatrix} k_1^* & k_2^* & \cdots & k_n^* \end{bmatrix}$ and k_j^*, $j \in [n]$, are determined by (5.66), the solvability of l_1-induced control problem of system (5.62) is presented in the following proposition.

Proposition 5.5 *For given a $\gamma > 0$, the problem of l_1-induced control of SIMO positive 2-D systems in the form of* (5.62) *is solvable if and only if the LP-based problem governed by* (5.65) *and the following condition*

$$\left(\begin{bmatrix}(A_s - I_n)\\ C\end{bmatrix}^\top - K_{op}^{*\top}\begin{bmatrix}B_s\\ D\end{bmatrix}^\top\right)\begin{bmatrix}\mu\\ \mathbf{1}_{n_z}\end{bmatrix} \prec 0 \tag{5.67}$$

is feasible for a vector $\mu \in \mathbb{R}^n$, $\mu \succ 0$. The optimal controller gain is obtained as

$$K_{op}^* = \begin{bmatrix}k_1^* & k_2^* & \cdots & k_n^*\end{bmatrix}, \tag{5.68}$$

where k_j^*, $j \in [n]$, are determined by (5.66).

Remark 5.12 Similar to Theorem 5.6, the result of Proposition 5.5 can be extended for MIMO 2-D systems in the form of (5.62), where the matrices B_1, B_2 and D have unitary rows. However, for general MIMO 2-D systems described by FMLSS model, analogous result of Theorem 5.7 is still left open although the feasible region of K in 5.63 can be completely described. The main difficulty is that the representation $K = \widetilde{H}^\dagger Z$ cannot induce the monotonicity of the resultant LP

$$\left(\begin{bmatrix}A_s - I_n\\ C\end{bmatrix}^\top - Z^\top \widetilde{H}^{\dagger\top}\begin{bmatrix}B_s\\ D\end{bmatrix}^\top\right)\begin{bmatrix}\mu\\ \mathbf{1}_{n_w}\end{bmatrix}$$

with respect to Z as (5.52) since the matrix $\begin{bmatrix}B_s\\ D\end{bmatrix}\widetilde{H}^\dagger$ is sign-indefinite.

5.5 Numerical Examples

In this section, we give some numerical examples and simulations to illustrate the effectiveness of the design method.

Example 5.1 Consider a control model governed the following Darboux-type partial differential equation

$$\begin{aligned}\frac{\partial T(x,t)}{\partial x} &= -\frac{\partial T(x,t)}{\partial t} - a_0 T(x,t) + a_1 T(x-\tau_x, t) + bu(x,t),\\ z(x,t) &= cT(x,t),\end{aligned} \tag{5.69}$$

where $T(x,t)$ is the unknown function representing temperature at (space) $x \in [0, x_f]$ and (time) $t \in [0, \infty)$, $u(x,t)$ is the control input, a_0, a_1, b and c are known real coefficients and τ_x is the space delay involving the heat propagation through material environment [27].

For given increments $\Delta x, \Delta t$, let $T(i, j) = T(i\Delta x, j\Delta t), u(i, j) = u(i\Delta x, j\Delta t)$ and $z(i, j) = z(i\Delta x, j\Delta t)$, $\tau_h = \lfloor \tau_x/\Delta x \rfloor$; $x^h(i, j) = T(i-1, j)$ and $x^v(i, j) = T(i, j)$. By a discretization process of (5.69), we obtain 2-D system (5.1)–(5.2) with

$$A = \begin{bmatrix} 0 & 1 \\ \frac{\Delta t}{\Delta x} & 1 - \frac{\Delta t}{\Delta x} - a_0\Delta t \end{bmatrix}, \quad A_d = \begin{bmatrix} 0 & 0 \\ a_1\Delta t & 0 \end{bmatrix}, \quad B = \begin{bmatrix} 0 \\ b\Delta t \end{bmatrix},$$

$$B_w = \begin{bmatrix} 0 \\ \Delta t \end{bmatrix}, \quad C = [0 \quad c], \quad C_d = 0, \quad D = D_w = 0.$$

For illustrative purpose, the system parameters are given as follows

Δx	Δt	a_0	a_1	b	c
0.1	0.05	1.0	2.0	5.0	0.9

It is clear that, for given parameters, system (5.1)–(5.2) is positive. In addition, we have

$$A = \begin{bmatrix} 0 & 1.0 \\ 0.5 & 0.45 \end{bmatrix}, \quad A_d = \begin{bmatrix} 0 & 0 \\ 0.1 & 0 \end{bmatrix}$$

and $\rho(A + A_d) = 1.0316$. Thus, by Theorem 5.1, the open-loop system is unstable. A norm-state trajectory $\|x(i, j)\|_1$ of the open-loop system with $\tau_h = 6$ (the system is independent of τ_v) and $w = 0$ is presented in Fig. 5.3. It can be seen that $\|x(i, j)\|_1$ does not converge to the plane $x = 0$, which demonstrates the unstable property of the open-loop system.

We now utilize the results presented in this chapter to check for the existence of an l_1-gain stabilizing controller. Taking (5.53) into account, we obtain $\delta_1^* = 2.0$, $\delta_2^* = 1.8$ and the LP-base problem governed by (5.48) and (5.54) is feasible if and

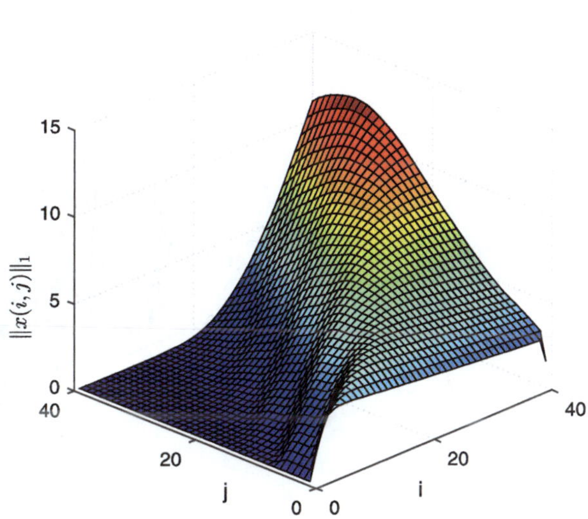

Fig. 5.3 A trajectory $\|x(i, j)\|_1$ of the open-loop system with $w = 0$ and horizontal delay $\tau_h = 6$

only if $\gamma > \gamma_{\min} = 0.05$. By Proposition 5.4, a desired stabilizing controller with optimal controller gain is obtained as

$$u_*(i, j) = -\begin{bmatrix} 2.0 & 1.8 \end{bmatrix} x(i, j).$$

A state trajectory $\|x(i, j)\|_1$ of the closed-loop system with $\tau_h = 6$ and random initial conditions and the output $z(i, j)$ with respect to $w(i, j) = \cos^2(0.1\pi(i + j))$ and zero initial condition are given in Figs. 5.4 and 5.5. The simulation results in Figs. 5.4 and 5.5 indicate that the closed-loop system is stable and $\|z\|_{l_1} < \gamma_{\min}\|w\|_{l_1}$ as revealed by the obtained theoretical results.

Example 5.2 Consider a 2-D system (5.1)–(5.2) with the system matrices

$$A = \begin{bmatrix} 0.3 & 0.1 & 0.2 \\ 0.5 & 0.2 & 0.1 \\ 0.2 & 0.3 & 0.5 \end{bmatrix}, \quad A_d = \begin{bmatrix} 0.3 & 0 & 0.1 \\ 0 & 0.1 & 0.1 \\ 0 & 0.5 & 0.1 \end{bmatrix}, \quad B = \begin{bmatrix} 0.1 & 0.1 \\ 0.1 & 0.0 \\ 0.1 & 0.2 \end{bmatrix}, \quad B_w = \begin{bmatrix} 0.1 \\ 0 \\ 0 \end{bmatrix},$$
$$C = \begin{bmatrix} 0.6 & 0.5 & 0.3 \end{bmatrix}, \quad C_d = \begin{bmatrix} 0.1 & 0 & 0.1 \end{bmatrix}, \quad D = \begin{bmatrix} 0.0 & 0.1 \end{bmatrix}, \quad D_w = 0.1.$$

It can be verify that $\rho(A + A_d) > 1$, and thus the open-loop system is unstable. We apply Theorem 5.7 to check for the existence of an l_1-induced stabilizing controller. It can be seen that the polyhedron $\mathscr{P}_1$ defined by (5.58) has three vertices that do not fall within the line $\mathbf{1}_4^\top Hx = 0$, which are indicated in Fig. 5.6 as

$$\begin{bmatrix} 5 \\ -3 \end{bmatrix}, \quad \begin{bmatrix} 5 \\ -2 \end{bmatrix} \text{ and } \begin{bmatrix} 4 \\ -1 \end{bmatrix}.$$

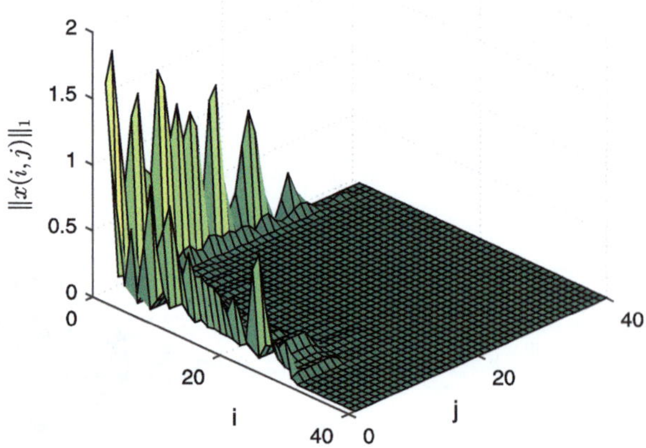

Fig. 5.4 A trajectory $\|x(i, j)\|_1$ of the closed-loop system with $w = 0$ and horizontal delay $\tau_h = 6$

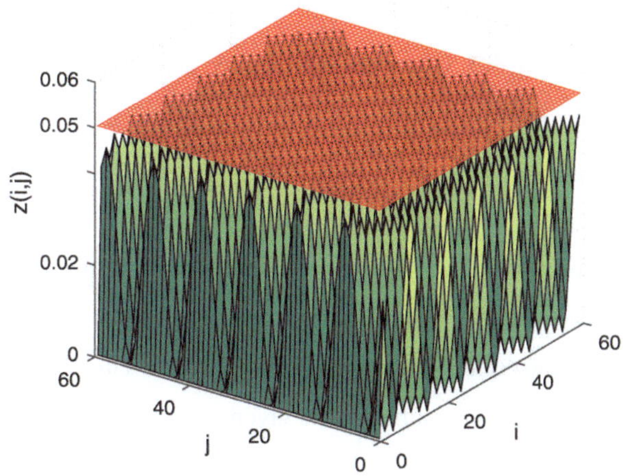

Fig. 5.5 Output trajectory $z(i, j)$ of the closed-loop system with $w(i, j) = \cos^2(0.1\pi(i + j))$ and horizontal delay $\tau_h = 6$

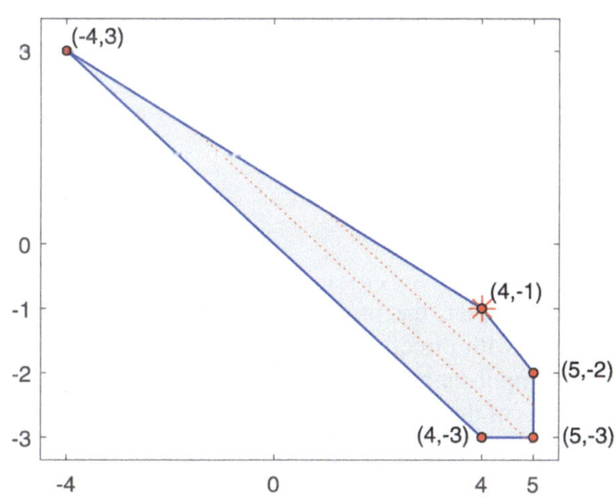

Fig. 5.6 Polyhedron $\mathscr{P}_1$

The vertex that maximizes the objective function $h_\mu(x) = \mathbf{1}_4^\top Hx$ is $x_1^* = \begin{bmatrix} 4 \\ -1 \end{bmatrix}$. Simiarly, the vertices that maximize $h_\mu(x)$ on $\mathscr{P}_2$ and $\mathscr{P}_3$ are $x_2^* = \begin{bmatrix} -1 \\ 2 \end{bmatrix}$ and $x_3^* = \begin{bmatrix} -1 \\ 3 \end{bmatrix}$, respectively. By solving the LP problem governed by (5.48) and (5.52) with respect to γ, where $Z = H \begin{bmatrix} x_1^* & x_2^* & x_3^* \end{bmatrix}$, an optimal feasible solution is obtained as

$$\gamma_{\min} = 0.2295, \quad \mu = \begin{bmatrix} 1.2862 & 0.9861 & 0.5833 \end{bmatrix}^\top .$$

By Theorem 5.7, the l_1-induced control problem of system (5.1)–(5.2) is solvable with $\gamma > \gamma_{\min}$. The controller gain K^s is given by

$$K^s = \begin{bmatrix} 4 & -1 & -1 \\ -1 & 2 & 3 \end{bmatrix}.$$

Example 5.3 Consider a delay-free 2-D system in the form of (5.1)–(5.2), where

$$A = \begin{bmatrix} 1 & 1 & 0 \\ 1 & 0 & 1 \\ 0 & 1 & 1 \end{bmatrix}, \quad B = \begin{bmatrix} 0 & 1 \\ 1 & 0 \\ 1 & 1 \end{bmatrix}, \quad B_w = \begin{bmatrix} 0 \\ 1 \\ 1 \end{bmatrix}, \quad C^\top = \begin{bmatrix} 1 \\ 1 \\ 1 \end{bmatrix}$$

and $A_d = 0$, $C_d = 0$, $D_w = 1$. Theorem 5.7 can help to find an l_1-induced stabilizing controller with optimal controller gain $K^s = \begin{bmatrix} -1 & 0 & 1 \\ 1 & 1 & 0 \end{bmatrix}$ and $\gamma_{\min} = 3.0$.

If the controller gain K is restricted to belonging the set of nonnegative matrices, then the best possible one is given by $K_+ = \begin{bmatrix} 0 & 0 & 1 \\ 0 & 1 & 0 \end{bmatrix}$. Clearly, condition

$$\mu^\top (A - I_3 - BK_+) + C = \begin{bmatrix} 1 + \mu_2 & 1 - \mu_2 + \mu_3 & 1 - \mu_3 \end{bmatrix} \prec 0$$

is not feasible for any $\mu \succ 0$. Thus, for MIMO 2-D systems, the problem of l_1-induced control cannot be always solvable only via nonnegative controller gains K.

5.6 Conclusion

In this chapter, the problem of l_1-gain control has been studied for positive 2-D systems with directional state delays. Exact value expressions of l_1-induced norm of the input-output operators are obtained using a 2-D Z-transform approach. On the basis of the obtained l_1-gain characterizations, necessary and sufficient LP-based conditions for l_1-induced performance with a prescribed attenuation level have been formulated. By utilizing some linear convex optimization procedures, necessary and sufficient conditions for the existence of a state-feedback controller that makes the closed-loop system positive, stable and has prescribed l_1-induced performance are formulated. Numerical examples with simulations have been presented to illustrate the effectiveness of the proposed method.

References

1. Farina, L., Rinaldi, S.: Positive Linear Systems: Theory and Applications. John Wiley & Sons, New York (2000)
2. Briat, C.: Stability and performance analysis of linear positive systems with delays using input-output methods. Int. J. Control **91**, 1669–1692 (2018)
3. Briat, C.: Hybrid $L_\infty \times \ell_\infty$-performance analysis and control of linear time-varying impulsive and switched positive systems. Nonlin. Anal. Hybrid Syst. **39**, 100980 (2021)

4. Smith, H.: Monotone Dynamical Systems: An Introduction to the Theory of Competitive and Cooperative Systems. American Mathematical Society, Providence, RI (2008)
5. Haddad, W.M., Chellaboina, V.S., Hui, Q.: Nonnegative and Compartmental Dynamical Systems. Princeton Univ Press, Princeton (NJ) (2010)
6. Altafini, C.: Consensus problems on networks with antagonistic interactions. IEEE Trans. Autom. Control **58**, 935–946 (2013)
7. Liu, X., Yu, W., Wang, L.: Necessary and sufficient asymptotic stability criterion for 2-D positive systems with time-varying state delays described by Roesser model. IET Control Theory Appl. **5**, 663–668 (2011)
8. Hien, L.V., Trinh, H.: Delay-dependent stability and stabilization of two-dimensional positive Markov jump systems with delays. IET Control Theory Appl. **11**, 1603–1610 (2017)
9. Shen, J., Wang, W.: Stability and positive observer design for positive 2D discrete-time system with multiple delays. Int. J. Syst. Sci. **48**, 1136–1145 (2017)
10. Hien, L.V., Trinh, H.: Exponential stability of two-dimensional homogeneous monotone systems with bounded directional delays. IEEE Trans. Autom. Control **63**, 2694–2700 (2018)
11. Bolajraf, M.: LP conditions for stability and stabilization of positive 2D discrete state-delayed Roesser models. Int. J. Control Auto. Syst. **16**, 2814–2821 (2018)
12. Hien, L.V., Trinh, H.: Observers design for 2-D positive time-delay Roesser systems. IEEE Trans. Circuit Syst: Expr. Brief **65**, 476–480 (2018)
13. Haddad, W.M., Chellaboina, V.S., August, E.: Stability and dissipativity theory for discrete-time non-negative and compartmental dynamical systems. Int. J. Control **76**, 1845–1861 (2003)
14. Haddad, W.M., Chellaboina, V.S., Rajpurohit, T.: Dissipativity theory for nonnegative and compartmental dynamical systems with time delay. IEEE Trans. Autom Control **49**, 747–751 (2004)
15. Briat, C.: Robust stability and stabilization of uncertain linear positive systems via integral linear constraints: L_1-gain and L_∞-gain characterization. Int. J. Robust Nonlinear Control **23**, 1932–1954 (2013)
16. Shen, J., Lam, J.: Input-output gain analysis for linear systems on cones. Automatica **77**, 44–50 (2017)
17. Xiang, W., Lam, J., Shen, J.: Stability analysis and $\mathscr{L}_1$-gain characterization for switched positive systems under dwell-time constraint. Automatica **85**, 1–8 (2017)
18. Shen, J., Lam, J.: Static output-feedback stabilization with optimal L_1-gain for positive linear systems. Automatica **63**, 248–253 (2016)
19. Zhang, J., Zhao, X., Zhang, R.: An improved approach to controller design of positive systems using controller gain decomposition. J. Frankl. Inst. **354**, 1356–1373 (2017)
20. Saeki, M.: L_1 synthesis of a static output controller for positive systems by LMI iteration. Int. J. Robust Nonlinear Control **28**, 1319–1333 (2018)
21. Hong, M.T., Hien, L.V.: Solvability of L_1-induced controller synthesis for positive systems with multiple delays. Int. J. Control Autom. Syst. **19**, 3569–3579 (2021)
22. Chen, X., Lam, J., Li, P., Shu, Z.: ℓ_1-induced norm and controller synthesis of positive systems. Automatica **43**, 1377–1385 (2013)
23. Fu, J., Duan, Z., Ghous, I.: ℓ_1-induced norm and controller synthesis for positive 2D systems with multiple delays. J. Frank. Inst. **357**, 7904–7920 (2020)
24. Duan, Z., Han, Y., Xiang, Z., Ghous, I.: On l_1-gain control for 2D delayed positive systems in FM LSS models: necessary and sufficient conditions. Int. J. Syst. Sci. **53**, 3449–3464 (2022)
25. Hien, L.V., Trinh, H., Pathirana, P.N.: On ℓ_1-gain control of 2-D positive Roesser systems with directional delays: Necessary and sufficient conditions. Automatica **112**, 108720 (2020)
26. Berman, A., Plemmons, R.J.: Nonnegative Matrices in the Mathematical Sciences. SIAM, Philadelphia (1994)
27. Hien, L.V., Trinh, H.: Stability of two-dimensional Roesser systems with time-varying delays via novel 2D finite-sum inequalities. IET Control Theory Appl. **10**, 1665–1674 (2016)
28. Hien, L.V., Trinh, H.: On reachable set estimation of two-dimensional systems described by the Roesser model with time-varying delays. Int. J. Robust Nonlinear Control **28**, 227–246 (2018)

29. Hien, L.V., Trinh, H.: Stability analysis and control of two-dimensional fuzzy systems with directional time-varying delays. IEEE Trans. Fuzzy Syst. **26**, 1550–1564 (2018)
30. Bliman, P.A.: Lyapunov equation for the stability of 2-D systems. Multidim. Syst. Sig. Process **13**, 201–222 (2002)
31. James, M.: The generalised inverse. Math. Gazet. **62**, 109–114 (1978)

Chapter 6
Peak-to-Peak Performance Analysis of Two-Dimensional Positive Time-Delay Systems

Abstract In this chapter, the analysis problem of peak-to-peak gain is studied for 2-D positive systems with interval time-varying delays. Directional delays in the state vectors are heterogeneous and time-varying, which are confined within different ranges with lower bounds other than zero. Novel comparison techniques involving interval state estimations subject to peak values of exogenous disturbances are presented and a characterization of l_∞-induced norm of the input-output operator is formulated. The obtained l_∞ gain characterization is then utilized to establish necessary and sufficient conditions in terms of linear programming (LP) settings for l_∞-induced performance with a prescribed attenuation level. Applications to l_∞-gain control problem via static output feedback controllers are also discussed. Numerical examples and simulations are given to illustrate the effectiveness of the proposed method.

Keywords 2-D systems · Roesser model · FMLSS model · Positive systems · Interval delays · Peak-to-peak gain · l_∞-induced · Linear programming

6.1 Introduction

The theory of positive systems [1] has become one of the most active research topics in recent years due to their elegant properties that have no counterpart in general dynamical systems [2–4]. Practical applications of positive systems have also been found in various areas such as economics, biology, ecology and epidemiology, chemistry, pharmacokinetics, population dynamics or communication [5–7]. This theory and many problems in control theory have also been developed for some classes of 2-D systems with delays, recently. We refer the reader to [8–13] for a few references.

In modeling dynamical systems in control engineering, it is well recognized that exogenous disturbances are unavoidable due to technical issues encountered in the data processing, operation and information transmission. Thus, the area

L. V. Hien, *Two-Dimensional Time-Delay Systems*, Studies in Systems, Decision and Control 605, https://doi.org/10.1007/978-3-031-95678-2_6

of robust control, which includes stability theory, performance analysis and controller/filter/observer design, plays a key role in design and applications of dynamical systems with noises and disturbances. To model and evaluate the effective of certain types of noise signals, in robust control, some performance indicators associated with signal energy or signal peak value such as L_1, L_2, H_∞ or L_∞ (respectively, l_1, l_2, H_∞, l_∞) for discrete-time systems) are widely used as most important tools [14–17]. For example, the H_∞ and l_2-l_∞ performance indices are used to deal with systems with the assumption of energy-bounded disturbance inputs. However, in many engineering systems as wind shear on aircraft wings or continuous road excitation on vehicle suspension systems, external disturbances are, in general, only persistent and amplitude-bounded rather than specifications on the total energy of a disturbance are required [18]. Thus, the l_∞-gain minimization is a more useful and effective approach to examining the responses of dynamic systems corrupted by persistently bounded disturbances [19, 20].

During the past two decades, the l_∞-induced theory, a natural counterpart control scheme to the existing H_∞ stability theory and which can deal with persistently bounded disturbances, has attracted considerable research attention due to its theoretical significance. Many results concerning with various control and filtering problems for both 1-D and 2-D systems have been published recently (see, e.g., [21–26] and the references therein). A common approach in the aforementioned works is the use of Lyapunov function scheme and its variants to derive sufficient conditions in the form of linear matrix inequalities (LMIs) that guarantee an upper bound of L_∞ or l_∞-induced gain. It should be noticed that the existing methods cannot help to derive tractable conditions for optimal peak-to-peak gain (l_∞-l_∞ index) as it usually produces much conservativeness in the established conditions, not only in terms of feasibility but also in terms of computational burden.

For positive systems, the use of linear co-positive Lyapunov functions [27, 28] is one of the most popular approaches, which stimulates L_∞ and l_∞ performance indices, especially, to deal with the attenuation levels of persistent peak-bounded disturbances. Up-to-date, considerably less attention has been devoted to formulating l_∞-gain characterization for 2-D positive systems with multiple interval delays. On one hand, the established analysis results for 1-D systems using co-positive Lyapunov functions, fundamental solution representation [29] or by utilizing the induced monotonicity [30] cannot help to derive necessary conditions for the existence of stabilizing controllers under l_∞ performance even for LTI systems without delay. On the other hand, due to technical challenges, one cannot directly extend such studies to 2-D processes using conventional 1-D systems theory.

In this chapter, we first formulate a characterization of l_∞-gain for a class of 2-D positive systems with heterogeneous interval delays in both Roesser model and Fornasini–Marchesini local state-space (FMLSS) model. The obtained characterizations are then utilized to derive necessary and sufficient LP-based conditions of l_∞-gain performance subject to prescribed attenuation level. As an application, a complete solution to static output-feedback l_∞-gain control problem is also presented.

6.2 Preliminaries

6.2.1 Problem Formulation

Consider a class of 2-D time-delay systems described by the Roesser model

$$\begin{bmatrix} x^h(i+1,j) \\ x^v(i,j+1) \end{bmatrix} = A_0 \begin{bmatrix} x^h(i,j) \\ x^v(i,j) \end{bmatrix} + \sum_{k=1}^{N} A_k \begin{bmatrix} x^h(i-\tau_{hk}(i),j) \\ x^v(i,j-\tau_{vk}(j)) \end{bmatrix} + B_w w(i,j), \tag{6.1}$$

$$z(i,j) = C_0 \begin{bmatrix} x^h(i,j) \\ x^v(i,j) \end{bmatrix} + \sum_{k=1}^{N} C_k \begin{bmatrix} x^h(i-\tau_{hk}(i),j) \\ x^v(i,j-\tau_{vk}(j)) \end{bmatrix} + D_w w(i,j), \tag{6.2}$$

where $x^h(i,j) \in \mathbb{R}^{n_h}$ and $x^v(i,j) \in \mathbb{R}^{n_v}$ are the horizontal and vertical state vectors; $z(i,j) \in \mathbb{R}^{n_z}$ and $w(i,j) \in \mathbb{R}^{n_w}$ are the regulated output and exogenous disturbance input vectors, respectively. A_0, A_k, B_w and C_0, C_k, D_w ($k \in [N]$) are given matrices of appropriate dimensions. $\tau_{hk}(i)$ and $\tau_{vk}(j)$ are time-varying directional delays which satisfy

$$\underline{\tau}_{hk} \le \tau_{hk}(i) \le \overline{\tau}_{hk}, \quad \underline{\tau}_{vk} \le \tau_{vk}(j) \le \overline{\tau}_{vk}, \tag{6.3}$$

where $\underline{\tau}_{hk}, \overline{\tau}_{vk}$ and $\underline{\tau}_{vk}, \overline{\tau}_{vk}$ are known positive integers involving lower bounds and upper bounds of delays along the horizontal and vertical directions.

Initial conditions of system (6.1) are specified as

$$\begin{aligned} x^h(i,j) &= \phi_h(i,j),\ i \in \mathbb{Z}[-\bar{\tau}_h, 0],\ j \in \mathbb{N}_0, \\ x^v(i,j) &= \phi_v(i,j),\ i \in \mathbb{N}_0,\ j \in \mathbb{Z}[-\bar{\tau}_v, 0], \end{aligned} \tag{6.4}$$

where $\bar{\tau}_h = \max_{1\le k\le N} \overline{\tau}_{hk}$ and $\bar{\tau}_v = \max_{1\le k\le N} \overline{\tau}_{vk}$. Moreover, the initial functions ϕ_h, ϕ_v in (6.4) are assumed to have finite support [31], that is, there exist positive integer T_h, T_v such that

$$\phi_h(i,j) = 0 \text{ for } j \ge T_h \text{ and } \phi_v(i,j) = 0 \text{ for } i \ge T_v. \tag{6.5}$$

This assumption is widely adopted in analysis and control of 2-D systems.

Similar to Chap. 4 (Definition 4.3) and Chap. 5, the concept of positivity of 2-D systems with inputs and outputs in the form of (6.1)–(6.2) is presented in following definition.

Definition 6.1 System (6.1)–(6.2) is said to be positive if for any nonnegative initial functions $\phi_h(i,j) \succeq 0$, $\phi_v(i,j)$ defined by (6.4)–(6.5) and nonnegative input $w(i,j) \succeq 0$, the state vector $x(i,j) = \begin{bmatrix} x^h(i,j) \\ x^v(i,j) \end{bmatrix}$ and the output vector $z(i,j)$ are always nonnegative, that is, $x(i,j) \succeq 0$ and $z(i,j) \succeq 0$ for all $i, j \in \mathbb{N}_0$.

Proposition 6.1 *System (6.1)–(6.2) is positive for any delays $\tau_{hk}(i)$, $\tau_{vk}(j)$, $k \in [N]$, if and only if the following condition is satisfied*

$$\mathscr{M} = \left[\begin{array}{c:c:c} A_0 & A_d & B_w \\ \hdashline C_0 & C_d & D_w \end{array}\right] \succeq 0, \tag{6.6}$$

where the augmented delay terms matrices A_d, C_d are defined as

$$A_d = \begin{bmatrix} A_1 \ A_2 \cdots A_N \end{bmatrix},$$
$$C_d = \begin{bmatrix} C_1 \ C_2 \cdots C_N \end{bmatrix}.$$

Proof The proof is similar to that of Proposition 5.1. We omit it here. □

Similar to Definition 4.2, the exponential stability of system (6.1) is given in the following definition.

Definition 6.2 System (6.1) without exogenous disturbance (i.e. $w = 0$) is said to be globally exponentially stable (GES) if there exist scalars $\alpha \in (0, 1)$ and $\beta > 0$ such that any solution $x(i, j)$ of system (6.1) subject to initial condition (6.4)–(6.5) satisfies the following inequality

$$\|x(i, j)\|_\infty \le \beta \Bigg(\sum_{k=0}^{i} \frac{\hat{\phi}_v(k)}{\alpha^{k+1}} + \sum_{l=0}^{j} \frac{\hat{\phi}_h(l)}{\alpha^{l+1}} \Bigg) \alpha^{i+j}, \quad i, j \in \mathbb{N}_0, \tag{6.7}$$

where $\hat{\phi}_v(k) = \sup_{-\bar{\tau}_v \le j \le 0} \|\phi_v(k, j)\|_\infty$ and $\hat{\phi}_h(l) = \sup_{-\bar{\tau}_h \le i \le 0} \|\phi_h(i, l)\|_\infty$.

As discussed in Chap. 4, Definition 6.2 is an extension of the conventional exponential stability concept developed for 1-D systems with delays.

A two-index sequence (function) $w : \mathbb{Z}_+ \times \mathbb{Z}_+ \to \mathbb{R}^{n_w}$ is said to be bounded if there exists a positive scalar $\overline{w}$ such that

$$\|w(k, l)\|_\infty \le \overline{w}, \quad \forall k, l \in \mathbb{Z}_+.$$

The l_∞-norm of a bounded sequence w is defined as

$$\|w\|_{l_\infty} = \sup_{k,l \in \mathbb{Z}_+} \|w(k, l)\|_\infty.$$

We denote by $l_\infty(\mathbb{R}^{n_w})$ the space of all bounded sequences w, that is,

$$l_\infty(\mathbb{R}^{n_w}) = \left\{ w : \mathbb{Z}_+ \times \mathbb{Z}_+ \to \mathbb{R}^{n_w} \middle| \|w\|_{l_\infty} < \infty \right\}.$$

In this chapter, the disturbance $w(i, j)$ is assumed to belong to $l_\infty(\mathbb{R}^{n_w})$.

Assume that system (6.1) without disturbance is GES. We define an input-output operator for system (6.1) as

$$\mathscr{L} : l_\infty(\mathbb{R}^{n_w}) \longrightarrow l_\infty(\mathbb{R}^{n_z}), \quad w \mapsto z$$

and l_∞-gain of system (6.1) under zero initial condition is defined by

$$\|\mathscr{L}\|_{l_\infty - l_\infty} = \sup_{\|w\|_{l_\infty} \neq 0} \frac{\|z\|_{l_\infty}}{\|w\|_{l_\infty}} = \sup_{\|w\|_{l_\infty}=1} \|z\|_{l_\infty}. \tag{6.8}$$

Definition 6.3 For a given $\gamma > 0$, system (6.1) is said to have l_∞-gain performance of level γ if, under zero initial condition, it holds that $\|\mathscr{L}\|_{l_\infty - l_\infty} < \gamma$. In other words, for any nonzero disturbance $w \in l_\infty(\mathbb{R}^{n_w})$, and under zero initial condition, the output trajectory $z(i, j)$ of system (6.1) satisfies

$$\|z(i, j)\|_\infty < \gamma \|w(i, j)\|_\infty, \quad \forall i, j \geq 0.$$

The main objective of this chapter is to formulate the value of l_∞-gain $\|\mathscr{L}\|_{l_\infty - l_\infty}$ for system (6.1) and characterize the l_∞-gain performance index by establishing necessary and sufficient conditions in terms of linear programming (LP) problems.

6.2.2 Auxiliary Results: Stability Analysis

In this section, we introduce equivalent (necessary and sufficient) conditions, by which the positive system (6.1) with $w = 0$ is GES.

Proposition 6.2 *The positive 2-D system* (6.1) *with* $w = 0$ *is GES if and only if the following LP-based condition*

$$\left(A_0 + \sum_{k=1}^{N} A_k\right) \chi - \chi \prec 0 \tag{6.9}$$

is feasible for a vector $\chi \in \mathbb{R}^n$, $\chi \succ 0$.

Proof See, Chap. 4, Theorem 4.3 for more details. □

By the properties of nonnegative matrices [32], it is found that there exists a positive vector $\chi \in \mathbb{R}^n$ satisfying condition (6.9) if and only if the spectral radius $\rho\left(A_0 + \sum_{k=1}^{N} A_k\right) < 1$ [1]. We summarize some stability criteria for system (6.1) in the following theorem.

Theorem 6.1 *For positive 2-D system* (6.1) *with* $w = 0$, *the following statements are equivalent.*

(i) *System* (6.1) *is GES.*

(ii) *There exists a vector* $\eta \in \mathbb{R}^n$, $\eta \succ 0$, *such that*

$$\eta^\top \left(A_0 + \sum_{k=1}^{N} A_k \right) - \eta^\top \prec 0. \tag{6.10}$$

(iii) *There exists a vector* $\chi \in \mathbb{R}^n$, $\chi \succ 0$, *such that*

$$\left(A_0 + \sum_{k=1}^{N} A_k \right) \chi - \chi \prec 0. \tag{6.11}$$

(iv) *The matrix* $A_0 + \sum_{k=1}^{N} A_k$ *is Schur stable. Equivalently, the spectral radius of the matrix* $A_0 + \sum_{k=1}^{N} A_k$ *satisfies*

$$\rho\left(A_0 + \sum_{k=1}^{N} A_k \right) < 1. \tag{6.12}$$

6.3 Performance Analysis Under Peak-to-Peak Gain

In this section, we present l_∞-gain analysis for positive 2-D systems with interval direction delays in the form of (6.1). Using the ideas of interval state and output estimations, a characterization of the exact value of l_∞-gain is first obtained. Then, necessary and sufficient LP-based conditions by which system (6.1)–(6.2) has l_∞-gain performance of level γ with prescribed threshold $\gamma > 0$ are formulated.

6.3.1 *Peak-to-Peak Gain Characterization*

Consider a 2-D positive system as given in (6.1) with zero initial condition (i.e. $\phi_h = 0$ and $\phi_v = 0$). Let $x_w(i, j)$ denote the corresponding solution of (6.1) with respect to input w. For convenience, we also define a partial order in $l_\infty(\mathbb{R}^{n_w})$ by

$$w_1 \preceq w_1 \text{ iff } w_1(i, j) \preceq w_2(i, j) \text{ for all } i, j \in \mathbb{Z}_+.$$

The following technical lemmas will be used to establish l_∞-gain of system (6.1).

Lemma 6.1 *For any* $w_1, w_2 \in l_\infty(\mathbb{R}^{n_w})$, *if* $w_1 \preceq w_2$ *then* $x_{w_1}(i, j) \preceq x_{w_2}(i, j)$ *for all* $i, j \geq 0$.

Proof Let $x_{\tilde{w}}(i, j) = x_{w_2}(i, j) - x_{w_1}(i, j)$. Then, it can be verified directly that $x_{\tilde{w}}(i, j)$ is a solution of system (6.1) with zero initial condition and nonnegative input

$$\tilde{w}(i, j) = w_2(i, j) - w_1(i, j) \succeq 0.$$

Thus, according to the positivity of system (6.1), $x_{\tilde{w}}(i, j) \succeq 0$ for all $i, j \geq 0$. □

For any $w \in l_\infty(\mathbb{R}^{n_w})$, it is obvious that

$$-|w(i, j)| \preceq w(i, j) \preceq |w(i, j)|, \quad \forall i, j \in \mathbb{Z}_+.$$

By Lemma 6.1, we have

$$-x_{|w|}(i, j) \preceq x_w(i, j) \preceq x_{|w|}(i, j).$$

This shows that $|x_w(i, j)| \preceq x_{|w|}(i, j)$. Moreover, for a $w \in l_\infty(\mathbb{R}^{n_w})$, we denote the upper bound vector of w by $\overline{w} = \|w\|_{l_\infty} \mathbf{1}_{n_w}$. Then, we have

$$|w(i, j)| \preceq \|w(i, j)\|_\infty \mathbf{1}_{n_w} \preceq \overline{w}.$$

To formulate an upper estimation for the state and output vectors $x(i, j)$ and $z(i, j)$ with respect to an input w, we consider the following associated positive 2-D system (also referred as upper scaled system)

$$\begin{bmatrix} \bar{x}^h(i+1, j) \\ \bar{x}^v(i, j+1) \end{bmatrix} = A_0 \begin{bmatrix} \bar{x}^h(i, j) \\ \bar{x}^v(i, j) \end{bmatrix} + \sum_{k=1}^{N} A_k \begin{bmatrix} \bar{x}^h(i - \underline{\tau}_{hk}, j) \\ \bar{x}^v(i, j - \underline{\tau}_{vk}) \end{bmatrix} + B_w \overline{w}, \tag{6.13}$$

$$\bar{z}(i, j) = C_0 \begin{bmatrix} \bar{x}^h(i, j) \\ \bar{x}^v(i, j) \end{bmatrix} + \sum_{k=1}^{N} C_k \begin{bmatrix} \bar{x}^h(i - \underline{\tau}_{hk}, j) \\ \bar{x}^v(i, j - \underline{\tau}_{vk}) \end{bmatrix} + D_w \overline{w}, \tag{6.14}$$

where $w \in l_\infty(\mathbb{R}^{n_w})$ and $\overline{w} = \|w\|_{l_\infty} \mathbf{1}_{n_w}$.

The following nondecreasing property of solutions of system (6.13) is useful for the next derivation.

Lemma 6.2 *Let $\bar{x}_w(i, j)$ be a solution of system* (6.13) *with zero initial condition. Then, for any positive integer k, l, it holds that*

$$\bar{x}_w(i+k, j+l) \succeq \bar{x}_w(i, j), \quad \forall i, j \in \mathbb{N}_0.$$

Proof Let $J_h = \begin{bmatrix} I_{n_h} & 0 \end{bmatrix}$ and $J_v = \begin{bmatrix} 0 & I_{n_v} \end{bmatrix}$. We define the following initial functions

$$\bar{\phi}_h(i, j) = J_h \bar{x}_w(i+k, j), \; i \in \mathbb{Z}[-\bar{\tau}_h, 0], \; j \in \mathbb{Z}[0, l],$$
$$\bar{\phi}_v(i, j) = J_v \bar{x}_w(i, j+l), \; i \in \mathbb{Z}[0, k], \; j \in \mathbb{Z}[-\bar{\tau}_v, 0].$$

It can be verified that $\bar{\xi}_{k,l}(i,j) = \bar{x}_w(i+k, j+l)$ is also a solution of system (6.13) with initial condition defined by functions $\bar{\phi}_h$ and $\bar{\phi}_v$. Let $e(i,j) = \bar{\xi}_{k,l}(i,j) - \bar{x}_w(i,j)$, $i, j \in \mathbb{N}_0$. Then, we have

$$\begin{bmatrix} e^h(i+1,j) \\ e^v(i,j+1) \end{bmatrix} = A_0 \begin{bmatrix} e^h(i,j) \\ e^v(i,j) \end{bmatrix} + \sum_{k=1}^{N} A_k \begin{bmatrix} e^h(i-\underline{\tau}_{hk}, j) \\ e^v(i, j-\underline{\tau}_{vk}) \end{bmatrix}. \tag{6.15}$$

Since $e^h(i,j) = \bar{\phi}_h(i,j) \succeq 0$, $e^v(i,j) = \bar{\phi}_v(i,j) \succeq 0$ and system (6.15) is positive, it is valid that $e(i,j) \succeq 0$ for all $i, j \geq 0$. This yields $\bar{x}_w(i+k, j+l) \succeq \bar{x}_w(i,j)$ as desired. □

Based on Lemma 6.2, the following lemmas are obtained.

Lemma 6.3 *For any $w \in l_\infty(\mathbb{R}^{n_w})$, $w \succeq 0$, let $x_w(i,j)$, $\bar{x}_w(i,j)$ and $z_w(i,j)$, $\bar{z}_w(i,j)$ be the state and output trajectories of systems* (6.1)–(6.2) *and* (6.13)–(6.14) *with zero initial condition, respectively. It holds that*

$$0 \preceq x_w(i,j) \preceq \bar{x}_w(i,j) \text{ and } 0 \preceq z_w(i,j) \preceq \bar{z}_w(i,j)$$

for all $i, j \geq 0$.

Proof Let $\bar{e}(i,j) = \bar{x}_w(i,j) - x_w(i,j)$. It follows from (6.1) and (6.13) that

$$\begin{aligned} \begin{bmatrix} \bar{e}^h(i+1,j) \\ \bar{e}^v(i,j+1) \end{bmatrix} &= A_0 \begin{bmatrix} \bar{e}^h(i,j) \\ \bar{e}^v(i,j) \end{bmatrix} + \sum_{k=1}^{N} A_k \begin{bmatrix} \bar{e}^h(i-\tau_{hk}(i), j) \\ \bar{e}^v(i, j-\tau_{vk}(j)) \end{bmatrix} \\ &\quad + \sum_{k=1}^{N} A_k \begin{bmatrix} \bar{x}^h(i-\underline{\tau}_{hk}, j) - \bar{x}^h(i-\tau_{hk}(i), j) \\ \bar{x}^v(i, j-\underline{\tau}_{vk}) - \bar{x}^v(i, j-\tau_{vk}(j)) \end{bmatrix} \\ &\quad + B_w(\overline{w} - w(i,j)). \end{aligned} \tag{6.16}$$

Since $i - \underline{\tau}_{hk} \geq i - \tau_{hk}(i)$ and $j - \underline{\tau}_{vk} \geq j - \tau_{vk}(j)$, by Lemma 6.2, we have

$$\begin{bmatrix} \bar{x}^h(i-\underline{\tau}_{hk}, j) - \bar{x}^h(i-\tau_{hk}(i), j) \\ \bar{x}^v(i, j-\underline{\tau}_{vk}) - \bar{x}^v(i, j-\tau_{vk}(j)) \end{bmatrix} \succeq 0.$$

This, together with $\overline{w} - w(i,j) \succeq 0$, yields that $\bar{e}(i,j) \succeq 0$ and, thus, $x_w(i,j) \preceq \bar{x}_w(i,j)$ for all $i, j \geq 0$.

On the other hand, from (6.2) and (6.14), we have

$$\begin{aligned}\bar{z}_w(i,j) - z_w(i,j) = C_0\bar{e}(i,j) + \sum_{k=1}^{N} C_k \begin{bmatrix} \bar{e}^h(i-\tau_{hk}(i),j) \\ \bar{e}^v(i,j-\tau_{vk}(j)) \end{bmatrix} \\ + \sum_{k=1}^{N} C_k \begin{bmatrix} \bar{x}^h(i-\underline{\tau}_{hk},j) - \bar{x}^h(i-\tau_{hk}(i),j) \\ \bar{x}^v(i,j-\underline{\tau}_{vk}) - \bar{x}^v(i,j-\tau_{vk}(j)) \end{bmatrix} \\ + D_w(\overline{w} - w(i,j)) \succeq 0. \end{aligned} \tag{6.17}$$

The last inequality in (6.17) ensures that $z_w(i,j) \preceq \bar{z}_w(i,j)$. □

To establish a lower bound for $x(i,j)$ and $z(i,j)$, similar to system (6.13)–(6.14), we consider the following scaled systems

$$\begin{bmatrix} \underline{x}^h(i+1,j) \\ \underline{x}^v(i,j+1) \end{bmatrix} = A_0 \begin{bmatrix} \underline{x}^h(i,j) \\ \underline{x}^v(i,j) \end{bmatrix} + \sum_{k=1}^{N} A_k \begin{bmatrix} \underline{x}^h(i-\overline{\tau}_{hk},j) \\ \underline{x}^v(i,j-\overline{\tau}_{vk}) \end{bmatrix} + B_w\underline{w}, \tag{6.18}$$

$$\underline{z}(i,j) = C_0 \begin{bmatrix} \underline{x}^h(i,j) \\ \underline{x}^v(i,j) \end{bmatrix} + \sum_{k=1}^{N} C_k \begin{bmatrix} \underline{x}^h(i-\overline{\tau}_{hk},j) \\ \underline{x}^v(i,j-\overline{\tau}_{vk}) \end{bmatrix} + D_w\underline{w}, \tag{6.19}$$

where $\underline{w} \in \mathbb{R}_+^{n_w}$ is a constant vector such that $w(i,j) \succeq \underline{w}$.

Lemma 6.4 *For any $w \in l_\infty(\mathbb{R}^{n_w})$ and a vector $\underline{w} \in \mathbb{R}_+^{n_w}$ such that $w(i,j) \succeq \underline{w}$, let $x_w(i,j)$, $\underline{x}_w(i,j)$ and $z_w(i,j)$, $\underline{z}_w(i,j)$ be the state and output trajectories of systems* (6.1)–(6.2) *and* (6.18)–(6.19) *with zero initial condition, respectively. It holds that*

$$0 \preceq \underline{x}_w(i,j) \preceq x_w(i,j) \text{ and } 0 \preceq \underline{z}_w(i,j) \preceq z_w(i,j)$$

for all $i, j \geq 0$.

Proof The proof is similar to that of Lemma 6.3. We omit it here. □

Next, we assume one of the stability conditions derived in Theorem 6.1 holds. Then, it can be deduced from (6.12) that the matrix $I_n - (A_0 + A_s)$ is invertible, where $A_s = \sum_{k=1}^{N} A_k$. Moreover,

$$(I_n - A_0 - A_s)^{-1} = \sum_{\nu=0}^{\infty} (A_0 + A_s)^\nu \succeq 0.$$

We say that a constant vector $x_e \in \mathbb{R}_+^n$ is a positive equilibrium point (EP$^+$) of system (6.13) if it satisfies the algebraic equation

$$x_e = (A_0 + A_s)x_e + B_w\overline{w}. \tag{6.20}$$

It can be verified from (6.13) and (6.20) that, subject to condition (6.12), the unique EP$^+$ of system (6.13) can be represented as

$$x_e = (I - A_0 - A_s)^{-1} B_w \overline{w}.$$

The following result gives a threshold for solutions of system (6.1).

Lemma 6.5 *For any $w \in l_\infty(\mathbb{R}^{n_w})$, the solution $x_w(i, j)$ of system* (6.1) *with zero initial condition satisfies*

$$|x_w(i, j)| \preceq \|w\|_{l_\infty} (I_n - A_0 - A_s)^{-1} B_w \mathbf{1}_{n_w} = x_e, \ \forall i, j \geq 0. \tag{6.21}$$

In particular, $x_w(i, j)$ also belongs to $l_\infty(\mathbb{R}^n)$.

Proof Since $|w(i, j)| \preceq \overline{w}$, by Lemmas 6.1 and 6.3, we have

$$|x_w(i, j)| \preceq x_{|w|}(i, j) \preceq \bar{x}_w(i, j), \ i, j \geq 0. \tag{6.22}$$

On the other hand, let us define the function $\hat{e}(i, j) = x_e - \bar{x}_w(i, j), i, j \geq 0$, then it follows from (6.13) and (6.20) that

$$\begin{bmatrix} \hat{e}^h(i+1, j) \\ \hat{e}^v(i, j+1) \end{bmatrix} = A_0 \begin{bmatrix} \hat{e}^h(i, j) \\ \hat{e}^v(i, j) \end{bmatrix} + \sum_{k=1}^{N} A_k \begin{bmatrix} \hat{e}^h(i - \underline{\tau}_{hk}, j) \\ \hat{e}^v(i, j - \underline{\tau}_{vk}) \end{bmatrix}. \tag{6.23}$$

System (6.23) is positive and $\hat{e}(0, 0) = x_e \succeq 0$, thus $\hat{e}(i, j) \succeq 0$. This, together with (6.22), gives $|x_w(i, j)| \preceq x_e$, by which we obtain

$$\|x_w\|_{l_\infty} = \sup_{i,j\geq 0} \|x_w(i, j)\|_\infty \leq \|x_e\|_\infty.$$

The proof is completed. □

Remark 6.1 It can be shown under equivalent conditions (6.10)–(6.12) that the positive 2-D system (6.23) is GES. Thus, the state trajectory $\bar{x}_w(i, j)$ of (6.13) increases exponentially to x_e. More precisely, there exists an $\alpha \in (0, 1)$ such that

$$0 \preceq x_e - \bar{x}_w(i, j) = \hat{e}(i, j) \preceq x_e \alpha^{i+j}, \ i, j \geq 0, \tag{6.24}$$

for each $w \in l_\infty(\mathbb{R}^{n_w})$. From (6.24), we obtain the interval estimate

$$\left(1 - \alpha^{i+j}\right) x_e \preceq \bar{x}_w(i, j) \preceq x_e, \ i, j \geq 0, \tag{6.25}$$

where $x_e = \|w\|_{l_\infty} (I - A_0 - A_s)^{-1} B_w \mathbf{1}_{n_w}$.

We now establish the following main result.

Theorem 6.2 *Assume that the 2-D system* (6.1) *is positive and GES. The value of l_∞-gain of system* (6.1) *under zero initial condition is expressed as*

$$\|\mathscr{L}\|_{l_\infty - l_\infty} = \left\| (C_0 + C_s) (I_n - A_0 - A_s)^{-1} B_w + D_w \right\|_\infty, \tag{6.26}$$

where $A_s = \sum_{k=1}^{N} A_k$ *and* $C_s = \sum_{k=1}^{N} C_k$.

Proof For any $w \in l_\infty(\mathbb{R}^{n_w})$, by Lemma 6.3, we have

$$|z_w(i, j)| \preceq z_{|w|}(i, j) \preceq \bar{z}(i, j),$$

where $\bar{z}(i, j)$ is determined by Eq. (6.14).

We will show that

$$\bar{z}(i, j) \preceq \|w\|_{l_\infty} \left[(C_0 + C_s)(I_n - A_0 - A_s)^{-1} B_w + D_w \right] \mathbf{1}_{n_w}.$$

For this, as shown in the proof of Lemma 6.5 that

$$\bar{x}(i, j) \preceq x_e, \quad \begin{bmatrix} \bar{x}^h(i - \underline{\tau}_{hk}, j) \\ \bar{x}^v(i, j - \underline{\tau}_{vk}) \end{bmatrix} \preceq x_e,$$

for all $i, j \geq 0$, where x_e is the EP defined in (6.20). This, together with (6.14), gives

$$\begin{aligned} \bar{z}(i, j) &\preceq (C_0 + C_s)x_e + D_w w \\ &= \Xi \overline{w} = \Xi \mathbf{1}_{n_w} \|w\|_{l_\infty}, \end{aligned} \tag{6.27}$$

where the matrix Ξ is given as

$$\Xi = (C_0 + C_s)(I_n - A_0 - A_s)^{-1} B_w + D_w.$$

From (6.27) and the fact $z_w(i, j) \preceq \bar{z}(i, j)$, we obtain

$$\begin{aligned} \|z_w(i, j)\|_\infty &\leq \|\bar{z}(i, j)\|_\infty \\ &\leq \left\| \Xi \mathbf{1}_{n_w} \right\|_\infty \|w\|_{l_\infty}, \quad i, j \geq 0. \end{aligned} \tag{6.28}$$

Taking supremum in i, j, from (6.28), we obtain

$$\begin{aligned} \|\mathscr{L}\|_{l_\infty - l_\infty} &= \sup_{w \neq 0} \frac{\|z_w\|_{l_\infty}}{\|w\|_{l_\infty}} \\ &\leq \left\| \Xi \mathbf{1}_{n_w} \right\|_\infty = \|\Xi\|_\infty. \end{aligned} \tag{6.29}$$

To complete the proof, we now show that $\|\mathscr{L}\|_{l_\infty - l_\infty} \geq \|\Xi\|_\infty$. Let $w_0 \in l_\infty(\mathbb{R}^{n_w})$ be a given input function with $\|w_0\|_{l_\infty} > 0$. We specify input disturbances $w(i, j)$ and $\underline{w}$ in (6.1) and (6.18) as

$$w(i, j) = \underline{w} = \|w_0\|_{l_\infty} \mathbf{1}_{n_w}, \quad i, j \geq 0.$$

By Lemma 6.4, we have

$$z_w(i, j) \succeq \underline{z}_w(i, j), \quad i, j \geq 0,$$

where $\underline{z}_w(i, j)$ is defined by (6.19). In addition, with the input $\underline{w}$, the unique EP of system (6.18) is given by

$$\underline{x}_e = (I_n - A_0 - A_s)^{-1} B_w \underline{w}.$$

Similar to the proof of Lemma 6.5 and Remark 6.1, we can conclude under condition (6.12) that there exists a scalar $\hat{\alpha} \in (0, 1)$ such that

$$\left(1 - \hat{\alpha}^{i+j}\right) \underline{x}_e \preceq \underline{x}_w(i, j) \preceq \underline{x}_e, \; i, j \geq 0. \tag{6.30}$$

Therefore, from (6.19) and (6.30), we obtain

$$\underline{z}_w(i, j) \succeq \left(\hat{\theta}_{i,j} C_0 + \sum_{k=1}^{N} \hat{\theta}_{i,j}^k C_k\right) \underline{x}_e + D_w \underline{w}, \tag{6.31}$$

where $\hat{\theta}_{i,j} = 1 - \hat{\alpha}^{i+j}$ and $\hat{\theta}_{i,j}^k = 1 - \hat{\alpha}^{i+j-\min(\overline{\tau}_{hk}, \overline{\tau}_{vk})}$.

From inequality (6.31), we get

$$\begin{aligned}
\sup_{i,j \geq 0} \|\underline{z}_w(i, j)\|_\infty &\geq \limsup_{i+j \to \infty} \left\| \left(\hat{\theta}_{i,j} C_0 + \sum_{k=1}^{N} \hat{\theta}_{i,j}^k C_k\right) \underline{x}_e + D_w \underline{w} \right\|_\infty \\
&= \left\| (C_0 + C_s)\underline{x}_e + D_w \underline{w} \right\|_\infty \\
&= \|\Xi \mathbf{1}_{n_w}\|_\infty \|w_0\|_{l_\infty} \\
&= \|\Xi\|_\infty \|w_0\|_{l_\infty}.
\end{aligned} \tag{6.32}$$

It follows from (6.32) and the fact $\|z_w\|_{l_\infty} \geq \sup_{i,j \geq 0} \|\underline{z}_w(i, j)\|_\infty$ that

$$\|\mathscr{L}\|_{l_\infty - l_\infty} = \sup_{\|w_0\|_{l_\infty} = 1} \|z_w\|_{l_\infty} \geq \|\Xi\|_\infty. \tag{6.33}$$

Finally, a combination of (6.29) and (6.33) gives $\|\mathscr{L}\|_{l_\infty - l_\infty} = \|\Xi\|_\infty$, which proves the equality (6.26). The proof is completed. □

Remark 6.2 Theorem 6.2 provides a characterization of l_∞-gain of positive 2-D systems with interval delays in the state and output vectors as presented in (6.1). It should be tressed here that, although the derived formula in (6.26) is delay-independent, this characterization, in general, is not an inherent property for 2-D time-delay systems. Thus, for 2-D systems with heterogeneous interval delays, how to formulate an l_∞-gain characterization proves to be an interesting and relevant issue. For this problem, the existing methods proposed for 1-D systems, for example, in [29, 33], which are based on lifting techniques and solution representations or the comparison method in [30], are not utilizable. In this chapter, a novel and systematic approach is presented. Specifically, by exploiting the monotonicity of state trajectories of systems (6.13)–(6.14) and (6.18)–(6.19), upper and lower bounds for state and output vectors of system (6.1) are obtained. Then, based on exponential convergence properties to equilibria of systems (6.13) and (6.18), an l_∞-gain characterization for system (6.1) is obtained. This result can be regarded as an extension of recent work [34] for 2-D time-delay systems. Noticing also that the method based on lifting techniques or solution representation can no longer be utilized for 2-D systems with interval time-varying delays.

6.3.2 l_∞-Gain Performance

In this section, based on the result of Theorem 6.2, we derive necessary and sufficient tractable conditions that are utilizable for the design problems involving l_∞-gain performance. Specifically, for a given $\gamma > 0$ referred to attenuation level, we formulate LP-based conditions by which the l_∞-gain satisfies $\|\mathscr{L}\|_{l_\infty - l_\infty} < \gamma$.

Theorem 6.3 *For a given $\gamma > 0$, the positive 2-D system* (6.1) *is GES and has l_∞-gain performance at level γ if and only if there exists a vector $\chi \in \mathbb{R}^n$, $\chi \succ 0$, that satisfies the following two LP-based conditions*

$$\left(A_0 + \sum_{k=1}^{N} A_k - I_n\right)\chi + B_w \mathbf{1}_{n_w} \prec 0, \tag{6.34}$$

$$\left(C_0 + \sum_{k=1}^{N} C_k\right)\chi + D_w \mathbf{1}_{n_w} - \gamma \mathbf{1}_{n_z} \prec 0. \tag{6.35}$$

Proof (Necessity) Let $\chi_s \in \mathbb{R}^n$, $\chi_s \succ 0$, be a positive vector that satisfies the stability condition (6.11). For a fixed $\varepsilon > 0$, we define

$$\chi = \varepsilon \chi_s + (I_n - A_0 - A_s)^{-1} B_w \mathbf{1}_{n_w},$$

where $A_s = \sum_{k=1}^{N} A_k$. It is clear that $\chi \succ 0$ and we have

$$(A_0 + A_s - I_n)\chi + B_w \mathbf{1}_{n_w} = \varepsilon (A_0 + A_s - I_n)\chi_s \prec 0. \tag{6.36}$$

On the other hand, by Theorem 6.2, $\|\mathscr{L}\|_{l_\infty - l_\infty} < \gamma$ if and only if

$$\Lambda = \left[(C_0 + C_s)(I_n - A_0 - A_s)^{-1} B_w + D_w\right] \mathbf{1}_{n_w} \prec \gamma \mathbf{1}_{n_z},$$

where $C_s = \sum_{k=1}^{N} C_k$. Thus, $\zeta = \gamma \mathbf{1}_{n_w} - \Lambda \succ 0$ and we have

$$(C_0 + C_s)\chi + D_w \mathbf{1}_{n_w} - \gamma \mathbf{1}_{n_z} = \varepsilon (C_0 + C_s)\chi_s - \zeta \prec 0 \tag{6.37}$$

for sufficiently small ε. The vector χ from (6.36) and (6.37) validates conditions (6.34) and (6.35).

(Sufficiency) Since $B_w\mathbf{1}_{n_w} \succeq 0$, condition (6.34) implies (6.11). By Theorem 6.1, system (6.1) with $w = 0$ is GES.

Let

$$\tilde{\chi} = (A_0 + A_s - I_n)\,\chi + B_w\mathbf{1}_{n_w}$$

then $\tilde{\chi} \prec 0$ and we have

$$\chi = (A_0 + A_d - I_n)^{-1}\left(\tilde{\chi} - B_w\mathbf{1}_{n_w}\right).$$

It follows from (6.35) that

$$\begin{aligned}\gamma\mathbf{1}_{n_z} &\succ (C_0 + C_s)\chi + D_w\mathbf{1}_{n_w}\\ &= (C_0 + C_s)(A_0 + A_s - I_n)^{-1}\tilde{\chi} + \Xi\mathbf{1}_{n_w},\end{aligned} \tag{6.38}$$

where the matrix Ξ is defined as in the proof of Theorem 6.2.

For any $w \in l_\infty(\mathbb{R}^{n_w})$ with $\|w\|_{l_\infty} = 1$, since

$$(A_0 + A_d - I_n)^{-1}\tilde{\chi} \succ 0,$$

from (6.28) and (6.38), we obtain

$$\|z\|_{l_\infty} \leq \|\Xi\mathbf{1}_{n_w}\|_\infty < \gamma.$$

The proof is completed. □

The following alternative performance conditions can be obtain by similar arguments used in the proof of Theorem 6.3.

Theorem 6.4 *For a given $\gamma > 0$, the positive 2-D system* (6.1) *is GES and has l_∞-gain performance at level γ if and only if the two following LP-based conditions are feasible for a vector $\eta \in \mathbb{R}^n$, $\eta \succ 0$*

$$\eta^\top\left(A_0 + \sum_{k=1}^{N} A_k - I_n\right) + \mathbf{1}_{n_z}^\top\left(C_0 + \sum_{k=1}^{N} C_k\right) \prec 0, \tag{6.39}$$

$$\eta^\top B_w + \mathbf{1}_{n_z}^\top D_w - \gamma\mathbf{1}_{n_w}^\top \prec 0. \tag{6.40}$$

6.3.3 *The Case of 2-D Time-Delay Systems Described by FMLSS Model*

In this section, we develop analogous results of the preceding section to the following 2-D systems described by the Fornasini–Marchesini local state-space model

$$\begin{aligned} x(i+1,j+1) &= A_{01}x(i+1,j) + A_{02}x(i,j+1) \\ &\quad + \sum_{k=1}^{N} \left[A_{d1k}x\left(i-\tau_{hk}(i), j+1\right) + A_{d2k}x\left(i+1, j-\tau_{vk}(j)\right)\right] \\ &\quad + B_{1w}w(i+1,j) + B_{2w}w(i,j+1), \\ z(i,j) &= Cx(i,j) + D_w w(i,j), \end{aligned} \tag{6.41}$$

where A_{01}, A_{02}, A_{d1k}, A_{d2k}, $k \in [N]$, and B_{1w}, B_{2w}, C, D_w are given matrices of appropriate dimensions. The initial condition of (6.41) is specified as

$$x(i,j) = \begin{cases} \phi_h(i,j), & i \in \mathbb{Z}[-\bar{\tau}_h, 0],\ j \in \mathbb{Z}[0, T_h], \\ \phi_v(i,j), & i \in \mathbb{Z}[0, T_v],\ j \in \mathbb{Z}[-\bar{\tau}_v, 0], \\ 0, & i > T_h \text{ and } j > T_v, \end{cases} \tag{6.42}$$

where $\bar{\tau}_h$ and $\bar{\tau}_v$ are positive integers such that $0 \leq \tau_{hk}(i) \leq \bar{\tau}_h$ and $0 \leq \tau_{vk}(j) \leq \bar{\tau}_v$.

As discussed in Chap. 4, system (6.41) is positive for any delays $\tau_{hk}(i)$ and $\tau_{vk}(j)$ if and only if the associated matrices A_{01}, A_{02}, A_{d1k}, A_{d2k}, $k \in [N]$, and B_{1w}, B_{2w}, C, D_w are all nonnegative. In addition, the positive system (6.41) with $w = 0$ is GES if and only if one of the following equivalent conditions holds

(i) There exists a vector $\chi \in \mathbb{R}^n$, $\chi \succ 0$, such that

$$\left(A_{01} + A_{02} + \tilde{A}_s - I_n\right)\chi \prec 0, \tag{6.43}$$

where $\tilde{A}_s = \sum_{k=1}^{N}(A_{d1k} + A_{d2k})$.

(ii) There exists a vector $\eta \in \mathbb{R}^n$, $\eta \succ 0$, such that

$$\eta^\top \left(A_{01} + A_{02} + \tilde{A}_s - I_n\right) \prec 0. \tag{6.44}$$

(iii) The spectral radius of the matrix $A_{01} + A_{02} + \tilde{A}_s$ satisfies

$$\rho(A_{01} + A_{02} + \tilde{A}_s) < 1. \tag{6.45}$$

The following result is obtained by a similar derivation process of Theorem 6.2.

Theorem 6.5 *Assume that the 2-D system* (6.41) *is positive and GES. The value of l_∞-gain of system* (6.41) *under zero initial condition is expressed as*

$$\|\mathscr{L}\|_{l_\infty - l_\infty} = \left\| C\left(I_n - A_{01} - A_{02} - \tilde{A}_s\right)^{-1}(B_{1w} + B_{2w}) + D_w \right\|_\infty, \tag{6.46}$$

where $\tilde{A}_s = \sum_{k=1}^{N}(A_{d1k} + A_{d2k})$.

Proof The proof is similar to that of Theorem 6.2. We omit it here. □

Based on the obtained result of Theorem 6.5, l_∞-gain performance of system (6.41) are derived in the following theorems.

Theorem 6.6 *For a given $\gamma > 0$, the positive 2-D system* (6.41) *is GES and has l_∞-gain performance at level γ if and only if one of the following conditions holds*

(i) *There exists a vector $\chi \in \mathbb{R}^n$, $\chi \succ 0$, that satisfies the following two LP-based conditions*

$$\begin{aligned} &\left(A_{01} + A_{02} + \tilde{A}_s - I_n\right)\chi + (B_{1w} + B_{2w})\mathbf{1}_{n_w} \prec 0, \\ &C\chi + D_w\mathbf{1}_{n_w} - \gamma\mathbf{1}_{n_z} \prec 0. \end{aligned} \tag{6.47}$$

(ii) *There exists a vector $\eta \in \mathbb{R}^n$, $\eta \succ 0$, that satisfies the following two LP-based conditions*

$$\begin{aligned} &\eta^\top\left(A_{01} + A_{02} + \tilde{A}_s - I_n\right) + \mathbf{1}_{n_z}^\top C \prec 0, \\ &\eta^\top(B_{1w} + B_{2w}) + \mathbf{1}_{n_z}^\top D_w - \gamma\mathbf{1}_{n_w}^\top \prec 0. \end{aligned} \tag{6.48}$$

6.4 Static Output-Feedback l_∞-Gain Control

As an application, in this section, we utilize the l_∞-gain performance analysis results presented in the preceding section to address the synthesis problem of an l_∞-gain static output-feedback control for the following positive 2-D systems with interval delays

$$\begin{aligned} \begin{bmatrix} x^h(i+1,j) \\ x^v(i,j+1) \end{bmatrix} &= A\begin{bmatrix} x^h(i,j) \\ x^v(i,j) \end{bmatrix} + A_d\begin{bmatrix} x^h(i-\tau_h(i),j) \\ x^v(i,j-\tau_v(j)) \end{bmatrix} + Bu(i,j) + B_w w(i,j), \\ z(i,j) &= C\begin{bmatrix} x^h(i,j) \\ x^v(i,j) \end{bmatrix} + C_d\begin{bmatrix} x^h(i-\tau_h(i),j) \\ x^v(i,j-\tau_v(j)) \end{bmatrix} + Du(i,j) + D_w w(i,j), \\ x_o(i,j) &= Mx(i,j) + Nw(i,j), \end{aligned} \tag{6.49}$$

where $u(i,j) \in \mathbb{R}^{n_u}$ is the control input and $x_o(i,j) \in \mathbb{R}^{n_o}$ is the measurement output vector, respectively. B, D, M, N are given real matrices and other notations are described as in system (6.1).

For positive system (6.49), a static output-feedback controller (SOFC) will be design in the form

$$u(i,j) = -Kx_o(i,j) \tag{6.50}$$

to make the closed-loop system positive, GES and possess a prescribed l_∞-gain performance, where $K \in \mathbb{R}^{n_u \times n_o}$ is the controller gain which will be determined. By incorporating the controller (6.50), the closed-loop system of (6.49) is obtained as

$$\begin{aligned}\begin{bmatrix} x^h(i+1,j) \\ x^v(i,j+1) \end{bmatrix} &= A_c \begin{bmatrix} x^h(i,j) \\ x^v(i,j) \end{bmatrix} + A_d \begin{bmatrix} x^h(i-\tau_h(i),j) \\ x^v(i,j-\tau_v(j)) \end{bmatrix} + B_{cw} w(i,j), \\ z(i,j) &= C_c \begin{bmatrix} x^h(i,j) \\ x^v(i,j) \end{bmatrix} + C_d \begin{bmatrix} x^h(i-\tau_h(i),j) \\ x^v(i,j-\tau_v(j)) \end{bmatrix} + D_{cw} w(i,j),\end{aligned} \tag{6.51}$$

where $A_c = A - BKM$, $B_{cw} = B_w - BKN$, $C_c = C - DKM$ and $D_{cw} = D_w - DKN$.

According to Proposition 6.1, the closed-loop system (6.51) is positive if and only if the following condition is satisfied

$$\left[\begin{array}{c|c} A & B_w \\ \hline C & D_w \end{array}\right] - \begin{bmatrix} B \\ D \end{bmatrix} K \begin{bmatrix} M & N \end{bmatrix} \succeq 0. \tag{6.52}$$

In addition, for a given $\gamma > 0$, by Theorem 6.1, system (6.51) is GES and has l_∞-gain performance at level γ if and only if the LP-based condition

$$\left(\mathscr{A} - \begin{bmatrix} B \\ D \end{bmatrix} K \begin{bmatrix} M & N \end{bmatrix} \right)^\top \begin{bmatrix} \eta \\ \mathbf{1}_{n_z} \end{bmatrix} \prec \gamma \begin{bmatrix} 0 \\ \mathbf{1}_{n_w} \end{bmatrix} \tag{6.53}$$

is feasible for a vector $\eta \in \mathbb{R}^n$, $\eta \succ 0$, where

$$\mathscr{A} = \left[\begin{array}{c|c} A + A_d - I_n & B_w \\ \hline C + C_d & D_w \end{array}\right].$$

We will determine whether a controller gain K satisfying (6.52) and (6.53) exists. To tackle this problem, we extend the ideas of vertex optimization proposed in [12]. In the following we illustrate the ideas by considering the case of single-input systems (i.e. $n_u = 1$).

6.4.1 Matrix Transformation Approach

The method of using certain type of matrix transformations is a very common approach in the existing literature to obtain a desired controller gain K (see, e.g., [29, 33, 35]). Note at first that, for a nontrivial control design, the matrix $\begin{bmatrix} B^\top & D^\top \end{bmatrix}$ is not identical zero. Thus, for any $\eta \in \mathbb{R}^n$, $\eta \succ 0$, it is clear that $B^\top \eta + D^\top \mathbf{1}_{n_z} > 0$. Thus, as revealed by (6.53), we define the transformation

$$K^\top \begin{bmatrix} B^\top & D^\top \end{bmatrix} \begin{bmatrix} \eta \\ \mathbf{1}_{n_z} \end{bmatrix} = Z \in \mathbb{R}^{n_o}. \tag{6.54}$$

Then, subject to (6.54), condition (6.52) holds if and only if

$$\left[\begin{array}{c:c} A & B_w \\ \hdashline C & D_w \end{array}\right] (B^\top \eta + D^\top \mathbf{1}_{n_z}) - \begin{bmatrix} B \\ D \end{bmatrix} Z^\top \begin{bmatrix} M & N \end{bmatrix} \succeq 0 \tag{6.55}$$

and condition (6.53) becomes

$$\mathscr{A}^\top \begin{bmatrix} \eta \\ \mathbf{1}_{n_z} \end{bmatrix} - \begin{bmatrix} M^\top \\ N^\top \end{bmatrix} Z \prec \gamma \begin{bmatrix} 0 \\ \mathbf{1}_{n_w} \end{bmatrix}. \tag{6.56}$$

Based on the above derivation, we have the following result.

Proposition 6.3 *For a given $\gamma > 0$, there exists an SOFC in the form of* (6.50) *that makes the closed-loop system* (6.51) *positive, GES and has l_∞-gain performance at level γ if the two LP-based conditions* (6.55)–(6.56) *is feasible for a vector $\eta \in \mathbb{R}^n$, $\eta \succ 0$, and a matrix $Z \in \mathbb{R}^{n_o}$. A desired controller gain is given by*

$$K = \frac{1}{B^\top \eta + D^\top \mathbf{1}_{n_z}} Z^\top. \tag{6.57}$$

Proposition 6.3 provides a scheme by which an explicit form of a desired controller gain can be obtained. However, this method only can help to derive tractable LP-based conditions subject to some constraints (matrix transformation equations).

6.4.2 Vertex Optimization Approach

In this section, we utilize vertex optimization techniques developed in [12] to tackle with the existence of a desired controller gain K subject to conditions (6.52)–(6.53). Let $G = \begin{bmatrix} B \\ D \end{bmatrix} = (g_i) \in \mathbb{R}_+^{n+n_o}$ and assume, without loss of generality, that $G \neq 0$ (i.e. at least one component $g_i > 0$). We also denote

$$\mathscr{M} = \left[\begin{array}{c:c} A & B_w \\ \hdashline C & D_w \end{array}\right] = (m_{ij}) \in \mathbb{R}_+^{(n+n_z)\times(n+n_w)}$$

and decompose the matrix

$$H = \begin{bmatrix} M & N \end{bmatrix} = \begin{bmatrix} h_1 \; h_2 \; \cdots \; h_{n+n_w} \end{bmatrix},$$

where $h_j \in \mathbb{R}^{n_o}$.

Since the controller gain K is a row matrix $K = k^\top$, $k \in \mathbb{R}^{n_o}$, condition (6.52) holds if and only if

$$k^\top h_j \le \sigma_j^* \triangleq \min_{1\le i\le n+n_z} \left\{ \frac{m_{ij}}{g_i} : g_i > 0 \right\}. \tag{6.58}$$

According to (6.58), we define the polyhedron Δ_P by

$$\Delta_P = \left\{ k \in \mathbb{R}^{n_o} \middle| k^\top H \preceq \sigma^* \right\}, \tag{6.59}$$

where $\sigma^* = \begin{bmatrix}\sigma_1^* \ \sigma_2^* \ \cdots \ \sigma_{n+n_w}^*\end{bmatrix}$. Assume that the matrix $H = \begin{bmatrix} M & N \end{bmatrix}$ has full-row rank, Eq. (6.59) defines a nonempty convex polyhedron. Let $\mathscr{V}$ be the set of vertices of Δ_P.

For a fixed vector $\eta \succ 0$, the function $\varphi_\eta(k) = k^\top H\eta$ is continuous and, thus, attains its maximum on the compact $\Omega = \Delta_P \cap \left\{k \in \mathbb{R}^{n_o} : \varphi_\eta(k) \ge 0\right\}$.

Lemma 6.6 *There exists a vertex $k_v \in \mathscr{V}$ such that $\varphi_\eta(k_v) = \max_{k\in\Delta_P} \varphi_\eta(k)$.*

Proof See, for example, Lemma 3 in [12]. □

The following theorem gives a criterion in terms of LP-based conditions for the existence of a desired l_∞-gain controller (6.50).

Theorem 6.7 *For a given $\gamma > 0$, there exists an SOFC in the form of* (6.50) *that makes the closed-loop system* (6.51) *positive, GES and has l_∞-gain performance at level γ if and only if there exists a vertex $k_v^* \in \mathscr{V}$ of Δ_P such that the LP-based problem*

$$\left(\begin{bmatrix} A + A_d - I_n & B_w \\ C + C_d & D_w \end{bmatrix} - \begin{bmatrix} B \\ D \end{bmatrix} k_v^{*\top} \begin{bmatrix} M & N \end{bmatrix} \right) \begin{bmatrix} \eta \\ \mathbf{1}_{n_w} \end{bmatrix} \prec \gamma \begin{bmatrix} 0 \\ \mathbf{1}_{n_z} \end{bmatrix} \tag{6.60}$$

is feasible for a vector $\eta \in \mathbb{R}^n$, $\eta \succ 0$.

Proof The *Sufficiency* is obvious. We now prove the *Necessity*. Let $K = k_0^\top$ be a desired controller gain. Then, it follows from (6.52) and (6.58), (6.59) that $k_0 \in \Delta_P$.

If $k_0 \notin \mathscr{V}$, we then activate the optimization problem

$$\texttt{maximize} \quad \varphi_\eta(k) = k^\top H \begin{bmatrix} \eta \\ \mathbf{1}_{n_w} \end{bmatrix} \quad \texttt{s.t.} \quad k \in \Delta_P. \tag{6.61}$$

By Lemma 6.6, there exists a vertex $k_v^* \in \mathscr{V}$ such that

$$\varphi_\eta(k_v^*) = \max_{k\in\Delta_P} \varphi_\eta(k).$$

Since $\varphi_\eta(k) \le \varphi_\eta(k_v^*)$ for all $k \in \Delta_P$, we have

$$\left(\mathscr{A} - Gk_v^{*\top} H\right)\begin{bmatrix}\eta \\ \mathbf{1}_{n_w}\end{bmatrix} \preceq (\mathscr{A} - GKH)\begin{bmatrix}\eta \\ \mathbf{1}_{n_w}\end{bmatrix}$$

for any K satisfying (6.52), where

$$\mathscr{A} = \left[\begin{array}{c:c} A + A_d - I_n & B_w \\ \hdashline C + C_d & D_w \end{array}\right] \text{ and } G = \begin{bmatrix} B \\ D \end{bmatrix}.$$

The above inequality validates condition (6.60). The proof is completed. □

For single-input single-output (SISO) systems, an optimal controller gain is explicitly obtained as

$$K = k_{\text{op}}^* = \min\left\{\frac{m_{ij}}{g_i h_j} : g_i h_j > 0\right\}$$

and a desired SOFC (6.50) exists if and only if there exists a vector $\eta \in \mathbb{R}^n$, $\eta \succ 0$, such that

$$\left(\mathscr{A} - k_{op}^* GH\right)\begin{bmatrix}\eta \\ \mathbf{1}_{n_w}\end{bmatrix} \prec \gamma \begin{bmatrix} 0 \\ \mathbf{1}_{n_z}\end{bmatrix}. \tag{6.62}$$

6.5 Numerical Examples

In this section, we give some numerical examples and simulations to illustrate the effectiveness of the design method.

Example 6.1 Consider a 2-D system as given in (6.1)–(6.2) with the system matrices

$$A_0 = \begin{bmatrix} 0.15 & 0.03 & 0.05 \\ 0.18 & 0.25 & 0.35 \\ 0.2 & 0.15 & 0.2 \end{bmatrix}, \quad A_1 = \begin{bmatrix} 0.2 & 0 & 0.1 \\ 0.1 & 0.15 & 0.25 \\ 0.15 & 0.1 & 0.2 \end{bmatrix}, \quad B_w = \begin{bmatrix} 0.25 \\ 0.15 \\ 0.05 \end{bmatrix},$$

$$C_0 = \begin{bmatrix} 0.25 & 0.35 & 0.48 \\ 0.5 & 0.6 & 0.15 \end{bmatrix}, \quad D_w = \begin{bmatrix} 0.35 \\ 0.16 \end{bmatrix}.$$

By using a search algorithm via the `linprog` toolbox in Matlab, it is found that the two LP-based conditions (6.34)–(6.35) is feasible with $\gamma \geq \gamma_{\min} \triangleq 1.9691$.

To support for the illustration of the analysis results, we take the interval delays $\tau_h(i) = 2 + 3|\sin(i\pi/2)|$ and $\tau_v(j) = 1 + 4|\cos(j\pi/2)|$. Some state trajectories of output $z(i, j)$ with respect to $\|w\|_{l_\infty} = 1$ and zero initial condition are presented in Figs. 6.1 and 6.2. In addition, the corresponding trajectory of input $w(i, j)$ is drawn in Fig. 6.3. The simulation results in Figs. 6.1 and 6.2 indicate that $\|z\|_{l_\infty} < \gamma_{\min}\|w\|_{l_\infty}$ as revealed by the obtained theoretical results.

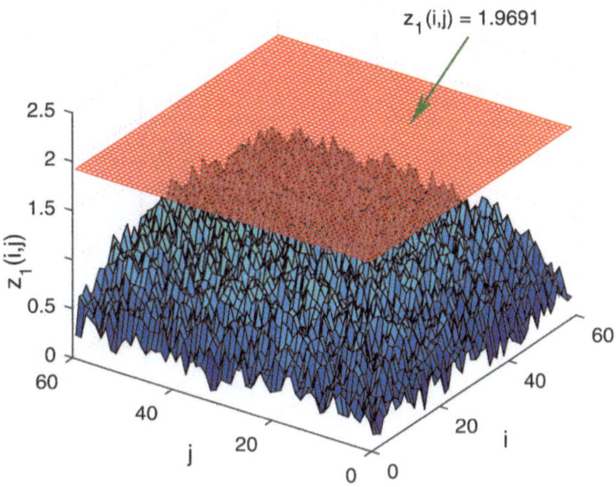

Fig. 6.1 A trajectory of $z_1(i, j)$ with zero initial condition, $\|w\|_{l_\infty} = 1$, and delays $\tau_h(i) = 2 + 3|\sin(i\pi/2)|$, $\tau_v(j) = 1 + 4|\cos(j\pi/2)|$

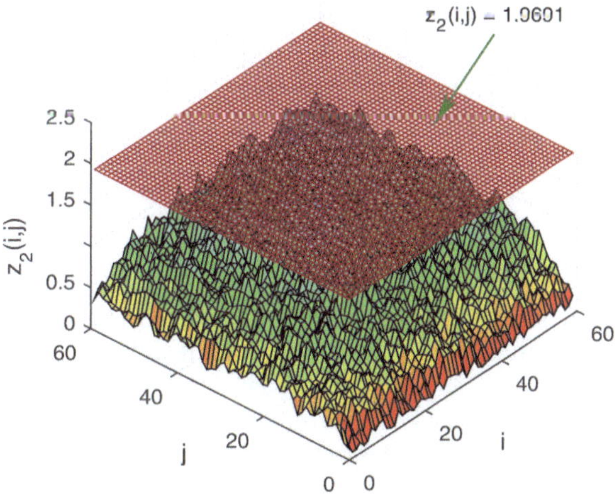

Fig. 6.2 A trajectory of $z_2(i, j)$ with zero initial condition, $\|w\|_{l_\infty} = 1$, and delays $\tau_h(i) = 2 + 3|\sin(i\pi/2)|$, $\tau_v(j) = 1 + 4|\cos(j\pi/2)|$

Example 6.2 Consider a 2-D control system represented as

$$\begin{aligned}
\begin{bmatrix} x^h(i+1, j) \\ x^v(i, j+1) \end{bmatrix} &= A \begin{bmatrix} x^h(i, j) \\ x^v(i, j) \end{bmatrix} + Bu(i, j) + B_w w(i, j), \\
z(i, j) &= C \begin{bmatrix} x^h(i, j) \\ x^v(i, j) \end{bmatrix} + C_d \begin{bmatrix} x^h(i - \tau_h(i), j) \\ x^v(i, j - \tau_v(j)) \end{bmatrix} + D_w w(i, j), \\
x_o(i, j) &= Mx(i, j) + Nw(i, j),
\end{aligned} \tag{6.63}$$

where the associated system matrices are given by

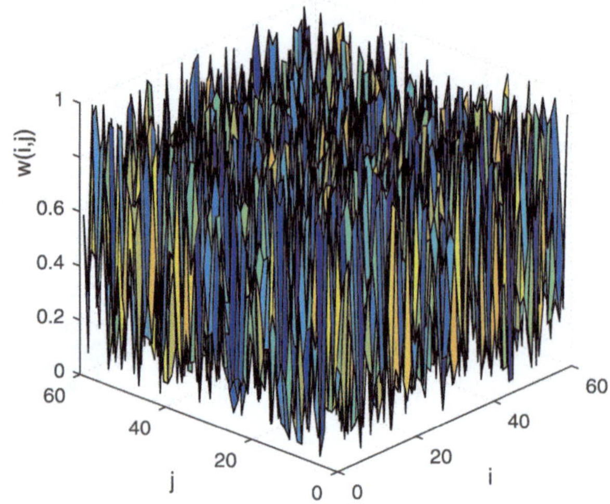

Fig. 6.3 A state trajectory of disturbance $w(i,j)$ with $\|w\|_{l_\infty}=1$

$$A=\begin{bmatrix}1.0 & 0.2 & 0.15\\ 0.2 & 0.4 & 0.45\\ 0.2 & 0.35 & 0.7\end{bmatrix},\quad B=\begin{bmatrix}1.0\\ 0\\ 1.0\end{bmatrix},\quad B_w=\begin{bmatrix}0.1\\ 0\\ 0.15\end{bmatrix}$$

$$C=\begin{bmatrix}0.2 & 0.25 & 0.26\\ 0.4 & 0.18 & 0.25\end{bmatrix},\quad C_d=\begin{bmatrix}0.1 & 0.1 & 0.05\\ 0.15 & 0.11 & 0.12\end{bmatrix},$$

$$D_w=\begin{bmatrix}0.1\\ 0.2\end{bmatrix},\quad M=\begin{bmatrix}1 & 1 & 0\end{bmatrix},\quad N=0.1.$$

System (6.63) is positive. However, the unforced system (i.e. with $u=0$) is unstable. Figure 6.4 presents a simulation result of $\|z(i,j)\|_\infty$ with respect to $\|w\|_{l_\infty}=1$, delays $\tau_h(i)=1+5|\sin(i\pi/2)|$, $\tau_v(j)=2+4|\cos(j\pi/2)|$ and under zero initial condition. It can be seen from Fig. 6.4 that the output $\|z(i,j)\|_\infty$ is unbounded.

Now, we apply the stabilization criterion in Theorem 6.7. The set of vertices $\mathscr{V}$ of the polyhedron Δ_P defined in (6.59) contains a unique element

$$k^*_{\mathrm{op}}=\min\left\{\frac{m_{ij}}{g_jh_j}\;\middle|\; g_ih_j\neq 0\right\}=0.2.$$

This is also the optimal controller gain. By this vertex, condition (6.60) or (6.62) is feasible for a vector $\eta\succ 0$ if and only if $\gamma>\gamma_{\min}=1.5316$. By Theorem 6.7, the closed-loop system of (6.63) is positive, GES and has l_∞-gain performance at level $\gamma>\gamma_{\min}$.

State trajectories $z_1(i,j)$ and $z_2(i,j)$ of the output vector of the closed-loop system with respect to $\|w\|_{l_\infty}=1$, delays $\tau_h(i)=1+5|\sin(i\pi/2)|$, $\tau_v(j)=2+4|\cos(j\pi/2)|$ and zero initial condition are presented in Figs. 6.5 and 6.6. Again, the simulation results in Figs. 6.5 and 6.6 show the effectiveness of the analysis results.

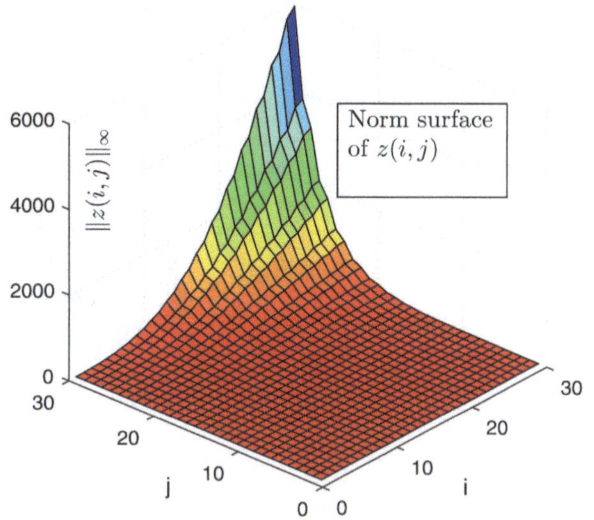

Fig. 6.4 An open-loop trajectory $\|z(i, j)\|_\infty$ with $\|w\|_{l_\infty} = 1$ and delays $\tau_h(i) = 1 + 5|\sin(i\pi/2)|$, $\tau_v(j) = 2 + 4|\cos(j\pi/2)|$

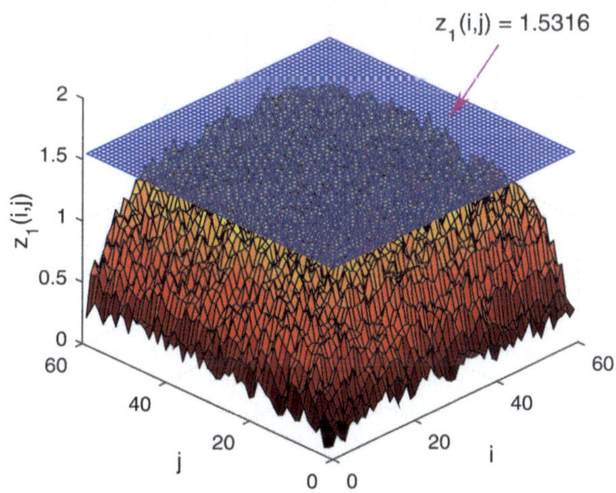

Fig. 6.5 State trajectory of $z_1(i, j)$ of the closed-loop system with zero initial condition, $\|w\|_{l_\infty} = 1$, and delays $\tau_h(i) = 1 + 5$ $|\sin(i\pi/2)|$, $\tau_v(j) = 2 + 4$ $|\cos(j\pi/2)|$

6.6 Conclusion

The problem of peak-to-peak performance analysis has been addressed in this chapter for 2-D positive systems with interval time-varying delays in both Roesser and Fornasini-Marchesini local state-space models. A characterization of l_∞-induced norm of the input-output operator has been formulated using novel comparison techniques involving interval state estimations subject to peak values of exogenous disturbances. On the basis of the obtained l_∞-gain characterization, necessary and sufficient LP-based conditions for l_∞-induced performance of the system with a prescribed

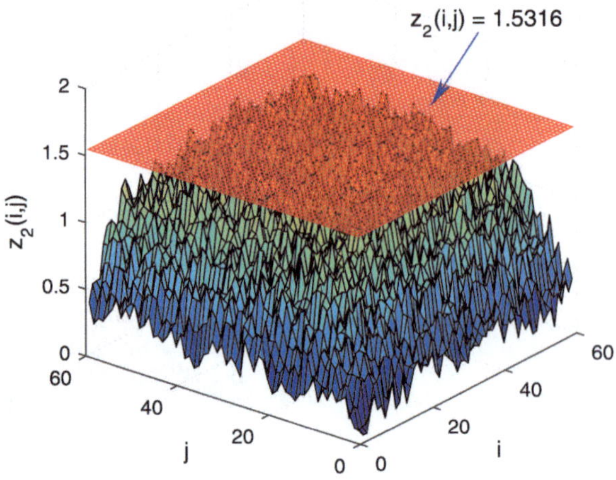

Fig. 6.6 State trajectory of $z_2(i, j)$ of the closed-loop system with zero initial condition, $\|w\|_{l_\infty} = 1$, and delays $\tau_h(i) = 1 + 5 |\sin(i\pi/2)|$, $\tau_v(j) = 2 + 4 |\cos(j\pi/2)|$

attenuation level have been formulated. As an application, the problem of l_∞-gain control via static output-feedback controllers has also been discussed. Numerical examples and simulations have been provided to illustrate the effectiveness of the proposed method.

References

1. Farina, L., Rinaldi, S.: Positive Linear Systems: Theory and Applications. John Wiley & Sons, New York (2000)
2. Briat, C.: Stability and performance analysis of linear positive systems with delays using input-output methods. Int J Control **91**, 1669–1692 (2018)
3. Wang, Z., Lim, C.C., Shen, Y.: Interval observer design for uncertain discrete-time linear systems. Syst Control Lett **116**, 41–46 (2018)
4. Briat, C.: Hybrid $L_\infty \times \ell_\infty$-performance analysis and control of linear time-varying impulsive and switched positive systems. Nonlin Anal Hybrid Syst **39**, 100980 (2021)
5. Jacquez, J.: Compartmental Analysis in Biology and Medicine. Univ Michigan Press, Ann Arbor (1985)
6. Smith, H.: Monotone Dynamical Systems: An Introduction to the Theory of Competitive and Cooperative Systems. American Mathematical Society, Providence, RI (2008)
7. Haddad, W.M., Chellaboina, V.S., Hui, Q.: Nonnegative and Compartmental Dynamical Systems. Princeton Univ Press, Princeton (NJ) (2010)
8. Hien, L.V., Trinh, H.: Delay-dependent stability and stabilization of two-dimensional positive Markov jump systems with delays. IET Control Theory Appl **11**, 1603–1610 (2017)
9. Hien, L.V., Trinh, H.: Exponential stability of two-dimensional homogeneous monotone systems with bounded directional delays. IEEE Trans Autom Control **63**, 2694–2700 (2018)
10. Hien, L.V., Trinh, H.: Observers design for 2-D positive time-delay Roesser systems. IEEE Trans Circuit Syst: Expr Brief **65**, 476–480 (2018)
11. Fu, J., Duan, Z., Ghous, I.: ℓ_1-induced norm and controller synthesis for positive 2D systems with multiple delays. J Frank Inst **357**, 7904–7920 (2020)

12. Hien, L.V., Trinh, H., Pathirana, P.N.: On ℓ_1-gain control of 2-D positive Roesser systems with directional delays: Necessary and sufficient conditions. Automatica **112**, 108720 (2020)
13. Duan, Z., Sun, Y., Ahn, C.K., Xiang, Z., Ghous, I.: L_1-gain control design for positive 2D continuous delayed systems. IEEE Trans Circuit Syst: Expr Brief **69**, 1317–1321 (2022)
14. Ahn, C.K., Wu, L., Shi, P.: Stochastic stability analysis for 2-D Roesser systems with multiplicative noise. Automatica **69**, 356–363 (2016)
15. Hien, L.V., Trinh, H., Lan-Huong, N.T.: Delay-dependent energy-to-peak stability of 2-D time-delay Roesser systems with multiplicative stochastic noises. IEEE Trans Autom Control **64**, 5066–5073 (2019)
16. Chang, X.H., Liu, R.R., Park, J.H.: A further study on output feedback H_∞ control for discrete-time systems. IEEE Trans Circuit Syst-II: Expr Brief **67**, 305–309 (2020)
17. Subramanian, K., Muthukumar, P., Trinh, H.: Nonfragile sampled-data H_∞ control design for high-speed train with parametric uncertainties. Int J Robust Nonlinear Control **31**, 1021–1034 (2021)
18. Coutinho, D., de Souza, C.E., Kinnaert, M., Schons, S.: Robust observer design for a class of discrete-time nonlinear singular systems with persistent disturbances. Int J Adapt Control Signal Process **35**, 51–68 (2021)
19. Dahleh, M., Shamma, J.: Rejection of persistent bounded disturbances: Nonlinear controllers. Syst Control Lett **18**, 245–252 (1992)
20. Ahn, C.K., Shmaliy, Y.S., Zhao, S., You, S.H., Sakthivel, R.: Further results on induced l_∞ RH FIR filtering. IEEE Trans Circuit Syst-II: Expr Brief **65**, 1124–1128 (2018)
21. Donkers, M.F., Heemels, W.H.: Output-based event-triggered control with guaranteed $\mathscr{L}_\infty$ gain and improved and decentralized event-triggering. IEEE Trans Autom Control **57**, 1362–1376 (2012)
22. Fan, C., Lam, J., Xie, X.C.: Peak-to-peak filtering for periodic piecewise linear polytopic systems. Int J Syst Sci **49**, 1997–2011 (2018)
23. Wang, H., Wang, C., Zheng, Y., Chen, Z., He, S.: Finite-time peak-to-peak gain minimization. Int J Control Autom Syst **17**, 319–326 (2019)
24. Hua, M., Bian, C., Deng, F., Fei, J., Chen, H., Cheng, P., Yao, F., Peng, Y.: Quantized peak-to-peak filtering for continuous-time nonhomogeneous Markov jump systems with structured uncertainty. J Frank Inst **357**, 13158–13188 (2020)
25. Ahn, C.K., Shi, P., Wu, L.: l_∞-gain performance analysis for two-dimensional Roesser systems with persistent bounded disturbance and saturation nonlinearity. Info Sci **333**, 126–139 (2016)
26. Ahn, C.K., Shi, P., Basin, M.V.: Two-dimensional peak-to-peak filtering for stochastic Fornasini-Marchesini systems. IEEE Trans Autom Control **63**, 1472–1479 (2018)
27. Knorn, F., Mason, O., Shorten, R.: On linear co-positive Lyapunov functions for sets of linear positive systems. Automatica **45**, 1943–1947 (2009)
28. Rami, M.A., Schönlein, M., Jordan, J.: Estimation of linear positive systems with unknown time-varying delays. Eur J Control **19**, 179–187 (2013)
29. Chen, X., Chen, M., Wang, L., Shen, J., Hu, J.: Static output-feedback controller synthesis for positive systems under ℓ_∞ performance. Int J Control Autom Syst **17**, 2871–2880 (2019)
30. Shen, J., Lam, J.: ℓ_∞/L_∞-gain analysis for positive linear systems with unbounded time-varying delays. IEEE Trans Autom Control **60**, 857–862 (2015)
31. Hien, L.V., Trinh, H.: Switching design for suboptimal guaranteed cost control of 2-D nonlinear switched systems in the Roesser model. Nonl Anal. Hybrid Syst **24**, 45–57 (2017)
32. Berman, A., Plemmons, R.J.: Nonnegative Matrices in the Mathematical Sciences. SIAM, Philadelphia (1994)
33. Chen, X., Wang, L., Chen, M., Shen, J.: ℓ_∞-induced output-feedback controller synthesis for positive nonlinear systems via T-S fuzzy model approach. Fuzzy Sets Syst **385**, 98–110 (2020)
34. Hong, M.T., Hien, L.V., Hang, T.T.M.: Static output-feedback peak-to-peak gain control of discrete-time positive linear systems with diverse interval delays. J Syst Control Engi **236**, 1578–1589 (2022)
35. Zhang, J., Zhao, X., Zhang, R.: An improved approach to controller design of positive systems using controller gain decomposition. J Frankl Inst **354**, 1356–1373 (2017)

Chapter 7
Stability of Two-Dimensional Time-Delay Roesser Systems with Markovian Switchings

Abstract In this chapter, the problem of stability analysis is presented for 2-D Markov jump systems in the Roesser model with interval delays and uncertain transition rates. Specifically, it is assumed that transition probabilities of the jumping process driven by a Markov chain are not exactly known but can be estimated by prescribed ranges. Inspired by the conventional Lyapunov–Krasovskii functional method, an analysis scheme for the stochastic stability property is first developed. The proposed scheme is then utilized in combination with weighted 2-D summation inequalities to derive delay-dependent stochastic stability conditions in terms of linear matrix inequalities for 2-D Markov jump Roesser systems with interval delays. Numerical examples are given to illustrate the effectiveness of the obtained results.

Keywords 2-D Markov jump systems · Roesser model · Stochastic stability · Uncertain transition probabilities · Directional delays

7.1 Introduction

As discussed in previous chapters, time-delay frequently occurs in control engineering systems due to technical limitations, for example, the constrained speed of signal processing and data transmission between various parts of the system [1, 2]. The existence of time-delay often degrades system performance and even causes the system instability [1]. Thus, the study regarding analysis and control of time-delay systems plays an essential role in applied models, which has attracted remarkable research attention in the past few decades. We refer the reader to recent works [3–7] for 1-D systems and [8–12] for 2-D systems with delays.

Beside the effect of time-delay, jumping phenomena are also very popular in control engineering. Among categories of hybrid dynamical systems (i.e. systems with switching parameters), Markov jump systems (MJSs) form one of the most important classes of stochastic hybrid systems in which the switching process is driven by a finite-state Markov chain [13]. Due to their efficacy and flexibility in modeling real-world problems [14–16], remarkable research attention has been devoted to study the

L. V. Hien, *Two-Dimensional Time-Delay Systems*, Studies in Systems, Decision and Control 605, https://doi.org/10.1007/978-3-031-95678-2_7

analysis and control of MJSs with or without delays. A very rich literature has been established for various classes of one-dimensional (1-D) Markov jump systems. We refer to [17–23] for a few references.

It is important to note that transition probabilities (TPs) of the switching process are essential and heavily affect the system performance. In the existing literature, a large number of analysis and synthesis results have been exploited based on the assumption that the TPs can be completely accessible [14, 20, 24, 25]. However, in practice, TPs of the jumping process are typically incompletely known, for instance, when adequate samples are costly or time-consuming to obtain. To this regard, from control perspectives, it is significant and challenging to study MJSs with deficient mode information (i.e. the TPs are not exactly known). Recently, the theory of 2-D Markov jump systems has also been extensively studied. For example, in [26, 27], the problems of H_2 fault detection and asynchronous observer-based control were studied for Markov jump 2-D systems without delays. Asynchronous control under conventional state-feedback and event-triggered state-feedback was also investigated for delay-free Markov jump Roesser systems with nonideal transition probabilities (TPs) in [28–30]. Delay-dependent H_∞ filtering and stability conditions were derived in [31, 32] for 2-D MJSs with delays and deficient transition rate information.

In this chapter, we study the problem of stochastic stability of 2-D Markov jump Roesser systems with interval time-varying delays and uncertain TPs. More precisely, the TPs of the jumping process driven by a Markov chain are not exactly known but confined within certain ranges. An analysis scheme for the stochastic stability property is extended from the conventional Lyapunov–Krasovskii functional (LKF) method. The proposed scheme is then utilized in combination with weighted 2-D summation inequalities to derive delay-dependent stochastic stability conditions in terms of linear matrix inequalities (LMIs) for 2-D Markov jump Roesser systems with interval delays. Numerical examples are given to illustrate the effectiveness of the obtained results.

7.2 Preliminaries

7.2.1 Model Description

Let $(\Omega, \mathscr{F}, \mathbb{P})$ be a complete probability space and $\theta : \mathbb{N}_0 \times \mathbb{N}_0 \to \mathscr{M}$ be a two-time scales homogeneous Markov process specifying the system mode which takes values in a finite set $\mathscr{M} = \{1, 2, \ldots, s\}$ with transition probabilities (transition rates)

$$\mathbb{P}\{\theta(i+1, j) = n|\theta(i, j) = m\} = \mathbb{P}\{\theta(i, j+1) = n|\theta(i, j) = m\} = \pi_{mn}$$

for all $n, m \in \mathscr{M}$, where $\pi_{mn} \geq 0$ and $\sum_{n=1}^{s} \pi_{mn} = 1, m \in \mathscr{M}$. The transition probability matrix is denoted as $\Pi = (\pi_{mn})$.

Consider the following Markov jump 2-D system in the Roesser model with time-varying delays

$$\begin{bmatrix} x^h(i+1,j) \\ x^v(i,j+1) \end{bmatrix} = A_{\theta(i,j)} \begin{bmatrix} x^h(i,j) \\ x^v(i,j) \end{bmatrix} + A_{d\theta(i,j)} \begin{bmatrix} x^h(i-d_h(i),j) \\ x^v(i,j-d_v(j)) \end{bmatrix}, \tag{7.1}$$

where $x^h(i,j) \in \mathbb{R}^{N_h}$ and $x^v(i,j) \in \mathbb{R}^{N_v}$ are the horizontal and the vertical state vectors, respectively, $A_{\theta(i,j)}$ and $A_{d\theta(i,j)}$ are mode-dependent real matrices of appropriate dimensions, $d_h(i)$ and $d_v(j)$ represent the time-delay along horizontal and vertical directions. It is assumed that the directional delays are confined within the ranges

$$d_1^h \le d_h(i) \le d_2^h, \quad d_1^v \le d_v(j) \le d_2^v, \quad \text{for all } i, j \in \mathbb{N}_0, \tag{7.2}$$

where d_1^h, d_1^v and d_2^h, d_2^v are known positive integers involving the lower bounds and upper bounds of delays. For the brevity of notation, when $\theta(i,j) = m \in \mathscr{M}$, the system matrices $A_{\theta(i,j)}$ and $A_{d\theta(i,j)}$ will be simply denoted as A_m and A_{dm}.

In system (7.1), initial mode $\theta_0 = \theta(0,0)$ of the Markov chain is defined by the initial probability distribution $\pi = (\pi_1, \pi_2, \ldots, \pi_s)$, where $\pi_m = \mathbb{P}\{\theta(0,0) = m\}$, and initial condition is defined as

$$\begin{aligned} x^h(i,j) &= \varphi_{i,j}, \ i \in \mathbb{Z}[-d_2^h, 0], \ j \in \mathbb{N}_0, \\ x^v(i,j) &= \psi_{i,j}, \ i \in \mathbb{N}_0, \ j \in \mathbb{Z}[-d_2^v, 0]. \end{aligned} \tag{7.3}$$

Moreover, we assume that the initial condition satisfies

$$\lim_{\kappa \to \infty} \sum_{k=0}^{\kappa} \mathbb{E}\left[\hat{\varphi}_k^2 + \hat{\psi}_k^2\right] < \infty, \tag{7.4}$$

where $\mathbb{E}[.]$ denotes the expectation, $\hat{\varphi}_k^2 = \sum_{i=-d_2^h}^{0} \varphi_{i,k}^\top \varphi_{i,k}$ and $\hat{\psi}_k^2 = \sum_{j=-d_2^v}^{0} \psi_{k,j}^\top \psi_{k,j}$.

It is assumed that, in model (7.1), the transition probabilities (TPs) are subject to uncertainties. More precisely, the transition probability matrix $\Pi = (\pi_{mn})$ can be represented as

$$\pi_{mn} = \hat{\pi}_{mn} + \Delta\pi_{mn}, \quad m, n \in \mathscr{M},$$

where $\Delta\pi_{mn} \in [-\delta_{mn}, \delta_{mn}]$ represents the error of uncertain transition rate π_{mn} and $\hat{\pi}_{mn}$, δ_{mn}, $m, n \in \mathscr{M}$, are known scalars. Without loss of generality, we assume in addition that $\hat{\pi}_{mn} - \delta_{mn} \ge 0$ for all $m, n \in \mathscr{M}$ and $\sum_{n=1}^{s} \hat{\pi}_{mn} \le 1$, $m \in \mathscr{M}$.

7.2.2 Problem Statement

Consider 2-D Markov jump systems with time-varying delays presented in (7.1). First, we introduce the following definition.

Definition 7.1 System (7.1) is said to be stochastically stable if

$$\mathbb{E}\left[\sum_{i,j=0}^{\infty}\|x(i,j)\|^2\Big|X_0,\theta_0\right]<\infty \tag{7.5}$$

for any state trajectory $x(i,j)=\begin{bmatrix}x^h(i,j)\\x^v(i,j)\end{bmatrix}$ and initial mode θ_0.

One of the main objectives of this chapter is to derive delay-dependent conditions which ensure the property of stochastic stability of system (7.1). An extended version of the conventional Lyapunov–Krasovskii functional (LKF) method is developed for 2-D Markov jump systems with delays. The proposed scheme is then utilized in combination with 2-D weighted summation inequalities to derive tractable LMI-based delay-dependent stability conditions.

7.3 Auxiliary Results

7.3.1 Mode-Dependent Functional Method

In this section, we formulate a technical lemma based on an extension of the 2-D LKF method. Let Θ be the set of finite-state functions $\theta:\mathbb{N}_0\times\mathbb{N}_0\to\mathscr{M}$, where $\mathscr{M}=\{1,2,\ldots,s\}$ with a given positive integer s. For a parameter-dependent functional $V:\Theta\times\mathbb{R}^N\to\mathbb{R}$, $(\theta,x)\mapsto V_\theta(x)$, defined by

$$V_\theta(x)=V_\theta^h(x^h)+V_\theta^v(x^v),$$

where $x^h\in\mathbb{R}^{N_h}$, $x^v\in\mathbb{R}^{N_v}$, $N=N_h+N_v$, $x=(x^h,x^v)$, and $V^h:\Theta\times\mathbb{R}^{N_h}\to\mathbb{R}$, $V^v:\Theta\times\mathbb{R}^{N_v}\to\mathbb{R}$ are some functionals. The weighted difference Δ_α of $V_\theta(x)$ along state trajectory $x(i,j)$ of system (7.1) is defined as

$$\Delta_\alpha V_{\theta(i,j)}(x(i,j))\triangleq\Delta_\alpha^h V_{\theta(i,j)}^h(x^h(i,j))+\Delta_\alpha^v V_{\theta(i,j)}^v(x^v(i,j)), \tag{7.6}$$

where

$$\Delta_\alpha^h V_{\theta(i,j)}^h(x^h(i,j))=\mathbb{E}[V_{\theta(i+1,j)}^h(x^h(i+1,j))|x(i,j),\theta(i,j)]-\alpha V_{\theta(i,j)}^h(x^h(i,j)),$$
$$\Delta_\alpha^v V_{\theta(i,j)}^v(x^v(i,j))=\mathbb{E}[V_{\theta(i,j+1)}^v(x^v(i,j+1))|x(i,j),\theta(i,j)]-\alpha V_{\theta(i,j)}^v(x^v(i,j)),$$

and α is a positive scalar.

The following lemma can be regarded as a 2-D mode-dependent functional-based scheme, which provides a useful tool to derive delay-dependent stability conditions.

Lemma 7.1 (Mode-dependent functional-based scheme) *Assume that there exist a mode-dependent functional* $V_\theta(x) = V_\theta^h(x^h) + V_\theta^v(x^v)$*, positive scalars* c_k *(*$k = 1, 2, 3, 4$*) and* $\alpha \in (0, 1)$ *such that*

(i) *For any* $x^h \in \mathbb{R}^{N_h}$*,* $x^v \in \mathbb{R}^{N_v}$*, and* $\theta \in \Theta$*,*

$$c_1\|x^h\|^2 \leq V_\theta^h(x^h) \leq c_2\|x^h\|^2, \quad c_3\|x^v\|^2 \leq V_\theta^v(x^v) \leq c_4\|x^h\|^2; \tag{7.7}$$

(ii) *The difference* $\Delta_\alpha V_{\theta(i,j)}(x(i, j))$ *defined by (7.6) satisfies*

$$\Delta_\alpha V_{\theta(i,j)}(x(i, j)) \leq 0, \quad (i, j) \in \mathbb{N}_0^2. \tag{7.8}$$

Then, the 2-D Markov jump system (7.1) is stochastically stable.

Proof For the brevity, along any state trajectory $x(i, j)$ of (7.1), we denote

$$\mathscr{V}^h(i, j) = V^h_{\theta(i,j)}(x^h(i, j)), \quad \mathscr{V}^v(i, j) = V^v_{\theta(i,j)}(x^v(i, j)).$$

It follows from condition (7.8) that

$$\begin{aligned}\mathbb{E}\left[\mathscr{V}^h(i+1, j)\middle|x(i, j), \theta(i, j)\right] &+ \mathbb{E}\left[\mathscr{V}^v(i, j+1)\middle|x(i, j), \theta(i, j)\right] \\ &\leq \alpha\left(\mathscr{V}^h(i, j) + \mathscr{V}^v(i, j)\right), \quad i, j \in \mathbb{N}_0. \end{aligned} \tag{7.9}$$

The following fact can be induced from basic properties of conditional expectation

$$\mathbb{E}\left[\mathbb{E}\left[\mathscr{V}^h(i+1, j)\middle|x(i, j), \theta(i, j)\right]\middle|X_0, \theta_0\right] = \mathbb{E}\left[\mathscr{V}^h(i+1, j)\middle|X_0, \theta_0\right].$$

By taking the expectation $\mathbb{E}[.|X_0, \theta_0]$ both sides of (7.9), and utilizing the above fact, we then obtain

$$\begin{aligned}\mathbb{E}\left[\mathscr{V}^h(i+1, j)\middle|X_0, \theta_0\right] &+ \mathbb{E}\left[\mathscr{V}^v(i, j+1)\middle|X_0, \theta_0\right] \\ &\leq \alpha\,\mathbb{E}\left[\mathscr{V}^h(i, j) + \mathscr{V}^v(i, j)\middle|X_0, \theta_0\right].\end{aligned}$$

This, together condition (7.7), yields

$$\begin{aligned}&\left(\mathbb{E}\left[\mathscr{V}^h(i+1, j)\middle|X_0, \theta_0\right] - \mathbb{E}\left[\mathscr{V}^h(i, j)\middle|X_0, \theta_0\right]\right) \\ &+ \left(\mathbb{E}\left[\mathscr{V}^v(i, j+1)\middle|X_0, \theta_0\right] - \mathbb{E}\left[\mathscr{V}^v(i, j)\middle|X_0, \theta_0\right]\right) \\ &\leq -(1-\alpha)\,\mathbb{E}\left[\mathscr{V}^h(i, j) + \mathscr{V}^v(i, j)\middle|X_0, \theta_0\right] \\ &\leq -(1-\alpha)c_{\min}\,\mathbb{E}\left[\|x(i, j)\|^2|X_0, \theta_0\right],\end{aligned} \tag{7.10}$$

where $c_{\min} = \min\{c_1, c_3\}$.

For any positive integers p and q, by summing up both sides of (7.10), we obtain

$$
\begin{aligned}
&\sum_{j=0}^{q-1} \left(\mathbb{E}\left[\mathscr{V}^h(p, j)\middle|X_0, \theta_0\right] - \mathbb{E}\left[\mathscr{V}^h(0, j)\middle|X_0, \theta_0\right]\right) \\
&+ \sum_{i=0}^{p-1} \left(\mathbb{E}\left[\mathscr{V}^v(i, q)\middle|X_0, \theta_0\right] - \mathbb{E}\left[\mathscr{V}^v(i, 0)\middle|X_0, \theta_0\right]\right) \\
&\leq -(1-\alpha)c_{\min}\,\mathbb{E}\left[\sum_{i=0}^{p-1}\sum_{j=0}^{q-1} \|x(i, j)\|^2 \middle| X_0, \theta_0\right]. \qquad (7.11)
\end{aligned}
$$

Let $p \to \infty$ and $q \to \infty$, it follows from (7.11) and condition (7.7) that

$$
\mathbb{E}\left[\sum_{i,j=0}^{\infty} \|x(i, j)\|^2 | X_0, \theta_0\right] \leq \frac{c_{\max}}{(1-\alpha)c_{\min}} \sum_{k=0}^{\infty} \mathbb{E}\left[\hat{\varphi}_k^2 + \hat{\psi}_k^2\right], \qquad (7.12)
$$

where $c_{\max} = \max\{c_2, c_4\}$. The result of (7.12) ensures that system (7.1) is stochastically stable. The proof is completed. □

Remark 7.1 Lemma 7.1 provides a useful tool to derive stability conditions using weighted LKF candidates. For $\alpha = 1$, the weighted difference defined in (7.6) is reduced to

$$
\begin{aligned}
\Delta V_{\theta(i,j)}(x(i, j)) &= \Delta_h V^h_{\theta(i,j)}(x^h(i, j)) + \Delta_v V^v_{\theta(i,j)}(x^v(i, j)) \\
&= \mathbb{E}[V^h_{\theta(i+1,j)}(x^h(i+1, j))|x(i, j), \theta(i, j)] - V^h_{\theta(i,j)}(x^h(i, j)) \\
&\quad + \mathbb{E}[V^v_{\theta(i,j+1)}(x^v(i, j+1))|x(i, j), \theta(i, j)] - V^v_{\theta(i,j)}(x^v(i, j)). \qquad (7.13)
\end{aligned}
$$

Note also that

$$
\Delta_\alpha V_{\theta(i,j)}(x(i, j)) = \Delta V_{\theta(i,j)}(x(i, j)) + (1-\alpha)V_{\theta(i,j)}(x(i, j)).
$$

This makes the process of manipulating weighted differences of LKF candidates more flexible. If condition (7.8) is modified as

$$
\Delta V_{\theta(i,j)} V(x(i, j)) \leq -c_0\|x(i, j)\|^2 \qquad (7.14)
$$

for some positive scalar c_0, the assertion of Lemma 7.1 is still valid.

7.3.2 2-D Summation Inequalities

For given integers $h \le k$, $p \le q$ and a function $w : \mathbb{Z}[h,k] \times [p,q] \to \mathbb{R}^N$, where N is a positive integer, we denote

$$\mathbb{A}_h(w) = \frac{1}{k-h+1}\sum_{i=h}^{k} w(i,.) \in \mathbb{R}^N, \quad \mathbb{A}_v(w) = \frac{1}{q-p+1}\sum_{j=p}^{q} w(.,j) \in \mathbb{R}^N$$

as the average value of w with respect to the first and the second arguments. It is fact that, for a given positive definite matrix $R \in \mathbb{S}_N^+$, the function $q_R : \mathbb{R}^N \to \mathbb{R}$, $x \mapsto q_R(x) = x^\top Rx$ is a convex function. Then, we have the following Jensen-type inequalities.

Lemma 7.2 (Jensen summation inequalities) *For a given matrix $R \in \mathbb{S}_N^+$, integers $h \le k$, $p \le q$, and a function $w : \mathbb{Z}[h,k] \times [p,q] \to \mathbb{R}^N$, the following inequalities hold*

$$\begin{aligned} \mathbb{A}_h(w^\top Rw) &\ge \mathbb{A}_h(w)^\top R\mathbb{A}_h(w), \\ \mathbb{A}_v(w^\top Rw) &\ge \mathbb{A}_v(w)^\top R\mathbb{A}_v(w). \end{aligned} \tag{7.15}$$

To sharpen the estimate in (7.15), it is necessary to find new bounds for the gaps $G_h = \mathbb{A}_h(w^\top Rw) - \mathbb{A}_h(w)^\top R\mathbb{A}_h(w)$ and $G_v = \mathbb{A}_v(w^\top Rw) - \mathbb{A}_v(w)^\top R\mathbb{A}_v(w)$ of (7.15). Note that when $h = k$, $G_h = 0$ and when $p = q$, $G_v = 0$. Thus, it is only necessary to find a new estimate for G_h for $h < k$.

Lemma 7.3 *For a given matrix $R \in \mathbb{S}_N^+$, integers $h < k$, $p \le q$, and a function $w : \mathbb{Z}[h,k] \times [p,q] \to \mathbb{R}^N$, the following inequality holds*

$$l_h^2\left[\mathbb{A}_h(w^\top Rw) - \mathbb{A}_h(w)^\top R\mathbb{A}_h(w)\right] \ge \frac{3(l_h+1)}{l_h-1}\chi_h^\top R\chi_h, \tag{7.16}$$

where $l_h = k - h + 1$ and $\chi_h = \sum_{i=h}^{k} w(i,j) - \frac{2}{l_h+1}\sum_{i=h}^{k}\sum_{l=h}^{i} w(l,j)$.

Proof The proof is similar to that of Lemma 2.3 in Chap. 2. We omit it here. □

Remark 7.2 By similar arguments used to derive (7.16) we obtain the following estimate

$$l_v^2\left[\mathbb{A}_v(w^\top Rw) - \mathbb{A}_v(w)^\top R\mathbb{A}_v(w)\right] \ge \frac{3(l_v+1)}{l_v-1}\chi_v^\top R\chi_v, \tag{7.17}$$

where $l_v = q - p + 1$ and $\chi_v = \sum_{j=p}^{q} w(i,j) - \frac{2}{l_v+1}\sum_{j=p}^{q}\sum_{l=p}^{j} w(i,l)$.

By simplifying the inequalities in (7.16) and (7.17), the following corollary is provided to make them easier to utilize in establishing stochastic stability conditions for 2-D time-delay systems in the form of (7.1).

Corollary 7.1 *Given positive integers h, k and a function* $w : \mathbb{Z}[-h, 0] \times [-k, 0] \to \mathbb{R}^N$. *Then, for any matrix* $R \in \mathbb{S}_N^+$, *the following inequalities hold*

$$\begin{aligned} h \sum_{i=-h}^{-1} \Delta_h w^\top(i,j) R \Delta_h w(i,j) &\geq \begin{bmatrix} w_{11} \\ w_{12} \end{bmatrix}^\top \begin{bmatrix} R & 0 \\ 0 & 3R \end{bmatrix} \begin{bmatrix} w_{11} \\ w_{12} \end{bmatrix}, \\ k \sum_{j=-k}^{-1} \Delta_v w^\top(i,j) R \Delta_v w(i,j) &\geq \begin{bmatrix} w_{21} \\ w_{22} \end{bmatrix}^\top \begin{bmatrix} R & 0 \\ 0 & 3R \end{bmatrix} \begin{bmatrix} w_{21} \\ w_{22} \end{bmatrix}, \end{aligned} \tag{7.18}$$

where

$$\begin{aligned} w_{11} &= w(0,j) - w(-h,j), \quad w_{21} = w(i,0) - w(i,-k), \\ w_{12} &= w(0,j) + w(-h,j) - \frac{2}{h+1} \sum_{i=-h}^{0} w(i,j), \\ w_{22} &= w(i,0) + w(i,-k) - \frac{2}{k+1} \sum_{j=-k}^{0} w(i,j). \end{aligned}$$

7.3.3 *Reciprocally Convex Inequalities*

To make this chapter self-contain, we recall here basic facts involving reciprocally convex inequalities and Schur complement lemma. As discussed in Chap. 2 [Sect. 2.3.3], to deal with with interval time-varying delays, it is necessary to bound the summation

$$S_R = \sum_{k=1}^{m} \frac{1}{\alpha_k} \chi_k^\top R_k \chi_k$$

for vectors $\chi_k \in \mathbb{R}^{N_k}$, matrices $R_k \in \mathbb{S}_{N_k}^+$ and scalars $\alpha_k > 0$ with $\sum_{k=1}^m \alpha_k = 1$. For the case $m = 2$, we have the following result.

Lemma 7.4 *For given matrices* $R_1 = R_1^\top > 0$, $R_2 = R_2^\top > 0$ *and* X_{12} *of appropriate dimensions, then we have*

$$\frac{1}{\alpha} \chi_1^\top R_1 \chi_1 + \frac{1}{1-\alpha} \chi_2^\top R_2 \chi_2 \geq \begin{bmatrix} \chi_1 \\ \chi_2 \end{bmatrix}^\top \begin{bmatrix} R_1 & X_{12} \\ X_{12}^\top & R_2 \end{bmatrix} \begin{bmatrix} \chi_1 \\ \chi_2 \end{bmatrix} \tag{7.19}$$

for any $\alpha \in (0, 1)$ *and vectors* χ_1, χ_2 *of appropriate dimensions provided that*

$$\begin{bmatrix} R_1 & X_{12} \\ X_{12}^\top & R_2 \end{bmatrix} \geq 0.$$

A generalized version of Lemma 7.4 can be obtained as follows.

Lemma 7.5 *Given matrices $R_k \in \mathbb{S}^+_{N_k}$, $k = 1, 2, \ldots, m$. For any matrices X_{jk}, $j, k = 1, 2, \ldots, m$, $j \neq k$, of appropriate dimensions, if*

$$\begin{bmatrix} R_j & X_{jk} \\ X_{jk}^\top & R_k \end{bmatrix} \geq 0,$$

then we have

$$\sum_{k=1}^{m} \frac{1}{\alpha_k} \chi_k^\top R_k \chi_k \geq \sum_{k=1}^{m} \chi_k^\top R_k \chi_k + \sum_{j \neq k} \begin{bmatrix} \chi_j \\ \chi_k \end{bmatrix}^\top \begin{bmatrix} R_j & X_{jk} \\ X_{jk}^\top & R_k \end{bmatrix} \begin{bmatrix} \chi_j \\ \chi_k \end{bmatrix} \tag{7.20}$$

for any $\alpha_k \in (0, 1)$ with $\sum_{k=1}^{m} \alpha_k = 1$ and $\chi_k \in \mathbb{R}^{N_k}$.

7.4 Stability Conditions

In this section, based on the scheme presented in Lemma 7.1 and by utilizing 2-D summation inequalities, we derive delay dependent stability conditions for 2-D Markov jump Roesser systems with interval time-varying delays as given in (7.1).

7.4.1 The Case of Uncertain Transition Probabilities

In light of the mode-dependent functional-based scheme proposed in Lemma 7.1, an augmented LKF candidate is constructed as follows

$$\underbrace{V_{\theta(i,j)}(x(i,j))}_{\mathscr{V}(i,j)} = \sum_{k=1}^{3} \Big(\underbrace{V^h_{\theta(i,j)k}(x^h(i,j))}_{\mathscr{V}^h_k(i,j)} + \underbrace{V^v_{\theta(i,j)k}(x^v(i,j))}_{\mathscr{V}^v_k(i,j)} \Big), \tag{7.21}$$

where

$$\mathscr{V}^h_1(i,j) = x^{h\top}(i,j) P^h_{\theta(i,j)} x^h(i,j), \quad \mathscr{V}^v_1(i,j) = x^{v\top}(i,j) P^v_{\theta(i,j)} x^v(i,j),$$

$$\mathscr{V}^h_2(i,j) = \sum_{k=i-d^h_1}^{i-1} \alpha^{i-k-1} x^{h\top}(k,j) Q^h_1 x^h(k,j) + \sum_{k=i-d^h_2}^{i-d^h_1-1} \alpha^{i-k-1} x^{h\top}(k,j) Q^h_2 x^h(k,j),$$

$$\mathscr{V}^v_2(i,j) = \sum_{k=j-d^v_1}^{j-1} \alpha^{j-k-1} x^{v\top}(i,k) Q^v_1 x^v(i,k) + \sum_{k=j-d^v_2}^{j-d^v_1-1} \alpha^{j-k-1} x^{v\top}(i,k) Q^v_2 x^v(i,k),$$

$$\mathscr{V}_3^h(i,j) = \sum_{k=-d_1^h}^{-1} \sum_{l=i+k}^{i-1} \alpha^{i-l-1} \eta_h^\top(l,j) R_1^h \eta_h(l,j) + \sum_{k=-d_2^h}^{-d_1^h-1} \sum_{l=i+k}^{i-1} \alpha^{i-l-1} \eta_h^\top(l,j) R_2^h \eta_h(l,j),$$

$$\mathscr{V}_3^v(i,j) = \sum_{k=-d_1^v}^{-1} \sum_{l=j+k}^{j-1} \alpha^{j-l-1} \eta_v^\top(i,l) R_1^v \eta_v(i,l) + \sum_{k=-d_2^v}^{-d_1^v-1} \sum_{l=j+k}^{j-1} \alpha^{j-l-1} \eta_v^\top(i,l) R_2^v \eta_v(i,l)$$

and $\eta_h(l,j) = \Delta_h(x^h(l,j))$, $\eta_v(i,l) = \Delta_v(x^v(i,l))$.

For the notational simplicity, we denote the block-row matrices

$$e_k = \begin{bmatrix} 0_{N\times(k-1)N} & I_N & 0_{N\times(7-k)N} \end{bmatrix}, \quad k = 1, 2, \ldots, 7,$$
$$\mathscr{A}_m = A_m e_1 + A_{dm} e_3, \quad \mathscr{D}_m = (A_m - I_N) e_1 + A_{dm} e_3, \quad m \in \mathscr{M},$$

and the following notations

$$\chi_0(i,j) = \begin{bmatrix} \chi_0^1(i,j) \\ \chi_0^2(i,j) \end{bmatrix}, \quad \chi_0^1(i,j) = \begin{bmatrix} x(i,j) \\ x_{dl}(i,j) \\ x_d(i,j) \\ x_{du}(i,j) \end{bmatrix}, \quad \chi_0^2(i,j) = \begin{bmatrix} x_{dl}^a(i,j) \\ x_d^a(i,j) \\ x_{du}^a(i,j) \end{bmatrix},$$

$$x_{dl}(i,j) = \begin{bmatrix} x^h(i-d_1^h, j) \\ x^v(i, j-d_1^v) \end{bmatrix}, \quad x_d(i,j) = \begin{bmatrix} x^h(i-d_h(i), j) \\ x^v(i, j-d_v(j)) \end{bmatrix},$$

$$x_{du}(i,j) = \begin{bmatrix} x^h(i-d_2^h, j) \\ x^v(i, j-d_2^v) \end{bmatrix}, \quad x_{dl}^a(i,j) = \begin{bmatrix} \xi_{11}(i,j) \\ \xi_{21}(i,j) \end{bmatrix},$$

$$x_d^a(i,j) = \begin{bmatrix} \xi_{12}(i,j) \\ \xi_{22}(i,j) \end{bmatrix}, \quad x_{du}^a(i,j) = \begin{bmatrix} \xi_{13}(i,j) \\ \xi_{23}(i,j) \end{bmatrix},$$

$$\xi_{11}(i,j) = \frac{1}{1+d_1^h} \sum_{k=i-d_1^h}^{i} x^h(k,j),$$

$$\xi_{12}(i,j) = \frac{1}{1+d_h(i)-d_1^h} \sum_{k=i-d_h(i)}^{i-d_1^h} x^h(k,j),$$

$$\xi_{13}(i,j) = \frac{1}{1+d_2^h-d_h(i)} \sum_{k=i-d_2^h}^{i-d_h(i)} x^h(k,j),$$

$$\xi_{21}(i,j) = \frac{1}{1+d_1^v} \sum_{l=j-d_1^v}^{j} x^v(i,l),$$

$$\xi_{22}(i,j) = \frac{1}{1+d_v(j)-d_1^v} \sum_{l=j-d_v(j)}^{j-d_1^v} x^v(i,l),$$

$$\xi_{23}(i,j) = \frac{1}{1+d_2^v-d_v(i)} \sum_{l=j-d_2^v}^{j-d_v(j)} x^v(i,l),$$

$$\Gamma_1 = \begin{bmatrix} e_1-e_2 \\ e_1+e_2-2e_5 \end{bmatrix}, \quad \Gamma_2 = \begin{bmatrix} e_2-e_3 \\ e_2+e_3-2e_6 \end{bmatrix},$$

$$\Gamma_3 = \begin{bmatrix} e_3-e_4 \\ e_3+e_4-2e_7 \end{bmatrix}, \quad \Gamma = \begin{bmatrix} \Gamma_2 \\ \Gamma_3 \end{bmatrix}, \quad \sigma_m = \sum_{n=1}^{s} \left(\hat{\pi}_{mn} - \delta_{mn}\right),$$

$$\Omega_{m,n}^0 = \mathscr{A}_m^\top \left(\tilde{P}_m + (1-\sigma_m)P_n\right)\mathscr{A}_m - \alpha e_1^\top P_m e_1, \quad \tilde{P}_m = \sum_{n=1}^{s} \left(\hat{\pi}_{mn} - \delta_{mn}\right) P_n,$$

$$\Omega_m^1 = \mathscr{D}_m^\top \left(\mathbb{I}_1^d R_1 + \mathbb{I}_2^d R_2\right)\mathscr{D}_m, \quad \mathbb{I}_1^d = \operatorname{diag}\left(d_1^h I_{N_h}, d_1^v I_{N_v}\right),$$
$$\mathbb{I}_2^d = \operatorname{diag}\left(d_{12}^h I_{N_h}, d_{12}^v I_{N_v}\right), \quad d_{12}^h = d_2^h - d_1^h, \quad d_{12}^v = d_2^v - d_1^v,$$

$$\Omega_2 = e_1^\top Q_1 e_1 - e_2^\top (\mathbb{I}_1^\alpha Q_1) e_2 + e_2^\top (\mathbb{I}_1^\alpha Q_2) e_2 - e_4^\top (\mathbb{I}_2^\alpha Q_2) e_4,$$
$$\mathbb{I}_1^\alpha = \operatorname{diag}\left(\alpha^{d_1^h} I_{N_h}, \alpha^{d_1^v} I_{N_v}\right), \quad \mathbb{I}_2^\alpha = \operatorname{diag}\left(\alpha^{d_2^h} I_{N_h}, \alpha^{d_2^v} I_{N_v}\right),$$
$$\Omega_3 = \Gamma_1^\top \Phi \Gamma_1, \quad \Phi = \operatorname{diag}\left(\mathbb{I}_1^\alpha (\mathbb{I}_1^d)^{-1} R_1, 3\Lambda R_1\right),$$
$$\Lambda = \operatorname{diag}\left(\frac{(d_1^h+1)\alpha^{d_1^h}}{d_1^h(d_1^h-1)} I_{N_h}, \frac{(d_1^v+1)\alpha^{d_1^v}}{d_1^v(d_1^v-1)} I_{N_v}\right),$$
$$\Omega_4 = \Gamma^\top \left(\frac{\alpha^{d_2^h}}{d_{12}^h} \mathbb{J}_h^T \Psi_h \mathbb{J}_h + \frac{\alpha^{d_2^v}}{d_{12}^v} \mathbb{J}_v^T \Psi_v \mathbb{J}_v\right)\Gamma,$$
$$\mathbb{J}_h = \operatorname{diag}\left(J_{N_h}, J_{N_h}, J_{N_h}, J_{N_h}\right), \quad J_{N_h} = \left[I_{N_h} \; 0_{N_h \times N_v}\right],$$
$$\mathbb{J}_v = \operatorname{diag}\left(J_{N_v}, J_{N_v}, J_{N_v}, J_{N_v}\right), \quad J_{N_v} = \left[0_{N_v \times N_h} \; I_{N_v}\right],$$
$$\widetilde{R}_2^h = \operatorname{diag}\left(R_2^h, 3R_2^h\right), \quad \widetilde{R}_2^v = \operatorname{diag}\left(R_2^v, 3R_2^v\right).$$

Delay-range-dependent stability for 2-D Markov jump systems with interval delays in the form of (7.1) is presented as the following theorem.

Theorem 7.1 *Consider system* (7.1) *with time-varying delays* $d_h(i) \in [d_1^h, d_2^h]$ *and* $d_v(j) \in [d_1^v, d_2^v]$. *Assume that there exist a scalar* $\alpha \in (0, 1)$, *positive definite matrices* $P_m = \operatorname{diag}(P_m^h, P_m^v) \in \mathbb{S}_N^+$, $m \in \mathscr{M}$, $Q_k = \operatorname{diag}(Q_k^h, Q_k^v) \in \mathbb{S}_N^+$, $R_k = \operatorname{diag}(R_k^h, R_k^v) \in \mathbb{S}_N^+$ $(k = 1, 2)$ *and matrices* $X \in \mathbb{R}^{2N_h \times 2N_h}$, $Y \in \mathbb{R}^{2N_v \times 2N_v}$ *such that*

$$\Psi_h = \begin{bmatrix} \widetilde{R}_2^h & X \\ X^\top & \widetilde{R}_2^h \end{bmatrix}, \quad \Psi_v = \begin{bmatrix} \widetilde{R}_2^v & Y \\ Y^\top & \widetilde{R}_2^v \end{bmatrix}, \tag{7.22}$$

and the following LMI-based condition holds for all $m, n \in \mathscr{M}$,

$$\Omega_{m,n} \triangleq \Omega^0_{m,n} + \Omega^1_m + \Omega_2 - \Omega_3 - \Omega_4 < 0. \tag{7.23}$$

Then, system (7.1) *with uncertain TPs is stochastically stable.*

Proof First, it can be verified from (7.21) that

$$\begin{aligned} c_1\|x^h(i,j)\|^2 \le \mathscr{V}^h(i,j) \le c_2\|x^h(i,j)\|^2, \\ c_3\|x^v(i,j)\|^2 \le \mathscr{V}^v(i,j) \le c_4\|x^v(i,j)\|^2, \end{aligned} \tag{7.24}$$

where $c_1 = \min_{m\in\mathscr{M}} \lambda_{\min}(P^h_m)$, $c_3 = \min_{m\in\mathscr{M}} \lambda_{\min}(P^v_m)$ and

$$\begin{aligned} c_2 = {} & \max_{m\in\mathscr{M}} \lambda_{\max}(P^h_m) + d^h_1\lambda_{\max}(Q^h_1) + (d^h_2 - d^h_1)\lambda_{\max}(Q^h_2) \\ & + \frac{1}{2}d^h_1(d^h_1+1)\lambda_{\max}(R^h_1) + \frac{1}{2}(d^h_2 - d^h_1)(d^h_2 + d^h_1 + 1)\lambda_{\max}(R^h_2), \\ c_4 = {} & \max_{m\in\mathscr{M}} \lambda_{\max}(P^v_m) + d^v_1\lambda_{\max}(Q^v_1) + (d^v_2 - d^v_1)\lambda_{\max}(Q^v_2) \\ & + \frac{1}{2}d^v_1(d^v_1+1)\lambda_{\max}(R^v_1) + \frac{1}{2}(d^v_2 - d^v_1)(d^v_2 + d^v_1 + 1)\lambda_{\max}(R^v_2). \end{aligned}$$

At the time $(i, j) \in \mathbb{N}^2_0$, $\theta(i, j) = m \in \mathscr{M}$, the system may jump to any mode $n \in \mathscr{M}$ along horizontal and vertical directions at times $(i+1, j)$ and $(i, j+1)$ with transition probability π_{mn}. To verify condition (7.8), we need to compute the weighted differences of $\Delta^h_\alpha \mathscr{V}^h(i,j)$ and $\Delta^v_\alpha \mathscr{V}^v(i,j)$. For the fist functional $\mathscr{V}^h_1(i,j)$ and $\mathscr{V}^v_1(i,j)$, we have

$$\begin{aligned} \Delta_h \mathscr{V}^h_1(i,j) &= \mathbb{E}\big[\mathscr{V}^h_1(i+1,j)|x(i,j), \theta(i,j)=m\big] - \mathscr{V}^h_1(i,j) \\ &= \mathbb{E}\big[x^{h\top}(i+1,j)P^h_{\theta(i+1,j)}x^h(i+1,j)\big|x(i,j),\theta(i,j)=m\big] \\ &\quad - x^{h\top}(i,j)P^h_{\theta(i,j)}x^h(i,j) \\ &= \sum_{n=1}^{s} \mathbb{P}\{\theta(i+1,j)=n|\theta(i,j)=m\}x^{h\top}(i+1,j)P^h_n x^h(i+1,j) \\ &\quad - x^{h\top}(i,j)P^h_m x^h(i,j) \\ &= x^{h\top}(i+1,j)\sum_{n=1}^{s}\pi_{mn}P^h_n x^h(i+1,j) - x^{h\top}(i,j)P^h_m x^h(i,j). \end{aligned} \tag{7.25}$$

Similarly, we also have

$$\Delta_v \mathscr{V}^v_1(i,j) = x^{v\top}(i,j+1)\sum_{n=1}^{s}\pi_{mn}P^v_n x^v(i,j+1) - x^{v\top}(i,j)P^v_m x^v(i,j). \tag{7.26}$$

A combination of (7.25) and (7.26) gives

$$\Delta\mathscr{V}_1(i,j)=\begin{bmatrix}x^h(i+1,j)\\x^v(i,j+1)\end{bmatrix}^\top\left(\sum_{n=1}^{s}\pi_{mn}P_n\right)\begin{bmatrix}x^h(i+1,j)\\x^v(i,j+1)\end{bmatrix}-x^\top(i,j)P_m x(i,j). \tag{7.27}$$

In system (7.1), since the transition probability matrix Π cannot be completely accessed, to derive tractable stability conditions, we need to eliminate the uncertainties from the combination $\sum_{n=1}^{s}\pi_{mn}P_n$. Let $\sigma_m=\sum_{n=1}^{s}(\hat{\pi}_{mn}-\delta_{mn})$. Then, according to the property of transition probabilities, $\sum_{n=1}^{s}\pi_{mn}=1$, we have

$$\sum_{n=1}^{s}(\Delta\pi_{mn}+\delta_{mn})=1-\sum_{n=1}^{s}\hat{\pi}_{mn}+\sum_{n=1}^{s}\delta_{mn}=1-\sigma_m\geq 0.$$

If $\sigma_m<1$, then we have

$$\begin{aligned}\sum_{n=1}^{s}\pi_{mn}P_n&=\sum_{n=1}^{s}(\hat{\pi}_{mn}-\delta_{mn})P_n+\sum_{n=1}^{s}(\Delta\pi_{mn}+\delta_{mn})P_n\\&=\underbrace{\sum_{n=1}^{s}(\hat{\pi}_{mn}-\delta_{mn})P_n}_{\tilde{P}_m}+(1-\sigma_m)\sum_{n=1}^{s}\frac{\Delta\pi_{mn}+\delta_{mn}}{1-\sigma_m}P_n\\&=\sum_{n=1}^{s}\rho_{mn}\left(\tilde{P}_m+(1-\sigma_m)P_n\right),\end{aligned} \tag{7.28}$$

where $\rho_{mn}=\frac{\Delta\pi_{mn}+\delta_{mn}}{1-\sigma_m}$. If $\sigma_m=1$ then all TPs at the row mth are precisely known and the term in the right-hand side of (7.28) is reduced to $\tilde{P}_m$. Therefore,

$$\begin{aligned}&\Delta\mathscr{V}_1(i,j)+(1-\alpha)\mathscr{V}_1(i,j)\\&\quad=\chi_0^\top(i,j)\Big[\sum_{n=1}^{s}\rho_{mn}\mathscr{A}_m^\top\left(\tilde{P}_m+(1-\sigma_m)P_n\right)\mathscr{A}_m-\alpha e_1^\top P_m e_1\Big]\chi_0(i,j)\\&\quad=\chi_0^\top(i,j)\sum_{n=1}^{s}\rho_{mn}\Omega_{m,n}^0\chi_0(i,j).\end{aligned} \tag{7.29}$$

Next, the difference $\Delta\mathscr{V}_2(i,j)$ can be manipulated as follows

$$\begin{aligned}\Delta\mathscr{V}_2^h(i,j)&=\mathbb{E}\left[\mathscr{V}_2^h(i+1,j)|x(i,j),\theta(i,j)=m\right]-\mathscr{V}_2^h(i,j)\\&=x^{h\top}(i,j)Q_1^h x^h(i,j)-\alpha^{d_1^h}x^{h\top}(i-d_1^h,j)Q_1^h x^h(i-d_1^h,j)\\&\quad+\alpha^{d_1^h}x^{h\top}(i-d_1^h,j)Q_2^h x^h(i-d_1^h,j)\\&\quad-\alpha^{d_2^h}x^{h\top}(i-d_2^h,j)Q_2^h x^h(i-d_2^h,j)-(1-\alpha)\mathscr{V}_2^h(i,j).\end{aligned} \tag{7.30}$$

Similarly, we also have

$$\begin{aligned}\Delta\mathscr{V}_2^v(i,j) &= \mathbb{E}\left[\mathscr{V}_2^v(i,j+1)|x(i,j),\theta(i,j)=m\right]-\mathscr{V}_2^v(i,j)\\ &= x^{v\top}(i,j)Q_1^v x^v(i,j)-\alpha^{d_1^v}x^{v\top}(i,j-d_1^v)Q_1^v x^v(i,j-d_1^v)\\ &\quad+\alpha^{d_1^v}x^{v\top}(i,j-d_1^v)Q_2^v x^v(i,j-d_1^v)\\ &\quad-\alpha^{d_2^v}x^{v\top}(i,j-d_2^v)Q_2^v x^v(i,j-d_2^v)-(1-\alpha)\mathscr{V}_2^v(i,j).\end{aligned}\tag{7.31}$$

Thus, it follows from (7.30) and (7.31) that

$$\begin{aligned}\Delta\mathscr{V}_2(i,j)+(1-\alpha)\mathscr{V}_2(i,j) &= \chi_0^\top(i,j)\Big[e_1^\top Q_1 e_1-e_2^\top(\mathbb{I}_1^\alpha Q_1)e_2\\ &\quad+e_2^\top(\mathbb{I}_2^\alpha Q_2)e_2-e_4^\top(\mathbb{I}_2^\alpha Q_2)e_4\Big]\chi_0(i,j)\\ &= \chi_0^\top(i,j)\Omega_2\chi_0(i,j).\end{aligned}\tag{7.32}$$

We now manipulate the differences $\Delta\mathscr{V}_3^h(i,j)$ and $\Delta\mathscr{V}_3^v(i,j)$. We have,

$$\begin{aligned}\Delta\mathscr{V}_3^h(i,j) &= \mathbb{E}[\mathscr{V}_3^h(i+1,j)|x(i,j),\theta(i,j)=m]-\mathscr{V}_3^h(i,j)\\ &= \sum_{k=-d_1^h}^{-1}\left(\sum_{l=i+1+k}^{i}\alpha^{i-l}\eta_h^\top(l,j)R_1^h\eta_h(l,j)-\sum_{l=i+k}^{i-1}\alpha^{i-l}\eta_h^\top(l,j)R_1^h\eta_h(l,j)\right)\\ &\quad+\sum_{k=-d_2^h}^{-d_1^h-1}\left(\sum_{l=i+k+1}^{i}\alpha^{i-l}\eta_h^\top(l,j)R_2^h\eta_h(l,j)\right.\\ &\quad\left.-\sum_{l=i+k}^{ki-1}\alpha^{i-l}\eta_h^\top(l,j)R_2^h\eta_h(l,j)\right)-(1-\alpha)\mathscr{V}_3^h(i,j)\\ &= \eta_h^\top(i,j)\left(d_1^hR_1^h+d_{12}^hR_2^h\right)\eta_h(i,j)-(1-\alpha)\mathscr{V}_3^h(i,j)\\ &\quad-\sum_{k=i-d_1^h}^{i-1}\alpha^{i-k}\eta_h^\top(k,j)R_1^h\eta_h(k,j)-\sum_{k=i-d_2^h}^{i-d_1^h-1}\alpha^{i-k}\eta_h^\top(k,j)R_2^h\eta_h(k,j)\\ &\le \eta_h^\top(i,j)\left(d_1^hR_1^h+d_{12}^hR_2^h\right)\eta_h(i,j)-(1-\alpha)\mathscr{V}_3^h(i,j)\\ &\quad-\alpha^{d_1^h}\sum_{k=i-d_1^h}^{i-1}\eta_h^\top(k,j)R_1^h\eta_h(k,j)-\alpha^{d_2^h}\sum_{k=i-d_2^h}^{i-d_1^h-1}\eta_h^\top(k,j)R_2^h\eta_h(k,j).\end{aligned}$$

Therefore,

$$\begin{aligned}\Delta\mathscr{V}_3^h(i,j)+(1-\alpha)\mathscr{V}_3^h(i,j) &\le \eta_h^\top(i,j)\left(d_1^hR_1^h+d_{12}^hR_2^h\right)\eta_h(i,j)\\ &\quad-\alpha^{d_1^h}\sum_{k=i-d_1^h}^{i-1}\eta_h^\top(k,j)R_1^h\eta_h(k,j)-\alpha^{d_2^h}\sum_{k=i-d_2^h}^{i-d_1^h-1}\eta_h^\top(k,j)R_2^h\eta_h(k,j).\end{aligned}\tag{7.33}$$

By similar lines used in deriving (7.33), we also have

$$\Delta\mathscr{V}_3^v(i,j)+(1-\alpha)\mathscr{V}_3^v(i,j)\le\eta_v^\top(i,j)\left(d_1^vR_1^v+d_{12}^vR_2^v\right)\eta_v(i,j)$$
$$-\alpha^{d_1^v}\sum_{l=j-d_1^v}^{j-1}\eta_v^\top(i,l)R_1^v\eta_v(i,l)-\alpha^{d_2^v}\sum_{l=j-d_2^v}^{j-d_1^v-1}\eta_v^\top(i,l)R_2^v\eta_v(i,l). \quad (7.34)$$

A combination of (7.33) and (7.34) gives

$$\Delta\mathscr{V}_3(i,j)+(1-\alpha)\mathscr{V}_3(i,j)\le\begin{bmatrix}\eta_h(i,j)\\\eta_v(i,j)\end{bmatrix}^\top\left(\mathbb{I}_1^dR_1+\mathbb{I}_2^dR_2\right)\begin{bmatrix}\eta_h(i,j)\\\eta_v(i,j)\end{bmatrix}$$
$$-\alpha^{d_1^h}\sum_{k=i-d_1^h}^{i-1}\eta_h^\top(k,j)R_1^h\eta_h(k,j)-\alpha^{d_1^v}\sum_{l=j-d_1^v}^{j-1}\eta_v^\top(i,l)R_1^v\eta_v(i,l)$$
$$-\alpha^{d_2^h}\sum_{k=i-d_2^h}^{i-d_1^h-1}\eta_h^\top(k,j)R_2^h\eta_h(k,j)-\alpha^{d_2^v}\sum_{l=j-d_2^v}^{j-d_1^v-1}\eta_v^\top(i,l)R_2^v\eta_v(i,l). \quad (7.35)$$

Observe that

$$\begin{bmatrix}\eta_h(i,j)\\\eta_v(i,j)\end{bmatrix}=\begin{bmatrix}x^h(i+1,j)-x^h(i,j)\\x^v(i,j+1)-x^v(i,j)\end{bmatrix}$$
$$=\left[(A_m-I_N)e_1+A_{dm}e_3\right]\chi_0(i,j)=\mathscr{D}_m\chi_0(i,j). \quad (7.36)$$

By utilizing Lemma 7.3, we have

$$-d_1^h\sum_{k=i-d_1^h}^{i-1}\eta_h^\top(k,j)R_1^h\eta_h(k,j)$$
$$\le-\left(x^h(i,j)-x^h(i-d_1^h,j)\right)^\top R_1^h\left(x^h(i,j)-x^h(i-d_1^h,j)\right)$$
$$-\frac{3(d_1^h+1)}{d_1^h-1}\left(x^h(i,j)+x^h(i-d_1^h,j)-2\xi_{11}(i,j)\right)^\top R_1^h$$
$$\times\left(x^h(i,j)+x^h(i-d_1^h,j)-2\xi_{11}(i,j)\right), \quad (7.37)$$

and

$$
\begin{aligned}
-d_1^v \sum_{l=j-d_1^v}^{j-1} & \eta_v^\top(i,l) R_1^v \eta_v(i,l) \\
& \leq -\left(x^v(i,j) - x^v(i,j-d_1^v)\right)^\top R_1^v \left(x^v(i,j) - x^v(i,j-d_1^v)\right) \\
& \quad - \frac{3(d_1^v+1)}{d_1^v-1} \left(x^v(i,j) + x^v(i,j-d_1^v) - 2\xi_{21}(i,j)\right)^\top R_1^v \\
& \quad \times \left(x^v(i,j) + x^v(i,j-d_1^v) - 2\xi_{21}(i,j)\right).
\end{aligned} \tag{7.38}
$$

From (7.37) and (7.38), we readily obtain

$$
\begin{aligned}
-\alpha^{d_1^h} \sum_{k=i-d_1^h}^{i-1} \eta_h^\top(k,j) R_1^h \eta_h(k,j) - \alpha^{d_1^v} \sum_{l=j-d_1^v}^{j-1} \eta_v^\top(i,l) R_1^v \eta_v(i,l) \\
\leq -\chi_0^\top(i,j) \Omega_3 \chi_0(i,j).
\end{aligned} \tag{7.39}
$$

Next, by utilizing the 2-D inequalities proposed in Corollary 7.1 and reciprocally convex inequalities, we have

$$
\begin{aligned}
d_{12}^h \sum_{k=i-d_2^h}^{i-d_h(i)-1} & \eta_h^\top(k,j) R_2^h \eta_h(k,j) + d_{12}^h \sum_{k=i-d_h(i)}^{i-d_1^h-1} \eta_h^\top(k,j) R_2^h \eta_h(k,j) \\
& \geq \frac{d_{12}^h}{d_h(i) - d_1^h} \omega_1^\top \widetilde{R}_2^h \omega_1 + \frac{d_{12}^h}{d_2^h - d_h(i)} \omega_2^\top \widetilde{R}_2^h \omega_2 \\
& \geq \begin{bmatrix} \omega_1 \\ \omega_2 \end{bmatrix}^\top \begin{bmatrix} \widetilde{R}_2^h & X \\ X^\top & \widetilde{R}_2^h \end{bmatrix} \begin{bmatrix} \omega_1 \\ \omega_2 \end{bmatrix},
\end{aligned}
$$

where

$$
\begin{aligned}
\omega_1 &= \begin{bmatrix} x^h(i-d_1^h,j) - x^h(i-d_h(i),j) \\ x^h(i-d_1^h,j) + x^h(i-d_h(i),j) - 2\xi_{12} \end{bmatrix}, \\
\omega_2 &= \begin{bmatrix} x^h(i-d_h(i),j) - x^h(i-d_2^h,j) \\ x^h(i-d_h(i),j) + x^h(i-d_2^h,j) - 2\xi_{13} \end{bmatrix}.
\end{aligned}
$$

Therefore,

$$
-d_{12}^h \sum_{k=i-d_2^h}^{i-d_1^h-1} \eta_h^\top(k,j) R_2^h \eta_h(k,j) \leq -\begin{bmatrix} \omega_1 \\ \omega_2 \end{bmatrix}^\top \begin{bmatrix} \widetilde{R}_2^h & X \\ X^\top & \widetilde{R}_2^h \end{bmatrix} \begin{bmatrix} \omega_1 \\ \omega_2 \end{bmatrix}. \tag{7.40}
$$

Note also that when $d_h(i) = d_1^h$ then $\omega_1 = 0$ and $d_h(i) = d_2^h$ then $\omega_2 = 0$, and the inequality in (7.40) is still valid. In addition, the augmented vector in (7.40) can be represented as

$$\begin{bmatrix} \omega_1 \\ \omega_2 \end{bmatrix} = (\mathbb{J}_h \Gamma)\chi_0(i, j).$$

Therefore,

$$-\alpha^{d_2^h} \sum_{k=i-d_2^h}^{i-d_1^h-1} \eta_h^\top(k, j) R_2^h \eta_h(k, j) \leq -\frac{\alpha^{d_2^h}}{d_{12}^h} \chi_0^\top(i, j) \left(\Gamma^\top \mathbb{J}_h^\top \Psi_h \mathbb{J}_h \Gamma\right) \chi_0(i, j). \tag{7.41}$$

By similar arguments, we obtain

$$-\alpha^{d_2^v} \sum_{l=j-d_2^v}^{j-d_1^v-1} \eta_v^\top(i, l) R_2^h \eta_v(i, l) \leq -\frac{\alpha^{d_2^v}}{d_{12}^v} \chi_0^\top(i, j) \left(\Gamma^\top \mathbb{J}_v^\top \Psi_v \mathbb{J}_v \Gamma\right) \chi_0(i, j). \tag{7.42}$$

It is consequent of (7.41) and (7.42),

$$\Delta \mathscr{V}_3(i, j) + (1-\alpha)\mathscr{V}_3(i, j) \leq \chi_0^\top(i, j) \left(\Omega_m^1 - \Omega_3 - \Omega_4\right) \chi_0(i, j). \tag{7.43}$$

From (7.29), (7.32) and (7.43), we finally obtain

$$\begin{aligned} \Delta_\alpha \mathscr{V}(i, j) &= \Delta \mathscr{V}(i, j) + (1-\alpha)\mathscr{V}(i, j) \\ &\leq \sum_{n=1}^{s} \rho_{mn} \chi_0^\top(i, j) \Omega_{m,n} \chi_0(i, j) \leq 0 \end{aligned}$$

according to condition (7.23). By Lemma 7.1, system (7.1) is stochastically stable. The proof is completed. □

Remark 7.3 In the proof of Theorem 7.1, reciprocally convex combination inequalities are used to tackle with the relationship between time-varying delays and their bounds. This approach can effectively reduce the conservatism of the proposed stability conditions both in terms of feasibility of LMI conditions and computational cost because no extra slack variable is introduced.

Remark 7.4 When the transition probability matrix $\Pi = (\pi_{mn})$ is completely accessed (i.e. π_{mn} are fully known), by assigning $\delta_{mn} = 0$, condition (7.23) is reduced to the following LMIs

$$\widehat{\Omega}_m = \widehat{\Omega}_m^0 + \Omega_m^1 + \Omega_2 - \Omega_3 - \Omega_4 < 0, \quad m \in \mathscr{M}, \tag{7.44}$$

where $\widehat{\Omega}_m^0 = \mathscr{A}_m^\top \left(\sum_{n=1}^s \pi_{mn} P_n\right) \mathscr{A}_m - \alpha e_1^\top P_n e_1$ and other matrices are defined as in Theorem 7.1.

Remark 7.5 For 2-D Markov jump systems with constant delays d_h and d_v, the selected mode-dependent LKF candidate (7.21) is reduced to the following simpler form

$$
\begin{aligned}
\widetilde{\mathscr{V}}(i,j) = {} & x^{\top}(i,j)P_{\theta(i,j)}x(i,j) + \sum_{k=i-d_h}^{i-1} \alpha^{i-k-1} x^{h\top}(k,j)Q^h x^h(k,j) \\
& + \sum_{l=j-d_v}^{j-1} \alpha^{j-l-1} x^{v\top}(i,l)Q^v x^v(i,l) \\
& + \sum_{k=-d_h}^{-1} \sum_{l=i+k}^{i-1} \alpha^{i-l-1} \eta_h^{\top}(l,j)R^h \eta_h(l,j) \\
& + \sum_{k=-d_v}^{-1} \sum_{l=j+k}^{j-1} \alpha^{j-l-1} \eta_v^{\top}(i,l)R^v \eta_v(i,l).
\end{aligned}
$$

Then, stability conditions of system (7.1) with constant delays can be derived as the following corollary.

Corollary 7.2 *System* (7.1) *with constant delays d_h, d_v is stochastically stable if there exist a scalar $\alpha \in (0,1)$, positive definite matrices $P_m^h \in \mathbb{S}_{N_h}^+$, $P_m^v \in \mathbb{S}_{N_v}^+$, $m \in \mathscr{M}$, $Q = \operatorname{diag}\left(Q^h, Q^v\right) \in \mathbb{S}_N^+$, $R = \operatorname{diag}\left(R^h, R^v\right) \in \mathbb{S}_N^+$ that satisfy the following LMI condition*

$$
\begin{aligned}
& \hat{\mathscr{A}}_m^{\top}\left(\tilde{P}_m + (1-\sigma_m)P_n\right)\hat{\mathscr{A}}_m - \alpha \hat{e}_1^{\top} P_m \hat{e}_1 + \hat{\mathscr{D}}_m^{\top}(\mathbb{I}_d R)\hat{\mathscr{D}}_m \\
& + \hat{e}_1^{\top} Q \hat{e}_1 - \hat{e}_2^{\top}(\mathbb{I}_\alpha Q)\hat{e}_2 - \hat{\Gamma}^{\top}\hat{\Phi}\hat{\Gamma} < 0, \quad m,n \in \mathscr{M},
\end{aligned} \tag{7.45}
$$

where $\hat{e}_k$ $(k = 1, 2, 3)$ are the block-rows of I_{3N} and other matrices are defined as

$$
\begin{aligned}
& \hat{\mathscr{A}}_m = A_m \hat{e}_1 + A_{dm}\hat{e}_2, \quad \hat{\mathscr{D}}_m = (A_m - I_N)\hat{e}_1 + A_{dm}\hat{e}_2, \quad m \in \mathscr{M}, \\
& \mathbb{I}_d = \operatorname{diag}\left(d_h I_{N_h}, d_v I_{N_v}\right), \quad \mathbb{I}_\alpha = \operatorname{diag}\left(\alpha^{d_h} I_{N_h}, \alpha^{d_v} I_{N_v}\right), \\
& \hat{\Phi} = \operatorname{diag}\left(\mathbb{I}_\alpha (\mathbb{I}_d)^{-1} R, 3\hat{\Lambda} R\right), \quad \hat{\Lambda} = \operatorname{diag}\left(\frac{(d_h+1)\alpha^{d_h}}{d_h(d_h-1)} I_{N_h}, \frac{(d_v+1)\alpha^{d_v}}{d_v(d_v-1)} I_{N_v}\right), \\
& \hat{\Gamma} = \left[(\hat{e}_1 - \hat{e}_2)^{\top}\ (\hat{e}_1 + \hat{e}_2 - 2\hat{e}_3)^{\top}\right]^{\top}.
\end{aligned}
$$

7.4.2 The Case of Complex Transition Probabilities

In this section, we consider 2-D Markov jump systems in the form of (7.1) with a more practical scenario for Markov chains $\theta(i,j)$. Typically, a transition probability π_{mn} is subject to one of three cases

(i) π_{mn} is precisely known;
(ii) π_{mn} is uncertain, that is, π_{mn} can be expressed as $\pi_{mn} = \hat{\pi}_{mn} + \Delta\pi_{mn}$ and $\Delta\pi_{mn} \in [-\delta_{mn}, \delta_{mn}]$ with known $\hat{\pi}_{mn} \geq \delta_{mn} \geq 0$;
(iii) π_{mn} is unknown, which is represented as #.

Precisely known TPs π_{mn} can be expressed as uncertain case $\hat{\pi}_{mn} + \Delta\pi_{mn}$, where $\delta_{mn} = 0$. For any $m \in \mathscr{M}$, $\mathscr{M} \triangleq \mathscr{N}_m^{\mathscr{K}} \cup \mathscr{N}_m^{\mathscr{U}}$, where, for $m \in \mathscr{M}$,

$$\mathscr{N}_m^{\mathscr{K}} = \{n \in \mathscr{M} : \pi_{mn} \text{ is known as } \hat{\pi}_{mn} + \Delta\pi_{mn}\},$$
$$\mathscr{N}_m^{\mathscr{U}} = \{n \in \mathscr{M} : \pi_{mn} \text{ is unknown }\}.$$

According to the proof of Theorem 7.1, to derive tractable stability conditions, we need to bound the matrix combination $P_m^c = \sum_{n \in \mathscr{M}} \pi_{mn} P_n$. If $\mathscr{N}_m^{\mathscr{K}} \neq \mathscr{M}$ (or equivalent $\mathscr{N}_m^{\mathscr{U}} \neq \emptyset$), we express this combination as

$$P_m^c = \underbrace{\sum_{n \in \mathscr{N}_m^{\mathscr{K}}} \pi_{mn} P_n}_{P_m^{\mathscr{K}}} + \sum_{n \in \mathscr{N}_m^{\mathscr{U}}} \pi_{mn} P_n. \tag{7.46}$$

Let $\pi_m^{\mathscr{K}} = \sum_{n \in \mathscr{N}_m^{\mathscr{K}}} \pi_{mn}$ then, we have

$$1 - \sum_{n \in \mathscr{M}} \pi_{mn} = \pi_m^{\mathscr{K}} + \sum_{n \in \mathscr{N}_m^{\mathscr{U}}} \pi_{mn}.$$

Since $\pi_m^{\mathscr{K}} < 1$, it can be verified that $\sum_{n \in \mathscr{N}_m^{\mathscr{U}}} \frac{\pi_{mn}}{1-\pi_m^{\mathscr{K}}} = 1$ and from (7.46), we have

$$P_m^c = \sum_{n \in \mathscr{N}_m^{\mathscr{U}}} \frac{\pi_{mn}}{1 - \pi_m^{\mathscr{K}}} \left(P_m^{\mathscr{K}} + \left(1 - \pi_m^{\mathscr{K}}\right) P_n\right).$$

Therefore,

$$\mathscr{A}_m^\top P_m^c \mathscr{A}_m = \sum_{n \in \mathscr{N}_m^{\mathscr{U}}} \frac{\pi_{mn}}{1 - \pi_m^{\mathscr{K}}} \mathscr{A}_m^\top \left(P_m^{\mathscr{K}} + \left(1 - \pi_m^{\mathscr{K}}\right) P_n\right) \mathscr{A}_m. \tag{7.47}$$

On the other hand, let

$$\hat{P}_m^{\mathscr{K}} = \sum_{n \in \mathscr{N}_m^{\mathscr{K}}} (\hat{\pi}_{mn} + \delta_{mn}) P_n \text{ and } \hat{\sigma}_m^{\mathscr{K}} = \sum_{n \in \mathscr{N}_m^{\mathscr{K}}} (\hat{\pi}_{mn} - \delta_{mn}),$$

then it can be verified that

$$P_m^{\mathscr{K}} \leq \hat{P}_m^{\mathscr{K}}, \quad 1 - \pi_m^{\mathscr{K}} \leq 1 - \hat{\sigma}_m^{\mathscr{K}}$$

for all $m \in \mathscr{M}$. This, together with (7.47), gives

$$\mathscr{A}_m^\top P_m^c \mathscr{A}_m \leq \sum_{n \in \mathscr{N}_m^{\mathscr{U}}} \frac{\pi_{mn}}{1 - \pi_m^{\mathscr{K}}} \mathscr{A}_m^\top \left(\hat{P}_m^{\mathscr{K}} + \left(1 - \hat{\sigma}_m^{\mathscr{K}}\right) P_n\right) \mathscr{A}_m. \tag{7.48}$$

In summary, we have the following result.

Theorem 7.2 *The 2-D time-varying delay system* (7.1) *with complex TPs is stochastically stable if there exist a scalar* $\alpha \in (0, 1)$ *and matrices* $P_m = \mathrm{diag}(P_m^h, P_m^v) \in \mathbb{S}_N^+$, $m \in \mathscr{M}$, $Q_k = \mathrm{diag}(Q_k^h, Q_k^v) \in \mathbb{S}_N^+$, $R_k = \mathrm{diag}(R_k^h, R_k^v) \in \mathbb{S}_N^+$ $(k = 1, 2)$, $X \in \mathbb{R}^{2N_h \times 2N_h}$, $Y \in \mathbb{R}^{2N_v \times 2N_v}$ *satisfying condition* (7.22) *such that the LMI-based condition*

$$\mathscr{A}_m^\top \left(\hat{P}_m^{\mathscr{K}} + \left(1 - \hat{\sigma}_m^{\mathscr{K}}\right) P_n\right) \mathscr{A}_m + \Xi_m < 0 \tag{7.49}$$

holds for all $m \in \mathscr{M}$ *and* $n \in \mathscr{N}_m^{\mathscr{K}}$, *where* $\Xi_m = \Omega_m^1 + \Omega_2 - \Omega_3 - \Omega_4$ *and other matrix notations are defined as in Theorem 7.1.*

7.5 Numerical Examples

In this section, we present some numerical examples to illustrate the efficacy of the obtained stability conditions.

Example 7.1 Consider system (7.1) with three modes (i.e. $s = 3$) and the system matrices

$$A_1 = \left[\begin{array}{cc|c} 0.1 & 0 & 0.05 \\ 0.2 & 0.09 & 0.1 \\ \hline 0.1 & 0 & 0.15 \end{array}\right], \quad A_2 = \left[\begin{array}{cc|c} 0.1 & 0 & 0 \\ 0 & -0.1 & 0.12 \\ \hline 0.15 & 0 & 0.16 \end{array}\right],$$

$$A_3 = \left[\begin{array}{cc|c} 0.1 & -0.1 & 0.15 \\ 0 & -0.1 & 0.1 \\ \hline 0.1 & 0.15 & 0.1 \end{array}\right], \quad A_{d1} = \left[\begin{array}{cc|c} -0.1 & 0 & 0.1 \\ 0.1 & -0.1 & 0 \\ \hline 0 & 0.1 & 0.15 \end{array}\right],$$

$$A_{d2} = \left[\begin{array}{cc|c} -0.1 & 0 & 0.1 \\ 0.1 & 0.09 & 0 \\ \hline 0 & 0.1 & -0.12 \end{array}\right], \quad A_{d3} = \left[\begin{array}{cc|c} 0.12 & 0.1 & 0.1 \\ 0.1 & -0.1 & 0 \\ \hline 0.1 & 0.1 & 0.11 \end{array}\right].$$

Case I: Uncertain transition probabilities

Assume that the transition probability matrix is given by $\Pi = \hat{\Pi} + \Delta\Pi$, where $\hat{\Pi} = (\hat{\pi}_{mn})$, $\Delta\Pi = (\Delta\pi_{mn})$, $m, n \in \mathscr{M} = \{1, 2, 3\}$, with the uncertain transition rates $\Delta\pi_{mn}$ which satisfy $|\Delta\pi_{mn}| \leq \delta_{mn}$ and

$$\hat{\Pi} = \begin{bmatrix} 0.3 & 0.51 & 0.16 \\ 0.4 & 0.25 & 0.3 \\ 0.5 & 0.25 & 0.2 \end{bmatrix}, \quad \delta_\Pi = (\delta_{mn}) = \begin{bmatrix} 0.05 & 0.01 & 0.02 \\ 0.1 & 0.05 & 0.1 \\ 0 & 0 & 0.01 \end{bmatrix}.$$

For interval directional delays $d_h(i) \in [2, 5]$, $d_v(j) \in [2, 6]$, by the Matlab LMI toolbox, it is found that conditions (7.22) and (7.23) are feasible with $\alpha = 0.992$ and (for short, some other matrices are omitted here)

$$P_1 = \text{diag}\left(\begin{bmatrix} 214.9063 & 3.5694 \\ 3.5694 & 204.1083 \end{bmatrix}, 261.9028\right),$$
$$P_2 = \text{diag}\left(\begin{bmatrix} 251.7818 & 36.9121 \\ 36.9121 & 247.8373 \end{bmatrix}, 209.4668\right),$$
$$P_3 = \text{diag}\left(\begin{bmatrix} 201.1407 & 75.0955 \\ 75.0955 & 311.3067 \end{bmatrix}, 240.6640\right).$$

By Theorem 7.1, system (7.1) with given parameters is stochastically stable.

To support the theoretical results, we make simulations with directional delays $d_h(i) = 2 + 3|\sin(i\pi/8)|$, $d_v(j) = 2 + 4|\cos(j\pi/6)|$ and randomly chosen initial sequences $x^h(i, j)$, $x^v(i, j)$ for $i, j \leq 20$ and $x^h(i, j) = 0$, $x^v(i, j) = 0$ elsewhere. State trajectories $x_1^h(i, j)$, $x_2^h(i, j)$ and $x^v(i, j)$ are presented in Figs. 7.1, 7.2 and 7.3. Bar mode transition is illustrated in Fig. 7.4 while diagonal mode transition is illustrated in Fig. 7.5. The simulation results presented in Figs. 7.1, 7.2, 7.3, 7.4 and 7.5 show that the system states converge as revealed by the theoretical results.

For the case of constant delays, Table 7.1 presents the upper bound of d_h for various values of α and $d_v = 10$. These numerical results illustrate the effectiveness of the proposed conditions in Theorem 7.1 and Corollary 7.2.

Cases II: Fully known transition probabilities

Assume that the transition probability matrix is fully accessed and given by

$$\Pi = \begin{bmatrix} 0.3 & 0.5 & 0.2 \\ 0.45 & 0.25 & 0.3 \\ 0.5 & 0.2 & 0.2 \end{bmatrix}.$$

Fig. 7.1 State trajectory x_1^h

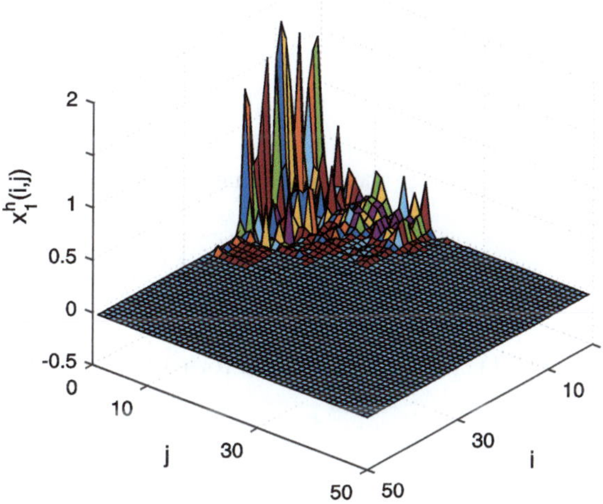

Fig. 7.2 State trajectory x_2^h

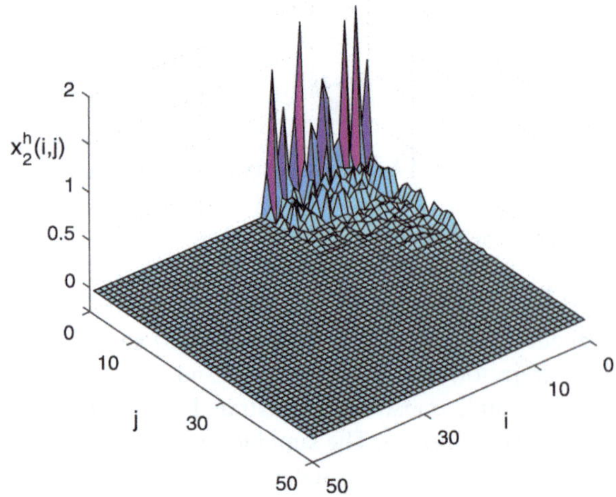

Fig. 7.3 State trajectory x^v

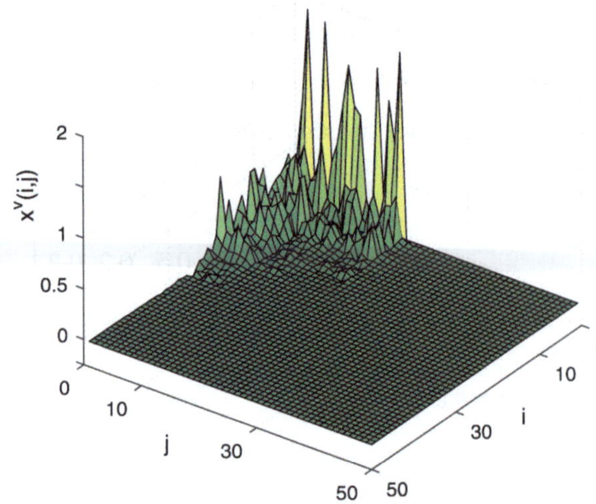

By decomposing the system matrices A_m, A_{dm}, $m = 1, 2, 3$, as

$$A_m = \left[\begin{array}{c|c} A_{11m} & A_{12m} \\ \hline A_{21m} & A_{22m} \end{array}\right], \quad A_{dm} = \left[\begin{array}{c|c} A_{d11m} & A_{d12m} \\ \hline A_{d21m} & A_{d22m} \end{array}\right],$$

system (7.1) can be written in the form of 2-D Markov jump systems in the FMLSS model

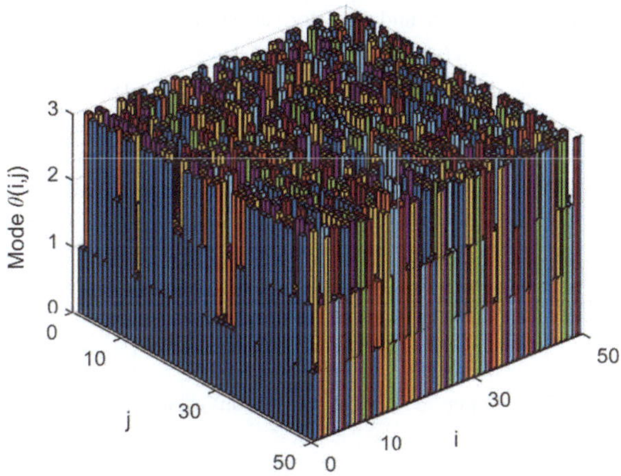

Fig. 7.4 Bar mode evolution

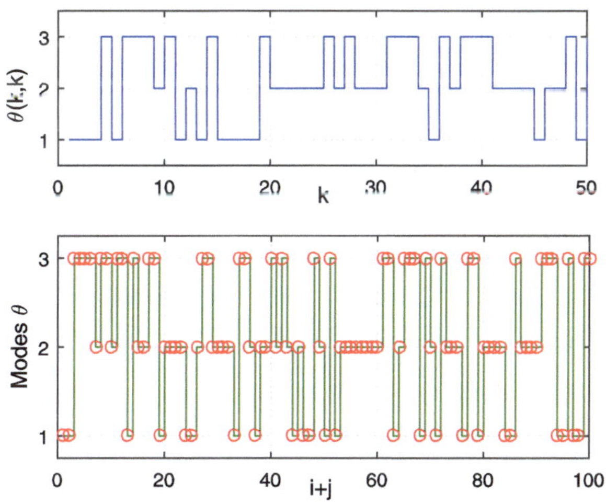

Fig. 7.5 Diagonal mode evolution

$$\begin{aligned} x(i+1, j+1) &= A_{1\theta(i,j)}x(i, j+1) + A_{2\theta(i,j)}x(i+1, j) \\ &\quad + A_{d1\theta(i,j)}x(i-d_h(i), j+1) + A_{d2\theta(i,j)}x(i+1, j-d_v(j)), \end{aligned} \tag{7.50}$$

where

$$A_{1m} = \begin{bmatrix} A_{11m} & A_{12m} \\ 0 & 0 \end{bmatrix}, \quad A_{2m} = \begin{bmatrix} 0 & 0 \\ A_{21m} & A_{22m} \end{bmatrix},$$

$$A_{d1m} = \begin{bmatrix} A_{d11m} & A_{d12m} \\ 0 & 0 \end{bmatrix}, \quad A_{d2m} = \begin{bmatrix} 0 & 0 \\ A_{d21m} & A_{d22m} \end{bmatrix}.$$

To illustrate the comparativeness of the obtained results, we apply Theorem 7.1 in this chapter for system (7.1) and Theorem 1 in [33] for system (7.50). The upper

Table 7.1 Upper bound of d_h for various α and $d_v = 10$

α	0.8	0.85	0.87	0.9	0.95
Allowable d_h	7	15	18	25	56

Table 7.2 Upper bound of d_2^h for various d_1^h and $d_1^v = 2$, $d_2^v = 4$

d_1^h	2	4	7	10
[33] (Theorem 1)	4	4	–	–
Theorem 7.1	7	8	11	14

Table 7.3 Upper bound of d_2^v for various d_1^v and $d_1^h = 2$, $d_2^h = 5$

d_1^h	2	5	7	10
[33] (Theorem 1)	2	–	–	–
Theorem 7.1	7	10	12	15

bounds of d_2^h and d_2^v delivered by Theorem 7.1 and [33] are given in Tables 7.2 and 7.3, respectively, where the symbol '–' indicates infeasible results. Clearly, the proposed method in this chapter delivers less conservatism than the use of free-weighting matrices and the classical Jensen inequality as in [33].

Example 7.2 Consider a model of thermal process in chemical reactors with two modes, which is expressed by the following Darboux-type equation [34]

$$\frac{\partial w(x,t)}{\partial x} = -\frac{\partial w(x,t)}{\partial t} - a_0^{\theta(x,t)} w(x,t) - a_1^{\theta(x,t)} w(x, t-\tau), \tag{7.51}$$

where $w(x,t)$ is the unknown function (e.g. temperature) at space $x \in [0, x_f]$ and time $t \in [0, \infty)$, $\tau > 0$ is a time-delay, $a_0^{\theta(x,t)}$, $a_1^{\theta(x,t)}$ are real coefficients and $\theta(x,t)$ represents the switching signal which specifies the working subsystem at (x,t).

We denote the following variables

$$\begin{aligned}
&w(i,j) = w(i\Delta x, j\Delta t), \quad \theta(i,j) = \theta(i\Delta x, j\Delta t),\\
&\frac{\partial w(x,t)}{\partial x} \simeq \frac{w(i,j) - w(i-1,j)}{\Delta x}, \quad \frac{\partial w(x,t)}{\partial t} \simeq \frac{w(i,j+1) - w(i,j)}{\Delta t},
\end{aligned}$$

then equation (7.51) can be represented as

$$\begin{aligned}
w(i,j+1) = {} & \left(1 - \frac{\Delta t}{\Delta x} - a_0^{\theta(i,j)} \Delta t\right) w(i,j)\\
& + \frac{\Delta t}{\Delta x} w(i-1,j) - a_1^{\theta(i,j)} \Delta t w(i, j-d_v).
\end{aligned} \tag{7.52}$$

Now, we define $x^h(i, j) = w(i-1, j)$ and $x^v(i, j) = w(i, j)$ then equation (7.52) can be recast into the form of 2-D system (7.1), where

$$A_{\theta(i,j)} = \begin{bmatrix} 0 & 1 \\ \frac{\Delta t}{\Delta x} & 1 - \frac{\Delta t}{\Delta x} - a_0^{\theta(i,j)} \Delta t \end{bmatrix}, \quad A_{d\theta(i,j)} = \begin{bmatrix} 0 & 0 \\ 0 & -a_1^{\theta(i,j)} \Delta t \end{bmatrix}.$$

The following parameters are taken from the literature [31]

$$\Delta t = 0.1, \quad \Delta x = 0.4, \quad a_0^1 = 1, \quad a_0^2 = 1.2, \quad a_1^1 = 1.2, \quad a_1^2 = 1.$$

Then, we have

$$A_1 = \begin{bmatrix} 0 & 1 \\ 0.25 & 0.65 \end{bmatrix}, \quad A_2 = \begin{bmatrix} 0 & 1 \\ 0.25 & 0.63 \end{bmatrix},$$
$$A_{d1} = \begin{bmatrix} 0 & 0 \\ 0 & -0.12 \end{bmatrix}, \quad A_{d2} = \begin{bmatrix} 0 & 0 \\ 0 & -0.1 \end{bmatrix}.$$

In this case the transition matrix is identity. By the Matlab LMI toolbox, it is found that condition (7.45) is feasible with $d_h \in (0, \infty)$ and $d_v \in (0, 14)$. By Corollary 7.2, the system is asymptotically stable with arbitrary switching signal $\theta(i, j)$.

7.6 Conclusion

In this chapter, the stochastic stability has been studied and presented for 2-D Markov jump Roesser systems with interval delays and uncertain transition rates. An analysis scheme for the stochastic stability property has been developed and utilized in combination with some 2-D summation inequalities to derive delay-dependent stochastic stability conditions. Numerical examples with simulations have been given to validate the effectiveness of the obtained results.

References

1. Sipahi, R., Niculescu, S.-I., Abdallah, C.T., Michiels, W., Gu, K.: Stability and stabilization of systems with time delay. IEEE Control Syst. Magaz. **31**, 38–65 (2011)
2. Witrant, E., Fridman, E., Sename, O., Dugard, L.: Recent Results on Time-Delay Systems: Analysis and Control. Springer, Basel (2016)
3. Hien, L.V., Trinh, H.: New finite-sum inequalities with applications to stability of discrete time-delay systems. Automatica **71**, 197–201 (2016)
4. Xiao, S., Xu, L., Zeng, H.B., Teo, K.L.: Improved stability criteria for discrete-time delay systems via novel summation inequalities. Int. J. Control Autom. Syst. **16**, 1592–1602 (2018)
5. Nam, P.T., Hiep, L.T.: A new delay-variation-dependent stability criterion for delayed discrete-time systems. J. Frankl. Inst. **357**, 6951–6967 (2020)

6. Zhang, C.K., Chen, W.H., Zhu, C., He, Y., Wu, M.: Stability analysis of discrete-time systems with time-varying delay via a delay-dependent matrix- separation-based inequality. Automatica **156**, 111192 (2023)
7. Sha, H., Park, J.H., Chen, J., Zhu, M., Nan, C.: Stability analysis of discrete-time systems with a time-varying delay via improved methods. IET Control Theory Appl. **18**, 951–959 (2024)
8. Ye, S., Li, J., Yao, J.: Robust H_∞ control for a class of 2-D discrete delayed systems. ISA Trans. **53**, 1456–1462 (2014)
9. Hien, L.V., Trinh, H.: Stability of two-dimensional Roesser systems with time-varying delays via novel 2D finite sum inequalities. IET Control Theory Appl. **10**, 1665–1674 (2016)
10. Huang, Y.B., An, J., He, Y., Wu, M.: Quasi-convex combination method and its application to the stability analysis of 2D discrete-time Roesser systems with time-varying delays. IET Control Theory Appl. **12**, 718–727 (2018)
11. Hien, L.V., Trinh, H.: Exponential stability of two-dimensional homogeneous monotone systems with bounded directional delays. IEEE Trans. Autom. Control **63**, 2694–2700 (2018)
12. Peng, D., Nie, H.: Abel lemma-based finite-sum inequality approach to stabilization for 2-D time-varying delay systems. Asian J. Control **23**, 1394–1406 (2021)
13. Costa, O.L.V., Fragoso, M.D., Marques, R.P.: Discrete-Time Markov Jump Linear Systems. Springer-Verlag, London (2005)
14. Geromel, J.C., Gonçalves, A.P., Fioravanti, A.R.: Dynamic output feedback control of discrete-time Markov jump linear systems through linear matrix inequalities. SIAM J. Control Optim. **48**, 573–593 (2009)
15. Guo, G., Lu, Z., Han, Q.L.: Control with Markov sensors/actuators assignment. IEEE Trans. Autom. Control **57**, 1799–1804 (2012)
16. Kao, Y., Xie, J., Wang, C., Karimi, H.R.: A sliding mode approach to H_∞ non-fragile observer-based control design for uncertain Markovian neutral-type stochastic systems. Automatica **52**, 218–226 (2015)
17. Oliveira, R.C., Vargas, A.N.: do Val JBR, Peres PLD,: Mode-independent H_2-control of a DC motor modeled as a Markov jump linear system. IEEE Trans. Control Syst. Tech. **22**, 1915–1919 (2014)
18. Hien, L.V., Dzung, N.T., Trinh, H.: Stochastic stability of nonlinear discrete-time Markovian jump systems with time-varying delay and partially unknown transition rates. Neurocomputing **175**, 450–458 (2016)
19. Li, X., Lam, J., Gao, H., Xiong, J.: H_∞ and H_2 filtering for linear systems with uncertain Markov transitions. Automatica **67**, 252–266 (2016)
20. Wu, Z.G., Shi, P., Shu, Z., Su, H., Lu, R.: Passivity-based asynchronous control for Markov jump systems. IEEE Trans. Autom Control **62**, 2020–2025 (2017)
21. Guerrero, J.C., Chávez-Fuentes, J.R., Casavilca, J.E., Costa, E.F.: Stability analysis of discrete-time Markov jump linear singular systems with partially known transition probabilities. Syst. Control Lett. **158**, 105057 (2021)
22. Tao, Y.Y., Wu, Z.G.: Asynchronous stabilization for hidden Markov jump linear systems with complex transition probabilities. Automatica **156**, 111200 (2023)
23. Kao, Y., Zhang, P., Wang, C., Xia, H.: Analysis and Design of Markov Jump Discrete Systems. Springer Nature, Singapore (2023)
24. Chen, Y., Zheng, W.X.: Exponential H_∞ filtering for stochastic Markovian jump systems with time delays. Int. J. Robust Nonliear Control **24**, 625–643 (2014)
25. Wang, Y., Zhang, S., Zhang, X., Wang, X.: Fault detection for a class of non-linear networked systems with Markovian transmission delays. IMA J. Math Control Inf. **32**, 141–159 (2015)
26. Wu, L., Yao, X., Zheng, W.X.: Generalized H_2 fault detection for two-dimensional Markovian jump systems. Automatica **48**, 1741–1750 (2012)
27. Hien, L.V., Trinh, H.: Observer-based control of 2-D Markov jump systems. IEEE Trans Circuit Syst-II **64**, 1322–1326 (2017)
28. Tao, Y.Y., Wu, Z.G.: Asynchronous control of two-dimensional Markov jump Roesser systems: An event-triggering strategy. IEEE Trans. Network Sci. Eng. **9**, 2278–2289 (2022)

29. Li, F., Shen, H.: Asynchronous control of 2-D Markov jump systems subject to general probabilities information. Nonlinear Anal. Hybrid Syst. **49**, 101357 (2023)
30. Tao, Y.Y., Che, W.W., Wu, Z.G., Xu, Y., Dong, S.: Asynchronous control of 2-D Markov jump Roesser systems with nonideal transition probabilities. IEEE Trans. Cynbern. **54**, 6643–6652 (2024)
31. Wei, Y., Qiu, J., Karimi, H.R., Wang, M.: Filtering design for two-dimensional Markovian jump systems with state-delays and deficient mode information. Info. Sci. **269**, 316–331 (2014)
32. Hien, L.V., Trinh, H.: Stability analysis of two-dimensional Markovian jump state-delayed systems in the Roesser model with uncertain transition probabilities. Info. Sci. **367**, 403–417 (2016)
33. Zhang, R., Zhang, Y., Hu, C., Meng, M.H., He, Q.: Delay-range-dependent H_∞ filtering for two-dimensional Markovain jump systems with interval delays. IET Control Theory Appl. **5**, 2191–2199 (2011)
34. Kaczorek, T.: Two-Dimensional Linear Systems. Springer, Berlin (1985)

Chapter 8
Analysis and Control of Two-Dimensional Roesser Systems with Stochastic Noises

Abstract In this chapter, the problems of stochastic stability, robust stabilization via synchronous state-feedback, and observer-based control are developed for some classes of two-dimensional (2-D) Roesser systems in the presence of state delays and multiplicative stochastic noises. An analysis scheme is first presented and utilized to derive delay-dependent stability conditions in terms of tractable linear matrix inequalities. Applications to the design problem of observer-based control for 2-D systems with exogenous disturbances and random packet dropouts in the control channel is then presented. As some extensions for 2-D systems with multiplicative noises, the stabilization problem via synchronous state-feedback control is also discussed for 2-D Roesser systems driven by inhomogeneous Markov chains. Numerical examples with simulations are given to illustrate the efficacy of the proposed methods.

Keywords 2-D systems · Multiplicative noises · Stochastic stability · Energy-to-peak stability · Synchronous control

8.1 Introduction

As discussed in previous chapters, subject to many practical reasons, time-delay phenomena frequently occur in engineering systems and industrial processes [1, 2]. A typical example can be found in the model of multi-agent systems (MASs). Due to the limited transmission rate of tele-communication channels, the data and signal exchange between agents is always infected by time delays, which heavily affects the consensus property of MASs [3]. Many other delay models can be found in chemical processes, transmission lines, or telecommunication networks. It is important to notice that the encounter of delays often degrades system performance and leads to unpredictable system performance [1]. Thus, the field of study in analysis and control of time-delay systems plays an essential role in applied models, which has attracted remarkable research attention over the past few decades. We refer the reader to recent works [4–13] for a few references.

L. V. Hien, *Two-Dimensional Time-Delay Systems*, Studies in Systems, Decision and Control 605, https://doi.org/10.1007/978-3-031-95678-2_8

Besides the effect of time-delay, dynamical systems with deterministic and/or stochastic exogenous disturbances are very popular in control engineering due to the inaccuracy of data processing and measurement errors [14, 15]. Among various ways to model external stochastic disturbances, additive noises [16] and multiplicative noises [17, 18] are most widely used. For example, multiplicative noisy systems (MNSs) can be used to describe dynamics of practical models in synthetic aperture radar, medical ultrasound or optical coherence tomography images [19]. In such a model, since stochastic signals get multiplied into states and outputs, the study on analysis and synthesis of MNSs is much more complicated, in particular, for MNSs with both delays and noises. A number of existing works concerning stability, performance analysis and controller/filter/observer design have been reported for one-dimensional (1-D) systems with certain types of multiplicative stochastic noises [20–24]. However, the theory for 2-D MNSs has received considerable less attention. Very few results devoted to extending the systems and control theory for 2-D MNSs have been reported. In particular, 2-D Roesser systems with stochastic noises get multiplied into system states were first investigated in [25]. Energy-to-peak stability (also known as l_2-l_∞ stability) was considered using linear matrix inequalities (LMIs) techniques. The problems of peak-to-peak (l_∞-l_∞) filtering and expected power bound were also studied for delay-free 2-D MNSs described by Fornasini–Marchesini local state-space (FMLSS) model in [26, 27]. The progress for 2-D MNSs with delays has not been well-developed so far. In particular, systematic approaches for the analysis and synthesis of 2-D systems subject to time-varying delays, exogenous disturbances, and multiplicative stochastic noises are of essential relevance, which motivate further development.

In this chapter, we first discuss the energy-to-peak stability for 2-D Roesser systems subject simultaneously to time-varying delays, external (deterministic) disturbances, and multiplicative stochastic noises in both state and output vectors. An analysis scheme is developed and utilized to derive delay-dependent stability conditions using LMI settings. Applications to the design problem of observer-based control for 2-D systems with exogenous disturbances and random packet dropouts in control channel is then presented. As other extensions, for 2-D systems with multiplicative noises and stochastic switching parameters, the stabilization problem via synchronous mode-dependent state-feedback controllers (SFCs) is also considered for 2-D Roesser systems driven by inhomogeneous Markov chains. Numerical examples are given to illustrate the efficacy of the proposed methods.

8.2 Preliminaries

8.2.1 Auxiliary Results

In this section, we recall some auxiliary results, which will be used in the presentation of the results in this chapter. First, we introduce the following properties of conditional expectation. For more details, we refer the reader to the monograph [28].

Lemma 8.1 *On a probability space $(\Omega, \mathscr{F}, \mathbb{P})$, let $\mathscr{G}$ be a sub-σ-algebra of $\mathscr{F}$ and X, Y be any random variables. We have the following properties.*

(i) $\mathbb{E}[X|\mathscr{G}] = X$ *and* $\mathbb{E}[XY|\mathscr{G}] = X\mathbb{E}[Y|\mathscr{G}]$ *if* X *is* $\mathscr{G}$*-measurable.*
(ii) *If* X *is independent of* $\mathscr{G}$ *then* $\mathbb{E}[X|\mathscr{G}] = \mathbb{E}[X]$.
(iii) $\mathbb{E}\left[\mathbb{E}[X|\mathscr{G}]\right] = \mathbb{E}[X]$.
(iv) $\mathbb{E}[\lambda X + \mu Y|\mathscr{G}] = \lambda\mathbb{E}[X|\mathscr{G}] + \mu\mathbb{E}[Y|\mathscr{G}]$, $\lambda, \mu \in \mathbb{R}$.
(v) *If* $X \leq Y$ *then* $\mathbb{E}[X|\mathscr{G}] \leq \mathbb{E}[Y|\mathscr{G}]$.

The following 2-D summation inequalities are useful for the manipulation of 2-D Lyapunov–Krasovskii functional (LKF) candidates (see, Chap. 2, Sect. 2.3.2).

Lemma 8.2 *Given positive integers h, k and a function $w : \mathbb{Z}[-h, 0] \times [-k, 0] \to \mathbb{R}^N$. For any matrix $R \in \mathbb{S}_n^+$, the following inequalities hold*

$$h \sum_{i=-h}^{-1} \Delta_h w^\top(i, j) R \Delta_h w(i, j) \geq \begin{bmatrix} w_{11} \\ w_{12} \end{bmatrix}^\top \begin{bmatrix} R & 0 \\ 0 & 3R \end{bmatrix} \begin{bmatrix} w_{11} \\ w_{12} \end{bmatrix},$$

$$k \sum_{j=-k}^{-1} \Delta_v w^\top(i, j) R \Delta_v w(i, j) \geq \begin{bmatrix} w_{21} \\ w_{22} \end{bmatrix}^\top \begin{bmatrix} R & 0 \\ 0 & 3R \end{bmatrix} \begin{bmatrix} w_{21} \\ w_{22} \end{bmatrix},$$

where $\Delta_h w(i, j) = w(i + 1, j) - w(i, j)$ and $\Delta_v w(i, j) = w(i, j + 1) - w(i, j)$ denote the horizontal and vertical differences of $w(i, j)$, respectively, and

$$w_{11} = w(0, j) - w(-h, j), \quad w_{21} = w(i, 0) - w(i, -k),$$

$$w_{12} = w(0, j) + w(-h, j) - \frac{2}{h+1} \sum_{i=-h}^{0} w(i, j),$$

$$w_{22} = w(i, 0) + w(i, -k) - \frac{2}{k+1} \sum_{j=-k}^{0} w(i, j).$$

To make this chapter self-contain, we also recall here basic facts involving reciprocally convex inequalities and Schur complement lemma. As discussed in Chap. 2 [Sect. 2.3.3], to deal with with interval time-varying delays, it is necessary to bound the summation

$$S_R = \sum_{k=1}^{m} \frac{1}{\alpha_k} \chi_k^\top R_k \chi_k$$

for vectors $\chi_k \in \mathbb{R}^{N_k}$, matrices $R_k \in \mathbb{S}_{N_k}^{+}$ and scalars $\alpha_k > 0$ with $\sum_{k=1}^{m} \alpha_k = 1$. For the case $m = 2$, we have the following result.

Lemma 8.3 *For given matrices $R_1 = R_1^\top > 0$, $R_2 = R_2^\top > 0$ and X_{12} of appropriate dimensions, if* $\begin{bmatrix} R_1 & X_{12} \\ X_{12}^\top & R_2 \end{bmatrix} \geq 0$, *then*

$$\frac{1}{\alpha} \chi_1^\top R_1 \chi_1 + \frac{1}{1-\alpha} \chi_2^\top R_2 \chi_2 \geq \begin{bmatrix} \chi_1 \\ \chi_2 \end{bmatrix}^\top \begin{bmatrix} R_1 & X_{12} \\ X_{12}^\top & R_2 \end{bmatrix} \begin{bmatrix} \chi_1 \\ \chi_2 \end{bmatrix}$$

for any $\alpha \in (0, 1)$ and vectors χ_1, χ_2 of appropriate dimensions.

The following result referred as Schur complement lemma will also be used to derive tractable design conditions.

Lemma 8.4 *For any matrices $X = X^\top$, $Z = Z^\top$ and a matrix Y of appropriate dimensions, we have*

$$\begin{bmatrix} X & Y \\ Y^\top & Z \end{bmatrix} > 0 \iff \begin{cases} Z > 0 \\ X - YZ^{-1}Y^\top > 0 \end{cases}$$

$$\iff \begin{cases} X > 0 \\ Z - Y^\top X^{-1} Y > 0. \end{cases}$$

8.3 Energy-to-Peak Stability of 2-D Time-Delay Roesser Systems

In this section, we study the energy-to-peak stability for 2-D Roesser systems subject simultaneously to time-varying delays, external disturbances, and multiplicative stochastic noises. An analysis scheme is developed and utilized to derive delay-dependent stability conditions.

8.3.1 Model Description

Consider a class of 2-D MNSs with time-varying delays described by the following Roesser-type model

$$\begin{aligned}\begin{bmatrix}x^h(i+1,j)\\x^v(i,j+1)\end{bmatrix} &= A\begin{bmatrix}x^h(i,j)\\x^v(i,j)\end{bmatrix} + A_d\begin{bmatrix}x^h(i-d_h(i),j)\\x^v(i,j-d_v(j))\end{bmatrix} + Bw(i,j)\\ &\quad + \Lambda(i,j)\left\{\hat{A}\begin{bmatrix}x^h(i,j)\\x^v(i,j)\end{bmatrix} + \hat{A}_d\begin{bmatrix}x^h(i-d_h(i),j)\\x^v(i,j-d_v(j))\end{bmatrix} + \hat{B}w(i,j)\right\},\\ z(i,j) &= C\begin{bmatrix}x^h(i,j)\\x^v(i,j)\end{bmatrix} + D\begin{bmatrix}x^h(i-d_h(i),j)\\x^v(i,j-d_v(j))\end{bmatrix} + Fw(i,j)\\ &\quad + \Theta(i,j)\left\{\hat{C}\begin{bmatrix}x^h(i,j)\\x^v(i,j)\end{bmatrix} + \hat{D}\begin{bmatrix}x^h(i-d_h(i),j)\\x^v(i,j-d_v(j))\end{bmatrix} + \hat{F}w(i,j)\right\},\end{aligned} \tag{8.1}$$

where $x^h(i,j) \in \mathbb{R}^{n_h}$ and $x^v(i,j) \in \mathbb{R}^{n_v}$ are the horizontal and vertical state vectors, respectively, $w(i,j) \in \mathbb{R}^{n_o}$ represents the exogenous disturbance belonging to l_2 and $z(i,j) \in \mathbb{R}^{n_z}$ represents the output vector. A, A_d, $\hat{A}$, $\hat{A}_d \in \mathbb{R}^{n\times n}$ ($n = n_h + n_v$) and B, $\hat{B} \in \mathbb{R}^{n\times n_o}$, C, D, $\hat{C}$, $\hat{D} \in \mathbb{R}^{n_z\times n}$, F, $\hat{F} \in \mathbb{R}^{n_z\times n_o}$ are given matrices.

In system (8.1), the two-parameter processes $\Lambda(i,j)$ and $\Theta(i,j)$ are defined by

$$\Lambda(i,j) = \text{diag}\left(\lambda_{ij}^{(1)}, \lambda_{ij}^{(2)}, \ldots, \lambda_{ij}^{(n)}\right) \text{ and } \Theta(i,j) = \text{diag}\left(\theta_{ij}^{(1)}, \theta_{ij}^{(2)}, \ldots, \theta_{ij}^{(n_z)}\right),$$

where $\lambda_{ij}^{(p)}$ ($p = 1, 2, \ldots, n$) and $\theta_{ij}^{(q)}$ ($q = 1, 2, \ldots, n_z$) are scalar-valued white noises defined on a complete probability space $(\Omega, \mathscr{F}, \mathbb{P})$. It is assumed that the random variables $\lambda_{ij}^{(p)}$ and $\theta_{ij}^{(q)}$ are independent with zero-mean and satisfy

$$\mathbb{E}\left[\lambda_{ij}^{(p)}\lambda_{kl}^{(p)}\right] = \sigma_p^2\delta_{ik}\delta_{jl}, \quad \mathbb{E}\left[\theta_{ij}^{(q)}\theta_{kl}^{(q)}\right] = \hat{\sigma}_q^2\delta_{ik}\delta_{jl}, \tag{8.2}$$

where σ_p and $\hat{\sigma}_q$ are given positive scalars, δ_{ik} denotes the Kronecker delta function.

The time-varying state delays $d_h(i)$ and $d_v(j)$ satisfy

$$d_1^h \le d_h(i) \le d_2^h, \quad d_1^v \le d_v(j) \le d_2^v, \tag{8.3}$$

where d_1^h, d_2^h and d_1^v, d_2^v are known positive integers involving the bounds of delays. We denote by $d_h^r = d_2^h - d_1^h$ and $d_v^r = d_2^v - d_1^v$ the variation range of delays $d_h(i)$ and $d_v(j)$. The initial condition of system (8.1) is specified as

$$\begin{aligned} x^h(i,j) &= \phi(i,j), \quad i \in \mathbb{Z}[-d_2^h, 0], \quad j \in \mathbb{N}_0,\\ x^v(i,j) &= \psi(i,j), \quad i \in \mathbb{N}_0, \quad j \in \mathbb{Z}[-d_2^v, 0], \end{aligned} \tag{8.4}$$

with $\phi(i,.)$ and $\phi(.,j)$ are mean-square summable sequences, that is,

$$M(\phi,\psi) \triangleq \sum_{k=0}^{\infty}\mathbb{E}[\hat{\phi}_k^2 + \hat{\psi}_k^2] < \infty, \tag{8.5}$$

where $\hat{\phi}_k = \max_{i\in\mathbb{Z}[-d_2^h,0]} \|\phi(i,k)\|$ and $\hat{\psi}_k = \max_{j\in\mathbb{Z}[-d_2^v,0]} \|\psi(k,j)\|$.

8.3.2 Problem Statement

Consider 2-D MNSs with time-varying delays presented in (8.1). First, we introduce the following definition.

Definition 8.1 (Stochastic stability) The 2-D MNS (8.1) with $w = 0$ is said to be stochastically stable if, for any solution $x(i, j)$ of (8.1) with respect to initial condition satisfying (8.4) and (8.5), it holds that

$$\mathbb{E}\left[\sum_{i,j=0}^{\infty} x^\top(i, j)x(i, j)\right] < \infty, \tag{8.6}$$

where, for convenience, we denote the whole state vector by $x(i, j) = \begin{bmatrix} x^h(i, j) \\ x^v(i, j) \end{bmatrix}$.

Remark 8.1 Condition (8.6) reveals that the total energy in mean-square of any solution of (8.1) is finite. In addition, (8.6) induces that $\lim_{i+j\to\infty} \mathbb{E}[x^\top(i, j)x(i, j)] = 0$. This ensures mean-square asymptotic stability of system (8.1).

The energy of a process $w \in l_2$ is associated with its l_2-norm $\|w\|_{l_2}^2 = \sum_{i,j=0}^{\infty} \|w(i, j)\|^2$. Meanwhile, if the output z belongs to l_∞, the value $\|z\|_{l_\infty}^2 = \sup_{i,j\geq 0} \mathbb{E}\left[z^\top(i, j)z(i, j)\right]$ is referred as the peak value of z in the sense of mean-square. The property energy-to-peak (EP) of system (8.1) is associated with these factors.

Definition 8.2 (Energy-to-peak performance) For given a scalar $\gamma > 0$, the 2-D MNS (8.1) is said to have an EP performance at level γ if the output $z \in l_\infty$ for any disturbance input $w \in l_2$ and there exists a constant $\beta > 0$ such that

$$\|z\|_{l_\infty}^2 < \gamma^2 \|w\|_{l_2}^2 + \beta M(\phi, \psi). \tag{8.7}$$

for all nonzero disturbance w.

Remark 8.2 Let $\mathscr{L}_{wz}$ denote the transfer function which maps the disturbance input w to output z under zero initial condition (i.e. $\phi = \psi = 0$). Then, for a given scalar $\gamma > 0$, and under zero initial condition, the l_2-l_∞ norm of $\mathscr{L}_{wz}$ satisfies

$$\|\mathscr{L}_{wz}\|_{l_2\text{-}l_\infty} = \sup_{0\neq w\in l_2} \frac{\|z\|_{l_\infty}}{\|w\|_{l_2}} < \gamma.$$

Definition 8.3 (Energy-to-peak stochastic stability) The 2-D MNS (8.1) is said to be energy-to-peak stochastically stable (EPSS) if the two following requirements hold

(i) System (8.1) with $w = 0$ is stochastically stable;
(ii) For a given $\gamma > 0$, system (8.1) has an EP performance at level γ.

The problem addressed in this section is to derive delay-dependent conditions by which 2-D MNSs in the form of (8.1) is EPSS. To tackle the problem, we first formulate an analysis scheme by extending the ideas of 2-D LKF method for stochastic time-delay systems. By constructing suitable LKF candidates and utilizing he proposed analysis scheme, delay-dependent stability conditions can be derived.

8.3.3 Functional-Based Analysis Scheme

In this section, we formulate a functional-based scheme to deal with the EPSS analysis problem of 2-D stochastic systems with delays. Let $V : \mathbb{R}^n \to \mathbb{R}$ be a functional defined by $V(x) = V^h(x^h) + V^v(x^v)$, where $x = (x^h, x^v)$, $x^h \in \mathbb{R}^{n_h}$, $x^v \in \mathbb{R}^{n_v}$, and $V^h : \mathbb{R}^{n_h} \to \mathbb{R}$, $V^v : \mathbb{R}^{n_v} \to \mathbb{R}$ are some functionals. The difference of $V(x)$ along state trajectory $x(i, j)$ of system (8.1) is defined as

$$\begin{aligned}\Delta V((x(i,j)) &\triangleq \Delta_h V^h(x^h(i,j)) + \Delta V^v(x^v(i,j)) \\ &= \mathbb{E}\left[V^h(x^h(i+1,j)) + V^v(x(i,j+1))\big|\mathscr{F}_{ij}^x\right] - V(x(i,j)),\end{aligned} \tag{8.8}$$

where $\mathscr{F}_{ij}^x$ is the σ-algebra generated by $\{x(k,l) : k \le i, l \le j\}$.

Theorem 8.1 *Given a scalar $\gamma > 0$. Assume that there exist a functional $V(x) = V^h(x^h) + V^v(x^v)$ and positive scalars c_0, c_1^h, c_2^h and c_1^v, c_2^v such that*

(i) *For any $x^h \in \mathbb{R}^{n_h}$, $x^v \in \mathbb{R}^{n_v}$,*

$$c_1^h\|x^h\|^2 \le V^h(x^h) \le c_2^h\|x^h\|^2, \quad c_1^v\|x^v\|^2 \le V^v(x^v) \le c_2^v\|x^h\|^2;$$

(ii) *The difference $\Delta V(x(i,j))$ of $V(x)$ along state trajectory $x(i,j)$ (8.1) satisfies*

$$\Delta V(x(i,j)) \le w^\top(i,j)w(i,j) - c_0\|x(i,j)\|^2, \quad (i,j) \in \mathbb{N}_0^2; \tag{8.9}$$

(iii) *For any nonzero disturbance w, it holds that*

$$\mathbb{E}\left[\frac{2}{\gamma^2}z^\top(i,j)z(i,j) - V(x(i,j))\big|\mathscr{F}_{ij}^x\right] \le w^\top(i,j)w(i,j). \tag{8.10}$$

Then, the 2-D MNS (8.1) is EPSS.

Proof For the brevity, along any state trajectory $x(i,j)$ of (8.1), we denote

$$\mathscr{V}^h(i,j) = V^h(x^h(i,j)), \quad \mathscr{V}^v(i,j) = V^v(x^v(i,j))$$

and $\mathscr{V}(i,j) = V(x(i,j))$. By taking expectation both sides of (8.9) and using the properties of conditional expectation (see, Lemma 8.1) [28], we have

$$\mathbb{E}[\mathscr{V}^h(i+1,j)] - \mathbb{E}[\mathscr{V}^h(i,j)] + \mathbb{E}[\mathscr{V}^v(i,j+1)] - \mathbb{E}[\mathscr{V}^v(i,j)] \\ \leq w^\top(i,j)w(i,j) - c_0\,\mathbb{E}[\|x(i,j)\|^2]. \quad (8.11)$$

Thus, for any integers $k \geq 1$ and $l \geq 1$, from (8.11) we get

$$\sum_{j=0}^{l-1}\big(\mathbb{E}[\mathscr{V}^h(k,j)] - \mathbb{E}[\mathscr{V}^h(0,j)]\big) + \sum_{i=0}^{k-1}\big(\mathbb{E}[\mathscr{V}^v(i,l)] - \mathbb{E}[\mathscr{V}^v(i,0)]\big) \\ \leq \sum_{i=0}^{k-1}\sum_{j=0}^{l-1} w^\top(i,j)w(i,j) - c_0\sum_{i=0}^{k-1}\sum_{j=0}^{l-1}\mathbb{E}[\|x(i,j)\|^2]. \quad (8.12)$$

When the exogenous disturbance w is ignored, it follows from (8.12) that

$$\sum_{i,j=0}^{\infty}\mathbb{E}[\|x(i,j)\|^2] \leq \frac{\max(c_2^h, c_2^v)}{c_0}M(\phi,\psi) < \infty,$$

which guarantees the stochastic stability of system (8.1).

On the other hand, from (8.12) we also have

$$\sum_{j=0}^{l-1}\mathbb{E}[\mathscr{V}^h(k,j)] \leq \sum_{i=0}^{k-1}\sum_{j=0}^{l-1} w^\top(i,j)w(i,j) + c_2M(\phi,\psi)$$

for any $k \geq 1$ and $l \geq 1$, where $c_2 = \max(c_2^h, c_2^v)$. Therefore,

$$\sum_{j=0}^{l}\mathbb{E}[\mathscr{V}^h(k,j)] \leq \sum_{i=0}^{k-1}\sum_{j=0}^{l} w^\top(i,j)w(i,j) + c_2M(\phi,\psi). \quad (8.13)$$

It can be deduced by similar arguments of (8.13) that

$$\sum_{i=0}^{k}\mathbb{E}[\mathscr{V}^v(i,l)] \leq \sum_{i=0}^{k}\sum_{j=0}^{l-1} w^\top(i,j)w(i,j) + c_2M(\phi,\psi). \quad (8.14)$$

A combination of (8.13) and (8.14), together with condition (iii), leads to

$$\frac{2}{\gamma^2}\mathbb{E}\left[z^\top(k,l)z(k,l)\right] < \mathbb{E}[\mathscr{V}(k,l)] + w^\top(k,l)w(k,l)$$

$$\leq 2\left(\sum_{i=0}^{k}\sum_{j=0}^{l} w^\top(i,j)w(i,j) + c_2 M(\phi,\psi)\right)$$

for any $(k, l) \in \mathbb{N}_0^2$, $k \geq 1$ and $l \geq 1$. It is clear that the last inequality is still valid for $k = 0$ or $l = 0$. Thus, for any $(k, l) \in \mathbb{N}_0^2$, we have

$$\frac{2}{\gamma^2}\mathbb{E}\left[z^\top(k,l)z(k,l)\right] < 2\left(\sum_{i,j=0}^{\infty} w^\top(i,j)w(i,j) + c_2 M(\phi,\psi)\right). \tag{8.15}$$

Finally, from (8.15), we readily obtain

$$\sup_{k,l\geq 0}\mathbb{E}\left[z^\top(k,l)z(k,l)\right] < \gamma^2\|w\|_{l_2}^2 + \beta M(\phi,\psi) \tag{8.16}$$

for a scalar $\beta \in (0, \gamma^2 c_2)$. The result of (8.16) shows that system (8.1) has an EP performance of level γ. The proof is completed. □

Remark 8.3 In the proof of Theorem 8.1, condition (8.9) can be replaced by the alternative condition that there exists a scalar $\alpha \in (0, 1)$ such that

$$\mathbb{E}\left[V^h(x^h(i+1,j)) + V^v(x(i,j+1))\big|\mathscr{F}_{ij}^x\right] - \alpha V(x(i,j)) \leq w^\top(i,j)w(i,j). \tag{8.17}$$

Indeed, from (8.17) and condition (i), we have

$$\begin{aligned}\mathbb{E}\Big[\mathscr{V}^h(i+1,j) + \mathscr{V}^v(i,j+1)\big|\mathscr{F}_{ij}^x\Big] - \mathscr{V}(i,j)&\\ \leq w^\top(i,j)w(i,j) - (1-\alpha)\mathscr{V}(i,j)&\\ \leq w^\top(i,j)w(i,j) - (1-\alpha)\min(c_1^h, c_1^v)\|x(i,j)\|^2.&\end{aligned}$$

This validates condition (8.9) and the conclusion of Theorem 8.1 still remains. By using the scalar α as a turning parameter, condition (8.17) can help to make the LMIzation process to derive stability conditions easier although, in general, it produces more conservatism in the resulting conditions than that of (8.9).

Remark 8.4 For other performance indices such as H_∞ (energy-to-energy l_2-l_2) or peak-to-peak (l_∞-l_∞), condition (iii) will be slightly modified and an analogous scheme can be obtained. In regard, the result of Theorem 8.1 can be effectively extended to various problems in the systems and control theory of 2-D Roesser systems with delays and multiplicative stochastic noises.

8.3.4 Energy-to-Peak Stability Conditions

For the brevity, we denote the delay terms $h_l(i) = d_h(i) - d_1^h$, $h_r(i) = d_2^h - d_h(i)$, $v_l(j) = d_v(j) - d_1^v$, $v_r(j) = d_2^v - d_v(j)$, block matrices

$$\begin{aligned} e_k &= \begin{bmatrix} 0_{n\times(k-1)n} & I_n & 0_{n\times[(7-k)n+n_o]} \end{bmatrix}, \quad k = 1, 2, \ldots, 7, \\ e_8 &= \begin{bmatrix} 0_{n_o\times 7n} & I_{n_o} \end{bmatrix}, \quad E_n^{(p)} = f_p f_p^\top \in \mathbb{R}^{n\times n}, \end{aligned}$$

where $\{f_1, f_2, \ldots, f_n\}$ is the canonical basis of $\mathbb{R}^n$, that means $E_n^{(p)}$ is the diagonal matrix with only pth entry equals one. We also denote the following vectors

$$\begin{aligned} \zeta_1(i,j) &= \begin{bmatrix} x(i,j) \\ x_l(i,j) \\ x_d(i,j) \\ x_u(i,j) \end{bmatrix}, \quad \zeta_2(i,j) = \begin{bmatrix} \bar{x}_l(i,j) \\ \bar{x}_d(i,j) \\ \bar{x}_u(i,j) \\ w(i,j) \end{bmatrix}, \\ \zeta(i,j) &= \begin{bmatrix} \zeta_1(i,j) \\ \zeta_2(i,j) \end{bmatrix}, \quad x_l(i,j) = \begin{bmatrix} x^h(i-d_1^h, j) \\ x^v(i, j-d_1^v) \end{bmatrix}, \\ x_u(i,j) &= \begin{bmatrix} x^h(i-d_2^h, j) \\ x^v(i, j-d_2^v) \end{bmatrix}, \quad \bar{x}_l(i,j) = \begin{bmatrix} \frac{1}{1+d_1^h}\sum_{k=i-d_1^h}^{i} x^h(k,j) \\ \frac{1}{1+d_1^v}\sum_{l=j-d_1^v}^{j} x^v(i,l) \end{bmatrix}, \\ \bar{x}_d(i,j) &= \begin{bmatrix} \frac{1}{1+h_l(i)}\sum_{k=i-d_h(i)}^{i-d_1^h} x^h(k,j) \\ \frac{1}{1+v_l(j)}\sum_{l=j-d_v(j)}^{j-d_1^v} x^v(i,l) \end{bmatrix}, \quad \bar{x}_u(i,j) = \begin{bmatrix} \frac{1}{1+h_r(i)}\sum_{k=i-d_2^h}^{i-d_h(i)} x^h(k,j) \\ \frac{1}{1+v_r(j)}\sum_{l=j-d_2^v}^{j-d_v(j)} x^v(i,l) \end{bmatrix}, \end{aligned}$$

and matrices

$$\begin{aligned} \mathscr{A} &= Ae_1 + A_d e_3 + Be_8, \quad \hat{\mathscr{A}} = \hat{A}e_1 + \hat{A}_d e_3 + \hat{B}e_8, \\ \Xi_1 &= \mathscr{A}^\top P \mathscr{A} + \sum_{p=1}^{n} \sigma_p^2 \hat{\mathscr{A}}^\top E_n^{(p)} P E_n^{(p)} \hat{\mathscr{A}} - e_1^\top P e_1 - e_8^\top e_8, \\ \Xi_2 &= e_1^\top Q e_1 + e_2^\top R e_2 - e_2^\top Q e_2 - e_4^\top R e_4, \\ \Xi_3 &= (\mathscr{A} - e_1)^\top (\mathbb{I}_1 W)(\mathscr{A} - e_1) + \sum_{p=1}^{n} \sigma_p^2 \hat{\mathscr{A}}^\top E_n^{(p)} (\mathbb{I}_1 W) E_n^{(p)} \hat{\mathscr{A}}, \\ \Xi_4 &= \begin{bmatrix} e_1 - e_2 \\ e_1 + e_2 - 2e_5 \end{bmatrix}^\top \operatorname{diag}(W, 3W) \begin{bmatrix} e_1 - e_2 \\ e_1 + e_2 - 2e_5 \end{bmatrix}, \\ \Xi_5 &= (\mathscr{A} - e_1)^\top (\mathbb{I}_2 Z)(\mathscr{A} - e_1) + \sum_{p=1}^{n} \sigma_p^2 \hat{\mathscr{A}}^\top E_n^{(p)} (\mathbb{I}_2 Z) E_n^{(p)} \hat{\mathscr{A}}, \\ \mathbb{I}_1 &= \operatorname{diag}\left((d_1^h)^2 I_{n_h}, (d_1^v)^2 I_{n_v}\right), \quad \mathbb{I}_2 = \operatorname{diag}\left((d_h^r)^2 I_{n_h}, (d_v^r)^2 I_{n_v}\right), \\ \Xi_6 &= \Gamma^\top \left(\mathbb{J}_h^{4\top} \Psi_X^h \mathbb{J}_h^4 + \mathbb{J}_v^{4\top} \Psi_Y^v \mathbb{J}_v^4\right) \Gamma, \end{aligned}$$

$$\Gamma_1 = \begin{bmatrix} e_2 - e_3 \\ e_2 + e_3 - 2e_6 \end{bmatrix}, \quad \Gamma_2 = \begin{bmatrix} e_3 - e_4 \\ e_3 + e_4 - 2e_7 \end{bmatrix}, \quad \Gamma = \begin{bmatrix} \Gamma_1 \\ \Gamma_2 \end{bmatrix},$$

$$\Psi_X^h = \begin{bmatrix} \widetilde{Z}^h & X \\ X^\top & \widetilde{Z}^h \end{bmatrix}, \quad \Psi_Y^v = \begin{bmatrix} \widetilde{Z}^v & Y \\ Y^\top & \widetilde{Z}^v \end{bmatrix}, \quad \widetilde{Z}^h = \operatorname{diag}(Z^h, 3Z^h),$$

$$\widetilde{Z}^v = \operatorname{diag}(Z^v, 3Z^v), \quad \mathbb{J}_h = \begin{bmatrix} I_{n_h} & 0_{n_h \times n_v} \end{bmatrix}, \quad \mathbb{J}_v = \begin{bmatrix} 0_{n_v \times n_h} & I_{n_v} \end{bmatrix},$$

$$\mathbb{J}_h^p = \underbrace{\operatorname{diag}(I_h, \ldots, I_h)}_{p \text{ blocks}}, \quad \mathbb{J}_v^q = \underbrace{\operatorname{diag}(I_v, \ldots, I_v)}_{q \text{ blocks}}.$$

We are now in a position to present delay-dependent stability conditions for 2-D MNSs with delays given in (8.1).

Theorem 8.2 *Given a scalar $\gamma > 0$ and assume that there exist matrices P, Q, R, W, Z that belong to $\mathbb{D}_n^+(n_h, n_v)$ and any matrices $X \in \mathbb{R}^{2n_h \times 2n_h}$, $Y \in \mathbb{R}^{2n_v \times 2n_v}$ such that the following LMI conditions are satisfied*

$$\Psi_X^h = \begin{bmatrix} \widetilde{Z}^h & X \\ X^\top & \widetilde{Z}^h \end{bmatrix} \geq 0, \quad \Psi_Y^v = \begin{bmatrix} \widetilde{Z}^v & Y \\ Y^\top & \widetilde{Z}^v \end{bmatrix} \geq 0, \tag{8.18}$$

$$\Xi_0 = \sum_{k=1}^{3} \Xi_k - \Xi_4 + \Xi_5 - \Xi_6 < 0, \tag{8.19}$$

$$\begin{bmatrix} -\Phi & \begin{bmatrix} \mathscr{H} \\ D_{\hat{\sigma}} \hat{\mathscr{H}} \end{bmatrix}^\top \\ \begin{bmatrix} \mathscr{H} \\ D_{\hat{\sigma}} \hat{\mathscr{H}} \end{bmatrix} & -\dfrac{\gamma^2}{2} I_{2n_z} \end{bmatrix} < 0, \tag{8.20}$$

where $\Phi = \operatorname{diag}(P, R, I_{n_o})$, $\mathscr{H} = \begin{bmatrix} C & D & F \end{bmatrix}$, $\hat{\mathscr{H}} = \begin{bmatrix} \hat{C} & \hat{D} & \hat{F} \end{bmatrix}$, and noise tensity matrix of output $D_{\hat{\sigma}} = \operatorname{diag}(\hat{\sigma}_1, \hat{\sigma}_2, \ldots, \hat{\sigma}_{n_z})$. Then, the 2-D MNS (8.1) *is EPSS for any directional time-varying delays within the ranges defined by* (8.3).

Proof By virtue of the functional-based scheme proposed in Theorem 8.1, we construct the following LKF candidate

$$\underbrace{V(x(i,j))}_{\mathscr{V}(i,j)} = \underbrace{x^\top(i,j) P x(i,j)}_{\mathscr{V}_0(i,j)} + \sum_{k=1}^{4} \Big(\underbrace{V_k^h(x^h(i,j))}_{\mathscr{V}_k^h(i,j)} + \underbrace{V_k^v(x^v(i,j))}_{\mathscr{V}_k^v(i,j)} \Big), \tag{8.21}$$

where

$$\mathscr{V}_1^h(i,j) = \sum_{k=i-d_1^h}^{i-1} x^{h\top}(k,j) Q^h x^h(k,j),$$

$$\mathscr{V}_2^h(i,j) = \sum_{k=i-d_2^h}^{i-d_1^h-1} x^{h\top}(k,j) R^h x^h(k,j),$$

$$\mathscr{V}_3^h(i,j) = d_1^h \sum_{k=-d_1^h}^{-1} \sum_{l=i+k}^{i-1} \eta_h^\top(l,j) W^h \eta_h(l,j)$$

$$\mathscr{V}_4^h(i,j) = d_h^r \sum_{k=-d_2^h}^{-d_1^h-1} \sum_{l=i+k}^{i-1} \eta_h^\top(l,j) Z^h \eta_h(l,j),$$

$$\mathscr{V}_1^v(i,j) = \sum_{l=j-d_1^v}^{j-1} x^{v\top}(i,l) Q^v x^v(i,l),$$

$$\mathscr{V}_2^v(i,j) = \sum_{l=j-d_2^v}^{j-d_1^v-1} x^{v\top}(i,l) R^v x^v(i,l),$$

$$\mathscr{V}_3^v(i,j) = d_1^v \sum_{k=-d_2^v}^{-1} \sum_{l=j+k}^{j-1} \eta_v^\top(i,l) W^v \eta_v(i,l),$$

$$\mathscr{V}_4^v(i,j) = d_v^r \sum_{k=-d_2^v}^{-d_1^v-1} \sum_{l=j+k}^{j-1} \eta_v^\top(i,l) Z^v \eta_v(i,l),$$

and $\eta_h(i,j) = \Delta_h(x^h(i,j))$, $\eta_v(i,j) = \Delta_v(x^v(i,j))$.

It can be verified from (8.21) that condition (i) of Theorem 8.1 is satisfied with $c_1^h = \lambda_{\min}(P^h)$, $c_1^v = \lambda_{\min}(P^v)$ and

$$\begin{aligned}
c_2^h &= \lambda_{\max}(P^h) + d_1^h \lambda_{\max}(Q^h) + d_h^r \lambda_{\max}(R^h) \\
&\quad + \frac{1}{2} d_1^h (d_1^h + 1) \lambda_{\max}(W^h) + \frac{1}{2} d_h^r (d_h^r + 1) \lambda_{\max}(Z^h), \\
c_2^v &= \lambda_{\max}(P^v) + d_1^v \lambda_{\max}(Q^v) + d_v^r \lambda_{\max}(R^v) \\
&\quad + \frac{1}{2} d_1^v (d_1^v + 1) \lambda_{\max}(W^v) + \frac{1}{2} d_v^r (d_v^r + 1) \lambda_{\max}(R_2^v).
\end{aligned}$$

We now compute the difference

$$\Delta \mathscr{V} = \mathbb{E}\left[\mathscr{V}^h(i+1,j) + \mathscr{V}^v(i,j+1) \middle| \mathscr{F}_{ij}^x\right] - \mathscr{V}(i,j)$$

and verify condition (ii) of Theorem 8.1. First, for the functional $\mathscr{V}_0$, we have

$$\begin{aligned}
\Delta \mathscr{V}_0 &= \mathbb{E}\left[\begin{bmatrix} x^h(i+1,j) \\ x^v(i,j+1) \end{bmatrix}^\top P \begin{bmatrix} x^h(i+1,j) \\ x^v(i,j+1) \end{bmatrix} \middle| \mathscr{F}_{ij}^x\right] - x^\top(i,j) P x(i,j) \\
&= \mathbb{E}\left[\mathscr{W}^\top(i,j) P \mathscr{W}(i,j) \middle| \mathscr{F}_{ij}^x\right] - x^\top(i,j) P x(i,j), \qquad (8.22)
\end{aligned}$$

where $\mathscr{W}(i, j)$ is the right-hand side of (8.1). Note from (8.1) that

$$\mathscr{W}(i, j) = \mathscr{A}\zeta(i, j) + \Lambda(i, j)\hat{\mathscr{A}}\zeta(i, j)$$

and $\Lambda(i, j) = \sum_{p=1}^{n} \lambda_{ij}^{(p)} E_n^{(p)}$. Therefore,

$$\begin{aligned}\mathscr{W}^\top(i, j)P\mathscr{W}(i, j) &= \zeta^\top(i, j)\mathscr{A}^\top P\mathscr{A}\zeta(i, j)\\ &+ 2\sum_{p=1}^{n} \lambda_{ij}^{(p)}\zeta^\top(i, j)\mathscr{A}^\top P E_n^{(p)}\hat{\mathscr{A}}\zeta(i, j)\\ &+ \left(\sum_{p=1}^{n} \lambda_{ij}^{(p)}\right)\left(\sum_{q=1}^{n} \lambda_{ij}^{(q)}\right)\zeta^\top(i, j)\hat{\mathscr{A}}^\top E_n^{(p)} P E_n^{(q)}\hat{\mathscr{A}}\zeta(i, j). \qquad (8.23)\end{aligned}$$

Using the properties of the process $\Lambda(i, j)$, from (8.22) and (8.23), we have

$$\begin{aligned}\Delta\mathscr{V}_0 &= \zeta^\top(i, j)\left(\mathscr{A}^\top P\mathscr{A} + \sum_{p=1}^{n} \sigma_p^2\hat{\mathscr{A}}^\top E_n^{(p)} P E_n^{(p)}\hat{\mathscr{A}} - e_1^\top P e_1\right)\zeta(i, j)\\ &= \zeta^\top(i, j)\left(\varXi_1 + e_8^\top e_8\right)\zeta(i, j). \qquad (8.24)\end{aligned}$$

Next, for functionals $\mathscr{V}_k = \mathscr{V}_k^h + \mathscr{V}_k^v$ $(k = 1, 2)$, we have

$$\begin{aligned}\Delta\mathscr{V}_1 &= \mathbb{E}\left[\mathscr{V}_1^h(i+1, j) + \mathscr{V}_1^v(i, j+1)\middle|\mathscr{F}_{ij}^x\right] - \mathscr{V}_1(i, j)\\ &= x^{h\top}(i, j)Q^h x^h(i, j) - x^{h\top}(i - d_1^h, j)Q^h x^h(i - d_1^h, j)\\ &\quad + x^{v\top}(i, j)Q^v x^v(i, j) - x^{v\top}(i, j - d_1^v)Q^v x^v(i, j - d_1^v)\\ &= x^\top(i, j)Qx(i, j) - x_l^\top(i, j)Qx_l(i, j)\\ &= \zeta^\top(i, j)\left(e_1^\top Q e_1 - e_2^\top Q e_2\right)\zeta(i, j). \qquad (8.25)\end{aligned}$$

Similarly, we also have

$$\Delta\mathscr{V}_2 = \zeta^\top(i, j)\left(e_2^\top R e_2 - e_4^\top R e_4\right)\zeta(i, j).$$

Therefore,

$$\Delta(\mathscr{V}_1 + \mathscr{V}_2) = \zeta^\top(i, j)\varXi_2\zeta(i, j). \qquad (8.26)$$

The difference of $\mathscr{V}_3^h$ is given by

$$
\begin{aligned}
\Delta\mathscr{V}_3^h &= \mathbb{E}\left[\mathscr{V}_3^h(i+1,j)\middle|\mathscr{F}_{ij}^x\right] - \mathscr{V}_3^h(i,j) \\
&= d_1^h\,\mathbb{E}\Big[\sum_{k=-d_1^h}^{-1}\sum_{l=i+k+1}^{i}\eta_h^\top(l,j)W^h\eta_h(l,j)\Big|\mathscr{F}_{ij}^x\Big] \\
&\quad - d_1^h\sum_{k=-d_1^h}^{-1}\sum_{l=i+k}^{i-1}\eta_h^\top(l,j)W^h\eta_h(l,j) \\
&= (d_1^h)^2\,\mathbb{E}\left[\eta_h^\top(i,j)W^h\eta_h(i,j)\middle|\mathscr{F}_{ij}^x\right] - d_1^h\sum_{k=i-d_1^h}^{i-1}\eta_h^\top(k,j)W^h\eta_h(k,j).
\end{aligned}
\tag{8.27}
$$

Similar to (8.27), we have

$$
\Delta\mathscr{V}_3^v = (d_1^v)^2\,\mathbb{E}\left[\eta_v^\top(i,j)W^v\eta_v(i,j)\middle|\mathscr{F}_{ij}^x\right] - d_1^v\sum_{l=j-d_1^v}^{j-1}\eta_v^\top(i,l)W^v\eta_v(i,l). \tag{8.28}
$$

A combination of (8.27) and (8.28) gives

$$
\begin{aligned}
\Delta\mathscr{V}_3 = {}& \mathbb{E}\left[\eta^\top(i,j)(\mathbb{I}_1W)\eta(i,j)\middle|\mathscr{F}_{ij}^x\right] - d_1^h\sum_{k=i-d_1^h}^{i-1}\eta_h^\top(k,j)W^h\eta_h(k,j) \\
& - d_1^v\sum_{l=j-d_1^v}^{j-1}\eta_v^\top(i,l)W^v\eta_v(i,l),
\end{aligned}
\tag{8.29}
$$

where $\eta(i,j) = [\eta_h^\top(i,j)\ \ \eta_v^\top(i,j)]^\top$. Since $\eta(i,j) = \mathscr{W}(i,j) - x(i,j)$, similar to (8.24), we have

$$
\begin{aligned}
\mathbb{E}\left[\eta^\top(i,j)(\mathbb{I}_1W)\eta(i,j)\middle|\mathscr{F}_{ij}^x\right] = {}& \zeta^\top(i,j)\Big[\big(\mathscr{A}-e_1\big)^\top(\mathbb{I}_1W)\big(\mathscr{A}-e_1\big) \\
& + \sum_{p=1}^{n}\sigma_p^2\hat{\mathscr{A}}^\top E_n^{(p)}(\mathbb{I}_1W)E_n^{(p)}\hat{\mathscr{A}}\Big]\zeta(i,j).
\end{aligned}
\tag{8.30}
$$

By Lemma 8.1, the summation terms in (8.29) will be manipulated as

$$
\begin{aligned}
& d_1^h\sum_{k=i-d_1^h}^{i-1}\eta_h^\top(k,j)W^h\eta_h(k,j) + d_1^v\sum_{l=j-d_1^v}^{j-1}\eta_v^\top(i,l)W^v\eta_v(i,l) \\
& \geq \begin{bmatrix} x(i,j)-x_l(i,j) \\ x(i,j)+x_l(i,j)-2\bar{x}_l(i,j)\end{bmatrix}^\top \widehat{W}\begin{bmatrix} x(i,j)-x_l(i,j) \\ x(i,j)+x_l(i,j)-2\bar{x}_l(i,j)\end{bmatrix},
\end{aligned}
\tag{8.31}
$$

where $\widehat{W} = \mathrm{diag}(W, 3W)$. Thus, from the estimations in (8.29)–(8.31), we obtain

$$\Delta \mathscr{V}_3 \leq \zeta^\top(i, j)\,(\Xi_3 - \Xi_4)\,\zeta(i, j). \tag{8.32}$$

Finally, for the functional $\mathscr{V}_4 = \mathscr{V}_4^h + \mathscr{V}_4^v$, similar to (8.29), we have

$$\begin{aligned}\Delta \mathscr{V}_4 = \zeta^\top(i, j)\Xi_5\zeta(i, j) - d_h^r \sum_{k=i-d_2^h}^{i-d_1^h-1} \eta_h^\top(k, j)Z^h\eta_h(k, j)\\ - d_v^r \sum_{l=j-d_2^v}^{j-d_1^v-1} \eta_v^\top(i, l)Z^v\eta_v(i, l).\end{aligned} \tag{8.33}$$

By utilizing Lemmas 8.1 and 8.2, if $d_1^h < d_h(i) < d_2^h$, then the first summation term in (8.33) will be manipulated as

$$\begin{aligned}d_h^r \sum_{k=i-d_2^h}^{i-d_1^h-1} \eta_h^\top(k, j)Z^h\eta_h(k, j) &\geq \frac{d_h^r}{h_l(i)}\chi_1^\top(\mathbb{J}_h^{2\top}\widetilde{Z}^h\mathbb{J}_h^2)\chi_1 + \frac{d_h^r}{h_r(i)}\chi_2^\top(\mathbb{J}_h^{2\top}\widetilde{Z}^h\mathbb{J}_h^2)\chi_2\\ &\geq \begin{bmatrix}\chi_1\\ \chi_2\end{bmatrix}^\top \mathbb{J}_h^{4\top}\Psi_X^h\mathbb{J}_h^4\begin{bmatrix}\chi_1\\ \chi_2\end{bmatrix},\end{aligned} \tag{8.34}$$

where

$$\begin{aligned}\chi_1 &= \begin{bmatrix}x_l(i, j) - x_d(i, j)\\ x_l(i, j) + x_d(i, j) - 2\bar{x}_d(i, j)\end{bmatrix} = \Gamma_1\zeta(i, j),\\ \chi_2 &= \begin{bmatrix}x_d(i, j) - x_u(i, j)\\ x_d(i, j) + x_u(i, j) - 2\bar{x}_u(i, j)\end{bmatrix} = \Gamma_2\zeta(i, j).\end{aligned}$$

It is clear that $\mathbb{J}_h^2\chi_1 = 0$ if $d_h(i) = d_1^h$ and $\mathbb{J}_h^2\chi_2 = 0$ if $d_h(i) = d_2^h$. Thus, the last inequality (8.34) is still valid when $d_h(i) = d_1^h$ and $d_h(i) = d_2^h$. Similar to (8.34), we also have

$$d_v^r \sum_{l=j-d_2^v}^{j-d_1^v-1} \eta_v^\top(i, l)Z^v\eta_v(i, l) \geq \begin{bmatrix}\chi_1\\ \chi_2\end{bmatrix}^\top \mathbb{J}_v^{4\top}\Psi_Y^v\mathbb{J}_v^4\begin{bmatrix}\chi_1\\ \chi_2\end{bmatrix}. \tag{8.35}$$

From (8.33) to (8.35), we readily obtain

$$\Delta \mathscr{V}_4 \leq \zeta^\top(i, j)\,(\Xi_5 - \Xi_6)\,\zeta(i, j). \tag{8.36}$$

It follows from (8.24), (8.26), (8.32), and (8.36) that

$$\Delta \mathscr{V} \leq w^\top(i, j)w(i, j) + \zeta^\top(i, j)\Xi_0\zeta(i, j). \tag{8.37}$$

Thus, by (8.19), there exists a scalar $c_0 > 0$ such that

$$\zeta^\top(i,j)\Xi_0\zeta(i,j) \le -c_0\|\zeta(i,j)\|^2$$

and therefore, from (8.37), we have

$$\Delta\mathscr{V} \le w^\top(i,j)w(i,j) - c_0\|x(i,j)\|^2.$$

The last inequality shows that condition (ii) of Theorem 8.1 is fulfilled.

As final step in the proof, we now verify condition (iii) of Theorem 8.1. By using the augmented vector $\hat{\zeta}(i,j) = \left[x^\top(i,j)\ x_d^\top(i,j)\ w^\top(i,j)\right]^\top$, the output $z(i,j)$ can be represented as

$$z(i,j) = \left(\mathscr{H} + \sum_{q=1}^{n_z}\theta_{ij}^{(q)}E_{n_z}^{(q)}\hat{\mathscr{H}}\right)\hat{\zeta}(i,j).$$

In addition, since $\sum_{q=1}^{n_z}\hat{\sigma}_q^2E_{n_z}^{(q)} = D_{\hat{\sigma}}^2$, we have

$$\mathbb{E}\left[z^\top(i,j)z(i,j)|\mathscr{F}_{ij}^x\right] = \hat{\zeta}^\top(i,j)\left(\mathscr{H}^\top\mathscr{H} + \hat{\mathscr{H}}^\top D_{\hat{\sigma}}^2\hat{\mathscr{H}}\right)\hat{\zeta}(i,j). \tag{8.38}$$

By incorporating the expression of functional $\mathscr{V}(i,j)$ in (8.21), from (8.38), we have

$$\begin{aligned}\mathbb{E}&\left[\frac{2}{\gamma^2}z^\top(i,j)z(i,j) - \mathscr{V}(i,j)|\mathscr{F}_{ij}^x\right] - w^\top(i,j)w(i,j)\\ &\le \hat{\zeta}^\top(i,j)\Big(\frac{2}{\gamma^2}\big[\mathscr{H}^\top\mathscr{H} + \hat{\mathscr{H}}^\top D_{\hat{\sigma}}^2\hat{\mathscr{H}}\big] - \Phi\Big)\hat{\zeta}(i,j).\end{aligned} \tag{8.39}$$

By the Schur complement (see, Lemma 8.4), condition (8.20) holds if and only if

$$-\Phi + \frac{2}{\gamma^2}\begin{bmatrix}\mathscr{H}\\ D_{\hat{\sigma}}\hat{\mathscr{H}}\end{bmatrix}^\top\begin{bmatrix}\mathscr{H}\\ D_{\hat{\sigma}}\hat{\mathscr{H}}\end{bmatrix} < 0. \tag{8.40}$$

It follows from (8.39) and (8.40) that

$$\mathbb{E}\left[\frac{2}{\gamma^2}z^\top(i,j)z(i,j) - \mathscr{V}(i,j)|\mathscr{F}_{ij}^x\right] - w^\top(i,j)w(i,j) \le 0,$$

which validates condition (iii) of Theorem 8.1. Thus, by Theorem 8.1, system (8.1) is EPSS. The proof is completed. □

Remark 8.5 The result of Theorem 8.2 can be extended to 2-D MNSs with delays where the system matrices are subject to different noise processes. In model (8.1), for the simplicity of notations in the presentation, we only develop the results for two different stochastic noises in the state and output as presented.

Remark 8.6 For 2-D MNSs with constant delays d_h and d_v, the functionals $\mathscr{V}_2(i, j)$ and $\mathscr{V}_4(i, j)$ are omitted and the LKF candidate (8.21) is reduced to the following simpler form

$$\begin{aligned}\widetilde{\mathscr{V}}(i, j) &= x^\top(i, j)Px(i, j) + \sum_{k=i-d_h}^{i-1} x^{h\top}(k, j)Q^h x^h(k, j)\\ &\quad + \sum_{l=j-d_v}^{j-1} x^{v\top}(i, l)Q^v x^v(i, l) + \sum_{k=-d_h}^{-1}\sum_{l=i+k}^{i-1} \eta_h^\top(l, j)W^h\eta_h(l, j)\\ &\quad + \sum_{k=-d_v}^{-1}\sum_{l=j+k}^{j-1} \eta_v^\top(i, l)W^v\eta_v(i, l).\end{aligned}$$

By similar lines used the proof of Theorem 8.2, EPSS conditions for system (8.1) with constant delays can be derived as the following corollary.

Corollary 8.1 *Given a scalar $\gamma > 0$ and assume that there exist matrices P, Q, $W \in \mathbb{D}_n^+(n_h, n_v)$ such that the following LMI conditions are satisfied*

$$\begin{aligned}&\mathscr{M} + \sum_{p=1}^{n} \sigma_p^2 \widehat{\Omega}^\top E_n^{(p)} (P + \mathbb{I}_d W)\, E_n^{(p)} \widehat{\Omega}\\ &\quad + \operatorname{diag}(-P + Q, -Q, 0, -I_{n_o}) - \digamma^\top \widetilde{W} \digamma < 0,\end{aligned} \tag{8.41}$$

$$\begin{bmatrix} -\widehat{\Phi} & \begin{bmatrix} \mathscr{H}^\top & \hat{\mathscr{H}}^\top D_{\hat{\sigma}} \end{bmatrix} \\ \begin{bmatrix} \mathscr{H} \\ D_{\hat{\sigma}}\hat{\mathscr{H}} \end{bmatrix} & -\dfrac{\gamma^2}{2} I_{2n_z} \end{bmatrix} < 0, \tag{8.42}$$

where $\mathscr{M} = \Omega_1^\top P \Omega_1 + \Omega_2^\top (\mathbb{I}_d W)\, \Omega_2$, $\widetilde{W} = \operatorname{diag}(W, 3W)$,

$$\begin{aligned}\Omega_1 &= \begin{bmatrix} A & A_d & 0_n & B \end{bmatrix}, \quad \Omega_2 = \begin{bmatrix} A - I_n & A_d & 0_n & B \end{bmatrix},\\ \widehat{\Omega} &= \begin{bmatrix} \hat{A} & \hat{A}_d & 0_n & \hat{B} \end{bmatrix}, \quad \mathbb{I}_d = \operatorname{diag}\left(d_h^2 I_{n_h}, d_v^2 I_{n_v}\right),\\ \digamma &= \begin{bmatrix} I_n & -I_n & 0_n & 0_{n\times n_o} \\ I_n & I_n & -2I_n & 0_{n\times n_o} \end{bmatrix}, \quad \widehat{\Phi} = \operatorname{diag}(P, Q, I_{n_o}).\end{aligned}$$

Then, the 2-D MNSs in the form of (8.1) *with constant delays d_h, d_v is EPSS.*

Remark 8.7 For unknown constant delays d_h and d_v satisfying $d_h \le d_h^*$ and $d_v \le d_v^*$, where d_h^* and d_v^* are known positive integers, the matrix $\mathbb{I}_d$ in (8.42) will be replaced by $\mathbb{I}_d^* = \operatorname{diag}\left((d_h^*)^2 I_{n_h}, (d_v^*)^2 I_{n_v}\right)$ and the conclusion of Corollary 8.1 still remains.

Remark 8.8 It is worth noting that the analysis scheme presented in Theorem 8.1 is utilizable for both 2-D MNSs with and without delays. More specifically, let us consider the following delay-free 2-D MNSs

$$\begin{aligned}\begin{bmatrix}x^h(i+1,j)\\x^v(i,j+1)\end{bmatrix} &= \left(A+\Lambda(i,j)\hat{A}\right)\begin{bmatrix}x^h(i,j)\\x^v(i,j)\end{bmatrix}+\left(B+\Lambda(i,j)\hat{B}\right)w(i,j),\\ z(i,j) &= \left(C+\Theta(i,j)\hat{C}\right)\begin{bmatrix}x^h(i,j)\\x^v(i,j)\end{bmatrix}+\left(F+\Theta(i,j)\hat{F}\right)w(i,j),\end{aligned} \tag{8.43}$$

where the stochastic noises $\Lambda(i, j)$ and $\Theta(i, j)$ are described as in (8.2). By following similar lines in the proof of Theorem 8.2, we have the following result.

Corollary 8.2 *Given a scalar $\gamma > 0$. The 2-D MNSs* (8.43) *is EPSS if there exists a matrix $P \in \mathbb{D}_n^+(n_h, n_v)$ such that*

$$\begin{bmatrix}\Pi_{11} & \Pi_{12}\\ \Pi_{12}^\top & \Pi_{22}\end{bmatrix}<0,\quad \begin{bmatrix}-P & * & * & *\\ 0 & -I_{n_o} & * & *\\ C & F & -\frac{\gamma^2}{2}I_{n_z} & *\\ D_{\hat{\sigma}}\hat{C} & D_{\hat{\sigma}}\hat{F} & 0 & -\frac{\gamma^2}{2}I_{n_z}\end{bmatrix}<0, \tag{8.44}$$

where

$$\Pi_{11} = -P + A^\top PA + \sum_{p=1}^{n}\sigma_p^2\hat{A}^\top E_n^{(p)}PE_n^{(p)}\hat{A},$$

$$\Pi_{12} = A^\top PB + \sum_{p=1}^{n}\sigma_p^2\hat{A}^\top E_n^{(p)}PE_n^{(p)}\hat{B},$$

$$\Pi_{22} = B^\top PB + \sum_{p=1}^{n}\sigma_p^2\hat{B}^\top E_n^{(p)}PE_n^{(p)}\hat{B} - I_{n_o},$$

and other notations are defined as in Theorem 8.2.

Proof The proof of Corollary 8.2 can be proceed similarly to Theorem 8.2 with the functional $V(x) = x^{h\top}P^hx^h + x^{v\top}P^vx^v$ and $P = \mathrm{diag}(P^h, P^v)$. We omit it here. □

8.3.5 *Observer-Based Control of 2-D Systems with Random Packet Dropouts*

Consider the following 2-D Roesser systems

$$\begin{aligned}\begin{bmatrix}x^h(i+1,j)\\x^v(i,j+1)\end{bmatrix} &= Ax(i,j)+Bu(i,j)+B_ww(i,j),\\ y(i,j) &= Cx(i,j)+Fw(i,j),\end{aligned} \tag{8.45}$$

where $u(i, j) \in \mathbb{R}^{n_u}$ is the control input and $y(i, j)$ is the measurement output vector. Other notations are described as in system (8.1).

In practice, full-state vector $x(i, j)$ is not always available to be accessed due to various technical reasons. Thus, conventional state-feedback controllers of the form $u(i, j) = Kx(i, j)$ are not always implementable. Alternatively, an observer-based controller $u(i, j) = K\hat{x}(i, j)$ can be used to stabilize system (8.45), where $\hat{x}(i, j)$ is observer-state vector. For illustrative representation, we consider the following Luenberger-type observer

$$\begin{aligned} \begin{bmatrix} \hat{x}^h(i+1, j) \\ \hat{x}^v(i, j+1) \end{bmatrix} &= A\hat{x}(i, j) + L\left(y(i, j) - \hat{y}(i, j)\right), \\ \hat{y}(i, j) &= C\hat{x}(i, j), \end{aligned} \tag{8.46}$$

where $L \in \mathbb{R}^{n \times n_o}$ is an observer gain being determined.

The data packet dropouts phenomena are frequently encountered in control engineering, which can be modeled using stochastic processes. In the schematic model presented in Fig. 8.1, where $x_{1,1}$ stands for the vector $\left[x^{h\top}(i+1, j)\ x^{v\top}(i, j+1)\right]^\top$, due to random packet dropouts, the actual control signal can be modeled as

$$u(i, j) = \gamma_{ij} K\hat{x}(i, j), \tag{8.47}$$

where γ_{ij} is a sequence of random variables with Bernoulli distribution which take values in $\{0, 1\}$ and have the following statistical probabilities

$$\mathbb{P}\{\gamma_{ij} = 1\} = \mathbb{E}[\gamma_{ij}] = \rho, \quad \mathbb{P}\{\gamma_{ij} = 0\} = 1 - \mathbb{E}[\gamma_{ij}] = 1 - \rho,$$

where ρ is a positive constant.

Let $\tilde{x}^h = \begin{bmatrix} x^h \\ \hat{x}^h \end{bmatrix}$, $\tilde{x}^v = \begin{bmatrix} x^v \\ \hat{x}^v \end{bmatrix}$, $\tilde{x} = \begin{bmatrix} \tilde{x}^h \\ \tilde{x}^v \end{bmatrix}$ and $\lambda_{ij} = \gamma_{ij} - \rho$. It can be verified that

$$\mathbb{E}[\lambda_{ij}] = 0 \text{ and } \mathbb{E}[\lambda_{ij}\lambda_{kl}] = \sigma^2\delta_{ik}\delta_{jl},$$

where $\sigma = \sqrt{\rho(1-\rho)}$. The resulting closed-loop system of (8.45) subject to observer-based controller defined by (8.46) and (8.47) is represented as

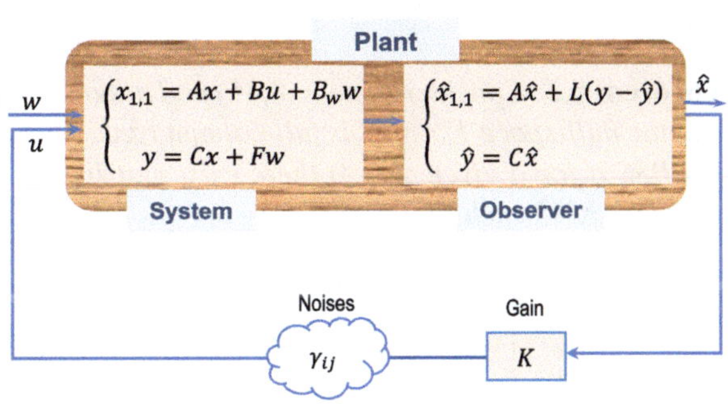

Fig. 8.1 A schematic random package dropouts model

$$\Pi \begin{bmatrix} \tilde{x}^h(i+1,j) \\ \tilde{x}^v(i,j+1) \end{bmatrix} = (\mathscr{A}_c + \lambda_{ij}\hat{\mathscr{A}_c})\Pi\tilde{x}(i,j) + \mathscr{B}w(i,j),$$
$$x(i,j) = \begin{bmatrix} J \; 0_{n\times n_v} \end{bmatrix}\tilde{x}(i,j), \tag{8.48}$$

where

$$\mathscr{A}_c = \begin{bmatrix} A & \rho BK \\ LC & A-LC \end{bmatrix}, \quad \hat{\mathscr{A}_c} = \begin{bmatrix} 0 & BK \\ 0 & 0 \end{bmatrix}, \quad \mathscr{B} = \begin{bmatrix} B_w \\ LF \end{bmatrix}$$
$$J = \begin{bmatrix} I_{n_h} & 0 & 0 \\ 0 & 0 & I_{n_v} \end{bmatrix}, \quad \Pi = \begin{bmatrix} J & 0_{n\times n_v} \\ 0_{n\times n_h} & J \end{bmatrix}.$$

The control objective is to design controller gain matrices K, L such that the closed-loop system (8.48) EPSS. By Corollary 8.2, the design parameters K, L are obtained by solving the following LMIs for $P \in \mathbb{D}^+_{2n}(2n_h, 2n_v)$

$$\begin{bmatrix} \mathscr{N} - \widetilde{P} & \mathscr{A}_c^\top \widetilde{P}\mathscr{B} \\ * & \mathscr{B}^\top \widetilde{P}\mathscr{B} - I_{n_o} \end{bmatrix} < 0, \quad \begin{bmatrix} -\widetilde{P} & [I_n \; 0]^\top \\ [I_n \; 0] & -\frac{\gamma^2}{2} I_n \end{bmatrix} < 0, \tag{8.49}$$

where $\mathscr{N} = \mathscr{A}_c^\top \widetilde{P}\mathscr{A}_c + \sigma^2 \hat{\mathscr{A}_c}^\top \widetilde{P}\hat{\mathscr{A}_c}$ and $\widetilde{P} = \Pi P \Pi^\top$. By Schur complement (Lemma 8.4), the first condition in (8.49) holds if and only if

$$\begin{bmatrix} -\operatorname{diag}(\widetilde{P}, I_{n_o}) & * & * \\ \widetilde{P}[\mathscr{A}_c \;\; \mathscr{B}] & -\widetilde{P} & 0 \\ \widetilde{P}[\sigma\hat{\mathscr{A}_c} \;\; 0] & 0 & -\widetilde{P} \end{bmatrix} < 0, \tag{8.50}$$

where the symbol $*$ indicates the symmetry terms. In addition, $P = \operatorname{diag}(P_h, P_v)$ with $P_h \in \mathbb{S}^+_{2n_h}$ and $P_v \in \mathbb{S}^+_{2n_v}$ if and only if

$$\widetilde{P} = \begin{bmatrix} P_1 & P_3 \\ * & P_2 \end{bmatrix},$$

where $P_1, P_2 \in \mathbb{D}^+_n(n_h, n_v)$ and P_3 is a diagonal block matrix. To simplify design conditions, we assign $P_3 = 0$. Based on (8.49) and (8.50), we have the following result.

Theorem 8.3 *Consider system* (8.45) *and assume that the matrix C is full-row rank and the null-space $F^\perp$ of F is full-column rank. Then, the closed-loop system* (8.48) *is EPSS if for a given $\gamma > 0$ there exist matrices $X \in \mathbb{D}^+_n(n_h, n_v)$ and $Y \in \mathbb{R}^{n_u\times n}$ and $Z \in \mathbb{R}^{n\times n}$ such that*

$$\begin{bmatrix} -\mathscr{X} & * & * & * \\ 0 & -(F^\perp)^\top F^\perp & * & * \\ \mathscr{W}_{13} & \mathscr{W}_{23} & -\mathscr{X} & * \\ \mathscr{W}_{14} & 0 & 0 & -X \end{bmatrix} < 0, \quad X < \gamma^2/2 I_n, \tag{8.51}$$

where $\mathscr{X} = \mathrm{diag}(X, X)$ *and*

$$\mathscr{W}_{13} = \begin{bmatrix} AX & \rho BY \\ Z & AX - Z \end{bmatrix}, \quad \mathscr{W}_{23} = \begin{bmatrix} B_w F^{\perp} \\ 0 \end{bmatrix}, \quad \mathscr{W}_{14} = \begin{bmatrix} 0 & \sigma BY \end{bmatrix}.$$

The controller gain and observer gain matrices are given by

$$K = YX^{-1}, \quad L = ZX^{-1}C^{\dagger}, \tag{8.52}$$

where $C^{\dagger} = C^{\top}(CC^{\top})^{-1}$ *is the pseudoinverse of* C.

Proof We define the matrices $P_1 = X^{-1}$ and $P_2 = X^{-1}$. By pre- and post-multiplying with $\mathscr{J}^{\top}$ and $\mathscr{J} = \mathrm{diag}(X, X, F^{\perp}, X, X, X, X)$, condition (8.50) holds if and only if

$$\begin{bmatrix} -\mathscr{X} & * & * & * \\ 0 & -(F^{\perp})^{\top} F^{\perp} & * & * \\ \widetilde{\mathscr{W}}_{13} & \mathscr{W}_{23} & -\mathscr{X} & * \\ \widetilde{\mathscr{W}}_{14} & 0 & 0 & -X \end{bmatrix} < 0, \tag{8.53}$$

where

$$\widetilde{\mathscr{W}}_{13} = \begin{bmatrix} AX & \rho BKX \\ LCX & AX - LCX \end{bmatrix}, \quad \widetilde{\mathscr{W}}_{14} = \begin{bmatrix} 0 & \sigma BKX \end{bmatrix}.$$

By the changing the variables

$$KX = Y \text{ and } LCX = Z,$$

condition (8.53) leads to the first condition in (8.51). On the other hand, by the Schur complement (Lemma 8.4), the second condition in (8.49) holds if and only if $-P_1 + (2/\gamma^2)I_n < 0$, which is equivalent to the second condition in (8.51). Therefore, by condition (8.51), the closed-loop system (8.48) is EPSS. A controller gain K and an observer gain L can be obtained as given in (8.52). The proof is completed. □

8.3.6 Numerical Examples

In this section, we give some numerical examples to illustrate the efficacy of the obtained stability conditions.

Example 8.1 Consider a 2-D MNS in the form of (8.1) with the matrices

$$A = \begin{bmatrix} 0.3 & -0.25 & 0.12 \\ -0.15 & 0.1 & 0 \\ 0.25 & 0.1 & -0.15 \end{bmatrix}, \quad A_d = \begin{bmatrix} 0.15 & -0.1 & 0 \\ 0.12 & -0.2 & -0.1 \\ 0.05 & -0.06 & 0.11 \end{bmatrix},$$

$$\hat{A} = \begin{bmatrix} 0.25 & 0.18 & 0.11 \\ 0 & 0.2 & 0.1 \\ 0.15 & 0.1 & 0.15 \end{bmatrix}, \quad \hat{A}_d = \begin{bmatrix} 0.14 & -0.1 & 0 \\ 0.12 & -0.2 & -0.1 \\ 0.05 & -0.12 & 0.1 \end{bmatrix},$$
$$B = \begin{bmatrix} 0.1 & 0.2 & 0.1 \end{bmatrix}^\top, \quad \hat{B} = \begin{bmatrix} 0.1 & 0.1 & 0 \end{bmatrix}^\top,$$
$$C = \begin{bmatrix} 1 & 1 & 0 \\ 0 & 1 & 1 \end{bmatrix}, \quad D = \begin{bmatrix} 0 & 0 & 1 \\ 1 & 0 & 0 \end{bmatrix}, \quad F = \begin{bmatrix} 0.15 \\ 0.2 \end{bmatrix},$$
$$\hat{C} = \begin{bmatrix} 0.2 & 0.1 & 0 \\ 0 & 0.1 & 0.1 \end{bmatrix}, \quad \hat{D} = \begin{bmatrix} 0 & 0 & 0.1 \\ 0.2 & 0 & 0 \end{bmatrix}, \quad \hat{F} = \begin{bmatrix} 0.1 \\ 0.1 \end{bmatrix}.$$

The white noises $\Lambda(i, j)$ and $\Theta(i, j)$ are defined by (8.2) with

$$D_\sigma = \text{diag}(0.4, 0.5, 1.2), \quad D_{\hat{\sigma}} = \text{diag}(1.1, 1.5),$$

and the time delays satisfy $2 \leq d_h(i) \leq 5$, $0 \leq d_v(j) \leq 4$.

By using Matlab LMI toolbox, the derived conditions (8.18)–(8.20) are feasible with minimal EPSS attenuation level $\gamma_* = 1.6202$. For example, with $\gamma = 1.8$, feasible solutions of (8.18)–(8.20) are given by

$$P = \text{diag}\left(\begin{bmatrix} 10.3907 & -3.4256 \\ -3.4256 & 5.5534 \end{bmatrix}, 4.2682\right),$$
$$Q = \text{diag}\left(\begin{bmatrix} 4.0245 & -2.1827 \\ -2.1827 & 2.5830 \end{bmatrix}, 4.0689\right),$$
$$R = \text{diag}\left(\begin{bmatrix} 3.1099 & -1.3183 \\ -1.3183 & 1.3591 \end{bmatrix}, 1.2650\right),$$
$$W = \text{diag}\left(\begin{bmatrix} 0.1902 & -0.0542 \\ -0.0542 & 0.0238 \end{bmatrix}, 416.4330\right),$$
$$Z = \text{diag}\left(\begin{bmatrix} 0.3997 & -0.2333 \\ -0.2333 & 0.2705 \end{bmatrix}, 0.0995\right),$$
$$X = \begin{bmatrix} -0.2475 & 0.1900 & 0.0003 & -0.0001 \\ 0.1855 & -0.2522 & -0.0001 & 0.0003 \\ -0.0004 & 0.0002 & -0.0004 & 0.0003 \\ 0.0001 & -0.0003 & 0.0003 & -0.0002 \end{bmatrix},$$
$$Y = \begin{bmatrix} -0.0954 & 0.0002 \\ -0.0001 & -0.0001 \end{bmatrix}.$$

By Theorem 8.2, the system is EPSS with level $\gamma = 1.8$.

In simulation results, the directional delays $d_h(i) \in [2, 5]$ and $d_v(j) \in [0, 4]$ are taken randomly. The white noise processes $\Lambda(i, j)$ and $\Theta(i, j)$ with variance matrices given by D_σ^2 and $D_{\hat{\sigma}}^2$ are presented in Figs. 8.2, 8.3, 8.4, 8.5 and 8.6. When the external disturbance is ignored (i.e. $w = 0$), Figs. 8.7, 8.8 and 8.9 show that the system with multiplicative noises specified above is stochastically stable. To further illustrate the

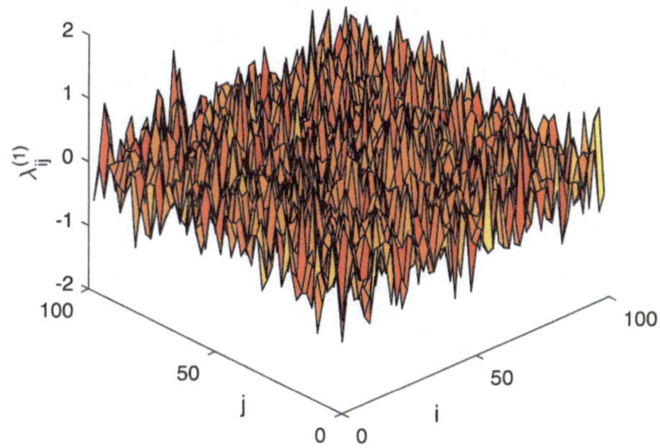

Fig. 8.2 Stochastic noise $\lambda_{ij}^{(1)}$

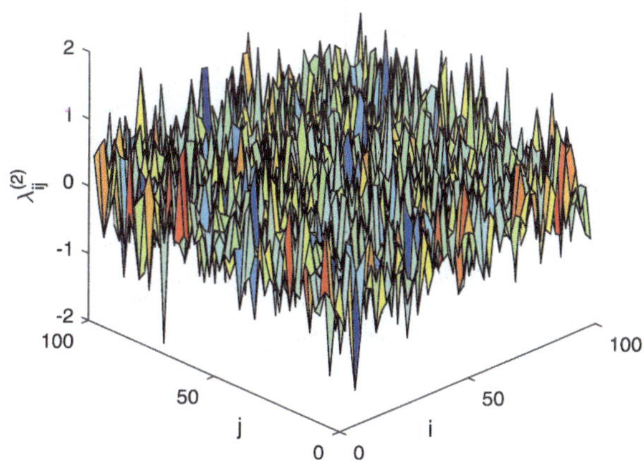

Fig. 8.3 Stochastic noise $\lambda_{ij}^{(2)}$

result, we set an l_2 input sequence as $w(i, j) = \frac{1}{j}\exp(-0.1i)$. The corresponding output trajectories $z_1(i, j)$ and $z_2(i, j)$ are presented in Figs. 8.10 and 8.11. It is found that

$$\sup_{i,j\geq 0} \mathbb{E}\left[z^\top(i, j)z(i, j)\right] \leq 6.0969,$$

$$\|w\|_{l_2} = \frac{\pi}{\sqrt{6(1-\exp(-0.2))}} \simeq 3.0124.$$

Clearly, $\|z\|_{l_\infty} < \gamma_* \|w\|_{l_2}$, where $\gamma_* = 1.6201$. The obtained simulation results demonstrate the effectiveness of our method.

Fig. 8.4 Stochastic noise $\lambda_{ij}^{(3)}$

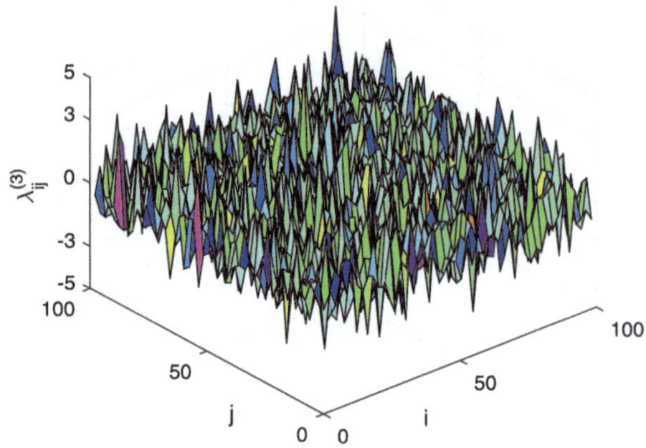

Fig. 8.5 Stochastic noise $\theta_{ij}^{(1)}$

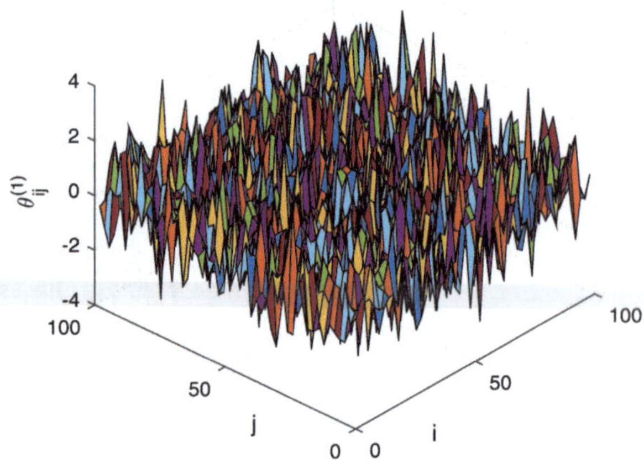

Example 8.2 Consider a control system described by the following Darboux-type equation [29]

$$\begin{aligned}\frac{\partial^2 T(x,t)}{\partial x \partial t} &= a_0 T(x,t) + a_1 \frac{\partial T(x,t)}{\partial t} + a_2 \frac{\partial T(x,t)}{\partial x} \\ &\quad + a_d T(x, t-\tau(t)) + b_1 u(x,t) + b_0 w(x,t), \\ z(x,t) &= cT(x,t) + b_2 u(x,t) + f w(x,t),\end{aligned} \tag{8.54}$$

where $T(x,t)$ is an unknown function at space $x \in [0, x_f]$ and time $t \in [0, \infty)$, $\tau(t)$ represents a time-varying delay, $u(x,t)$ is the control input, $w(x,t)$ is the disturbance and a_k, b_k, $k = 0, 1, 2$, a_d, c, f are real coefficients. Denote

Fig. 8.6 Stochastic noise $\theta_{ij}^{(2)}$

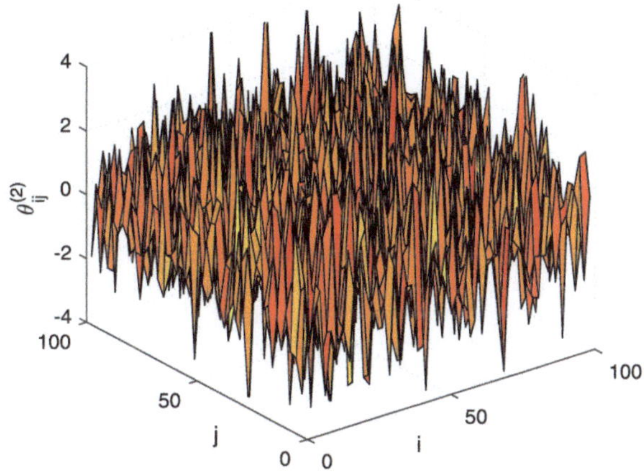

Fig. 8.7 State trajectory $x_1^h(i, j)$ of the system with $w = 0$

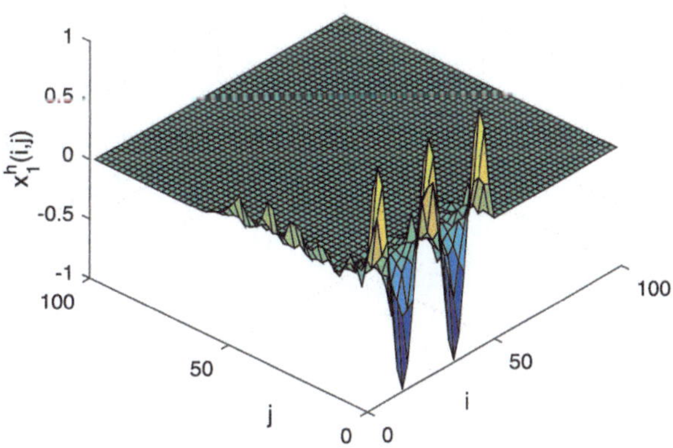

$$x^h(x, t) = \frac{\partial T(x, t)}{\partial t} - a_2 T(x, t), \quad x^v(x, t) = T(x, t),$$

then equation (8.54) can be written as

$$\begin{bmatrix} \frac{\partial x^h}{\partial x} \\ \frac{\partial x^v}{\partial t} \end{bmatrix} = \begin{bmatrix} a_1 & a_0 + a_1 a_2 \\ 1 & a_2 \end{bmatrix} \begin{bmatrix} x^h \\ x^v \end{bmatrix} + \begin{bmatrix} 0 & a_d \\ 0 & 0 \end{bmatrix} \begin{bmatrix} 0 \\ x_d^v \end{bmatrix} + \begin{bmatrix} b_1 \\ 0 \end{bmatrix} u + \begin{bmatrix} b_0 \\ 0 \end{bmatrix} w, \tag{8.55}$$

where $x_d^v = T(x, t - \tau(t))$. For given increments Δx and Δt, we denote $x^h(i, j) = x^h(i\Delta x, j\Delta t)$, $x^v(i, j) = x^v(i\Delta x, j\Delta t)$, $u(i, j) = u(i\Delta x, j\Delta t)$, $w(i, j) = w(i\Delta x, j\Delta t)$ and $d_v(j) = \lfloor \tau(j\Delta t) \rfloor$. By using Euler approximation scheme, from (8.55), we have

Fig. 8.8 State trajectory $x_2^h(i, j)$ of the system with $w = 0$

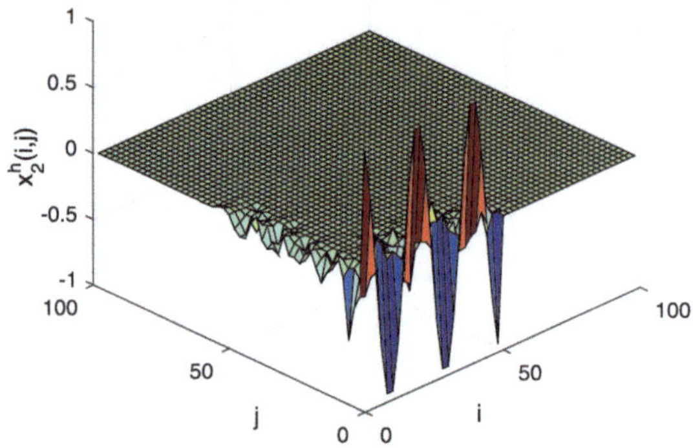

Fig. 8.9 State trajectory $x^v(i, j)$ of the system with $w = 0$

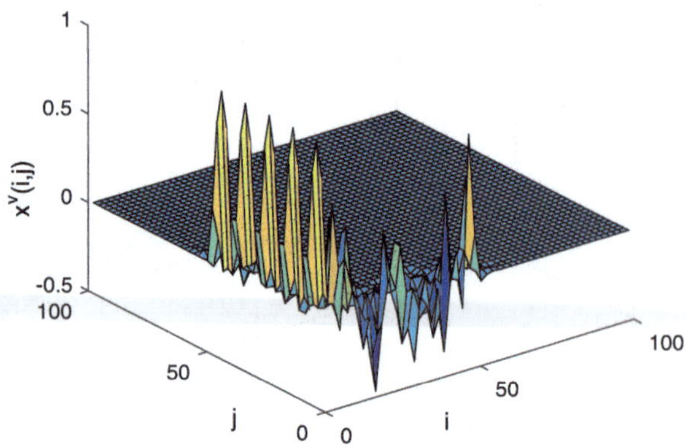

$$\begin{bmatrix} x^h(i+1, j) \\ x^v(i, j+1) \end{bmatrix} = A \begin{bmatrix} x^h(i, j) \\ x^v(i, j) \end{bmatrix} + A_d \begin{bmatrix} x^h(i - d_h(i), j) \\ x^v(i, j - d_v(j)) \end{bmatrix} + B_c u(i, j) + B_w w(i, j),$$
$$z(i, j) = Cx(i, j) + b_2 u(i, j) + f w(i, j), \tag{8.56}$$

with the system matrices

$$A = \begin{bmatrix} 1 + a_1 \Delta x & (a_0 + a_1 a_2)\Delta x \\ \Delta t & 1 + a_2 \Delta t \end{bmatrix}, \quad A_d = \begin{bmatrix} 0 & a_d \Delta x \\ 0 & 0 \end{bmatrix},$$
$$B_c = \begin{bmatrix} b_1 \Delta x \\ 0 \end{bmatrix}, \quad B_w = \begin{bmatrix} b_0 \Delta x \\ 0 \end{bmatrix}, \quad C = \begin{bmatrix} 0 & c \end{bmatrix}.$$

A conventional state-feedback controller for (8.56) is given as $u(i, j) = Kx(i, j)$, $K \in \mathbb{R}^{n_u \times n}$ is the controller gain. By incorporating noises in random packet dropouts, actual control can be modeled as

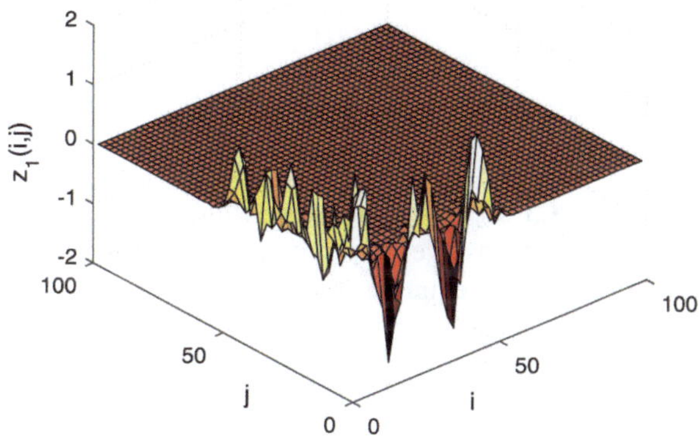

Fig. 8.10 Output trajectory $z_1(i, j)$ with $w(i, j) = \frac{1}{j}\exp(-0.1i)$

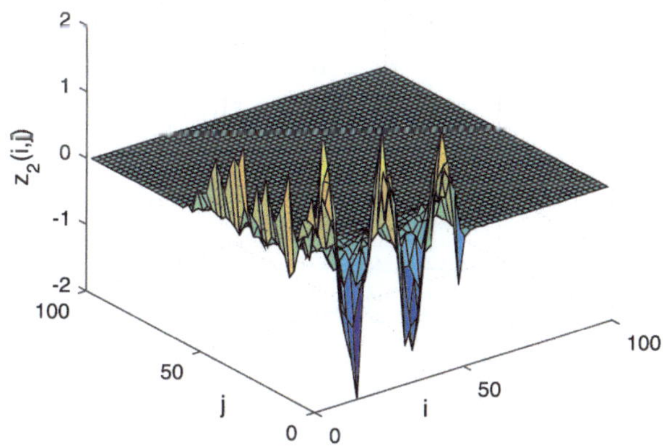

Fig. 8.11 Output trajectory $z_2(i, j)$ with $w(i, j) = \frac{1}{j}\exp(-0.1i)$

$$u(i, j) = \lambda_{ij} K x(i, j). \tag{8.57}$$

The closed-loop system is in the form of (8.1), where $\hat{A} = B_c K$, $\hat{C} = b_2 K$, $F = f$.

For illustrative purpose, the parameters are given as

$$\begin{aligned}
&a_0 = 0.2, \quad a_1 = a_2 = -1, \quad a_d = -0.8, \\
&b_k = 1, \ k = 0, 1, 2, \quad c = 1, \quad f = 0.5, \quad \Delta x = 0.1, \quad \Delta t = 0.05.
\end{aligned}$$

The white noises in (8.2) are defined by λ_{ij} with $\sigma_p = \hat{\sigma}_q = \sigma = 1.5$. The controller gain is fixed, for example, $K = \begin{bmatrix} 0 & 0.2 \end{bmatrix}$ and time delay $0 \leq d_v(j) \leq 10$. By using Matlab LMI toolbox, it is found that the derived conditions (8.18)–(8.20) in Theorem 8.2 are feasible with minimal level $\gamma_* = 0.8165$. By Theorem 8.2, the closed-loop system of (8.56) subject to (8.57) is EPSS with level $\gamma_{\min} = 0.8165$.

With l_2 disturbance sequence $w(i, j) = \frac{1}{j}\exp(-0.1i)$, we have

$$\sup_{i,j\geq 0} \mathbb{E}\left[z^\top(i, j)z(i, j)\right] \leq 2.25,$$

whereas $\|w\|_{l_2} = \frac{\pi}{\sqrt{6(1-\exp(-0.2))}} \simeq 3.0124$. Thus, $\|z\|_{l_\infty} < \gamma_* \|w\|_{l_2}$. This illustrates the effectiveness of the analysis result.

Example 8.3 In this example, we illustrate the effectiveness of the proposed design method of an observer-based controller for the following control model described by Darboux-type equation

$$\frac{\partial^2 T(x,t)}{\partial x \partial t} = a_0 s(x,t) + a_1 \frac{\partial T(x,t)}{\partial t} + a_2 \frac{\partial T(x,t)}{\partial x} + bu(x,t), \tag{8.58}$$

where $u(x, t)$ is the control input and a_0, a_1, a_2, b are real coefficients.

By a similar discretization process for (8.56), we obtain model (8.45), where

$$A = \begin{bmatrix} 1 + a_1 \Delta x & (a_0 + a_1 a_2)\Delta x \\ \Delta t & 1 + a_2 \Delta t \end{bmatrix}, \quad B = \begin{bmatrix} b\Delta x \\ 0 \end{bmatrix},$$
$$C = \begin{bmatrix} 0 & 1 \end{bmatrix}, \quad F = \begin{bmatrix} 0 & 1 \end{bmatrix}.$$

Given parameters

$$a_0 = -1.2, \quad a_1 = a_2 = -1, \quad b = 1, \quad \Delta x = 0.1, \quad \Delta t = 0.05,$$

and $B_w = \begin{bmatrix} 0.1 & 0.1 \\ 0 & 0.2 \end{bmatrix}$. It can be verified that $\text{rank}(C) = 1$, $\text{rank}(F^\perp) = 1$, and thus C is full-row rank and $F^\perp$ is full-column rank.

For $\rho = 0.6$, the proposed LMI conditions in (8.51) are feasible with minimum allowable $\gamma_{\min} = 0.3321$. For example, with $\gamma = 0.5$, a feasible solution of (8.51) is given by

$$X = \text{diag}(0.1074, 0.1057), \quad Y = \begin{bmatrix} -0.1444 & -0.0192 \end{bmatrix},$$
$$Z = \begin{bmatrix} 0.0112 & -0.0022 \\ -0.0004 & 0.0059 \end{bmatrix}.$$

By (8.52), the gain matrices are obtained as

$$K = \begin{bmatrix} -1.3446 & -0.1821 \end{bmatrix}, \quad L^\top = \begin{bmatrix} -0.0204 & 0.0561 \end{bmatrix}.$$

By Theorem 8.3, the closed-loop system (8.48) is EPSS at level $\gamma \geq 0.3321$.

A simulation result for norm state trajectory $\|x(i, j)\|$ of the closed-loop system (8.48) conducted with $w(i, j) = \exp(-0.2(i + j))$ is given in Fig. 8.12. The stochastic noise γ_{ij} is also depicted in Fig. 8.13. It is found that

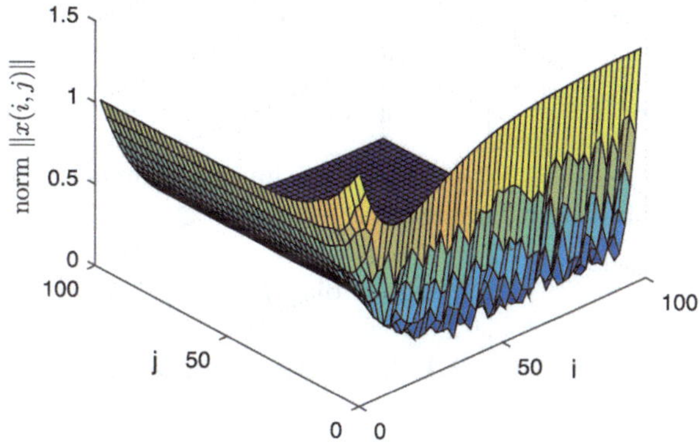

Fig. 8.12 State trajectory $\|x(i, j)\|$

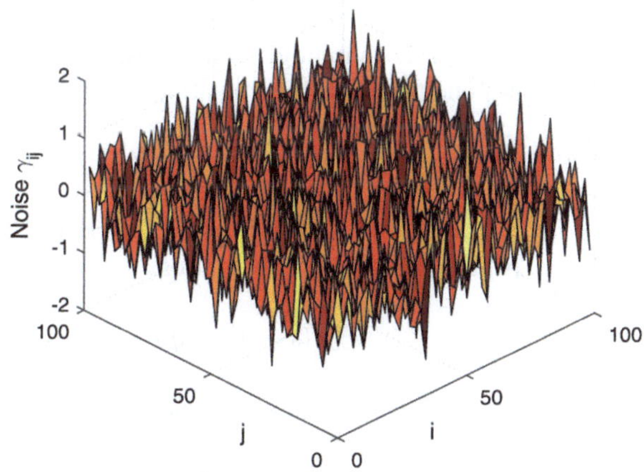

Fig. 8.13 Stochastic noise γ_{ij}

$$\|w\|_{l_2} = 5.5167, \quad \|x\|_{l_\infty} < 1.5.$$

Thus, $\|x\|_{l_\infty} < \gamma_* \|w\|_{l_2}$ as revealed by the obtained theoretical result.

8.4 Stabilization of Inhomogeneous Markov Jump 2-D Systems with Multiplicative Noises

In this section, we present design conditions of a synchronous state-feedback controller (SSFC) for 2-D Roesser systems with multiplicative noises

$$\begin{bmatrix} x^h(i+1,j) \\ x^v(i,j+1) \end{bmatrix} = A_{\theta(i,j)} \begin{bmatrix} x^h(i,j) \\ x^v(i,j) \end{bmatrix} + B_{\theta(i,j)}u(i,j)$$
$$+ \beta(i,j) \left(\hat{A}_{\theta(i,j)} \begin{bmatrix} x^h(i,j) \\ x^v(i,j) \end{bmatrix} + \hat{B}_{\theta(i,j)}u(i,j) \right), \ i,j \in \mathbb{N}_0, \tag{8.59}$$

where $x = \begin{bmatrix} x^h \\ x^v \end{bmatrix} \in \mathbb{R}^d$ is the state vector, $u \in \mathbb{R}^{d_c}$ is the control input, A_θ, $\hat{A}_\theta \in \mathbb{R}^{d \times d}$ and B_θ, $\hat{B}_\theta \in \mathbb{R}^{d \times d_c}$ are given parameter-dependent system matrices. The multiplicative parameter

$$\beta(i,j) = \mathrm{diag}(\beta_1(i,j), \beta_2(i,j), \ldots, \beta_d(i,j)),$$

where $\beta_k(i,j)$ $(k = 1, 2, \ldots, d)$ are stochastic noises defined on a probability space $(\Omega, \mathscr{F}, \mathbb{P})$, which are independent 2-D random variables with zero-mean and satisfy

$$\mathbb{E}\left[\beta_p(i,j)\beta_p(k,l)\right] = \rho_p^2 \delta_{ik}\delta_{jl}, \tag{8.60}$$

where ρ_p, $p = 1, 2, \ldots, d$, are known scalars and δ_{ik} is the Kronecker notation. $\theta(i,j)$ is a 2-D inhomogeneous Markov chain with finite state-space $\mathscr{M} = \{1, 2, \ldots, M\}$ and time-varying transition probability specified as

$$\mathbb{P}[\theta(i+1,j) = n | \theta(i,j) = m] = \mathbb{P}[\theta(i,j+1) = n | \theta(i,j) = m] = \pi_{mn}^{(\sigma)}, \tag{8.61}$$

where $\sigma = \sigma(s)$, $s = i + j + 1$, is a Markov chain with state-space $\mathscr{N} = \{1, 2, \ldots, N\}$ and transition probabilities

$$\mathbb{P}[\sigma(s+1) = q | \sigma(s) = p] = \upsilon_{pq}. \tag{8.62}$$

We denote by $\Upsilon = (\upsilon_{pq})_{N \times N}$ the probability transition matrix of switching signal σ.

Remark 8.9 The parameter σ in (8.61) represents a switching process, which is a piecewise constant function $\sigma : \mathbb{N}_0 \to \mathscr{N}$. As discussed in existing literatures, the 2-D system causality imposes an increment depending only on $i + j$, by which the switching can be assumed to occur only at each sampling point of i or j, that is, $\sigma(i,j) = \sigma(k,l)$ if $i + j = k + l$. We denote $\sigma(i,j) = \sigma(s)$ for $s = i + j + 1$. This assumption is widely adopted in analysis and control of 2-D switched systems.

Assumption 8.4.1 The processes $\theta(i,j)$ and $\beta(i,j)$ are independent and the stochastic variation of $\sigma(s)$ is independent with respect to the σ-algebra generated by $\{\beta(i,j) : i, j \leq s - 1\}$.

We consider the problem of designing an SSFC for system (8.59) in the form

$$u(i,j) = K_{\theta(i,j)}x(i,j) \tag{8.63}$$

to stabilize system (8.59), where $K_{\theta(i,j)}$ is the controller gain, which will be determined. The closed-loop system of (8.59) is expressed as

$$\begin{bmatrix} x^h(k+1,l) \\ x^v(k,l+1) \end{bmatrix} = \left(A^c_{\theta(i,j)} + \beta(i,j)A^c_{\theta(i,j)}\right) x(i,j), \tag{8.64}$$

where $A^c_\theta = A_\theta + B_\theta K_\theta$ and $\hat{A}^c_\theta = \hat{A}_\theta + \hat{B}_\theta K_\theta$.

8.4.1 Stability Analysis

We recall here that [12], system (8.64) is said to be stochastically stable if any solution $x(i,j)$ of (8.64) satisfies

$$\mathbb{E}\left[\sum_{i,j=0}^{\infty} |x(i,j)|^2\right] < \infty.$$

Moreover, system (8.59) is robustly stabilizable via SSFCs with respect to multiplicative stochastic noises $\beta(i,j)$ if there exists an SSFC (8.63) such that the closed-loop system (8.64) is stochastically stable.

Theorem 8.4 *Assume that there exist matrices* $P_n = \mathrm{diag}(P^h_n, P^v_n) \in \mathbb{D}^+_d(d_h, d_v)$, *where* $d = d_h + d_v$ *and* $n \in \mathscr{M}$, *such that the following LMI condition is satisfied*

$$A^{c\top}_m \Phi_{m,p} A^c_m + \sum_{k=1}^{d} \rho_k^2 \hat{A}^{c\top}_m \digamma_k \Phi_{m,p} \digamma_k \hat{A}^c_m - P_m < 0, \quad m \in \mathscr{M},\ p \in \mathscr{N}, \tag{8.65}$$

where $\digamma_k = e_k e_k^\top \in \mathbb{R}^{d\times d}$, e_k *is the kth vector in the canonical basis of* $\mathbb{R}^d$ *and*

$$\Phi_{m,p} = \sum_{q=1}^{N} \upsilon_{pq} \sum_{n=1}^{M} \pi^{(q)}_{mn} P_n.$$

Then, the closed-system (8.64) *is stochastically stable.*

Proof Consider the following parameter-dependent Lyapunov functional

$$V_\theta(x) = x^\top P_\theta x, \tag{8.66}$$

where $P_\theta = P_m$ whenever $\theta(i,j) = m \in \mathscr{M}$. Let $\lambda_1 = \min_{m\in\mathscr{M}} \left\{\lambda_{\min}(P^h_m), \lambda_{\min}(P^v_m)\right\}$ and $\lambda_2 = \max_{m\in\mathscr{M}} \left\{\lambda_{\max}(P^h_m), \lambda_{\max}(P^v_m)\right\}$, then it can be seen that

$$\lambda_1 I_d \le P_m \le \lambda_2 I_d, \quad m \in \mathscr{M},$$

and hence

$$\lambda_1\|x\|^2 \le V_\theta(x) \le \lambda_2\|x\|^2$$

for all $\theta \in \mathscr{M}$ and $x \in \mathbb{R}^d$. We denote $V^h(x^h) = x^{h\top} P_\theta^h x^h$, $V^v(x^v) = x^{v\top} P_\theta^v x^v$ and compute the difference ΔV of the functional $V = V_\theta(x)$ along state trajectory $x(i, j)$. Assume the system is active at mode $\theta(i, j) = m \in \mathscr{M}$ at the point (i, j) and $\sigma(s) = p$ for some $p \in \mathscr{N}$. Note that the difference of $V_\theta(x)$ along state trajectory $x(i, j)$ is given by

$$\begin{aligned}\Delta V = &\underbrace{\mathbb{E}\left[V^h(x^h(i+1, j))\middle|x(i, j), m, p\right] - V^h(x^h(i, j))}_{\Delta V^h} \\ &+ \underbrace{\mathbb{E}\left[V^v(x^v(i, j+1))\middle|x(i, j), m, p\right] - V^v(x^v(i, j))}_{\Delta V^v}.\end{aligned}$$

By the total probability properties, we have

$$\begin{aligned}\Delta V^h = &\sum_{n\in\mathscr{M}, q\in\mathscr{N}} \mathbb{E}\left[x^{h\top}(i+1, j) P_n^h x^h(i+1, j)\right] \\ &\times \mathbb{P}\left[\theta(i+1, j) = n, \sigma(s+1) = q\middle|\theta(i, j) = m, \sigma(s) = p\right] \\ &- x^{h\top}(i, j) P_m^h x^h(i, j).\end{aligned} \tag{8.67}$$

In addition, according to Assumption 8.4.1,

$$\begin{aligned}&\mathbb{P}\left[\theta(i+1, j) = n, \sigma(s+1) = q\middle|\theta(i, j) = m, \sigma(s) = p\right] \\ &= \mathbb{P}\left[\theta(i+1, j) = n\middle|\theta(i, j) = m, \sigma(s+1) = q, \sigma(s) = p\right] \\ &\quad\times \mathbb{P}\left[\sigma(s+1) = q\middle|\theta(i, j) = m, \sigma(s) = p\right] \\ &= \mathbb{P}\left[\theta(i+1, j) = n\middle|\theta(i, j) = m\right]\mathbb{P}\left[\sigma(s+1) = q\middle|\sigma(s) = p\right] \\ &= \upsilon_{pq}\pi_{mn}^{(q)}.\end{aligned}$$

Therefore,

$$\Delta V^h = \mathbb{E}\left[x^{h\top}(i+1, lj)\Phi_{m,p}^h x^h(i+1, j)\right] - x^{h\top}(i, j) P_m^h x^h(i, j), \tag{8.68}$$

where $\Phi_{m,p}^h = \sum_{q=1}^N \upsilon_{pq} \sum_{n=1}^M \pi_{mn}^{(q)} P_n^h$. By similar arguments, we also have

$$\Delta V^v = \mathbb{E}\left[x^{v\top}(i, j+1)\Phi_{m,p}^v x^v(i, j+1)\right] - x^{v\top}(i, j) P_m^v x^v(i, j), \tag{8.69}$$

where $\Phi_{m,p}^v = \sum_{q=1}^N \upsilon_{pq} \sum_{n=1}^M \pi_{mn}^{(q)} P_n^v$. A combination of (8.68) and (8.69) gives

$$\Delta V = \mathbb{E}\left[x_{1,1}^\top(i, j)\Phi_{m,p} x_{1,1}(i, j)\right] - x^\top(i, j) P_m x(i, j), \tag{8.70}$$

where

$$x_{1,1}(i,j) = \begin{bmatrix} x^h(i+1,j) \\ x^v(i,j+1) \end{bmatrix} = \left(A^c_m + \beta(i,j)\hat{A}^c_m\right) x(i,j)$$
$$= \left(A^c_m + \sum_{k=1}^{d} \beta_k(i,j) F_k \hat{A}^c_m\right) x(i,j).$$

Therefore, similar to the proof of Theorem 8.2, from (8.70), we get

$$\Delta V = x^\top(i,j)\Xi_{m,p}x(i,j), \tag{8.71}$$

where

$$\Xi_{m,p} = A^{c\top}_m \Phi_{m,p} A_m + \sum_{k=1}^{d} \rho_k^2 \hat{A}^{c\top}_m F_k \Phi_{m,p} F_k \hat{A}^c_m - P_m.$$

Due to (8.65), there exists a scalar $\lambda_3 > 0$ such that $\Delta V \leq -\lambda_3 |x(i,j)|^2$, by which we can conclude that the closed-loop system (8.64) is stochastically stable. The proof is completed. □

Remark 8.10 For homogeneous Markov chains, the process $\sigma(s)$ is constant ($\sigma(s)$ takes only one mode). To this case, condition (8.65) is reduced to the following simplified LMI

$$A^{c\top}_m \widetilde{P}_m A^c_m + \sum_{k=1}^{d} \rho_k^2 \hat{A}^{c\top}_m F_k \widetilde{P}_m F_k \hat{A}^c_m - P_m < 0, \tag{8.72}$$

where $\widetilde{P}_m = \sum_{n=1}^{M} \pi_{mn} P_n$.

8.4.2 Controller Design

In this section, we present the design problem of an SSFC in the form of (8.63). By Theorem 8.4, the closed-loop system (8.64) is stochastically stable if there exist matrices $P_n = \text{diag}(P^h_n, P^v_n) \in \mathbb{D}^+_d(d_h, d_v)$, $n \in \mathscr{M}$, such that

$$A^{c\top}_m \Phi_{m,p} A^c_m + \sum_{k=1}^{d} \rho_k^2 \hat{A}^{c\top}_m F_k \Phi_{m,p} F_k \hat{A}^c_m - P_m < 0, \quad m \in \mathscr{M}, \tag{8.73}$$

where $A^c_m = A_m + B_m K_m$ and $\hat{A}^c_m = \hat{A}_m + \hat{B}_m K_m$. We denote

$$G^m_q = \left[\sqrt{\pi^{(q)}_{m1}}\ \sqrt{\pi^{(q)}_{m2}}\ \cdots\ \sqrt{\pi^{(q)}_{mM}}\right],$$

$$H_p^m = \left[\sqrt{\upsilon_{p1}} \otimes G_1^m \cdots \sqrt{\upsilon_{pN}} \otimes G_N^m\right] \otimes I_d,$$
$$\mathscr{D}_P = \mathrm{diag}(P_1, P_2, \ldots, P_M), \quad \mathcal{F}_\rho = \left[\rho_1 \mathcal{F}_1 \ \rho_2 \mathcal{F}_2 \cdots \rho_d \mathcal{F}_d\right],$$

where $\otimes$ denotes the Kronecker product. Then, it can be verified that

$$\begin{aligned}
\Phi_{m,p} &= \sum_{q=1}^{N} \upsilon_{pq} \sum_{n=1}^{M} \pi_{mn}^{(q)} P_n \\
&= \sum_{q=1}^{N} \upsilon_{pq} \left(G_q^m \otimes I_d\right) \mathscr{D}_P \left(G_q^{m\top} \otimes I_d\right) \\
&= H_p^m \left(I_N \otimes \mathscr{D}_P\right) H_p^{m\top}.
\end{aligned}$$

In addition to this,

$$\sum_{k=1}^{d} \rho_k^2 \mathcal{F}_k \Phi_{m,p} \mathcal{F}_k = \mathcal{F}_\rho \left(I_d \otimes \Phi_{m,p}\right) \mathcal{F}_\rho^\top.$$

Therefore, condition (8.73) can be written as

$$\begin{aligned}
&A_m^{c\top} H_p^m \left(I_N \otimes \mathscr{D}_P\right) H_p^{m\top} A_m^c \\
&+ \hat{A}_m^{c\top} \mathcal{F}_\rho \left(I_d \otimes H_p^m\right) \left(I_{dN} \otimes \mathscr{D}_P\right) \left(I_d \otimes H_p^{m\top}\right) \mathcal{F}_\rho^\top \hat{A}_m^c - P_m < 0.
\end{aligned} \tag{8.74}$$

By the Schur complement (Lemma 8.4), condition (8.74) holds if and only if

$$\begin{bmatrix} -P_m & A_m^{c\top} H_p^m & \hat{A}_m^{c\top} \mathcal{F}_\rho \left(I_d \otimes H_p^m\right) \\ * & -I_N \otimes \mathscr{D}_P^{-1} & 0 \\ * & * & -I_{dN} \otimes \mathscr{D}_P^{-1} \end{bmatrix} < 0, \tag{8.75}$$

By defining the matrices $X_m = P_m^{-1} \in \mathbb{D}_d(d_h, d_v)$, we have

$$X_m A_m^{c\top} = X_m A_m^\top + X_m K_m^\top B_m^\top,$$
$$X_m \hat{A}_m^{c\top} = X_m \hat{A}_m^\top + X_m K_m^\top \hat{B}_m^\top.$$

In addition, we change the variables $Y_m = K_m X_m$, then

$$X_m A_m^{c\top} = \Omega_{X_m,Y_m}^\top \text{ and } X_m \hat{A}_m^{c\top} = \hat{\Omega}_{X_m,Y_m}^\top,$$

where

$$\Omega_{X_m,Y_m} = A_m X_m + B_m Y_m, \quad \hat{\Omega}_{X_m,Y_m} = \hat{A}_m X_m + \hat{B}_m Y_m.$$

In summary, design conditions of a desired SSFC (8.63) are given in the following theorem.

Theorem 8.5 *System* (8.59) *is robustly stabilizable via SSFCs subject to multiplicative stochastic noises defined by* (8.60) *if there exist matrices* $X_m \in \mathbb{D}_d^+(d_h, d_v)$ *and* $Y_m \in \mathbb{R}^{d_c \times d}$ $(m \in \mathscr{M})$ *that satisfy the following LMI conditions*

$$\begin{bmatrix} -X_m & \Xi_m^\top H_{m,p} & \hat{\Xi}_m^\top F_Q\left(I_d \otimes H_{m,p}\right) \\ * & -I_N \otimes \mathscr{D}_X & 0 \\ * & * & -I_{dN} \otimes \mathscr{D}_X \end{bmatrix} < 0,\ m \in \mathscr{M},\ p \in \mathscr{N}. \tag{8.76}$$

The controller gains are obtained as

$$K_m = Y_m X_m^{-1},\ m \in \mathscr{M}.$$

Remark 8.11 Theorem 8.5 in this chapter deals with the design of a synchronous mode-dependent SFC for 2-D systems with multiplicative stochastic noises. By similar lines used, for example, in [30–32], and the proof of Theorems 8.4, 8.5, analogous LMI conditions of (8.76) can be formulated to design an asynchronous state-feedback controller (ASFC) of the form

$$u(i, j) = K_{\gamma(i,j)} x(i, j),$$

where $\gamma(i, j)$ is a Markov chain, which determines the switching process of controller, with finite state space $\mathscr{M}_c = \{1, 2, \ldots, M_c\}$. Note in addition that, the design procedure proposed in this section can be extended to 2-D systems with state time-delays by employing some arguments in the proof of Theorems 8.1 and 8.2. We omit details here.

Remark 8.12 For 2-D systems in the form of (8.59) with nonideal transition probabilities, that is, some entries of the matrices $\Pi_{(\sigma)} = (\pi_{mn}^{(\sigma)})$ and $\Upsilon = (\upsilon_{pq})$ are unknown or uncertain, by similar methods presented in [33] and Chap. 7, design conditions of an SSFC (8.60) can be obtained from (8.76).

8.4.3 An Illustrative Example

Example 8.4 Consider the following control model described by a Darboux equation

$$\frac{\partial T(x,t)}{\partial x} + \frac{\partial T(x,t)}{\partial t} = -a_{\theta(x,t)} T(x,t) + b_{\theta(x,t)} u(x,t) \tag{8.77}$$

where $T(x, t)$ is the unknown function, $u(x, t)$ is the control input and a_θ, b_θ are parameter-dependent real coefficients, $\theta = \theta(x, t)$ is a switching parameter. Similar to Sect. 8.3.6, for given increments Δx and Δt, we denote

Table 8.1 System parameters

Parameters	$m = 1$	$m = 2$	$m = 3$
a_m	1.25	1.58	0.86
b_m	1.0	0.5	1.0
Increments	$\Delta t = 0.1$		$\Delta x = 0.1$

$$T(i, j) = T(i\Delta x, j\Delta t), \quad \theta(i, j) = \theta(i\Delta x, j\Delta t).$$

Then, by using Euler approximation method, from (8.77), we obtain

$$\begin{aligned} T(i, j+1) = &\left(1 - \frac{\Delta t}{\Delta x} - a_{\theta(i,j)}\Delta t\right) T(i, j) \\ &+ \frac{\Delta t}{\Delta x} T(i-1, j) + b_{\theta(i,j)} u(i, j). \end{aligned} \tag{8.78}$$

Define state variables $x^h(i, j) = T(i-1, j)$ and $x^v(i, j) = T(i, j)$. By incorporating stochastic noises subject to noisy environment effects, Eq. (8.78) can be recast in the form of (8.59) with

$$A_\theta = \hat{A}_\theta = \begin{bmatrix} 0 & 1 \\ \frac{\Delta t}{\Delta x} & 1 - \frac{\Delta t}{\Delta x} - a_\theta \Delta t \end{bmatrix}, \quad B_\theta = \hat{B}_\theta = \begin{bmatrix} 0 \\ b_\theta \Delta t \end{bmatrix}.$$

We assume the jumping process $\theta(i, j)$ is driven by a three-mode inhomogeneous Markov chain with transition probabilities given by matrices

$$\Pi_{\{\sigma=1\}} = \begin{bmatrix} 0.8 & 0.1 & 0.1 \\ 0.36 & 0.54 & 0.1 \\ 0.2 & 0.05 & 0.75 \end{bmatrix}, \quad \Pi_{\{\sigma=2\}} = \begin{bmatrix} 0.6 & 0.15 & 0.25 \\ 0.3 & 0.5 & 0.2 \\ 0.25 & 0.1 & 0.65 \end{bmatrix},$$

and second-order probability matrix

$$\Upsilon = (\upsilon_{pq}) = \begin{bmatrix} 0.8 & 0.2 \\ 0.7 & 0.3 \end{bmatrix}.$$

The system matrices are specified by the following parameters (Table 8.1).

Simulation results with given transition rates and parameters show that the open system (i.e. without control) is unstable. State trajectories of the open system with scalar-valued stochastic noise $\beta(i, j)$ whose intensity $\rho = 0.1$ are presented in Figs. 8.14 and 8.15. It can be seen from conducted simulation results that the unforced system of (8.59) is unstable.

We now apply the design method proposed in this section. By using the Matlab LMI toolbox, it is found that condition (8.76) is feasible with $\rho \leq 2.22$. For $\rho = 2.22$, a feasible solution of (8.76) is defined by

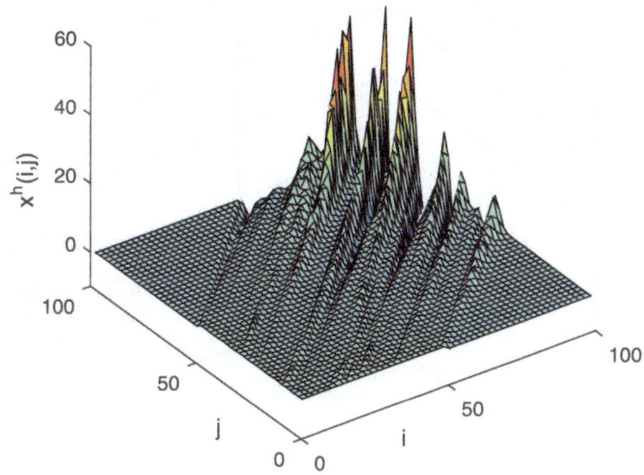

Fig. 8.14 A state trajectory $x^h(i, j)$ of the open-loop system with $\rho = 0.1$

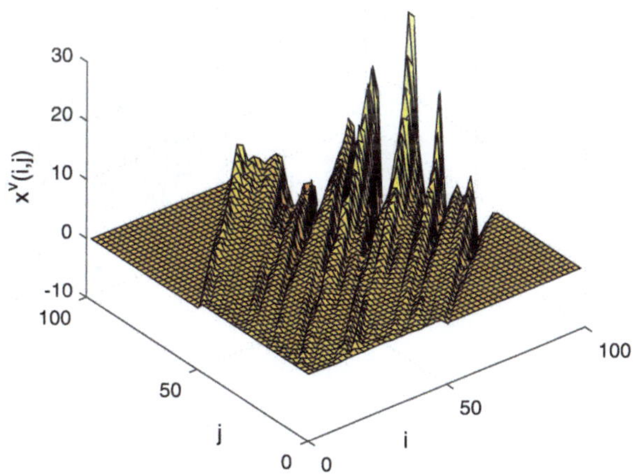

Fig. 8.15 A state trajectory $x^v(i, j)$ of the open-loop system with $\rho = 0.1$

$$X_1 = \text{diag}(9.1239, 1.1605), \quad X_2 = \text{diag}(2.7375, 0.5403), \quad X_3 = \text{diag}(11.0303, 1.6494),$$
$$Y_1 = \begin{bmatrix} -88.5907 & 1.4072 \end{bmatrix}, \quad Y_2 = \begin{bmatrix} -52.2842 & 1.5081 \end{bmatrix}, \quad Y_3 = \begin{bmatrix} -107.1481 & 1.4016 \end{bmatrix}.$$

Then, a desired SSFC (8.63) is composed of the following controller gains

$$K_1 = \begin{bmatrix} -9.7097 & 1.2126 \end{bmatrix}, \quad K_2 = \begin{bmatrix} -19.0994 & 2.7910 \end{bmatrix},$$
$$K_3 = \begin{bmatrix} -9.7139 & 0.8498 \end{bmatrix}.$$

By Theorem 8.5, the 2-D system (8.59) is robustly stabilizable via SSFC (8.63) with respect to stochastic noise $\beta(i, j)$.

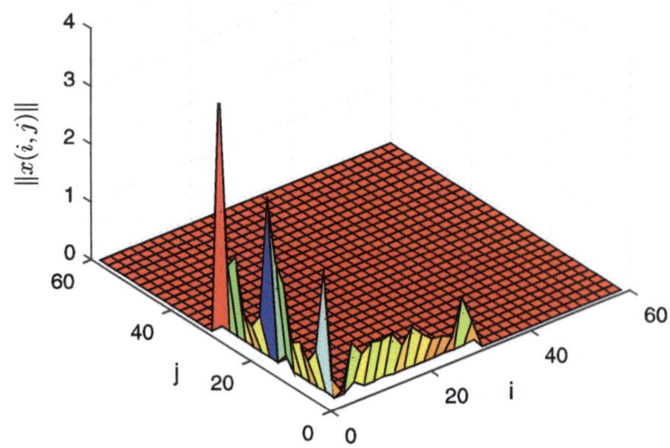

Fig. 8.16 A state trajectory $\|x(i, j)\|$ of the closed-loop system with $\rho = 2.22$

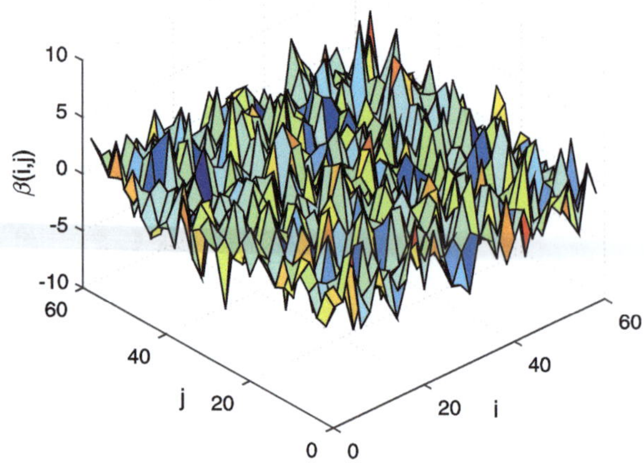

Fig. 8.17 Stochastic noise $\beta(i, j)$

A closed-loop norm-state trajectory $\|x(i, j)\|$ with $\rho = 2.22$ is depicted in Fig. 8.16. Stochastic noise $\beta(i, j)$ is presented in Fig. 8.17 and mode transitions $\theta(i, j)$, $\sigma(s)$ are given in Fig. 8.18, respectively. The obtained simulation results demonstrate the effectiveness of the proposed design method.

8.5 Conclusion

In this chapter, the problems of energy-to-peak stochastic stability, robust stochastic stabilization via synchronous state-feedback and observer-based control have been developed for 2-D Roesser systems with multiplicative stochastic noises. Systematic

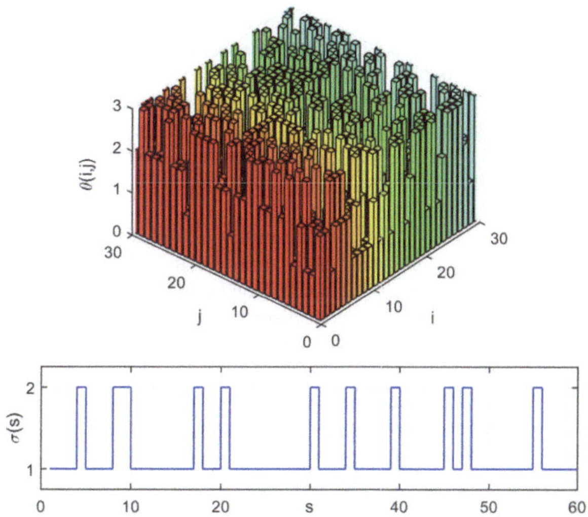

Fig. 8.18 Mode evolutions of $\theta(i, j)$ and $\sigma(s)$

approaches in stability analysis and controller design have been presented. Extensions to observer-based control of 2-D systems with random packet dropouts and 2-D systems with multiplicative noises driven by inhomogeneous Markov chains have also been discussed. Numerical examples with simulations have been provided to illustrate the efficacy of the proposed methods.

References

1. Sipahi, R., Niculescu, S.-I., Abdallah, C.T., Michiels, W., Gu, K.: Stability and stabilization of systems with time delay. IEEE Control Syst Magaz **31**, 38–65 (2011)
2. Witrant, E., Fridman, E., Sename, O., Dugard, L.: Recent Results on Time-Delay Systems: Analysis and Control. Springer, Basel (2016)
3. Zhang, H., Yue, D., Zhao, W., Hu, S., Dou, C.: Distributed optimal consensus control for multiagent systems with input delay. IEEE Trans Cybern **48**, 1747–1759 (2018)
4. Ge, X., Hoi, K., Vong, S.W.: A delay-variation-dependent stability criterion for discrete-time systems via a bivariate quadratic function negative-determination lemma. J Frankl Inst **359**, 4976–4996 (2022)
5. Hien, L.V.: Delay-dependent positivity and stability analysis of discrete-time systems with delay. IEEE Control Syst Lett **6**, 2575–2580 (2022)
6. Zhang, C.K., Chen, W.H., Zhu, C., He, Y., Wu, M.: Stability analysis of discrete-time systems with time-varying delay via a delay-dependent matrix- separation-based inequality. Automatica **156**, 111192 (2023)
7. Sha, H., Park, J.H., Chen, J., Zhu, M., Nan, C.: Stability analysis of discrete-time systems with a time-varying delay via improved methods. IET Control Theory Appl **18**, 951–959 (2024)
8. Hien, L.V., Trinh, H.: Stability of two-dimensional Roesser systems with time-varying delays via novel 2D finite sum inequalities. IET Control Theory Appl **10**, 1665–1674 (2016)
9. Zhao, D., Shen, D., Wang, Y.: Fault diagnosis and compensation for two-dimensional discrete time systems with sensor faults and time-varying delays. Int J Robust Nonlinear Control **27**, 3296–3320 (2017)

10. Huang, Y.B., An, J., He, Y., Wu, M.: Quasi-convex combination method and its application to the stability analysis of 2D discrete-time Roesser systems with time-varying delays. IET Control Theory Appl **12**, 718–727 (2018)
11. Hien, L.V., Trinh, H.: On reachable set estimation of two-dimensional systems described by the Roesser model with time-varying delays. Int J Robust Nonlinear Control **28**, 227–246 (2018)
12. Hien, L.V., Lan-Huong, N.T., Hieu, T.: Delay-dependent energy-to-peak stability of 2-D time-delay Roesser systems with multiplicative stochastic noises. IEEE Trans Autom Control **64**, 5066–5073 (2019)
13. Hien, L.V., Vu, L.H.: Finite-region boundedness and (Q, S, R)-dissipativity of 2-D singular Roesser systems with mixed time-varying delays. J Frank Inst **361**, 1416–1435 (2024)
14. Gray, W.S., González, O.R., Doğan, M.: Stability analysis of digital linear flight controllers subject to electromagnetic disturbances. IEEE Trans Aero Electron Syst **36**, 1204–1218 (2000)
15. Fernández, J.H., Speyer, J.L., Idan, M.: Stochastic estimation for two-state linear dynamic systems with additive Cauchy noises. IEEE Trans Autom Control **60**, 3367–3372 (2015)
16. Mao, Z., Zhan, Y., Tao, G., Jiang, B., Yan, X.G.: Sensor fault detection for rail vehicle suspension systems with disturbances and stochastic noises. IEEE Vehicle Technol **66**, 4691–4705 (2017)
17. Li, L., Zhang, H.: Stabilization of discrete-time systems with multiplicative noise and multiple delays in the control variable. SIAM J Control Optim **54**, 894–917 (2016)
18. Li, L., Zhang, H., Fu, M.: Linear quadratic regulation for discrete-time systems with multiplicative noise and multiple input delays. Optim Control Appl Method **38**, 295–316 (2017)
19. Petrou, M., Petrou, C.: Image Processing: The Fundamentals, 2nd edn. John Wiley & Sons, Chichester, UK (2010)
20. Li, G., Zhang, W., Chen, M.: Robust H_2/H_∞ control for periodic stochastic difference systems with multiplicative noise. IET Control Theory Appl **9**, 2451–2457 (2015)
21. Zhang, H., Li, L., Xu, J., Fu, M.: Linear quadratic regulation and stabilization of discrete-time systems with delay and multiplicative noise. IEEE Trans Autom Control **60**, 2599–2613 (2015)
22. Liang, X., Xu, J., Zhang, H.: Discrete-time LQG control with input delay and multiplicative noise. IEEE Trans Aero Elect Syst **53**, 3079–3090 (2017)
23. Qi, Q., Zhang, H.: Output feedback control and stabilization for multiplicative noise systems with intermittent observations. IEEE Trans Cybern **48**, 2128–2138 (2018)
24. Jia, R., Zong, X.: Time-varying formation control of linear multiagent systems with time delays and multiplicative noises. Int J Robust Nonlinear Control **31**, 9008–9025 (2021)
25. Ahn, C.K., Wu, L., Shi, P.: Stochastic stability analysis for 2-D Roesser systems with multiplicative noise. Automatica **69**, 356–363 (2016)
26. Ahn, C.K., Shi, P., Basin, M.V.: Two-dimensional peak-to-peak filtering for stochastic Fornasini-Marchesini systems. IEEE Trans Autom Control **63**, 1472–1479 (2018)
27. Wang, B., Zhu, Q., Xu, L.: Expected power bound and stability of two-dimensional digital filters with multiplicative noise in the FMLSS model. J Frank Inst **358**, 2500–2514 (2021)
28. Kallenberg, O.: Foundations of Modern Probability, 2nd edn. York, Springer (2002)
29. Kaczorek, T.: Two-Dimensional Linear Systems. Springer, Berlin (1985)
30. Hien, L.V., Trinh, H.: Observer-based control of 2-D Markov jump systems. IEEE Trans Circuit Syst-II **64**, 1322–1326 (2017)
31. Hien, L.V., Dzung, N.T.: Asynchronous control of discrete-time stochastic bilinear systems with Markovian switchings. Int J Syst Sci **50**, 23–34 (2019)
32. Zhang, M., Shen, C., Wu, Z.G.: Asynchronous observer-based control for exponential stabilization of Markov jump systems. IEEE Trans Circuit Syst-II **67**, 2039–2043 (2020)
33. Zhou, M., Fu, Y.: Stability and stabilization for discrete-time Markovian jump stochastic systems with piecewise homogeneous transition probabilities. Int J Control Autom Syst **17**, 2165–2173 (2019)

Chapter 9
Stability and Stabilization of Two-Dimensional Fuzzy Systems with Time-Varying Delays

Abstract This chapter deals with the stability and stabilization via fuzzy rule-based state-feedback controllers (SFCs) problems for two-dimensional (2-D) Takagi-Sugeno fuzzy systems in the Roesser model with interval state delays. In light of a 2-D Lyapunov–Krasovskii functional (LKF) method, an augmented fuzzy parameter-dependent LKF candidate is constructed and employed to establish delay-dependent stability conditions. On the basis of the proposed analysis results, the design problem of a stabilizing rule-based SFC for 2-D fuzzy time-delay systems is addressed. Numerical examples are given to show the efficacy and advantage of the obtained results.

Keywords 2-D fuzzy systems · Roesser model · 2-D summation inequalities · Directional delays · LMIs

9.1 Introduction

Time-delay phenomena are omnipresent in practical systems and industrial processes [1, 2]. For instance, the reaction of engineering systems to exogenous signals is never instantaneous and always infected by time delays [3]. In the effect of delays, system behavior and performance is usually unpredictable and even chaotic. Thus, the study concerning analysis and control of time-delay systems plays essential role in control engineering [4]. Despite a long history, this research area is still one of the most active topics in the past few decades. For a few references, we refer the reader to recent works [5–9] on one-dimensional (1-D) time-delay systems and [10–15] for various classes of 2-D systems with delays.

It can be realized through the existing literature that the approach based on the celebrated Krasovskii theorem, widely known as Lyapunov–Krasovskii functional (LKF) method, is one of the most efficient approaches to derive tractable stability conditions for time-delay systems [6]. Such conditions for LTI delay systems are derived in certain types of linear matrix inequalities (LMIs) [9] which can be classified into two kinds, say delay-dependent stability conditions and delay-independent ones. Roughly speaking, delay-dependent stability conditions, which depend explicitly on

L. V. Hien, *Two-Dimensional Time-Delay Systems*, Studies in Systems, Decision and Control 605, https://doi.org/10.1007/978-3-031-95678-2_9

the magnitude of delays, are more reasonable and less conservative than delay-independent ones. Remarkable research attention has been devoted to developing methods in order to derive as less conservative as possible delay-dependent stability conditions not only in enlarging the allowable delay range but also in reducing computational burden.

Another important aspect beside the effect of time-delay is that actual engineering systems are often modeled by complex nonlinear systems which pose much difficulty and challenging in analysis and control [16]. It has been well recognized since its inception in the pioneering work [17] that fuzzy rule-based modeling has become a successful approach for complex nonlinear systems. Among popular rule-based models, Takagi–Sugeno (T-S) approach [18] is the most effective way to represent nonlinear systems by fuzzy sets and fuzzy reasoning. Over the past two decades, the topic of analysis and control has been extensively developed for 1-D T-S fuzzy systems with delays (see, for example, [19–28] and the references therein). In particular, the H_∞ filter design problem was investigated for T-S fuzzy systems with unknown membership functions [22]. Based on fuzzy parameter-dependent LKF approach, and by utilizing some generalized Jensen-type summation inequalities or free-weighting matrices, delay-dependent stability conditions were derived for T-S fuzzy systems with time-varying delay and known membership functions [24, 28]. The design problem of fuzzy rule-based state feedback controllers was also discussed in [24]. Dissipativity analysis and sampled-data control were also developed for various classes of T-S fuzzy systems with delays in recent works [25–27].

It is noticing that all the aforementioned works only concern with the analysis and control of 1-D fuzzy systems. The theory for 2-D T-S fuzzy systems has received growing research attention. Only a few results concerning stability analysis and synthesis of 2-D fuzzy systems have been reported. For example, the problems of stability and H_∞ stabilization via state feedback controllers (SFCs) or H_∞ filter design were studied for T-S delay-free systems described by the FMLSS model in [29–31]. In [32], H_∞ control problem was studied for T-S fuzzy systems in FMLSS model with delays and missing measurements. By using a common LKF, that does not depend membership functions, and utilizing the free-weighting matrix method, delay-dependent stability conditions were derived and employed into designing fuzzy rule-based static output feedback controllers. In [33], improved fuzzy parameter-dependent LKFs were constructed and new 2-D summation inequalities were utilized to derive less conservative delay-dependent stability and stabilization via fuzzy rule-based SFCs for 2-D T-S fuzzy systems in the Roesser model with interval directional delays. Sliding mode control and filtering were also investigated for some classes of 2-D fuzzy-affine systems with delays in recent works [34–36].

In this chapter, we consider the problems of stability and stabilization via fuzzy-rule based SFCs 2-D T-S fuzzy systems described by the Roesser model with interval state delays. An analysis scheme extended from the conventional LKF method is first developed for fuzzy parameter-dependent functionals. By using some improved fuzzy parameter-dependent LKFs, delay-dependent stability conditions are derived for T-S fuzzy Roesser systems with interval delays using the LMIs framework.

On the basis of the derived stability conditions, the design problem of fuzzy rule-based SFCs is also discussed. Numerical examples are given to show the efficacy of the obtained results.

9.2 Preliminaries

9.2.1 Model Description

Consider the following rule-based model of 2-D Roesser systems with delays

RULE k: IF (ϑ_1 is Θ_1^k) $\wedge$ (ϑ_2 is Θ_2^k) $\wedge$ ... $\wedge$ (ϑ_p is Θ_p^k) THEN

$$\begin{bmatrix} x^h(i+1,j) \\ x^v(i,j+1) \end{bmatrix} = A_k \begin{bmatrix} x^h(i,j) \\ x^v(i,j) \end{bmatrix} + A_{dk} \begin{bmatrix} x^h(i-d_h(i),j) \\ x^v(i,j-d_v(j)) \end{bmatrix} + B_k u(i,j), \tag{9.1}$$

where $k \in \mathscr{R} = \{1, 2, \ldots, N_r\}$, N_r is the number of IF-THEN rules, $x^h(i,j) \in \mathbb{R}^{n_h}$, $x^v(i,j) \in \mathbb{R}^{n_v}$ are the horizontal and vertical state vectors, respectively, $u(i,j) \in \mathbb{R}^{n_u}$ is the control input, $\vartheta_l = \vartheta_l(i,j), l = 1, 2, \ldots, p$, are the premise variables and Θ_l^k are the fuzzy sets associated with the kth IF-THEN rule. A_k, $A_{dk} \in \mathbb{R}^{n\times n}$ ($n = n_h + n_v$) and $B_k \in \mathbb{R}^{n\times n_u}$ are given real system matrices with appropriate dimensions. $d_h(i)$ and $d_v(j)$ represent the time-varying delays along horizontal and vertical directions. It is assumed that the directional delays are being confined within prescribed ranges

$$d_1^h \le d_h(i) \le d_2^h, \quad d_1^v \le d_v(j) \le d_2^v, \tag{9.2}$$

where d_1^h, d_1^v and d_2^h, d_2^v are give positive integers involving the lower bounds and upper bounds of delays.

The initial condition of (9.1) is defined by sequences $\phi_h(i,j)$ and $\phi_v(i,j)$ as

$$\begin{aligned} x^h(i,j) &= \phi_h(i,j), \ i \in \mathbb{Z}[-d_2^h, 0], \ j \in \mathbb{N}_0, \\ x^v(i,j) &= \phi_v(i,j), \ i \in \mathbb{N}_0, \ j \in \mathbb{Z}[-d_2^v, 0]. \end{aligned} \tag{9.3}$$

A vector-valued sequence $\phi : \mathbb{N}_0 \to \mathbb{R}^q$ is said to belong to the space $l_2(\mathbb{R}^q)$ if the series $\sum_{k=0}^{\infty} \phi^\top(k)\phi(k)$ is finite. The sequence ϕ is said to have finite support if there exists a positive integer T such that $\phi(k) = 0$ for all $k \ge T$. In this chapter, we consider 2-D T-S fuzzy systems (9.1) with initial sequences $\phi_h(i,.) \in l_2(\mathbb{R}^{n_h})$, $i \in \mathbb{Z}[-d_2^h, 0]$, and $\phi_v(.,j) \in l_2(\mathbb{R}^{n_v})$, $j \in \mathbb{Z}[-d_2^v, 0]$. This assumption it is widely adopted in the area of analysis and control of 2-D systems [10, 11]. We define

$$\hat{\phi}_h(j) = \sup_{i\in\mathbb{Z}[-d_2^h,0]} \|\phi_h(i,j)\| \text{ and } \hat{\psi}_v(i) = \sup_{j\in\mathbb{Z}[-d_2^v,0]} \|\phi_v(i,j)\|.$$

It can be verified that

$$\lim_{T\to\infty}\sum_{k=0}^{T}\left(\hat{\phi}_h^2(k)+\hat{\phi}_v^2(k)\right)<\infty. \tag{9.4}$$

Condition (9.4) is trivial if the initial sequences $\phi_h(i,.)$, $\phi_v(.,j)$ have finite supports.

Assume that the premise variables and control inputs are independent (that is, for any $l \in \{1, 2, \dots, p\}$, ϑ_l does not depend on $u(i,j)$) [24]. Then, the defuzzified dynamic output of system (9.1) using the common center-average defuzzifier approach can be inferred as

$$\begin{aligned}\begin{bmatrix}x^h(i+1,j)\\x^v(i,j+1)\end{bmatrix}=\sum_{k=1}^{N_r}\mu_k(\vartheta(i,j))\Big\{&A_k\begin{bmatrix}x^h(i,j)\\x^v(i,j)\end{bmatrix}\\&+A_{dk}\begin{bmatrix}x^h(i-d_h(i),j)\\x^v(i,j-d_v(j))\end{bmatrix}+B_ku(i,j)\Big\},\end{aligned} \tag{9.5}$$

where $\vartheta(i,j)=(\vartheta_1(i,j),\dots,\vartheta_p(i,j))$,

$$\mu_k(\vartheta)=\frac{\prod_{l=1}^{p}m_{kl}(\vartheta_l)}{\sum_{k=1}^{N_r}\prod_{l=1}^{p}m_{kl}(\vartheta_l)}, \tag{9.6}$$

and $m_{kl}(\vartheta_l)$ represents the grade membership of ϑ_l in Θ_l^k. For the brevity, we denote $\mu_k(\vartheta)$ as μ_k if it does not cause any confusion. Note also that the fuzzy basic functions $\mu_k(\vartheta)$, $k=1,2,\dots,N_r$, satisfy

$$0\le\mu_k(\vartheta)\le 1,\quad \sum_{k=1}^{N_r}\mu_k(\vartheta)=1.$$

The overall fuzzy model (9.1) can be written as

$$\begin{bmatrix}x^h(i+1,j)\\x^v(i,j+1)\end{bmatrix}=A(\mu)\begin{bmatrix}x^h(i,j)\\x^v(i,j)\end{bmatrix}+A_d(\mu)\begin{bmatrix}x^h(i-d_h(i),j)\\x^v(i,j-d_v(j))\end{bmatrix}+B(\mu)u(i,j), \tag{9.7}$$

where

$$A(\mu)=\sum_{k=1}^{N_r}\mu_kA_k,\quad A_d(\mu)=\sum_{k=1}^{N_r}\mu_kA_{dk},\quad B(\mu)=\sum_{k=1}^{N_r}\mu_kB_k.$$

9.2.2 Problem Formulation

Definition 9.1 The open system (without control input u) of (9.7) is said to be globally asymptotically stable (GAS) if, for any initial sequences satisfying (9.3)–(9.4), the corresponding state trajectory of the system satisfies

$$\lim_{T\to\infty}\sup_{(i,j)\in\Gamma_T}\left\|\begin{bmatrix}x^h(i,j)\\x^v(i,j)\end{bmatrix}\right\|=0,$$

where Γ_T denotes the diagonal segment $\{(i,j)\in\mathbb{N}_0^2 : i+j=T\}$.

Definition 9.1 indicates that, for any state trajectory $x(i,j)=\begin{bmatrix}x^{h\top}(i,j)\\x^{v\top}(i,j)\end{bmatrix}$, $\|x(i,j)\|$ tends to zero as $i+j\to\infty$ and the convergence is uniformly to the sum $i+j$.

To stabilize the 2-D fuzzy system (9.7), a fuzzy rule-based SFC will be designed in the following law.

RULE k: IF $(\vartheta_1$ is $\Theta_1^k)\ \wedge\ (\vartheta_2$ is $\Theta_2^k)\ \wedge\ \ldots\ \wedge (\vartheta_p$ is $\Theta_p^k)$ THEN

$$u(i,j)=F_kx(i,j), \tag{9.8}$$

where, $F_k\in\mathbb{R}^{n_u\times n}$, $k\in\mathscr{R}$, are the controller gains, which will be determined. By fuzzy blending similar to (9.7), the overall state-feedback controller (SFCL) is represented as

$$u(i,j)=\sum_{k=1}^{N_r}\mu_kF_kx(i,j). \tag{9.9}$$

By integrating the SFCL (9.9), the closed-loop system of (9.7) is obtained as

$$\begin{bmatrix}x^h(i+1,j)\\x^v(i,j+1)\end{bmatrix}=A^c(\mu)\begin{bmatrix}x^h(i,j)\\x^v(i,j)\end{bmatrix}+A_d(\mu)\begin{bmatrix}x^h(i-d_h(i),j)\\x^v(i,j-d_v(j))\end{bmatrix}, \tag{9.10}$$

where $F(\mu)=\sum_{k=1}^{N_r}\mu_kF_k$ and

$$\begin{aligned}A^c(\mu)&=A(\mu)+B(\mu)F(\mu)\\&=\sum_{k=1}^{N_r}\mu_k\left(A_k+\sum_{l=1}^{N_r}\mu_k\mu_lB_kF_l\right).\end{aligned}$$

The main objectives of this chapter is to derive delay-dependent conditions that ensure the GAS property of the open-loop system (9.7). We then design an SFCL in the form of (9.8) that makes the closed-loop system (9.10) asymptotically stable. To tackle with this, we first develop an analysis scheme by extending the LKF method. By reformulating some 2-D summation inequalities, we then utilize the proposed scheme and efficient tools to derive less conservative delay-dependent stability conditions for system (9.10) in the form of LMI settings. Finally, based on the analysis results, a design procedure is proposed to obtain desired stabilizing controller gains for 2-D T-S fuzzy control systems (9.1).

9.3 Auxiliary Results

9.3.1 Parameter-Dependent Functional-Based Method

In this section, we present an analysis scheme based on an extension of the 2-D LKF method. Let $\mathscr{P}_V$ denote the set of admissible premise variables $\mu = (\mu_1, \mu_2, \ldots, \mu_{N_r})$ defined by (9.6). Consider a (rule-based) parameter-dependent functional $V : \mathscr{P}_V \times \mathbb{R}^n \to \mathbb{R}$, $(\mu, x) \mapsto V_\mu(x)$, defined by

$$V_\mu(x) = V_\mu^h(x^h) + V_\mu^v(x^v),$$

where $x = (x^h, x^v)$, $x^h \in \mathbb{R}^{n_h}$, $x^v \in \mathbb{R}^{n_v}$ and $V^h : \mathscr{P}_V \times \mathbb{R}^{n_h} \to \mathbb{R}$, $V^v : \mathscr{P}_V \times \mathbb{R}^{n_v} \to \mathbb{R}$ are some functionals. The difference ΔV along state trajectory $x(i, j)$ of system (9.7) is defined as

$$\Delta V \triangleq \underbrace{\Delta_h V^h_{\mu(i,j)}(x^h(i, j)) + \Delta_v V^v_{\mu(i,j)}(x^v(i, j))}_{\Delta V_{\mu(i,j)}(x(i,j))}, \tag{9.11}$$

where

$$\begin{aligned}
\Delta_h V^h_{\mu(i,j)}(x^h(i, j)) &= V^h_{\mu(i+1,j)}(x^h(i+1, j)) - V^h_{\mu(i,j)}(x^h(i, j)),\\
\Delta_v V^v_{\mu(i,j)}(x^v(i, j)) &= V^v_{\mu(i,j+1)}(x^v(i, j+1)) - V^v_{\mu(i,j)}(x^v(i, j)).
\end{aligned}$$

The following functional-based scheme is given to deal with fuzzy rule-based 2-D systems as given in (9.7).

Lemma 9.1 (Parameter-dependent functional-based scheme) *Assume that there exist a rule-based functional $V_\mu(x) = V_\mu^h(x^h) + V_\mu^v(x^v)$ and positive scalars c_0, c_k ($k = 1, 2, 3, 4$) such that*

(i) *For any $x^h \in \mathbb{R}^{n_h}$, $x^v \in \mathbb{R}^{n_v}$, and $\mu \in \mathscr{P}_V$,*

$$c_1\|x^h\|^2 \le V_\mu^h(x^h) \le c_2\|x^h\|^2, \quad c_3\|x^v\|^2 \le V_\mu^v(x^v) \le c_4\|x^h\|^2; \tag{9.12}$$

(ii) *The difference ΔV along any state trajectory satisfies*

$$\Delta V_{\mu(i,j)}(x(i, j)) \le -c_0\|x(i, j)\|^2. \tag{9.13}$$

Then, the open-loop system of (9.7) *is GAS. Moreover, along any state trajectory* $x(i, j)$ *of system* (9.7) *with* $u = 0$*, we also have*

$$\lim_{T\to\infty} \sum_{i+j=T} V_{\mu(i,j)}(x(i, j)) = 0.$$

Proof Denote

$$E(k) = \sum_{(i,j)\in\Gamma_k} V_{\mu(i,j)}(x(i, j))$$

as the energy of the functional V stored in the diagonal line Γ_k and $E(0) = 0$. It follows from (9.13) that

$$\begin{aligned} V^h_{\mu(i+1,j)}(x^h(i + 1, j)) + V^v_{\mu(i,j+1)}(x^v(i, j + 1)) \\ \leq V_{\mu(i,j)}(x(i, j)) - c_0 x^\top(i, j)x(i, j), \quad (i, j) \in \mathbb{N}_0^2. \end{aligned} \tag{9.14}$$

For any integer $q \geq 1$, by taking (9.14) into account, we then obtain

$$\begin{aligned} E(q + 1) &= \sum_{i+j=q+1} V_{\mu(i,j)}(x(i, j)) \\ &= \sum_{i+j=q} \left[V^h_{\mu(i+1,j)}(x^h(i + 1, j)) + V^v_{\mu(i,j+1)}(x^v(i, j + 1))\right] \\ &\quad + V^h_{\mu(0,q+1)}(x^h(0, q + 1)) + V^v_{\mu(q+1,0)}(x^v(q + 1, 0)) \\ &\leq \sum_{i+j=q} V_{\mu(i,j)}(x(i, j)) - c_0 \sum_{i+j=q} \|x(i, j)\|^2 \\ &\quad + V^h_{\mu(0,q+1)}(x^h(0, q + 1)) + V^v_{\mu(q+1,0)}(x^v(q + 1, 0)) \\ &\leq E(q) + F(q + 1) - c_0 \sum_{i+j=q} \|x(i, j)\|^2, \end{aligned}$$

where $F(q) = V^h_{\mu(0,q)}(x^h(0, q)) + V^v_{\mu(q,0)}(x^v(q, 0))$. By using the variation formula of linear difference equations, it follows from the inequality

$$E(k + 1) \leq E(k) + F(k + 1) - c_0 \sum_{i+j=k} \|x(i, j)\|^2$$

that

$$E(q) \leq E(0) + \sum_{k=1}^{q} F(k) - c_0 \sum_{k=0}^{q-1} \sum_{i+j=k} \|x(i, j)\|^2. \tag{9.15}$$

for any integer $q \geq 1$. From (9.12) and (9.15), we get

$$c_0 \sum_{k=0}^{q-1} \sum_{i+j=k} \|x(i,j)\|^2 + E(q) \le \max(c_2, c_4) \sum_{k=0}^{q} \left(\hat{\phi}_h^2(k) + \hat{\phi}_v^2(k) \right)$$

and therefore,

$$\min\left(1, \frac{c_0}{\max(c_2, c_4)}\right) \sum_{k=0}^{q} E(k) \le \max(c_2, c_4) \sum_{k=0}^{\infty} \left(\hat{\phi}_h^2(k) + \hat{\phi}_v^2(k) \right) < \infty. \tag{9.16}$$

It can be deduced from (9.16) that $\sum_{k=0}^{\infty} E(k)$ is finite which implies

$$\lim_{T\to\infty} \sum_{i+j=T} V_{\mu(i,j)}(x(i,j)) = 0$$

and the 2-D fuzzy system (9.7) is GAs. The proof is completed. □

Remark 9.1 Lemma 9.1 provides a useful tool to derive stability conditions for 2-D fuzzy systems with delays in the form of (9.7). As discussed in Chap. 2, to derive less conservative delay-dependent stability conditions, not only rule-based LKF candidates $V_\mu(x)$ but also the bounding techniques used to manipulate the difference of $V_\mu(x)$ along state trajectories of system (9.7) to meet the derived conditions in Lemma 9.1 are essential.

Remark 9.2 As a special case, for common LKF, that is, $V_\mu(x) = V(x)$, the result of Lemma 9.1 is reduced to that of Lemma 2.1 in Chap. 2. In this regard, Lemma 9.1 can be viewed as an extension of Lemma 2.1 to adapt with parameter-dependent 2-D systems with delays in the form of (9.7).

9.3.2 Summation Inequalities

In this section, we present a novel approach to derive some useful summation inequalities, which are utilized to derive stability and stabilization conditions for 2-D time-delay systems. For given integers $a < b$ and matrices $R_k \in \mathbb{S}_+^n$, $k \in \mathbb{Z}[a,b]$, since $\sum_{k=a}^{b} R_k$ is also a positive definite matrix, we can define the matrix R_H as

$$R_H = \left(\sum_{k=a}^{b} R_k^{-1} \right)^{-1},$$

which also belongs to $\mathbb{S}_n^+$. Now, for vectors $\chi_k \in \mathbb{R}^n$, we denote the vectors

$$\chi_s = \sum_{k=a}^{b} \chi_k, \quad \hat{\chi}_k = \chi_k - R_k^{-1} R_H \chi_s.$$

It is easy to verify that

$$\sum_{k=a}^{b} \chi_k^\top R_k \chi_k = \chi_s^\top R_H \chi_s + \sum_{k=a}^{b} \hat{\chi}_k^\top R_k \hat{\chi}_k. \tag{9.17}$$

As a consequence of (9.17), we have the following results.

Lemma 9.2 (Generalized Jensen inequality) *For any vector-valued function $\chi : \mathbb{Z}[a,b] \to \mathbb{R}^n$, $k \mapsto \chi_k$, and matrix-valued function $R : \mathbb{Z}[a,b] \to \mathbb{S}_n^+$, $k \mapsto R_k$, the following inequality holds*

$$\sum_{k=a}^{b} \chi_k^\top R_k \chi_k \geq \left(\sum_{k=a}^{b} \chi_k\right)^\top R_H \left(\sum_{k=a}^{b} \chi_k\right). \tag{9.18}$$

Proof The proof is straightforward from (9.17). We omit it here. □

Remark 9.3 It is interesting to note that the summation inequality (9.18) encompasses the Jensen-based summation inequality as a special case. More specifically, for a constant matrix-valued function $R_k = R$, $k \in \mathbb{Z}[a,b]$, we have

$$R_H = \left((b-a+1)R^{-1}\right)^{-1} = \frac{1}{b-a+1} R.$$

Thus, it follows from (9.18) that

$$\sum_{k=a}^{b} \chi_k^\top R \chi_k \geq \left(\sum_{k=a}^{b} \chi_k\right) R \left(\sum_{k=a}^{b} \chi_k\right) \text{ (Jensen's form).} \tag{9.19}$$

Corollary 9.1 (Jensen summation inequalities) *For a vector-valued function $w : \mathbb{Z}[a,b] \times [c,d] \to \mathbb{R}^n$ and a matrix $R \in \mathbb{S}_n^+$, the following summation inequalities hold*

$$\begin{aligned} \sum_{i=a}^{b} w^\top(i,j) R w(i,j) &\geq \frac{1}{b-a+1} \left(\sum_{i=a}^{b} w(i,j)\right)^\top R \left(\sum_{i=a}^{b} w(i,j)\right), \\ \sum_{j=c}^{d} w^\top(i,j) R w(i,j) &\geq \frac{1}{d-c+1} \left(\sum_{j=c}^{d} w(i,j)\right)^\top R \left(\sum_{j=a}^{b} w(i,j)\right). \end{aligned} \tag{9.20}$$

Lemma 9.3 (Multiple kernels summation inequality) *With the assumption of Lemma 9.2, the following inequality holds*

$$\sum_{k=a}^{b} \chi_k^\top R_k \chi_k \geq \chi_s^\top R_H \chi_s + \frac{3(l+1)}{l(l-1)} \widetilde{\chi}_s^\top \widetilde{\chi}_s, \tag{9.21}$$

where $l = l_{\mathbb{Z}}(a,b) = b - a + 1$ *and*

$$\widetilde{\chi}_s = \sum_{k=a}^{b} R_k^{\frac{1}{2}} \hat{\chi}_k - \frac{2}{l+1} \sum_{k=a}^{b} \sum_{l=a}^{k} R_l^{\frac{1}{2}} \hat{\chi}_l.$$

Proof Let $Z_k = R_k^{\frac{1}{2}} \hat{\chi}_k$ and $Z_s = \sum_{k=a}^{b} Z_k$. We define the vector

$$\hat{Z}_k = Z_k - \frac{1}{l} Z_s - \frac{6}{l(l-1)}(k - \frac{a+b}{2})\widetilde{\chi}_s.$$

Then, it can be verified by some direct computation that

$$\sum_{k=a}^{b} \|\hat{Z}_k\|^2 = \sum_{k=a}^{b} \|Z_k\|^2 - \|Z_s\|^2 - \frac{3(l+1)}{l(l-1)} \|\widetilde{\chi}_s\|^2.$$

This fact yields

$$\sum_{k=a}^{b} \hat{\chi}_k^{\top} R_k \hat{\chi}_k = \sum_{k=a}^{b} \|Z_k\|^2 \geq \frac{3(l+1)}{l(l-1)} \widetilde{\chi}_s^{\top} \widetilde{\chi}_s,$$

which, together (9.17), validates (9.21). The proof is completed. □

For a constant valued function $R_k = R$, we have

$$\widetilde{\chi}_s = R^{\frac{1}{2}} \left(\chi_s - \frac{2}{l+1} \chi_d \right),$$

where $\chi_d = \sum_{k=a}^{b} \sum_{t=a}^{k} \chi_t$. From (9.21) we have the following result.

Lemma 9.4 *For a matrix* $R \in \mathbb{S}_+^n$, *integers* $a < b$ *and vectors* $\chi_k \in \mathbb{R}^n$, $k \in \mathbb{Z}[a,b]$, *the following inequality holds*

$$\sum_{k=a}^{b} \chi_k^{\top} R \chi_k \geq \frac{1}{l} \left(\sum_{k=a}^{b} \chi_k \right)^{\top} R \left(\sum_{k=a}^{b} \chi_k \right) + \frac{3(l+1)}{l(l-1)} \check{\chi}^{\top} R \check{\chi}, \tag{9.22}$$

where $l = b - a + 1$ *and* $\check{\chi} = \sum_{k=a}^{b} \chi_k - \frac{2}{l+1} \sum_{k=a}^{b} \sum_{t=a}^{k} \chi_t$.

Remark 9.4 Lemma 9.4 encompasses the summation inequalities derived from discrete Wirtinger inequality [37, 38], Abel-lemma or Cauchy-Schwarz inequality. As shown in Lemma 9.4, an alternative simple proof is presented for a more general version of summation inequality with multiple kernels (9.21) which induces (9.22) as a very special case.

By simplifying the inequality (9.22) we obtain the following corollary, which makes the results of (9.22) easier to utilize for 2-D systems with delays.

Corollary 9.2 *Given positive integers d_h, d_v. For any matrix $R \in \mathbb{S}_n^+$ and function $w : \mathbb{Z}[-d_h, 0] \times [-d_v, 0] \to \mathbb{R}^n$, the following inequalities hold*

$$\begin{aligned} \sum_{i=-d_h}^{-1} \Delta_h w^\top(i, j) R \Delta_h w(i, j) &\geq \frac{1}{d_h} \begin{bmatrix} w_{h1} \\ w_{h2} \end{bmatrix}^\top \begin{bmatrix} R & 0 \\ * & 3R \end{bmatrix} \begin{bmatrix} w_{h1} \\ w_{h2} \end{bmatrix}, \\ \sum_{j=-d_v}^{-1} \Delta_v w^\top(i, j) R \Delta_v w(i, j) &\geq \frac{1}{d_v} \begin{bmatrix} w_{h1} \\ w_{h2} \end{bmatrix}^\top \begin{bmatrix} R & 0 \\ * & 3R \end{bmatrix} \begin{bmatrix} w_{h1} \\ w_{h2} \end{bmatrix}, \end{aligned} \tag{9.23}$$

where

$$\begin{aligned} w_{h1} &= w(0, j) - w(-d_h, j), \\ w_{h2} &= w(0, j) + w(-d_h, j) - \frac{2}{d_h + 1} \sum_{i=-d_h}^{0} w(i, j), \\ w_{v1} &= w(i, 0) - w(i, -d_v), \\ w_{v2} &= w(i, 0) + w(i, -d_v) - \frac{2}{d_v + 1} \sum_{j=-d_v}^{0} w(i, j). \end{aligned}$$

9.3.3 Reciprocally Convex Inequalities

The following matrix inequality is a useful tool to deal with the interval time-varying delays. For given matrices $R_k \in \mathbb{S}_n^+$, $k = 1, 2$, a matrix X of appropriate dimension and vectors ζ_1, ζ_2, it is easy to see that

$$\frac{1}{\alpha} \zeta_1^\top R_1 \zeta_1 + \frac{1}{\beta} \zeta_2^\top R_2 \zeta_2 = \zeta^\top \begin{bmatrix} R_1 & X \\ X^\top & R_2 \end{bmatrix} \zeta + \zeta_{\alpha,\beta}^\top \begin{bmatrix} R_1 & X \\ X^\top & R_2 \end{bmatrix} \zeta_{\alpha,\beta} \tag{9.24}$$

holds for all $\alpha, \beta \in (0, 1)$, $\alpha + \beta = 1$, where

$$\zeta = \begin{bmatrix} \zeta_1 \\ \zeta_2 \end{bmatrix}, \quad \zeta_{\alpha\beta} = \begin{bmatrix} \sqrt{\frac{\beta}{\alpha}} \zeta_1 \\ -\sqrt{\frac{\alpha}{\beta}} \zeta_2 \end{bmatrix}.$$

Consequently, if $\begin{bmatrix} R_1 & X \\ X^\top & R_2 \end{bmatrix} \geq 0$ then

$$\frac{1}{\alpha}\zeta_1^\top R_1\zeta_1 + \frac{1}{1-\alpha}\zeta_2^\top R_2\zeta_2 \geq \begin{bmatrix}\zeta_1\\ \zeta_2\end{bmatrix}^\top \begin{bmatrix}R_1 & X\\ X^\top & R_2\end{bmatrix}\begin{bmatrix}\zeta_1\\ \zeta_2\end{bmatrix} \tag{9.25}$$

for all $\alpha \in (0, 1)$.

9.3.4 Matrix Inequalities

The following equivalent matrix inequalities will be utilized for zero-type equations.

Lemma 9.5 *For any matrices $A^\top = A \in \mathbb{R}^{n\times n}$, $B \in \mathbb{R}^{m\times n}$ with* $\mathrm{rank}(B) < n$, *$C \in \mathbb{S}_m^+$ and any matrix $X \in \mathbb{R}^{n\times m}$, the following statements are equivalent*

(i) *Finsler's lemma [40]: $x^\top Ax < 0$ for all $x \neq 0$ subject to $Bx = 0$ if and only if*

$$(B^\perp)^\top AB^\perp < 0 \Longleftrightarrow A + XB + B^\top X^\top < 0.$$

(ii) *$A - B^\top CB < 0$ if and only if*

$$\begin{bmatrix}A + B^\top X + X^\top B & *\\ X & -C\end{bmatrix} < 0.$$

Based on Schur decomposition, the following lemma provides a useful tool to linearize matrix inequalities,

Lemma 9.6 (Schur complement lemma) *For given matrices $X = X^\top$, $Z = Z^\top$, and a matrix Y of appropriate dimensions, it holds that*

$$\begin{bmatrix}X & Y\\ Y^\top & Z\end{bmatrix} > 0 \Longleftrightarrow \begin{cases}Z > 0\\ X - YZ^{-1}Y^\top > 0\end{cases}$$
$$\Longleftrightarrow \begin{cases}X > 0\\ Z - Y^\top X^{-1}Y > 0.\end{cases}$$

9.4 Stability Analysis

In this section, based on the scheme presented in Lemma 9.1 and by utilizing the formulated summation inequalities, we derive delay-dependent stability conditions for the following 2-D fuzzy systems with interval delays

$$\begin{bmatrix}x^h(i+1, j)\\ x^v(i, j+1)\end{bmatrix} = A(\mu)\begin{bmatrix}x^h(i, j)\\ x^v(i, j)\end{bmatrix} + A_d(\mu)\begin{bmatrix}x^h(i - d_h(i), j)\\ x^v(i, j - d_v(j))\end{bmatrix}. \tag{9.26}$$

According to the analysis scheme proposed in Lemma 9.1, we construct the following fuzzy parameter-dependent LKF candidate

$$V_\mu(x) = \underbrace{\tilde{x}^\top P(\mu)\tilde{x}}_{V_\mu^0(x)} + \sum_{k=1}^{4} V_k(x), \tag{9.27}$$

where $\tilde{x} = \left[\tilde{x}^{h\top}\ \tilde{x}^{v\top}\right]^\top$ and

$$P(\mu) = \sum_{k=1}^{N_r} \mu_k P_k = \sum_{k=1}^{N_r} \mu_k \operatorname{diag}(P_k^h, P_k^v) = \operatorname{diag}(P^h(\mu), P^v(\mu)),$$

$$\tilde{x}^h(i,j) = \begin{bmatrix} x^h(i,j) \\ \sum_{s=i-d_1^h}^{i-1} x^h(s,j) \\ \sum_{t=i-d_2^h}^{i-d_1^h-1} x^h(t,j) \end{bmatrix}, \quad \tilde{x}^v(i,j) = \begin{bmatrix} x^v(i,j) \\ \sum_{s=j-d_1^v}^{j-1} x^v(i,s) \\ \sum_{t=j-d_2^v}^{j-d_1^v-1} x^v(i,t) \end{bmatrix},$$

$$V_1(x(i,j)) = \sum_{k=i-d_1^h}^{i-1} x^{h\top}(k,j) Q_1^h x^h(k,j) + \sum_{l=j-d_1^v}^{j-1} x^{v\top}(i,l) Q_1^v x^v(i,l),$$

$$V_2(x(i,j)) = \sum_{k=i-d_2^h}^{i-d_1^h-1} x^{h\top}(k,j) Q_2^h x^h(k,j) + \sum_{l=j-d_2^v}^{j-d_1^v-1} x^{v\top}(i,l) Q_2^v x^v(i,l),$$

$$V_3(x(i,j)) = d_1^h \sum_{k=-d_1^h}^{-1} \sum_{l=i+k}^{i-1} \eta_h^\top(l,j) R_1^h \eta_h(l,j) + d_1^v \sum_{k=-d_1^v}^{-1} \sum_{l=j+k}^{j-1} \eta_v^\top(i,l) R_1^v \eta_v(i,l),$$

$$V_4(x(i,j)) = d_r^h \sum_{k=-d_2^h}^{-d_1^h-1} \sum_{l=i+k}^{i-1} \eta_h^\top(l,j) R_2^h \eta_h(l,j) + d_r^v \sum_{k=-d_2^v}^{-d_1^v-1} \sum_{l=j+k}^{j-1} \eta_v x^\top(i,l) R_2^v \eta_v(i,l),$$

$$\eta_h(i,j) = \Delta_h(x^h(i,j)), \quad \eta_v(i,j) = \Delta_v(x^v(i,j)), \quad d_r^h = d_2^h - d_1^h, \quad d_r^v = d_2^v - d_1^v.$$

For the notational simplicity, we denote the matrices $e_k = \left[0_{n\times(k-1)n}\ I_n\ 0_{n\times(8-k)n}\right]$ $(k = 1, 2, \ldots, 8)$, delay terms $r_{h1}(i) = d_h(i) - d_1^h$, $r_{h2}(i) = d_2^h - d_h(i)$, $r_{v1}(j) = d_v(j) - d_1^v$, $r_{v2}(j) = d_2^v - d_v(j)$ and augmented vectors

$$\chi_x(i,j) = \begin{bmatrix} x(i,j) \\ x_l(i,j) \\ x_d(i,j) \\ x_u(i,j) \end{bmatrix}, \quad \chi_s(i,j = \begin{bmatrix} \eta(i,j) \\ x_l^s(i,j) \\ x_d^s(i,j) \\ x_u^s(i,j) \end{bmatrix}, \quad \chi(i,j) = \begin{bmatrix} \chi_x(i,j) \\ \chi_s(i,j) \end{bmatrix},$$

$$x_l(i,j) = \begin{bmatrix} x^h(i-d_1^h, j) \\ x^v(i, j-d_1^v) \end{bmatrix}, \quad x_d(i,j) = \begin{bmatrix} x^h(i-d_h(i), j) \\ x^v(i, j-d_v(j)) \end{bmatrix},$$

$$x_u(i,j) = \begin{bmatrix} x^h(i-d_2^h, j) \\ x^v(i, j-d_2^v) \end{bmatrix}, \quad \eta(i,j) = \begin{bmatrix} \Delta_h x^h(i,j) \\ \Delta_v x^v(i,j) \end{bmatrix}, \quad x_l^s(i,j) = \begin{bmatrix} \zeta_{11}(i,j) \\ \zeta_{21}(i,j) \end{bmatrix},$$

$$x_d^s(i,j) = \begin{bmatrix} \zeta_{12}(i,j) \\ \zeta_{22}(i,j) \end{bmatrix}, \quad x_u^s(i,j) = \begin{bmatrix} \zeta_{13}(i,j) \\ \zeta_{23}(i,j) \end{bmatrix},$$

$$\zeta_{11}(i,j) = \frac{1}{1+d_1^h} \sum_{l=i-d_1^h}^{i} x^h(l,j), \quad \zeta_{12}(i,j) = \frac{1}{1+r_{h1}(i)} \sum_{l=i-d_h(i)}^{i-d_1^h} x^h(l,j),$$

$$\zeta_{13}(i,j) = \frac{1}{1+r_{h2}(i)} \sum_{l=i-d_2^h}^{i-d_h(i)} x^h(l,j), \quad \zeta_{21}(i,j) = \frac{1}{1+d_1^v} \sum_{l=j-d_1^v}^{j} x^v(i,l),$$

$$\zeta_{22}(i,j) = \frac{1}{1+r_{v1}(j)} \sum_{l=j-d_v(j)}^{j-d_1^v} x^v(i,l), \quad \zeta_{23}(i,j) = \frac{1}{1+r_{v2}(j)} \sum_{l=j-d_2^v}^{j-d_v(j)} x^v(i,l).$$

We also denote $\mathbb{I}(\alpha,\beta) = \mathrm{diag}\left((1+\alpha)I_{n_h}, (1+\beta)I_{n_v}\right)$ for scalars α, β, and matrices

$$\Gamma = \begin{bmatrix} I_{n_h} & 0 & 0 & 0 & 0 & 0 \\ 0 & 0 & 0 & I_{n_v} & 0 & 0 \\ 0 & I_{n_h} & 0 & 0 & 0 & 0 \\ 0 & 0 & 0 & 0 & I_{n_v} & 0 \\ 0 & 0 & I_{n_h} & 0 & 0 & 0 \\ 0 & 0 & 0 & 0 & 0 & I_{n_v} \end{bmatrix} \in \mathbb{R}^{3n\times 3n},$$

$$\Pi_1(d_h, d_v) = \begin{bmatrix} e_1 + e_5 \\ \mathbb{I}(d_1^h, d_1^v)e_6 - e_2 \\ \mathbb{I}(r_{h1}, r_{v1})e_7 + \mathbb{I}(r_{h2}, r_{v2})e_8 - e_3 - e_4 \end{bmatrix},$$

$$\Pi_2(d_h, d_v) = \begin{bmatrix} e_1 \\ \mathbb{I}(d_1^h, d_1^v)e_6 - e_1 \\ \mathbb{I}(r_{h1}, r_{v1})e_7 + \mathbb{I}(r_{h2}, r_{v2})e_8 - e_2 - e_3 \end{bmatrix},$$

$$\Omega_{k,q}^0 = \Pi_1^\top(d_h, d_v)\Gamma \mathscr{P}_{k,q}\Gamma^\top \Pi_1(d_h, d_v), \quad \mathscr{P}_{k,q} = \mathrm{diag}(P_k^h, P_q^v),$$

$$\Lambda_m = \Pi_2^\top(d_h, d_v)\Gamma P_m \Gamma^\top \Pi_2(d_h, d_v), \quad P_m = \mathrm{diag}(P_m^h, P_m^v),$$

$$\Upsilon_m = (A_m - I_n)e_1 + A_{dm}e_3 - e_5,$$

$$\Omega_1 = e_1^\top Q_1 e_1 - e_2^\top Q_1 e_2 + e_2^\top Q_2 e_2 - e_4^\top Q_2 e_4, \quad \Omega_2 = e_5^\top \widehat{R}_1 e_5 - F_1^\top \widetilde{R}_1 F_1,$$

$$\Omega_3 = e_5^\top \widehat{R}_2 e_5 - F^\top \left(\mathbb{J}_{n_h}^{4\top} \Psi_X^h \mathbb{J}_{n_h}^4 + \mathbb{J}_{n_v}^{4\top} \Psi_Y^v \mathbb{J}_{n_v}^4\right) F,$$

$$\widehat{R}_1 = \mathrm{diag}\left((d_1^h)^2 R_1^h, (d_1^v)^2 R_1^v\right), \quad \widetilde{R}_1 = \mathrm{diag}(R_1, 3R_1),$$

$$\widehat{R}_2 = \mathrm{diag}\left((d_r^h)^2 R_2^h, (d_r^v)^2 R_2^v\right), \quad \widetilde{R}_2^h = \mathrm{diag}(R_2^h, 3R_2^h), \quad \widetilde{R}_2^v = \mathrm{diag}(R_2^v, 3R_2^v),$$

$$F_1 = \begin{bmatrix} e_1 - e_2 \\ e_1 + e_2 - 2e_6 \end{bmatrix}, \quad F_2 = \begin{bmatrix} e_2 - e_3 \\ e_2 + e_3 - 2e_7 \end{bmatrix}, \quad F_3 = \begin{bmatrix} e_3 - e_4 \\ e_3 + e_4 - 2e_8 \end{bmatrix},$$

$$F = \begin{bmatrix} F_2 \\ F_3 \end{bmatrix}, \quad \Psi_X^h = \begin{bmatrix} \widetilde{R}_2^h & X \\ * & \widetilde{R}_2^h \end{bmatrix}, \quad \Psi_Y^v = \begin{bmatrix} \widetilde{R}_2^v & Y \\ * & \widetilde{R}_2^v \end{bmatrix},$$

$$\mathbb{J}^k_{n_h} = \underbrace{\operatorname{diag}\left([I_{n_h} \quad 0_{n_h \times n_v}], \ldots, [I_{n_h} \quad 0_{n_h \times n_v}]\right)}_{k \text{ blocks}},$$
$$\mathbb{J}^k_{n_v} = \underbrace{\operatorname{diag}\left([0_{n_v \times n_h} \quad I_{n_v}], \ldots, [0_{n_v \times n_h} \quad I_{n_v}]\right)}_{k \text{ blocks}}$$

By using the fuzzy rule-based functional (9.27), and by utilizing the scheme presented in Lemma 9.1, delay-dependent stability conditions for system (9.26) are given as in the following theorem.

Theorem 9.1 *Consider fuzzy systems in* (9.26) *with interval delays* $d_h(i) \in [d_1^h, d_2^h]$ *and* $d_v(j) \in [d_1^v, d_2^v]$. *Assume that there exist matrices* $P_k = \operatorname{diag}\left(P_k^h, P_k^v\right)$, $P_k^h \in \mathbb{S}^+_{3n_h}$, $P_k^v \in \mathbb{S}^+_{3n_v}$ $(k \in \mathscr{R})$, $Q_s = \operatorname{diag}(Q_s^h, Q_s^v) \in \mathbb{S}^+_n$, $R_s = \operatorname{diag}(R_s^h, R_s^v) \in \mathbb{S}^+_n$ $(s = 1, 2)$, *real matrices* $X \in \mathbb{R}^{2n_h \times 2n_h}$, $Y \in \mathbb{R}^{2n_v \times 2n_v}$ *and a matrix* $Z \in \mathbb{R}^{3n \times 7n}$ *such that*

$$\Psi_X^h = \begin{bmatrix} \widetilde{R}_2^h & X \\ X^\top & \widetilde{R}_2^h \end{bmatrix} \geq 0, \quad \Psi_Y^v = \begin{bmatrix} \widetilde{R}_2^v & Y \\ Y^\top & \widetilde{R}_2^v \end{bmatrix} \geq 0, \tag{9.28}$$

and the following LMI-based condition holds for $d_h \in \{d_1^h, d_2^h\}$, $d_v \in \{d_1^v, d_2^v\}$,

$$\left[\begin{array}{c:c} \begin{array}{c} \left(\Upsilon_m^\perp\right)^\top \Phi_{k,q} \Upsilon_m^\perp \\ + \operatorname{Sym}\left\{Z^\top \Pi_2(d_h, d_v) \Upsilon_m^\perp\right\} \end{array} & Z^\top \\ \hdashline * & -\Gamma P_m \Gamma^\top \end{array}\right] < 0, \tag{9.29}$$

where $\Phi_{k,q} = \Omega_{k,q}^0 + \sum_{s=1}^3 \Omega_s$. *Then, system* (9.26) *is GAS.*

Proof It is clear that condition (i) of Lemma 9.1 is fulfilled. We only need to validate condition (ii). Let $V_\mu^{0h} = \tilde{x}^{h\top}(i, j) P^h(\mu) \tilde{x}^h(i, j)$ and $V_\mu^{0v} = \tilde{x}^{v\top}(i, j) P^v(\mu) \tilde{x}^v(i, j)$. The difference of V_μ^0 along state trajectory $x(i, j)$ of system (9.26) is given by

$$\Delta V_\mu^0 \triangleq \Delta V^0_{\mu(i,j)}(x(i, j)) = \underbrace{\Delta_h \left(V^{0h}_{\mu(i,j)}(x^h(i, j))\right)}_{\Delta_h V_\mu^{0h}} + \underbrace{\Delta_v \left(V^{0v}_{\mu(i,j)}(x^v(i, j))\right)}_{\Delta_v V_\mu^{0v}}, \tag{9.30}$$

where

$$\Delta_h V_\mu^{0h} = \tilde{x}^{h\top}(i+1, j) P_1^h(\mu) \tilde{x}^h(i+1, j) - \tilde{x}^{h\top}(i, j) P^h(\mu) \tilde{x}^h(i, j),$$
$$\Delta_v V_\mu^{0v} = \tilde{x}^{v\top}(i, j+1) P_2^v(\mu) \tilde{x}^h(i, j+1) - \tilde{x}^{v\top}(i, j) P^v(\mu) \tilde{x}^h(i, j),$$

and

$$P_1^h(\mu) = \sum_{k=1}^{N_r} \mu_{1k} P_k^h, \quad P_2^v(\mu) = \sum_{k=1}^{N_r} \mu_{2k} P_k^v,$$
$$\mu_{1k} = \mu_k(\vartheta(i+1, j)), \quad \mu_{2k} = \mu_k(\vartheta(i, j+1)).$$

Hereafter, for simplicity, we will write the vector $\chi(i, j)$ as χ. Note at first that the state transition matrix Γ satisfies $\Gamma^\top = \Gamma^{-1}$ and we have the following vector representation

$$\Gamma \begin{bmatrix} \tilde{x}^h(i+1, j) \\ \tilde{x}^v(i, j+1) \end{bmatrix} = \Pi_1(d_h, d_v)\chi, \quad \Gamma \begin{bmatrix} \tilde{x}^h(i, j) \\ \tilde{x}^v(i, j) \end{bmatrix} = \Pi_2(d_h, d_v)\chi. \tag{9.31}$$

Therefore, from (9.30) and (9.31), we have

$$\Delta V_\mu^0 = \sum_{k=1}^{N_r}\sum_{q=1}^{N_r}\sum_{m=1}^{N_r} \mu_{1k}\mu_{2q}\mu_m \chi^\top \left(\Omega_{k,q}^0 - \Lambda_m\right)\chi. \tag{9.32}$$

Next, the difference $\Delta V_1(x)$ is given by

$$\begin{aligned} \Delta V_1 &= x^{h\top}(i, j) Q_1^h x^h(i, j) - x^{h\top}(i - d_1^h, j) Q_1^h x^h(i - d_1^h, j) \\ &\quad + x^{v\top}(i, j) Q_1^v x^v(i, j) - x^{v\top}(i, j - d_1^v) Q_1^v x^v(i, j - d_1^v) \\ &= \chi^\top \left(e_1^\top Q_1 e_1 - e_2^\top Q_1 e_2\right)\chi. \end{aligned} \tag{9.33}$$

Similarly, we also have

$$\begin{aligned} \Delta V_2 &= x^{h\top}(i - d_1^h, j) Q_2^h x^h(i - d_1^h, j) - x^{h\top}(i - d_2^h, j) Q_2^h x^h(i - d_2^h, j) \\ &\quad + x^{v\top}(i, j - d_1^v) Q_2^v x^v(i, j - d_1^v) - x^{v\top}(i, j - d_2^v) Q_2^v x^v(i, j - d_2^v) \\ &= \chi^\top \left(e_2^\top Q_2 e_2 - e_4^\top Q_2 e_4\right)\chi. \end{aligned} \tag{9.34}$$

A combination of (9.33) and (9.34) gives

$$\Delta(V_1 + V_2) = \chi^\top \Omega_1 \chi. \tag{9.35}$$

We now compute the difference ΔV_3 and utilize the inequalities derived in Corollary 9.2 to manipulate ΔV_3. Observe that

$$\eta(i, j) = \begin{bmatrix} \Delta_h x^h(i, j) \\ \Delta_v x^v(i, j) \end{bmatrix} = e_5 \chi(i, j).$$

Therefore,

$$\Delta V_3 = \chi^\top e_5^\top \widehat{R}_1 e_5 \chi - d_1^h \sum_{l=i-d_1^h}^{i-1} \eta_h^\top(l, j) R_1^h \eta_h(l, j) - d_1^v \sum_{l=j-d_1^v}^{j-1} \eta_v^\top(i, l) R_1^v \eta_v(i, l). \tag{9.36}$$

By utilizing Corollary 9.2, the summation terms in (9.36) will be manipulated as

$$\begin{aligned} -d_1^h \sum_{l=i-d_1^h}^{i-1} \eta_h^\top(l, j) R_1^h \eta_h(l, j) &\le -\xi_{h1}^\top R_1^h \xi_{h1} - 3\xi_{h2}^\top R_1^h \xi_{h2}, \\ -d_1^v \sum_{l=j-d_1^v}^{j-1} \eta_v^\top(i, l) R_1^v \eta_v(i, l) &\le -\xi_{v1}^\top R_1^v \xi_{v1} - 3\xi_{v2}^\top R_1^v \xi_{v2}, \end{aligned} \tag{9.37}$$

where

$$\begin{aligned} \xi_{h1} &= x^h(i, j) - x^h(i - d_1^h, j), \quad \xi_{h2} = x^h(i, j) + x^h(i - d_1^h, j) - 2\zeta_{11}(i, j), \\ \xi_{v1} &= x^v(i, j) - x^v(i, j - d_1^v), \quad \xi_{v2} = x^v(i, j) + x^v(i, j - d_1^v) - 2\zeta_{21}(i, j). \end{aligned}$$

It follows from (9.36) and (9.37) that

$$\Delta V_3 \le \chi^\top \left(e_5^\top \widehat{R}_1 e_5 - F_1 \widetilde{R}_1 F_1\right) \chi. \tag{9.38}$$

Similar to (9.36), the difference of V_4 along state trajectory $x(i, j)$ is given by

$$\Delta V_4 = \chi^\top e_5^\top \widehat{R}_2 e_5 \chi - d_r^h \sum_{l=i-d_2^h}^{i-d_1^h-1} \eta_h^\top(l, j) R_2^h \eta_h(l, j) - d_r^v \sum_{l=j-d_2^v}^{j-d_1^v-1} \eta_v^\top(i, l) R_2^v \eta_v(i, l). \tag{9.39}$$

If $d_1^h < d_h(i) < d_2^h$ then, by utilizing the reciprocally convex inequality (see, Sect. 9.3.3), we have

$$
\begin{aligned}
-d_r^h \sum_{l=i-d_2^h}^{i-d_1^h-1} \eta_h^\top(l,j) R_2^h \eta_h(l,j) &= -d_r^h \sum_{l=i-d_2^h}^{i-d_h(i)-1} \eta_h^\top(l,j) R_2^h \eta_h(l,j) \\
&\quad - d_r^h \sum_{l=i-d_h(i)}^{i-d_1^h-1} \eta_h^\top(l,j) R_2^h \eta_h(l,j) \\
&\leq -\frac{d_r^h}{r_{h1}(i)} \tilde{\xi}_{h1}^\top \widetilde{R}_2^h \tilde{\xi}_{h1} - \frac{d_r^h}{r_{h2}(i)} \tilde{\xi}_{h2}^\top \widetilde{R}_2^h \tilde{\xi}_{h2} \\
&\leq -\begin{bmatrix} \tilde{\xi}_{h1} \\ \tilde{\xi}_{h2} \end{bmatrix}^\top \begin{bmatrix} \widetilde{R}_2^h & X \\ * & \widetilde{R}_2^h \end{bmatrix} \begin{bmatrix} \tilde{\xi}_{h1} \\ \tilde{\xi}_{h2} \end{bmatrix},
\end{aligned} \tag{9.40}
$$

where

$$
\begin{aligned}
\tilde{\xi}_{h1} &= \begin{bmatrix} x^h(i-d_1^h, j) - x^h(i-d_h(i), j) \\ x^h(i-d_1^h, j) + x^h(i-d_h(i), j) - 2\zeta_{12} \end{bmatrix}, \\
\tilde{\xi}_{h2} &= \begin{bmatrix} x^h(i-d_h(i), j) - x^h(i-d_2^h, j) \\ x^h(i-d_h(i), j) + x^h(i-d_2^h, j) - 2\zeta_{13} \end{bmatrix}.
\end{aligned}
$$

If $d_h(i) = d_1^h$ the $\tilde{\xi}_{h1} = 0$ and if $d_h(i) = d_2^h$, $\tilde{\xi}_{h2} = 0$. This indicates that the last inequality in (9.40) is still valid.

Similarly, we also have

$$
- d_r^v \sum_{l=j-d_2^v}^{j-d_1^v-1} \eta_v^\top(i,l) R_2^v \eta_v(i,l) \leq -\begin{bmatrix} \tilde{\xi}_{v1} \\ \tilde{\xi}_{v2} \end{bmatrix}^\top \begin{bmatrix} \widetilde{R}_2^v & Y \\ * & \widetilde{R}_2^v \end{bmatrix} \begin{bmatrix} \tilde{\xi}_{v1} \\ \tilde{\xi}_{v2} \end{bmatrix}, \tag{9.41}
$$

where

$$
\begin{aligned}
\tilde{\xi}_{v1} &= \begin{bmatrix} x^v(i, j-d_1^v) - x^v(i, j-d_v(j)) \\ x^v(i, j-d_1^v) + x^v(i, j-d_v(j)) - 2\zeta_{22} \end{bmatrix} \\
\tilde{\xi}_{v2} &= \begin{bmatrix} x^v(i, j-d_v(j)) - x^v(i, j-d_2^v) \\ x^v(i, j-d_v(j)) + x^v(i, j-d_2^v) - 2\eta_{23} \end{bmatrix}.
\end{aligned}
$$

Note also that

$$
\begin{aligned}
&\begin{bmatrix} \tilde{\xi}_{h1} \\ \tilde{\xi}_{h2} \end{bmatrix} = (J_{n_h} F)\chi, \quad \begin{bmatrix} \tilde{\xi}_{v1} \\ \tilde{\xi}_{v2} \end{bmatrix} = (J_{n_v} F)\chi, \\
&J_{n_h} = [I_{n_h} \quad 0_{n_h \times n_v}], \quad J_{n_v} = [0_{n_v \times n_h} \quad I_{n_v}].
\end{aligned}
$$

Therefore, from (9.39)–(9.41), we have

$$
\Delta V_4 \leq \chi^\top \left[e_5^\top \widehat{R}_2 e_5 - F^\top \left(\mathbb{J}_{n_h}^{4\top} \Psi_X^h \mathbb{J}_{n_h}^4 + \mathbb{J}_{n_v}^{4\top} \Psi_Y^v \mathbb{J}_{n_v}^4 \right) F \right] \chi. \tag{9.42}
$$

From (9.32), (9.35), (9.38), and (9.42), and by using S-procedure [41], we obtain the estimate of the difference ΔV

$$\Delta V_{\mu}(x) \leq \sum_{k=1}^{N_r}\sum_{q=1}^{N_r}\sum_{m=1}^{N_r} \mu_{1k}\mu_{2l}\mu_m \chi^\top \big(\Phi_{k,q} - \Lambda_m\big)\chi. \tag{9.43}$$

Note from (9.26) that

$$\Upsilon_m \chi = 0, \quad m \in \mathscr{R}.$$

Thus, by Lemma 9.1, stability conditions for system (9.26) are derived from the following condition

$$\sum_{k=1}^{N_r}\sum_{q=1}^{N_r}\sum_{m=1}^{N_r} \mu_{1k}\mu_{2l}\mu_m \chi^\top \left(\Phi_{k,q} - \Lambda_m\right)\chi < 0 \;\; s.t. \sum_{m=1}^{N_r} \mu_m \Upsilon_m \chi = 0. \tag{9.44}$$

By Finsler lemma, condition (9.44) holds if and only if

$$\sum_{k=1}^{N_r}\sum_{q=1}^{N_r}\sum_{m=1}^{N_r} \mu_{1k}\mu_{2l}\mu_m \left(\Phi_{k,q} - \Lambda_m + \mathrm{Sym}\{\mathscr{F}\Upsilon_m\}\right) < 0 \tag{9.45}$$

for any matrix $\mathscr{F}$ with appropriate dimension. By the properties of fuzzy basic functions μ_k, condition (9.45) holds if for any k, q, m,

$$\Phi_{k,q} - \Lambda_m + \mathrm{Sym}\{\mathscr{F}\Upsilon_m\} < 0,$$

which, by Lemma 9.5, is equivalent to the following condition

$$\left(\Upsilon_m^\perp\right)^\top \left(\Phi_{k,q} - \Lambda_m\right)\Upsilon_m^\perp < 0. \tag{9.46}$$

It should be pointed out that condition (9.46) is still not tractable in the sense of computation as its left-hand side is not convex with respect to delay variables d_h, d_v according to the term

$$-\left(\Upsilon_m^\perp\right)^\top \Lambda_m \Upsilon_m^\perp.$$

In order to formulate tractable condition from (9.46), we will utilize Lemma 9.5 to convert (9.46) into LMIs that are convex with respect to d_h and d_v. For this, we rewrite the inequality (9.46) in the form

$$\left(\Upsilon_m^\perp\right)^\top \Phi_{k,q}\Upsilon_m^\perp - \left(\Pi_2(d_h, d_v)\Upsilon_m^\perp\right)^\top (\Gamma P_m \Gamma^\top)\Pi_2(d_h, d_v)\Upsilon_m^\perp < 0. \tag{9.47}$$

By Lemma 9.5 (the equivalence in (ii)), condition (9.47) holds if and only if there exists a matrix Z of appropriate dimension such that

$$\mathscr{M}_{k,q,m}(d_h, d_v) \triangleq \left[\begin{array}{c:c} \begin{array}{c}\left(\Upsilon_m^{\perp}\right)^{\top} \Phi_{k,q} \Upsilon_m^{\perp} \\ + \operatorname{Sym}\left\{Z^{\top} \Pi_2(d_h, d_v) \Upsilon_m^{\perp}\right\}\end{array} & Z^{\top} \\ \hdashline * & -\Gamma P_m \Gamma^{\top} \end{array}\right] < 0.$$

It is clear that the matrix $\mathscr{M}_{k,q,m}(d_h, d_v)$ is affine and thus convex with respect to d_h and d_v. Therefore,

$$\mathscr{M}_{k,q,m}(d_h, d_v) < 0, \quad (d_h, d_v) \in \mathbb{Z}[d_1^h, d_2^h] \times [d_1^v, d_2^v]$$

if and only if the corresponding LMIs at vertices (d_h, d_v), that is,

$$\mathscr{M}_{k,q,m}(d_h, d_v)\Big|_{\substack{d_h \in \{d_1^h, d_2^h\} \\ d_v \in \{d_1^v, d_2^v\}}} < 0. \tag{9.48}$$

According to (9.48), if condition (9.29) holds for $d_h \in \{d_1^h, d_2^h\}$ and $d_v \in \{d_1^v, d^v\}$ then, together (9.28), by Lemma 9.1, the fuzzy 2-D system (9.26) is GAS. The proof is completed. □

Remark 9.5 In Theorem 9.1, the decision matrix variables are directly derived from augmented parameter-dependent LKF candidates as (9.27) and reciprocally convex inequality. Thus, not only the feasibility domain, this approach can also help to reduce the computational complexity of the derived stability conditions in comparison to, for example, the method of free-weighting matrices [24]. More precisely, the number of decision variables of the derived conditions in Theorem 9.1 is $21n^2 + (2 + 1.5N_r)n + (6 + 4.5N_r)(n_h^2 + n_v^2)$, where $n = n_h + n_v$. Thus, the proposed approach for 1-D fuzzy systems, this number is $(27 + 4.5N_r)n^2 + (2 + 1.5N_r)n$, which is clearly fewer than $(62.5 + 4.5N_r)n^2 + (4.5 + 1.5N_r)n$ delivered by the result of [24].

Remark 9.6 In the proof of Theorem 9.1, if the functional (9.27) is independent of fuzzy parameters, that is $\operatorname{diag}(P_k^h, P_k^v) = P$ for all $k \in \mathscr{R}$, condition (9.29) is reduced to the following condition

$$\left(\Upsilon_m^{\perp}\right)^{\top} \left(\widetilde{\Omega} - \widetilde{\Lambda}\right) \Upsilon_m^{\perp} < 0, \tag{9.49}$$

where

$$\widetilde{\Omega} = \Pi_1^{\top}(d_h, d_v)\Gamma P \Gamma^{\top} \Pi_1(d_h, d_v) + \sum_{s=2}^{3} \Omega_s,$$
$$\widetilde{\Lambda} = \Pi_2^{\top}(d_h, d_v)\Gamma P \Gamma^{\top} \Pi_2(d_h, d_v).$$

Similar to Theorem 9.1, from (9.49), we have the following corollary.

Corollary 9.3 *Asusme that there exist positive definite matrices* $P = \mathrm{diag}(P^h, P^v) \in \mathbb{S}^+_{3n}$, $Q_s = \mathrm{diag}(Q^h_s, Q^v_s) \in \mathbb{S}^+_n$, $R_s = \mathrm{diag}(R^h_s, R^v_s) \in \mathbb{S}^+_n$ $(s = 1, 2)$, *and real matrices* $X \in \mathbb{R}^{2n_h \times 2n_h}$, $Y \in \mathbb{R}^{2n_v \times 2n_v}$, *and* $Z \in \mathbb{R}^{3n \times 7n}$ *that satisfy condition* (9.28) *and the following LMI holds for* $m \in \mathscr{R}$, $d_h \in \{d^h_1, d^h_2\}$ *and* $d_v \in \{d^v_1, d^v_2\}$,

$$\left[\begin{array}{c|c} \begin{array}{c}\left(\Upsilon_m^{\perp}\right)^{\top} \widetilde{\Omega} \Upsilon_m^{\perp} \\ + \mathrm{Sym}\left\{Z^{\top} \Pi_2(d_h, d_v) \Upsilon_m^{\perp}\right\}\end{array} & Z^{\top} \\ \hline * & -\Gamma P \Gamma^{\top} \end{array}\right] < 0. \tag{9.50}$$

Then, system (9.26) *is GAS for any delays* $d_h(i) \in [d^h_1, d^h_2]$ *and* $d_v(j) \in [d^v_1, d^v_2]$.

Remark 9.7 To further illustrate the superior of the method presented in Theorem 9.1, we formulate an analogous result using the conventional Jensen summation inequalities. Indeed, if the Jensen inequalities in (9.20) are employed to manipulate the summations in (9.36) and (9.39), condition (9.50) is reduced to the following LMI-based condition

$$\left[\begin{array}{c|c} \begin{array}{c}\left(\Upsilon_m^{\perp}\right)^{\top} \Xi \Upsilon_m^{\perp} \\ + \mathrm{Sym}\left\{Z^{\top} \Pi_2(d_h, d_v) \Upsilon_m^{\perp}\right\}\end{array} & Z^{\top} \\ \hline * & -\Gamma P \Gamma^{\top} \end{array}\right] < 0, \tag{9.51}$$

where

$$\begin{aligned}
\Xi &= \Pi_1^{\top}(d_h, d_v) \Gamma P \Gamma^{\top} \Pi_1(d_h, d_v) + \Omega_1 + e_5^{\top}(\widehat{R}_1 + \widehat{R}_2) e_5 \\
&\quad - (e_1 - e_2)^{\top} R_1 (e_1 - e_2) - \varkappa_h^{\top} \hat{\Psi}^h_X \varkappa_h - \varkappa_v^{\top} \hat{\Psi}^v_Y \varkappa_v, \\
\varkappa_h &= \begin{bmatrix} J_{n_h}(e_2 - e_3) \\ J_{n_h}(e_3 - e_4) \end{bmatrix}, \quad \varkappa_v = \begin{bmatrix} J_{n_v}(e_2 - e_3) \\ J_{n_v}(e_3 - e_4) \end{bmatrix}, \\
\hat{\Psi}^h_X &= \begin{bmatrix} R^h_2 & \hat{X} \\ * & R^h_2 \end{bmatrix}, \quad \hat{\Psi}^v_Y = \begin{bmatrix} R^v_2 & \hat{Y} \\ * & R^v_2 \end{bmatrix}.
\end{aligned}$$

Similar to Corollary 9.3, system (9.26) is GAS if $\hat{\Psi}^h_X \geq 0$, $\hat{\Psi}^v_Y \geq 0$ and condition (9.51) holds for $d_h \in \{d^h_1, d^h_2\}$, $d_v \in \{d^v_1, d^v_2\}$,

Remark 9.8 For 2-D fuzzy systems in the form of (9.26) with constant delays, the functionals V_2 and V_4 in (9.27) will be omitted and V^0_μ is simplified as

$$\hat{V}^0_\mu = \begin{bmatrix} \hat{x}^h(i, j) \\ \hat{x}^v(i, j) \end{bmatrix}^{\top} P(\mu) \begin{bmatrix} \hat{x}^h(i, j) \\ \hat{x}^v(i, j) \end{bmatrix},$$

where

$$\hat{x}^h(i,j) = \begin{bmatrix} x^h(i,j) \\ \sum_{k=i-d_h}^{i-1} x^h(k,j) \end{bmatrix}, \quad \hat{x}^v(i,j) = \begin{bmatrix} x^v(i,j) \\ \sum_{l=j-d_v}^{v-1} x^v(i,l) \end{bmatrix}.$$

By similar lines used in the proof of Theorem 9.1, we obtain the following result.

Corollary 9.4 *For given positive integers d_h and d_v, assume that there exist matrices $P_k = \mathrm{diag}(P_k^h, P_k^v) \in \mathbb{S}_{2n}^+$ $(k \in \mathscr{R})$, $Q = \mathrm{diag}(Q^h, Q^v) \in \mathbb{S}_n^+$ and $R = \mathrm{diag}(R^h, R^v) \in \mathbb{S}_n^+$ such that the following condition holds for $k, q, m \in \mathscr{R}$*

$$\left(\hat{\Upsilon}_m^{\perp}\right)^{\top} \left(\hat{\Pi}_1^{\top} \hat{\Gamma} \mathscr{P}_{k,q} \hat{\Gamma}^{\top} \hat{\Pi}_1 - \hat{\Pi}_2^{\top} \hat{\Gamma} P_m \hat{\Gamma}^{\top} \hat{\Pi}_2 + \mathscr{H}\right) \hat{\Upsilon}_m^{\perp} < 0, \tag{9.52}$$

where

$$\hat{\Pi}_1 = \begin{bmatrix} \hat{e}_1 + \hat{e}_3 \\ \mathbb{I}(d_h, d_v)\hat{e}_4 - \hat{e}_2 \end{bmatrix}, \quad \hat{\Pi}_2 = \begin{bmatrix} \hat{e}_1 \\ \mathbb{I}(d_h, d_v)\hat{e}_4 - \hat{e}_1 \end{bmatrix},$$

$$\mathscr{H} = \hat{e}_1^{\top} Q \hat{e}_1 - \hat{e}_2^{\top} Q \hat{e}_2 + \hat{e}_3^{\top} \hat{R} \hat{e}_3 - \hat{F}^{\top} \tilde{R} \hat{F},$$

$$\hat{\Upsilon}_m = (A_m - I_n)\hat{e}_1 + A_{dm}\hat{e}_2 - \hat{e}_3,$$

$$\hat{R} = \mathrm{diag}\left(d_h^2 R^h, d_v^2 R^v\right), \quad \tilde{R} = \mathrm{diag}(R, 3R),$$

$$\hat{F} = \begin{bmatrix} \hat{e}_1 - \hat{e}_2 \\ \hat{e}_1 + \hat{e}_2 - 2\hat{e}_4 \end{bmatrix}, \quad \hat{\Gamma} = \begin{bmatrix} I_{n_h} & 0 & 0 & 0 \\ 0 & 0 & I_{n_v} & 0 \\ 0 & I_{n_h} & 0 & 0 \\ 0 & 0 & 0 & I_{n_v} \end{bmatrix},$$

$$\hat{e}_k = \left[0_{n\times(k-1)n} \; I_n \; 0_{n\times(4-k)n}\right] \in \mathbb{R}^{n\times 4n} (k = 1, 2, 3, 4).$$

Then, system (9.26) *with constant delays d_h, d_v is GAS.*

9.5 Fuzzy Rule-Based State Feedback Controller Design

In this section, the results of Theorem 9.1 and Corollary 9.3 will be utilized to derive stabilization conditions via rule-based controllers (9.9). In comparison to system (9.26), the closed-loop system (9.10) is much more complicated as the matrix $A_\mu^c = A_\mu + B_\mu F_\mu$ is involved double summation

$$A_\mu^c = \sum_{m=1}^{N_r} \sum_{s=1}^{N_r} \mu_m \mu_s \left(A_m + B_m F_s\right).$$

To obtain desired controller gains F_s, instead of using zero-type Eq. (9.44) for Υ_m, we employ the following identity

$$2\sum_{m=1}^{N_r}\sum_{s=1}^{N_r}\mu_m\mu_s\chi^\top(e_1+e_5)^\top\mathscr{W}\Upsilon_{m,s}\chi=0, \tag{9.53}$$

where

$$\Upsilon_{m,s}=(A_m+B_mF_s-I_n)e_1+A_{dm}e_3-e_5$$

and $\mathscr{W}$ is a free-weighting matrix in the form $\mathscr{W}=\operatorname{diag}(\mathscr{W}^h,\mathscr{W}^v)$. Following similar lines in the proof of Theorem 9.1, stabilization conditions for 2-D fuzzy systems (9.1) are presented in the following theorem.

Theorem 9.2 *Assume that there exist matrices $P_k=\operatorname{diag}(P_k^h,P_k^v)\in\mathbb{S}_{3n}^+$ $(k\in\mathscr{R})$, $Q_l=\operatorname{diag}(Q_l^h,Q_l^v)$, $R_l=\operatorname{diag}(R_l^h,R_l^v)\in\mathbb{S}_n^+$ $(l=1,2)$, real matrices $X\in\mathbb{R}^{2n_h\times 2n_h}$, $Y\in\mathbb{R}^{2n_v\times 2n_v}$, $\widetilde{Z}\in\mathbb{R}^{3n\times 8n}$, $G_k\in\mathbb{R}^{n_u\times n}$ $(k\in\mathscr{R})$ and a nonsingular matrix $\mathscr{U}=\operatorname{diag}(\mathscr{U}^h,\mathscr{U}^v)\in\mathbb{R}^{n\times n}$ that satisfy condition (9.28) and the following condition holds for $d_h\in\{d_1^h,d_2^h\}$, $d_v\in\{d_1^v,d_2^v\}$ and $k,q,m,s\in\mathscr{R}$*

$$\left[\begin{array}{c|c}\Omega_{k,q}+\operatorname{Sym}\{\widetilde{Z}^\top\Pi_2(d_h,d_v)\} & \\ \quad+\operatorname{Sym}\{(e_1+e_5)^\top\widetilde{\Upsilon}_{m,s}\} & \widetilde{Z}^\top \\ \hline * & -\Gamma P_m\Gamma^\top\end{array}\right]<0, \tag{9.54}$$

where

$$\widetilde{\Upsilon}_{m,s}=(A_m\mathscr{U}+B_mG_s-\mathscr{U})e_1+A_{dm}\mathscr{U}e_3-\mathscr{U}e_5,$$

and other matrix notations are defined as in Theorem 9.1. Then, there exists a fuzzy rule-based SFC as in (9.8) *that makes the closed-loop system* (9.10) *asymptotically stable for any delays $d_h(i)\in[d_1^h,d_2^h]$ and $d_v(j)\in[d_1^v,d_2^v]$. The corresponding controller gains are obtained as*

$$F_k=G_k\mathscr{U}^{-1},\quad k\in\mathscr{R}. \tag{9.55}$$

Proof Let $\mathscr{W}=\mathscr{U}^{-1}$ and $\mathscr{W}_{[k]}=\operatorname{diag}\underbrace{(\mathscr{W},\mathscr{W},\ldots,\mathscr{W})}_{k\text{ blocks}}$. We define the matrices

$$\begin{aligned}&\widetilde{P}_k=\mathscr{W}_{[3]}^\top P_k\mathscr{W}J_{[3]},\quad k\in\mathscr{R},\quad \widetilde{P}(\mu)=\sum_{k=1}^{N_r}\mu_k\widetilde{P}_k,\\ &\widetilde{Q}_l=\mathscr{W}^\top Q_l\mathscr{W},\quad \widetilde{R}_l=\mathscr{W}^\top R_l\mathscr{W},\quad l=1,2,\\ &\widetilde{X}=\mathscr{W}_{[2]}^\top X\mathscr{W}_{[2]},\quad \widetilde{Y}=\mathscr{W}_{[2]}^\top Y\mathscr{W}_{[2]}.\end{aligned} \tag{9.56}$$

By employing the functional $\widetilde{V}_\mu(x)$ obtained from (9.27) for the matrices $\widetilde{P}(\mu)$, $\widetilde{Q}_l$, $\widetilde{R}_l$ $(l=1,2)$, and similar to Theorem 9.1, we can conclude that the closed-loop system (9.10) is GAS if $\Psi_{\widetilde{X}}^h\geq 0$, $\Psi_{\widetilde{Y}}^v\geq 0$, and the following condition holds

$$\sum_{k=1}^{N_r}\sum_{q=1}^{N_r}\sum_{m=1}^{N_r}\mu_{1k}\mu_{2q}\mu_m\left(\breve{\Omega}_{k,q}-\widehat{\Lambda}_m\right) + \mathtt{Sym}\left\{\sum_{m=1}^{N_r}\sum_{s=1}^{N_r}\mu_m\mu_s(e_1+e_5)^\top \mathscr{W}\Upsilon_{m,s}\right\}<0, \tag{9.57}$$

where $\breve{\Omega}_{k,q}$ and $\widehat{\Lambda}_m$ are defined as in Theorem 9.1 with the matrices given in (9.56).

The following identities will be used to transform condition (9.57)

$$\begin{aligned} &e_k\mathscr{W}_{[8]}^{-1}=\mathscr{U}e_k,\quad \Pi_l(d_h,d_v)\mathscr{W}_{[8]}^{-1}=\mathscr{W}_{[3]}^{-1}\Pi_l(d_h,d_v),\quad l=1,2,\\ &\mathscr{W}_{[3]}^{-\top}\Gamma\widetilde{P}_{k,q}\Gamma^\top\mathscr{W}_{[3]}^{-1}=\Gamma P_{k,q}\Gamma^\top,\quad \mathscr{W}_{[3]}^{-\top}\Gamma\widetilde{P}_m\Gamma^\top\mathscr{W}_{[3]}^{-1}=\Gamma P_m\Gamma^\top. \end{aligned}$$

By pre- and post-multiplying with $\mathscr{W}_{[8]}^{-\top}$ and $\mathscr{W}_{[8]}^{-1}$, it is found that

$$\Psi_{\widetilde{X}}^h\geq 0 \text{ and } \Psi_{\widetilde{Y}}^v\geq 0$$

if and only if condition (9.28) holds. Moreover, condition (9.57) is satisfied if and only if

$$\sum_{k=1}^{N_r}\sum_{q=1}^{N_r}\sum_{m=1}^{N_r}\mu_{1k}\mu_{2l}\mu_m\left(\Omega_{k,q}-\Lambda_m\right) + \mathtt{Sym}\left\{\sum_{m=1}^{N_r}\sum_{s=1}^{N_r}\mu_m\mu_s(e_1+e_5)^\top\left(\Upsilon_m\mathscr{W}_{[8]}^{-1}+B_mF_s\mathscr{U}e_1\right)\right\}<0, \tag{9.58}$$

where $\Upsilon_m=(A_m-I_n)e_1+A_{dm}e_3-e_5$. It is clear that

$$\Upsilon_m\mathscr{W}_{[8]}^{-1}=(A_m\mathscr{U}-\mathscr{U})e_1+A_{dm}\mathscr{U}e_3-\mathscr{U}e_5 \text{ and } F_s\mathscr{U}=G_s.$$

Thus, condition (9.58) can be written as

$$\sum_{k=1}^{N_r}\sum_{q=1}^{N_r}\sum_{m=1}^{N_r}\mu_{1k}\mu_{2l}\mu_m\mu_s\left[\left(\Omega_{k,q}-\Lambda_m\right)+\mathtt{Sym}\left\{(e_1+e_5)^\top\widetilde{\Upsilon}_{m,s}\right\}\right]<0. \tag{9.59}$$

On the other hand, similar to (9.47), it can be verified that

$$\begin{aligned} &\Omega_{k,q}-\Lambda_m+\mathtt{Sym}\left\{(e_1+e_5)^\top\widetilde{\Upsilon}_{m,s}\right\}\\ &=\Omega_{k,q}+\mathtt{Sym}\left\{(e_1+e_5)^\top\widetilde{\Upsilon}_{m,s}\right\}-\Pi_2^\top(d_h,d_v)(\Gamma P_m\Gamma^\top)\Pi_2(d_h,d_v)<0 \end{aligned}$$

if and only if there exists a matrix $\widetilde{Z}$ such that

$$\left[\begin{array}{c:c} \Omega_{k,q} + \mathrm{Sym}\left\{\widetilde{Z}^\top \Pi_2(d_h, d_v)\right\} & \widetilde{Z}^\top \\ + \mathrm{Sym}\left\{(e_1 + e_5)^\top \widetilde{\Upsilon}_{m,s}\right\} & \\ \hdashline * & -\Gamma P_m \Gamma^\top \end{array}\right] < 0.$$

Thus, (9.59) can be deduced by the condition (9.54). As a consequence, the closed-loop system (9.10) is GAS. The proof is completed. □

Remark 9.9 Under the analysis scheme presented in Lemma 9.1 with common LKF (independent of fuzzy rules), on the basis of stability conditions derived in Corollary 9.3, design conditions of a fuzzy rule-based SFC for system (9.1) are obtained as in the following corollary.

Corollary 9.5 *Assume that there exist positive definite matrices* $P = \mathrm{diag}(P^h, P^v) \in \mathbb{S}^+_{3n}$, $Q_l = \mathrm{diag}(Q^h_l, Q^v_l)$, $R_l = \mathrm{diag}(R^h_l, R^v_l) \in \mathbb{S}^+_n$ $(l = 1, 2)$, *real matrices* $X \in \mathbb{R}^{2n_h \times 2n_h}$, $Y \in \mathbb{R}^{2n_v \times 2n_v}$, $Z \in \mathbb{R}^{3n \times 8n}$, $G_k \in \mathbb{R}^{n_u \times n}$ $(k \in \mathscr{R})$ *and a nonsingular matrix* $\mathscr{U} = \mathrm{diag}(\mathscr{U}^h, \mathscr{U}^v) \in \mathbb{R}^{n \times n}$ *that satisfy condition* (9.28) *and the following condition holds for* $d_h \in \{d^h_1, d^h_2\}$, $d_v \in \{d^v_1, d^v_2\}$ *and* $m, s \in \mathscr{R}$

$$\left[\begin{array}{c:c} \widetilde{\Omega} + \mathrm{Sym}\left\{\widetilde{Z}^\top \Pi_2(d_h, d_v)\right\} & \widetilde{Z}^\top \\ + \mathrm{Sym}\left\{(e_1 + e_5)^\top \widetilde{\Upsilon}_{m,s}\right\} & \\ \hdashline * & -\Gamma P \Gamma^\top \end{array}\right] < 0, \tag{9.60}$$

where $\widetilde{\Omega}$ *and* $\widetilde{\Upsilon}_{m,s}$ *are defined in Corollary 9.3 and Theorem 9.2, respectively. Then, the closed-loop system* (9.10) *subject to fuzzy rule-based SFC* (9.8) *is GAS.*

The following result can be obtained as a special case of Corollary 9.5 for 2-D fuzzy systems in the form of (9.1) with constant delays.

Corollary 9.6 *Given positive integers* d_h *and* d_v. *Assume that there exist positive definite matrices* $P = \mathrm{diag}(P^h, P^v) \in \mathbb{S}^+_{2n}$, $Q = \mathrm{diag}(Q^h, Q^v)$, $R = \mathrm{diag}(R^h, R^v) \in \mathbb{S}^+_n$, $G_k \in \mathbb{R}^{n_u \times n}$ $(k \in \mathscr{R})$ *and a nonsingular matrix* $\mathscr{U} = \mathrm{diag}(\mathscr{U}^h, \mathscr{U}^v)$ *satisfying the following condition for* $m, s \in \mathscr{R}$

$$\hat{\Pi}_1^\top \check{P} \hat{\Pi}_1 - \hat{\Pi}_2^\top \check{P} \hat{\Pi}_2 + \mathscr{H} + \mathrm{Sym}\left\{(e_1 + e_3)^\top \widehat{\Upsilon}_{m,s}\right\} < 0, \tag{9.61}$$

where

$$\check{P} = \hat{\Gamma} P \hat{\Gamma}^\top, \quad \widehat{\Upsilon}_{m,s} = (A_m \mathscr{U} + B_m G_s - \mathscr{U}) e_1 + A_{dm} e_2 - \mathscr{U} e_3,$$

and other matrix notations are defined in Corollary 9.4 and Theorem 9.2. Then, the closed-loop system (9.10) *with constant delays* d_h, d_v *is asymptotically stable subject to fuzzy rule-based SFC* (9.8) *with the controller gains given in* (9.55).

Remark 9.10 For T-S fuzzy systems described by the FMLSS model, a fuzzy rule-based functional mentioned in Lemma 9.1 is defined by $V_\mu(x) = V_\mu^h(x) + V_\mu^v(x)$. If the difference of $V_\mu(x))$ along state trajectory is negative definite, that is, there exists a scalar $\tilde{c}_0 > 0$ such that

$$
\begin{aligned}
&V_{\mu(i+1,j+1)}(x(i+1, j+1)) - V^h_{\mu(i,j+1)}(x(i, j+1)) \\
&\quad - V^v_{\mu(i+1,j)}(x(i+1, j)) \leq -\tilde{c}_0 \|x(i, j)\|^2,
\end{aligned}
\tag{9.62}
$$

then the conclusion of Lemma 9.1 is still valid. By suitably modifying the LKF candidate (9.27), analogous results can be obtained.

9.5.1 Numerical Examples

In this section, we present some numerical examples to demonstrate the effectiveness of the analysis and design conditions proposed in this chapter.

Example 9.1 Consider a model of thermal process described by the following nonlinear partial differential equation

$$
\frac{\partial w(x,t)}{\partial x} + \frac{\partial w(x,t)}{\partial t} = -a_0(x, t, w)w(x, t) - a_1(x, t, w)w(x, t - \tau_t), \tag{9.63}
$$

where $w(x, t)$ is the unknown function representing (e.g., temperature) at (space) x and time t, τ_t represents the time-delay and $a_0 = a_0(x, t, w)$, $a_1 = a_1(x, t, w)$ are state-dependent coefficients.

For given increments Δx, Δt, we approximate Eq. (9.63) by

$$
\begin{aligned}
&w(i, j) = w(i\Delta x, j\Delta t), \\
&\frac{\partial w(x,t)}{\partial x} \simeq \frac{w(i, j) - w(i-1, j)}{\Delta x}, \\
&\frac{\partial w(x,t)}{\partial t} \simeq \frac{w(i, j+1) - w(i, j)}{\Delta t},
\end{aligned}
$$

then Eq. (9.63) can be written in the form

$$
\begin{aligned}
w(i, j+1) = &\left(1 - \frac{\Delta t}{\Delta x} - a_0(.)\Delta t\right) w(i, j) \\
&+ \frac{\Delta t}{\Delta x} w(i-1, j) - a_1(.)\Delta t w(i, j - d_v(j)),
\end{aligned}
\tag{9.64}
$$

where $\lfloor . \rfloor$ denotes the floor function and $d_v(j) = \lfloor \frac{\tau_t}{\Delta t} \rfloor$.

We define state vectors $x^h(i, j) = w(i-1, j)$ and $x^v(i, j) = w(i, j)$. Then, Eq. (9.64) can be recast into the following system

$$\begin{bmatrix} x^h(i+1,j) \\ x^v(i,j+1) \end{bmatrix} = \mathscr{A}(i,j,x)\begin{bmatrix} x^h(i,j) \\ x^v(i,j) \end{bmatrix} + \mathscr{A}_d(i,j,x)\begin{bmatrix} x^h(i-d_1(i),j) \\ x^v(i,j-d_2(j)) \end{bmatrix}, \tag{9.65}$$

where

$$\mathscr{A}(i,j,x) = \begin{bmatrix} 0 & 1 \\ \frac{\Delta t}{\Delta x} & 1-\frac{\Delta t}{\Delta x} - a_0(.)\Delta t \end{bmatrix}, \quad \mathscr{A}_d(i,j,x) = \begin{bmatrix} 0 & 0 \\ 0 & -a_1(.)\Delta t \end{bmatrix}.$$

Assume that the coefficients a_0, a_1 are confined within certain ranges of the form

$$a_0^l \le a_0 \le a_0^u, \quad a_1^l \le a_1 \le a_1^u,$$

where a_0^l, a_1^l and a_0^u, a_1^u are given scalars involving the lower bounds and upper bounds of the coefficients a_0, a_1. Noticing also that when the bounds a_0^l, a_1^l and a_0^u, a_1^u are unknown or partially unknown, we can use the proposed method of [22] to present system (9.65) in the T-S form. According to the above assumption, we can get the following representation

$$\begin{aligned} a_0(.) &= \mu_{11}(a_0)a_0^l + \mu_{12}(a_0)a_0^u, \\ a_1(.) &= \mu_{21}(a_1)a_1^l + \mu_{22}(a_1)a_1^u, \end{aligned} \tag{9.66}$$

where $\mu_{1k}(a_0) \ge 0$, $\mu_{2k}(a_1) \ge 0$ $(k = 1, 2)$ and

$$\mu_{11}(a_0) + \mu_{12}(a_0) = 1, \quad \mu_{21}(a_1) + \mu_{22}(a_1) = 1.$$

Based on (9.66), system (9.65) can be represented by a four-rule fuzzy system (9.26), where

$$A_1 = \begin{bmatrix} 0 & 1 \\ \frac{\Delta t}{\Delta x} & 1-\frac{\Delta t}{\Delta x} - a_0^l\Delta t \end{bmatrix} = A_3, \quad A_2 = \begin{bmatrix} 0 & 1 \\ \frac{\Delta t}{\Delta x} & 1-\frac{\Delta t}{\Delta x} - a_0^u\Delta t \end{bmatrix} = A_4,$$
$$A_{d1} = \begin{bmatrix} 0 & 0 \\ 0 & -a_1^l\Delta t \end{bmatrix} = A_{d2}, \quad A_{d3} = \begin{bmatrix} 0 & 0 \\ 0 & -a_1^u\Delta t \end{bmatrix} = A_{d4}.$$

Let $\Delta t = 0.05$, $\Delta x = 0.1$, $a_0^l = 1$, $a_0^u = 2$ and $a_1^l = 0$, $a_1^u = 1$. By using Matlab LMI toolbox to test the derived conditions of Corollary 9.3 (common Lyapunov scheme). Allowable results of d_2^v for various d_1^v and the number of iterations (NoI) are given in Table 9.1 Condition (9.51) in Remark 9.7 is also conducted. It is found that the derived conditions mentioned in Remark 9.7 do not generate any feasible solution. This result demonstrates the merit of the 2-D summation inequalities proposed in Lemma 9.4, Corollary 9.2 in comparison to the common Jensen inequalities.

To simulate the result, we take $a_0(.) = 1 + 2\sin^2(x^v(i,j))$, $a_1(.) = \cos^2(x^v(i,j))$ and $d_v(j) = \text{round}(30|\sin(j\pi/8)|)$. The conducted state trajectories $x^h(i,j)$, $x^v(i,j)$ with finite support initial sequences $x^h(0,j) = 1$ for $j \le 100$ and $x^v(i,j) = -1$ for $i \le 100$, $j \in \mathbb{Z}[-30,0]$ are presented in Figs. 9.1 and 9.2. The results of

Table 9.1 Allowable bound of d_2^v for various d_1^v

$NoI \backslash d_1^v$	0	5	10	15	20
20	20	24	28	34	40
30	31	35	38	42	48
50	38	43	47	52	59

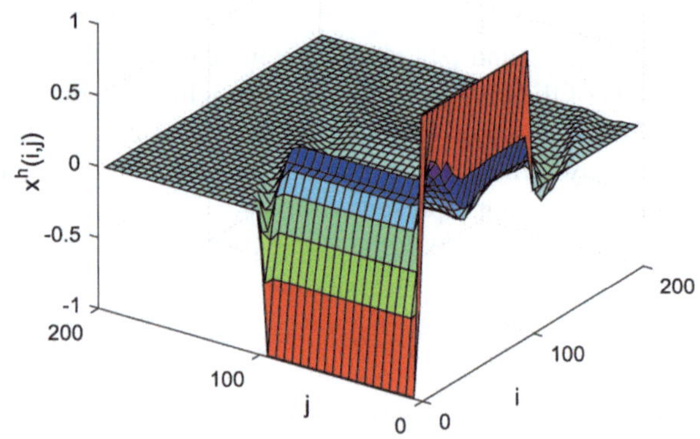

Fig. 9.1 State trajectory $x^h(i, j)$ with coefficients $a_0(.) = 1 + 2\sin^2(x^v(i, j))$ and $a_1(.) = \cos^2(x^v(i, j))$

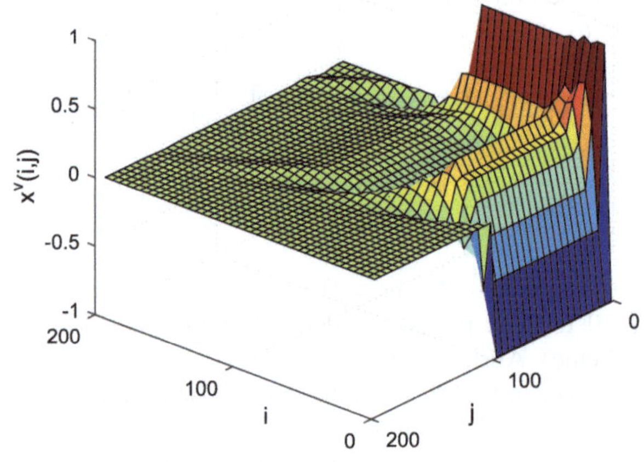

Fig. 9.2 State trajectory $x^h(i, j)$ with coefficients $a_0(.) = 1 + 2\sin^2(x^v(i, j))$ and $a_1(.) = \cos^2(x^v(i, j))$

Table 9.1 and Figs. 9.1 and 9.2 illustrate the effectiveness of the obtained stability conditions.

Example 9.2 Consider the following 2-D T-S fuzzy control system

$$\begin{bmatrix} x^h(i+1,j) \\ x^v(i,j+1) \end{bmatrix} = \sum_{k=1}^{3} \mu_k(\vartheta) \left\{ A_k \begin{bmatrix} x^h(i,j) \\ x^v(i,j) \end{bmatrix} + A_{dk} \begin{bmatrix} x^h(i-d_h,j) \\ x^v(i,j-d_v) \end{bmatrix} + B_k u(i,j) \right\}, \tag{9.67}$$

where $\vartheta(i,j) = x^h(i,j) + x^v(i,j)$ and constant delays d_h, d_v. Membership functions are given by

$$\mu_k(\vartheta(i,j)) = \frac{\phi_k(\vartheta(i,j))}{\sum_{k=1}^{3} \phi_k(\vartheta(i,j))}, \quad k = 1, 2, 3,$$

where

$$\phi_1(\vartheta)) = \exp(-\frac{1}{4}(\vartheta+4)^2), \quad \phi_2(\vartheta) = \exp(-\frac{1}{4}\vartheta^2), \quad \phi_3(\vartheta) = \exp(-\frac{1}{4}(\vartheta-4)^2).$$

The system matrices of two fuzzy rules are given as follows

$$\begin{aligned}
(A_1, A_{d1}, B_1) &= \left(\begin{bmatrix} 0 & 1 \\ 0.05 & 1 \end{bmatrix}, \begin{bmatrix} -0.1 & 0 \\ 0.1 & -0.5 \end{bmatrix}, \begin{bmatrix} 1 \\ 0.5 \end{bmatrix} \right), \\
(A_2, A_{d2}, B_2) &= \left(\begin{bmatrix} 0 & 1 \\ 0.1 & 0.95 \end{bmatrix}, \begin{bmatrix} 0.1 & 0 \\ 0.1 & -0.6 \end{bmatrix}, \begin{bmatrix} 0.5 \\ 1 \end{bmatrix} \right), \\
(A_3, A_{d3}, B_3) &= \left(\begin{bmatrix} 0 & 1 \\ -0.1 & 1.2 \end{bmatrix}, \begin{bmatrix} 0 & 0.1 \\ 0.2 & -0.25 \end{bmatrix}, \begin{bmatrix} 0.8 \\ 0.6 \end{bmatrix} \right).
\end{aligned}$$

The behaviors of membership functions along state trajectories $x(i,j)$ are depicted in Figs. 9.3, 9.4 and 9.5.

Moreover, with the specified data, it is found that the derived conditions of Corollary 9.4 do not deliver any feasible solution. Thus, stability of the open-loop (unforced) system of (9.67) is not assured. Simulation results given in Figs. 9.6 and 9.7 taken with $d_h = 0$, $d_v = 5$ reveal that the open-loop system is unstable.

Let $d_h = 10$, $d_v = 10$. By solving condition (9.61) via Matlab LMI toolbox we obtain

$$\mathcal{U} = \mathrm{diag}(0.1402, 0.1076)$$

and the controller gains are obtained as

$$F_1 = F_2 = F_3 = \begin{bmatrix} -0.0149 & -1.0957 \end{bmatrix}.$$

According to Corollary 9.6, the closed-loop system of (9.67) is GAS. State trajectories of the corresponding closed-loop system are presented in Figs. 9.8 and 9.9. The presented simulations indicate the effectiveness of the design scheme.

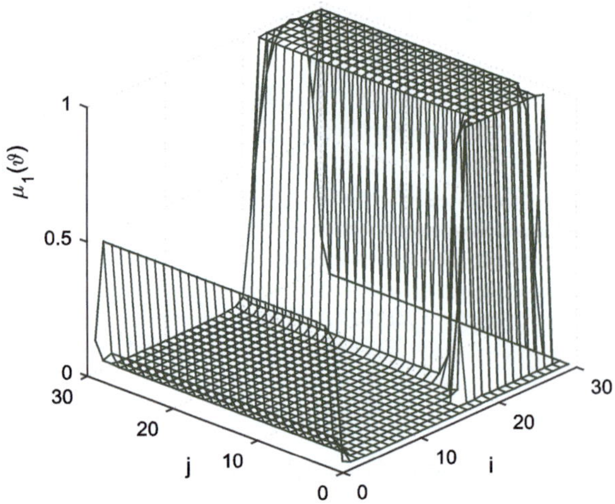

Fig. 9.3 Membership function $\mu_1(\vartheta(i, j))$

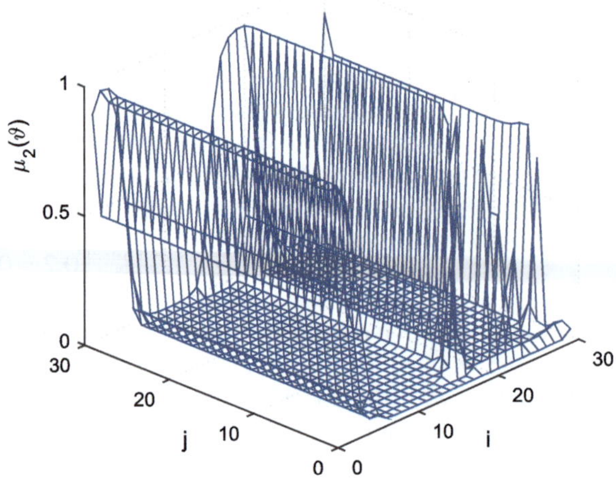

Fig. 9.4 Membership function $\mu_2(\vartheta(i, j))$

Example 9.3 Consider a 2-D -T fuzzy system as (9.26) with the following data taken from [24]

$$(A_1, A_{d1}) = \left(\begin{bmatrix} -0.291 & 1 \\ 0 & 0.95 \end{bmatrix}, \begin{bmatrix} 0.012 & 0.014 \\ 0 & 0.015 \end{bmatrix} \right)$$

$$(A_2, A_{d2}) = \left(\begin{bmatrix} -0.1 & 0 \\ 1 & -0.2 \end{bmatrix}, \begin{bmatrix} 0.01 & 0 \\ 0.01 & 0.015 \end{bmatrix} \right).$$

To make a comparison, we assume that the information propagation occurs only in j-axis. Let $d_h(i) = 0$ and solving the LMIs (9.28) and (9.44) for various d_1^v. The allowable bounds of d_2^v are given in Table 9.2. Clearly, our approach based

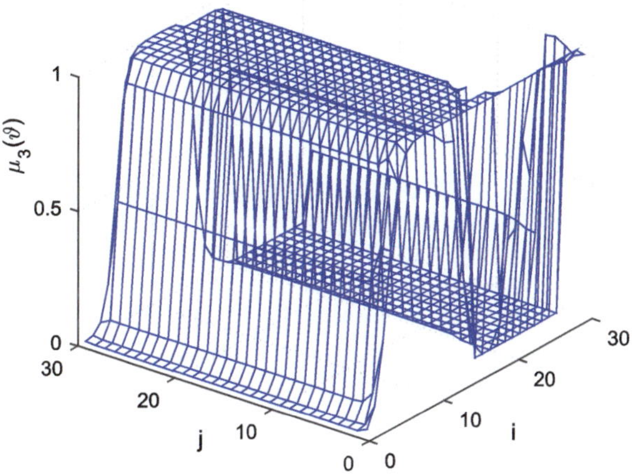

Fig. 9.5 Membership function $\mu_3(\vartheta(i, j))$

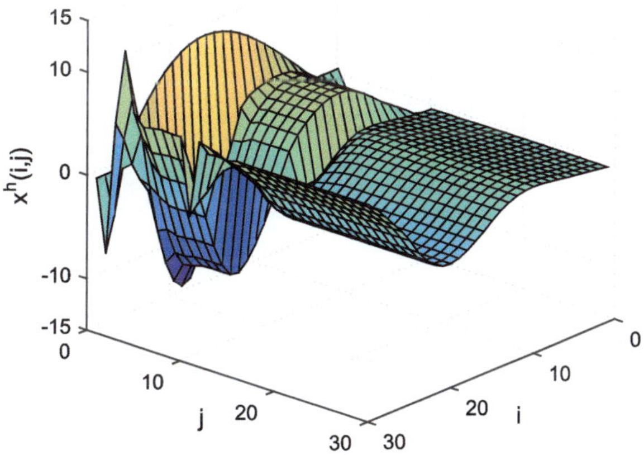

Fig. 9.6 State trajectory $x^h(i, j)$ of the unforced system

on improved rule-based LKFs and new summation inequalities can deliver better results over the existing literature not only the feasibility of conditions but also the computational complexity.

9.6 Conclusion

In this chapter, the problems of stability and stabilization via fuzzy rule-based state-feedback controllers have been discussed for a class of 2-D fuzzy systems described by the Roesser model with directional time-varying delays. The common center-average defuzzifier approach has been utilized to present the defuzzified output

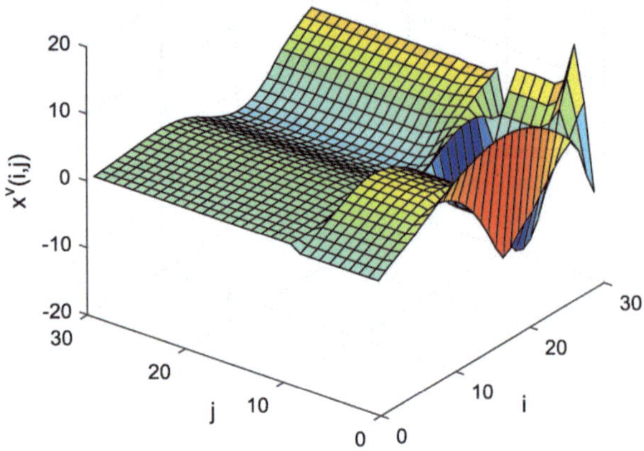

Fig. 9.7 State trajectory $x^v(i, j)$ of the unforced system

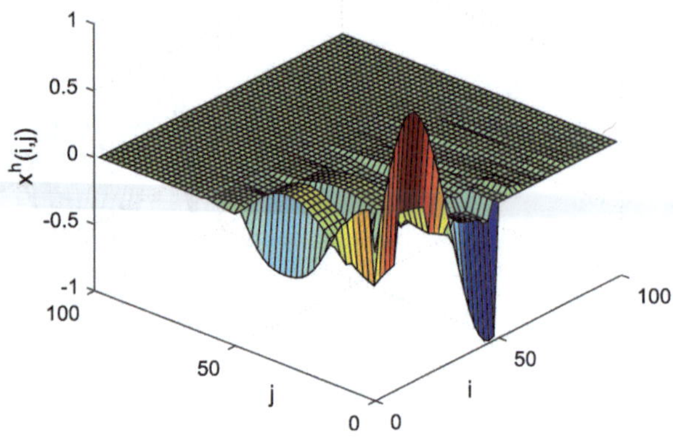

Fig. 9.8 State trajectory $x^h(i, j)$ of the closed-loop system

dynamics. A functional-based scheme has been proposed and employed to derive delay-dependent stability conditions based on augmented fuzzy parameter-dependent LKF candidates. On the basis of the obtained analysis results, delay-dependent tractable conditions have been formulated to design a rule-based stabilizing state-feedback controller. Numerical examples have been provided to demonstrate the efficacy of the obtained results.

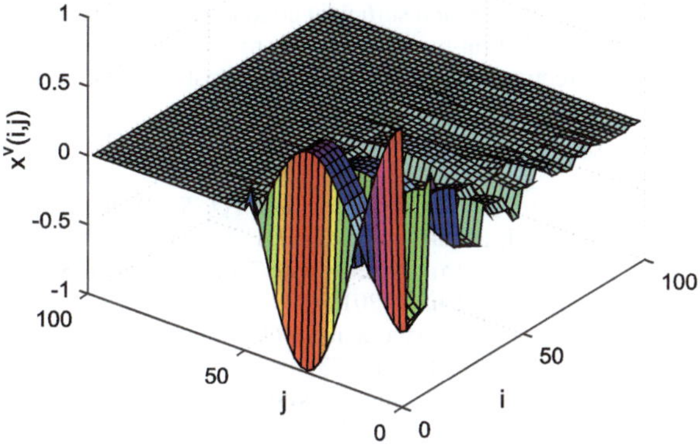

Fig. 9.9 State trajectory $x^{\upsilon}(i, j)$ of the closed-loop system

Table 9.2 Allowable d_2^{υ} for various d_1^{υ}

d_1^{υ}	3	5	10	12	NoDV
[21]	24	26	31	33	690
[42]	38	40	45	47	6165
[24]	52	54	58	60	301
Corollary 9.5	58	61	66	69	124

References

1. Witrant, E., Fridman, E., Sename, O., Dugard, L.: Recent Results on Time-Delay Systems: Analysis and Control. Springer, Basel (2016)
2. Atay, F.M.: Complex Time-Delay Systems: Theory and Applications. Springer, Berlin (2016)
3. Pham, T.N., Trinh, H., Hien, L.V., Wong, K.P.: Integration of electric vehicles for load frequency output feedback H_∞ control of smart grids. IET Gener. Trans. Distrib. **10**, 3341–3352 (2016)
4. Sipahi, R., Niculescu, S.-I., Abdallah, C.T., Michiels, W., Gu, K.: Stability and stabilization of systems with time delay. IEEE Control Syst. Magaz. **31**, 38–65 (2011)
5. Hien, L.V., Hieu, T.: New finite-sum inequalities with applications to stability of discrete time-delay systems. Automatica **71**, 197–201 (2016)
6. Zhang, X.M., Han, Q.L., Seuret, A., Gouaisbaut, F., He, Y.: Overview of recent advances in stability of linear systems with time-varying delays. IET Control Theory Appl. **13**, 1–16 (2019)
7. Nam, P.T., Hiep, L.T.: A new delay-variation-dependent stability criterion for delayed discrete-time systems. J. Frankl. Inst. **357**, 6951–6967 (2020)
8. Zhang, C.K., Chen, W.H., Zhu, C., He, Y., Wu, M.: Stability analysis of discrete-time systems with time-varying delay via a delay-dependent matrix- separation-based inequality. Automatica **156**, 111192 (2023)
9. Sha, H., Park, J.H., Chen, J., Zhu, M., Nan, C.: Stability analysis of discrete-time systems with a time-varying delay via improved methods. IET Control Theory Appl. **18**, 951–959 (2024)
10. Hien, L.V., Trinh, H.: Stability of two-dimensional Roesser systems with time-varying delays via novel 2D finite sum inequalities. IET Control Theory Appl. **10**, 1665–1674 (2016)

11. Huang, Y.B., An, J., He, Y., Wu, M.: Quasi-convex combination method and its application to the stability analysis of 2D discrete-time Roesser systems with time-varying delays. IET Control Theory Appl. **12**, 718–727 (2018)
12. Hien, L.V., Trinh, H.M., Pathirana, P.N.: On l_1-gain control of 2-D positive Roesser systems with directional delays: Necessary and sufficient conditions. Automatica **112**, 108720 (2020)
13. Trinh, H., Hien, L.V.: Prediction-based approach to stabilization of 2-D continuous-time Roesser systems with directional input delays. J. Frank Inst. **357**, 4779–4794 (2020)
14. Chelliq, E.H., Alfidi, M., Chalh, Z.: Admissibility and robust H_∞ controller design for uncertain 2D singular discrete systems with interval time-varying delays. Optim. Control Appl. Meth. **44**, 2476–2503 (2023)
15. Hien, L.V., Vu, L.H.: Finite-region boundedness and (Q, S, R)-dissipativity of 2-D singular Roesser systems with mixed time-varying delays. J. Frankl. Inst. **361**, 1416–1435 (2024)
16. Wu, L., Su, X., Shi, P.: Fuzzy Control Systems with Time-Delay and Stochastic Perturbation: Analysis and Synthesis. Springer, Cham (2015)
17. Zadeh, L.: Outline of a new approach to the analysis of complex systems and decision processes. IEEE Trans. Syst. Man Cybern. **3**, 28–44 (1973)
18. Takagi, T., Sugeno, M.: Fuzzy identification of systems and its applications to modeling and control. IEEE Trans. Syst. Man Cybern. **15**, 116–132 (1985)
19. Lin, C., Wang, Q.G., Lee, T.H., He, Y.: LMI Approach to Analysis and Control of Takagi-Sugeno Fuzzy Systems with Time Delay. Springer-Verlag, Berlin (2007)
20. Hien, L.V.: Exponential stability and stabilisation of fuzzy time-varying delay systems. Int. J. Syst. Sci. **41**, 1155–1161 (2010)
21. Wu, L., Su, X., Shi, P., Qiu, J.: A new approach to stability analysis and stabilization of discrete-time T-S fuzzy time-varying delay systems. IEEE Trans. Syst. Man Cybern: Cybern. **41**, 273–286 (2011)
22. Li, X.J., Yang, G.H.: Fault detection for T-S fuzzy systems with unknown membership functions. IEEE Trans. Fuzzy Syst. **22**, 139–152 (2014)
23. Yin, S., Shi, P., Yang, H.: Adaptive fuzzy control of strict-feedback nonlinear time-delay systems with unmodeled dynamics. IEEE Trans. Cybern. **46**, 1926–1938 (2016)
24. Park, M.J., Kwon, O.M.: Stability and stabilization of discrete-time T-S fuzzy systems with time-varying delay via Cauchy-Schwartz-based summation inequality. IEEE Trans. Fuzzy Syst. **25**, 128–140 (2017)
25. Sakthivel, R., Karthick, S.A., Kaviarasan, B., Alzahrani, F.: Dissipativity-based non-fragile sampled-data control design of interval type-2 fuzzy systems subject to random delays. ISA Trans. **83**, 154–164 (2018)
26. Wang, X., Park, J.H., Yang, H., Zhao, G., Zhong, S.: An improved fuzzy sampled-data control to stabilization of T-S fuzzy systems with state delays. IEEE Trans. Cybern. **50**, 3125–3135 (2020)
27. Ge, C., Park, J.H., Hua, C., Guan, X.: Dissipativity analysis for T-S fuzzy system under memory sampled-data control. IEEE Trans. Cybern. **51**, 961–969 (2021)
28. Li, Y., He, Y., Yang, Y.: Stability analysis for delayed T-S fuzzy systems: A compensation Lyapunov-Krasovskii functional method combined with free-weighting matrices. ISA Trans. **142**, 12–19 (2023)
29. Chen, X., Lam, J., Gao, H.: Zhou,: Stability analysis and control design for 2-D fuzzy systems via basis-dependent Lyapunov functions. Multidim. Syst. Sig. Process **24**, 395–415 (2013)
30. Boukili, B., Hmamed, A., Benzaouia, A., Hajjaji, A.E.: H_∞ filtering of two-dimensional T-S fuzzy systems. Circuit Syst. Sig. Process **33**, 1737–1761 (2014)
31. Li, L.: Observer-based H_∞ controller for 2-D T-S fuzzy model. Int. J. Syst. Sci. **47**, 3455–3464 (2016)
32. Luo, Y., Wang, Z., Liang, J., Wei, G., Alsaadi, F.E.: H_∞ control for 2-D fuzzy systems with interval time-varying delays and missing measurements. IEEE Trans. Cybern. **47**, 365–377 (2017)
33. Hien, L.V., Trinh, H.: Stability analysis and control of two-dimensional fuzzy systems with directional time-varying delays. IEEE Trans. Fuzzy. Syst. **26**, 1550–1564 (2018)

34. Ji, W., Qiu, J., Lam, H.K.: Fuzzy-affine-model-based sliding-mode control for discrete-time nonlinear 2-D systems via output feedback. IEEE Trans. Cybern. **53**, 979–987 (2023)
35. Wang, M., Lam, H.K., Qiu, J., Yan, H., Li, Z.: Fuzzy-affine-model-based filtering design for continuous-time Roesser-type 2-D nonlinear systems. IEEE Trans. Cybern. **53**, 3220–3230 (2023)
36. Chen, W.H., Zhang, C.K., Liu, Z.Z., Wang, L., He, Y.: Stability and filtering for delayed discrete-time T-S fuzzy systems via membership-dependent approaches. IEEE Trans. Fuzzy Syst. **32**, 6101–6111 (2024)
37. Nam, P.T., Pubudu, P.N., Trinh, H.: Discrete Wirtinger-based inequality and its application. J. Frankl. Inst. **352**, 1893–1905 (2015)
38. Seuret, A., Gouaisbaut, F., Fridman, E.: Stability of discrete-time systems with time-varying delays via a novel summation inequality. IEEE Trans. Autom. Control **60**, 2740–2745 (2015)
39. Zhang, X.M., Han, Q.L.: Abel lemma-based finite-sum inequality and its application to stability analysis for linear discrete time-delay systems. Automatica **57**, 199–202 (2015)
40. Skelton, R.E., Iwasaki, T., Grigoradis, K.M.: A Unified Algebraic Approach to Linear Control Design. Taylor & Francis, New York (1997)
41. Boyd, S., Ghaoui, L.E., Feron, E., Balakrishnan, V.: Linear Matrix Inequalities in System and Control Theory. SIAM, Philadelphia (1994)
42. Li, Z., Gao, H., Agarwal, R.K.: Stability analysis and controller synthesis for discrete-time delayed fuzzy systems via small gain theorem. Info. Sci. **226**, 93–104 (2013)

Chapter 10
Stability and Stabilization via Event-Triggered Control of Two-Dimensional Singular Systems with Delays

Abstract In this chapter, the problems of stability and stabilization via event-triggered state-feedback controllers are presented for two-dimensional (2-D) singular Roesser systems with directional time-varying delays. Based on a 2-D Lyapunov–Krasovskii functional (LKF) method, and by utilizing free matrix equations, delay-dependent conditions are first derived in terms of linear matrix inequalities (LMIs) to ensure the regularity, causality and internal stability of 2-D singular Roesser systems with time-varying delays. The derived stability conditions are then employed to derive tractable conditions for the design of a stabilizing event-triggered state-feedback controller (ETC). Numerical examples are given to illustrate the effectiveness of the proposed analysis and design conditions.

Keywords 2-D singular systems · Roesser model · Event-triggered mechanism · Admissibility

10.1 Introduction

Singular systems form an important category of dynamical systems, which are widely used to model a large number of practical phenomena in circuits networks, power systems, aerospace engineering or multibody mechanics [1–4]. This type of systems is also referred as descriptor systems or differential-algebraic equations [5–7]. One of the major differences involving their dynamic behavior is that, in singular systems, state and output variables are involved in both differential/difference equations (dynamic part) and algebraic constraints. This fact results in many different features of singular systems in comparison to normal systems such as impulsive behavior in the state-response, non-properness of the transfer matrix or non-causality between states and input/outputs [6]. Besides the singularity governed by algebraic constraints of relevant states and outputs, time-delay phenomenon is omnipresent in control engineering [8, 9]. Due to theoretical and practical issues, the study on analysis and control of time-delay systems, in general, and singular systems with delays, in particular, has attracted remarkable research attention in the past few decades. We refer the reader to recent works [10–17] for a few references.

L. V. Hien, *Two-Dimensional Time-Delay Systems*, Studies in Systems, Decision and Control 605, https://doi.org/10.1007/978-3-031-95678-2_10

As mentioned in previous chapters, the theory of two-dimensional (2-D) systems has been extensively studied and developed due to a wide range of applications, for example, in electrical circuits, image processing, data transmission or multi-dimensional digital filtering. A large number of interesting results concerning the problems of stability analysis and controller design for various classes of 2-D systems with and without delays have been reported in the literature [18–23]. However, very little attempt of scholars has been successfully extended to 2-D singular systems with delays. For example, the stability analysis and state-feedback controller design problems were studied for 2-D linear singular Roesser systems without delay in [24]. Based on some decomposing techniques [25] and by using 2-D quadratic Lyapunov functions, stability and stabilization conditions in the type of linear matrix inequalities (LMIs) were obtained. The H_∞ control problem was also investigated for 2-D singular Roesser systems with constant delays [26, 27] using a bounded real lemma and delay-independent LMI-based approach. Unfortunately, the proposed methods in [26, 27] cannot be extendable to 2-D singular systems with time-varying delays. Very recently, some extended results on stability, H_∞ stabilization or finite-region dissipativity analysis were developed for 2-D singular Roesser systems with directional time-varying delays in [28–30] and [31], respectively.

On the other hand, it has been well recognized as one of the most advanced control techniques, event-triggered control (ETC) scheme has been extensively studied for various types of one-dimensional (1-D) time-delay systems [32–36]. For 2-D systems, this topic has received growing research attention in recent years. In particular, the stabilization problem, sliding mode control via ETC scheme were developed for nominal delay-free 2-D systems in [37, 38] and [39, 40]. The design problem of an H_∞ asynchronous ETC was also studied for 2-D Markov jump systems [41]. Up to date, this topic has not been fully developed for 2-D systems, in general, and 2-D singular systems with delays, in particular. It should be emphasized that according to the singularity coupled with delayed states, the proposed methods in existing literature [37–41] cannot be simply extended to 2-D singular systems with time-varying delays.

In this chapter, we consider the problems of stability analysis and controller design for some classes of 2-D singular systems in the Roesser model with extended time-varying delays. We first study the stability of 2-D descriptor systems with mixed discrete and distributed (summation) interval delays. A systematic approach based on a 2-D Lyapunov–Krasovskii functional (LKF) method and zero-type free matrix equations is presented to derive delay-dependent stability conditions in terms of LMIs which ensure the system regularity, causality and internal stability. We then address the stabilization problem via event-triggered state-feedback controllers (ETCs) for 2-D singular systems with state delays. The effectiveness of the obtained results is illustrated by numerical examples.

10.2 Preliminaries

10.2.1 Auxiliary Results

Let $E \in \mathbb{R}^{n\times n}$ and $A \in \mathbb{R}^{n\times n}$ be given matrices with $\text{rank}(E) = r < n$. Assume that the dimension n is split as $n = n_h + n_v$ and $E = \text{diag}(E_h, E_v)$, where $E_h \in \mathbb{R}^{n_h\times n_h}$, $E_v \in \mathbb{R}^{n_v\times n_v}$, with $\text{rank}(E_h) = r_h \le n_h$, $\text{rank}(E_v) = r_v \le n_v$, and $r = r_h + r_v$. Then, there exist nonsingular matrices M_h, $N_h \in \mathbb{R}^{n_h\times n_h}$ and M_v, $N_v \in \mathbb{R}^{n_v\times n_v}$ such that

$$M_h E_h N_h = \left[\begin{array}{c:c} I_{r_h} & 0 \\ \hdashline 0 & 0 \end{array}\right] \in \mathbb{R}^{n_h\times n_h}, \quad M_v E_v N_v = \left[\begin{array}{c:c} I_{r_v} & 0 \\ \hdashline 0 & 0 \end{array}\right] \in \mathbb{R}^{n_v\times n_v},$$

where, for a given positive integer m, I_m denotes the identity matrix in $\mathbb{R}^{m\times m}$. Denote $\tilde{M} = \text{diag}(M_h, M_v)$ and $\tilde{N} = \text{diag}(N_h, N_v)$. It is easy to verify that

$$\tilde{M} E \tilde{N} - \text{diag}(I_{r_h}, 0, I_{r_v}, 0).$$

In addition, let $\Pi \in \mathbb{R}^{n\times n}$ be the transition matrix given by

$$\Pi = \begin{bmatrix} I_{r_h} & 0 & 0 & 0 \\ 0 & 0 & I_{r_v} & 0 \\ 0 & I_{n_h-r_h} & 0 & 0 \\ 0 & 0 & 0 & I_{n_v-r_v} \end{bmatrix}.$$

Then, we have the following decomposition

$$\bar{E} \triangleq MEN = \left[\begin{array}{c:c} I_r & 0_{r\times(n-r)} \\ \hdashline 0_{(n-r)\times r} & 0_{(n-r)\times(n-r)} \end{array}\right] \text{ and } MAN = \left[\begin{array}{c:c} A_{11} & A_{12} \\ \hdashline A_{21} & A_{22} \end{array}\right], \tag{10.1}$$

where $0_{p\times q}$ denotes the zero matrix of $p \times q$ dimension and $M = \Pi\tilde{M}$, $N = \tilde{N}\Pi$.

First, similar to [25], we give the following definition.

Definition 10.1 The pair of matrices (E, A) is said to be regular if the polynomial $\det(E\mathbb{I}(\lambda, \mu) - A)$ is not identically zero, where $\mathbb{I}(\lambda, \mu) = \text{diag}(\lambda I_{n_h}, \mu I_{n_v})$. The pair (E, A) is said to be causal if $\deg(\det(sE - A)) = \text{rank}(E)$.

Lemma 10.1 *If the pair (E, A) is regular then the nonsingular matrices M, N in (10.1) can be selected such that the matrix A_{22} is nonsingular.*

Proof Let $\lambda, \mu \in \mathbb{C}$ such that $\det(E\mathbb{I}(\lambda, \mu) - A) \neq 0$. We define the matrices

$$\hat{A} = \mathscr{F}_{\lambda,\mu}^{-1} A, \quad \hat{E} = \mathscr{F}_{\lambda,\mu}^{-1} E,$$

where $\mathscr{F}_{\lambda,\mu} = E\mathbb{I}(\lambda, \mu) - A$. Then, we have

$$\hat{A} = I_n - \hat{E}\mathbb{I}(\lambda, \mu).$$

Since $\text{rank}(\hat{E}) = r$, we can find a nonsingular matrix S such that $S\hat{E}S^{-1} = \text{diag}(E_r, 0)$, where $E_r \in \mathbb{R}^{r\times r}$ is a nonsingular matrix. We now define the nonsingular matrices

$$M = \text{diag}(E_r^{-1}, I_{n-r})S\mathscr{F}_{\lambda,\mu}^{-1}, \quad N = S^{-1},$$

then it can be verified that $MEN = \text{diag}(I_r, 0)$ and

$$\begin{aligned} MAN &= \text{diag}(E_r^{-1}, I_{n-r})S\left(I_n - \hat{E}\mathbb{I}(\lambda, \mu)\right)S^{-1} \\ &= \text{diag}(E_r^{-1}, I_{n-r})\left(I_n - \text{diag}(E_r, 0)S\mathbb{I}(\lambda, \mu)S^{-1}\right) \\ &= \left[\begin{array}{c|c} E_r^{-1} - S_{11} & -S_{12} \\ \hline 0 & I_{n-r} \end{array}\right], \end{aligned}$$

where $S\mathbb{I}(\lambda, \mu)S^{-1} = \left[\begin{array}{c|c} S_{11} & S_{12} \\ \hline S_{21} & S_{22} \end{array}\right]$. With the selected matrices, clearly the matrix $A_{22} = I_{n-r}$ is nonsingular. The proof is completed. □

Lemma 10.2 *The pair* (E, A) *is regular and causal if and only if there exist nonsingular matrices* M, N *in* (10.1) *such that the matrix* A_{22} *is nonsingular.*

Proof The necessity is deduced from Lemma 10.1. We now prove the sufficiency. Assume that $\det(A_{22}) \neq 0$. We denote

$$\widehat{M} = \begin{bmatrix} I_r & -A_{12}A_{22}^{-1} \\ 0_{(n-r)\times r} & I_{n-r} \end{bmatrix} M, \quad \widehat{N} = N\begin{bmatrix} I_r & 0_{r\times(n-r)} \\ -A_{22}^{-1}A_{21} & A_{22}^{-1} \end{bmatrix}.$$

Then, it can be checked by direct computation that

$$\widehat{E} \triangleq \widehat{M}E\widehat{N} = \text{diag}(I_r, 0_{n-r}), \quad \widehat{A} \triangleq \widehat{M}A\widehat{N} = \text{diag}(\hat{A}_{11}, I_{n-r}),$$

where $\hat{A}_{11} = A_{11} - A_{12}A_{22}^{-1}A_{21}$. It follows from the above decomposition that

$$\begin{aligned} \widehat{M}\left(E\mathbb{I}(\lambda, \mu) - A\right)\widehat{N} &= \text{diag}\left(\begin{bmatrix} \lambda I_{r_h} & 0 \\ 0 & \mu I_{r_v} \end{bmatrix}, 0_{n-r}\right) - \begin{bmatrix} \hat{A}_{11} & 0 \\ 0 & I_{n-r} \end{bmatrix} \\ &= \text{diag}\left(\begin{bmatrix} \lambda I_{r_h} & 0 \\ 0 & \mu I_{r_v} \end{bmatrix} - \hat{A}_{11}, -I_{n-r}\right). \end{aligned}$$

Based on this representation, we have

$$\det\left(E\mathbb{I}(\lambda, \mu) - A\right) = (-1)^{n-r}\det(\widehat{M}^{-1})\det(\widehat{N}^{-1})\det\left(\begin{bmatrix} \lambda I_{r_h} & 0 \\ 0 & \mu I_{r_v} \end{bmatrix} - \hat{A}_{11}\right).$$

Thus, $\det\left(E\mathbb{I}(\lambda, \mu) - A\right)$ is not identically zero and the pair (E, A) is regular.

On the other hand, we also have

$$\begin{aligned}\det(sE - A) &= \det\left(\widehat{M}^{-1}(s\widehat{E} - \widehat{A})\widehat{N}^{-1}\right)\\ &= (-1)^{n-r}\det(\widehat{M}^{-1})\det(\widehat{N}^{-1})\det(sI_r - \hat{A}_{11}).\end{aligned}$$

This shows that $\deg(\det(sE - A)) = r = \operatorname{rank}(E)$, which proves the causality of the pair (E, A). □

10.3 Stability of Singular Roesser Systems with Extended Delays

10.3.1 Model Description

Consider the following 2-D descriptor system with mixed delays

$$\begin{aligned}\begin{bmatrix} E_h x^h(i+1, j) \\ E_v x^v(i, j+1) \end{bmatrix} &= A \begin{bmatrix} x^h(i, j) \\ x^v(i, j) \end{bmatrix} + A_d \begin{bmatrix} x^h(i - d_h(i), j) \\ x^v(i, j - d_v(j)) \end{bmatrix}\\ &\quad + A_s \begin{bmatrix} \sum_{l=1}^{s_h(i)} x^h(i - l, j) \\ \sum_{l=1}^{s_v(j)} x^v(i, j - l) \end{bmatrix},\end{aligned} \tag{10.2}$$

where $x^h(i, j) \in \mathbb{R}^{n_h}$ and $x^v(i, j) \in \mathbb{R}^{n_v}$ denote the horizontal and vertical state vectors, respectively, $E_h \in \mathbb{R}^{n_h \times n_h}$, $E_v \in \mathbb{R}^{n_v \times n_v}$ and A, A_d, $A_s \in \mathbb{R}^{n \times n}$ $(n = n_h + n_v)$ are given real matrices with $\operatorname{rank}(E) = r < n$, $E = \operatorname{diag}(E_h, E_v)$. In model (10.2), $d_h(i)$, $s_h(i)$ and $d_v(j)$, $s_v(j)$ represent respectively the time-delays along the horizontal and vertical directions, which are confined in prescribed ranges

$$\begin{aligned}&d_{hm} \le d_h(i) \le d_{hM}, \quad d_{vm} \le d_v(j) \le d_{vM},\\ &s_{hm} \le s_h(i) \le s_{hM}, \quad s_{vm} \le s_v(j) \le s_{vM}, \quad \text{for all } i, j \in \mathbb{N}_0,\end{aligned} \tag{10.3}$$

where d_{hm}, d_{hM}, d_{vm}, d_{vM}, s_{hm}, s_{hM}, s_{vm} and s_{vM} are known nonnegative integers involving the upper and the lower bounds of delays. Denote $\sigma_h = \max(d_{hM}, s_{hM})$ and $\sigma_v = \max(d_{vM}, s_{vM})$ as the upper bounds of delays. The overall state vector of (10.2) is defined as $x(i, j) = \operatorname{col}\{x^h(i, j), x^v(i, j)\} \in \mathbb{R}^n$.

Initial condition of system (10.2) is specified by

$$\begin{aligned}x^h(k, j) &= \phi(k, j), \quad k \in \mathbb{Z}[-\sigma_h, 0], \quad j \in \mathbb{Z}^+,\\ x^v(i, l) &= \psi(i, l), \quad l \in \mathbb{Z}[-\sigma_v, 0], \quad i \in \mathbb{Z}^+,\end{aligned} \tag{10.4}$$

where $\phi(k, .) \in l_2(\mathbb{Z}^+)$ for $k \in \mathbb{Z}[-\sigma_h, 0]$ and $\psi(., l) \in l_2(\mathbb{Z}^+)$ for $l \in \mathbb{Z}[-\sigma_v, 0]$.

We recall here that, the matrix pair (E, A) is regular if $\det[E\mathbb{I}(z, w) - A]$ is not identically zero and (E, A) is causal if $\deg(\det(sE - A)) = \mathrm{rank}(E)$. We give the following definition for 2-D singular systems as presented in (10.2).

Definition 10.2 (i) The 2-D singular system (10.2) is said to be acceptable if the pair (E, A) is regular and causal.

(ii) The acceptable 2-D singular system (10.2) is said to be internally stable if for any compatible initial condition (10.4), the corresponding solution satisfies

$$\lim_{q\to\infty} \sup_{(i,j)\in\Gamma_q} \left\| \begin{bmatrix} x^h(i, j) \\ x^v(i, j) \end{bmatrix} \right\| = 0,$$

where Γ_q denotes the diagonal line $\Gamma_q = \{(i, j) \in \mathbb{N}_0^2 : i + j = q\}$.

(iii) System (10.2) is said to be admissible if it is acceptable and internally stable.

Similar to 1-D singular systems, regularity and causality of the pair (E, A) guarantee system (10.2) to have a decomposition into slow- and fast-subsystems, which deduces the global existence and uniqueness of solutions [4]. Our main aim in this section is to derive delay-dependent conditions which ensure the system regularity, causality and internal stability. By utilizing a systematic approach based on 2-D LKF method and zero-type free matrix equations, LMI-based admissibility conditions are obtained.

10.3.2 Stability Conditions

For the brevity, we denote the following augmented vectors

$$x(i, j) = \begin{bmatrix} x^h(i, j) \\ x^v(i, j) \end{bmatrix}, \quad x_{1,1}(i, j) = \begin{bmatrix} x^h(i+1, j) \\ x^v(i, j+1) \end{bmatrix},$$
$$x_d(i, j) = \begin{bmatrix} x^h(i - d_h(i), j) \\ x^v(i, j - d_v(j)) \end{bmatrix}, \quad x_{dM}(i, j) = \begin{bmatrix} x^h(i - d_{hM}, j) \\ x^v(i, j - d_{vM}) \end{bmatrix},$$
$$x_{dm}(i, j) = \begin{bmatrix} x^h(i - d_{hm}, j) \\ x^v(i, j - d_{vm}) \end{bmatrix}, \quad x_s(i, j) = \begin{bmatrix} \sum_{l=1}^{s_h(i)} x^h(i-l, j) \\ \sum_{l=1}^{s_v(j)} x^v(i, j-l) \end{bmatrix},$$
$$\eta_h(i, j) = x^h(i+1, j) - x^h(i, j), \quad \eta_v(i, j) = x^v(i, j+1) - x^v(i, j),$$
$$\chi(i, j) = \mathrm{col}\left\{x(i, j), x_d(i, j), x_{dm}(i, j), x_{dM}(i, j), x_s(i, j)\right\}.$$

To derive LMI-based admissibility conditions for system (10.2), we construct the following functional candidate

$$V(i,j) \triangleq V(x(i,j)) = \underbrace{\sum_{k=1}^{7} V_k^h(x^h(i,j))}_{V_h(i,j)} + \underbrace{\sum_{k=1}^{7} V_k^v(x^v(i,j))}_{V_v(i,j)}, \tag{10.5}$$

where

$$
\begin{aligned}
V_1^h(i,j) &= x^{h\top}(i,j)E_h^\top P^h E_h x^h(i,j),\\
V_2^h(i,j) &= \sum_{l=i-d_{hm}}^{i-1} x^{h\top}(l,j)Q_1^h x^h(l,j),\\
V_3^h(i,j) &= \sum_{l=i-d_h(i)}^{i-1} x^{h\top}(l,j)Q_2^h x^h(l,j) + \sum_{k=-d_{hM}}^{-d_{hm}} \sum_{l=i+k}^{i-1} x^{h\top}(l,j)Q_2^h x^h(l,j),\\
V_4^h(i,j) &= d_{hm} \sum_{k=-d_{hm}}^{-1} \sum_{l=i+k}^{i-1} \eta_h^\top(l,j)E_h^\top R_1^h E_h \eta_h(l,j),\\
V_5^h(i,j) &= d_{hM} \sum_{k=-d_{hM}}^{-1} \sum_{l=i+k}^{i-1} \eta_h^\top(l,j)E_h^T R_2^h E_h \eta_h(l,j),\\
V_6^h(i,j) &= (d_{hM}-d_{hm}) \sum_{k=-d_{hM}}^{-d_{hm}-1} \sum_{l=i+k}^{i-1} \eta_h^\top(l,j)E_h^\top R_3^h E_h \eta_h(l,j),\\
V_7^h(i,j) &= s_{hM} \sum_{k=s_{hm}}^{s_{hM}} \sum_{l=1}^{k} \sum_{q=i-l}^{i-1} x^{h\top}(q,j)Z^h x^h(q,j),
\end{aligned}
$$

and

$$
\begin{aligned}
V_1^v(i,j) &= x^{v\top}(i,j)E_v^\top P^v E_v x^v(i,j),\\
V_2^v(i,j) &= \sum_{l=j-d_{vm}}^{j-1} x^{v\top}(i,l)Q_1^v x^v(i,l),\\
V_3^v(i,j) &= \sum_{l=j-d_v(j)}^{j-1} x^{v\top}(i,l)Q_2^v x^v(i,l) + \sum_{k=-d_{vM}}^{-d_{vm}} \sum_{l=j+k}^{j-1} x^{v\top}(i,l)Q_2^v x^v(i,l),\\
V_4^v(i,j) &= d_{vm} \sum_{k=-d_{vm}}^{-1} \sum_{l=j+k}^{j-1} \eta_v^\top(i,l)E_v^\top R_1^v E_v \eta_v(i,l),\\
V_5^v(i,j) &= d_{vM} \sum_{k=-d_{vM}}^{-1} \sum_{l=j+k}^{j-1} \eta_v^\top(i,l)E_v^T R_2^v E_v \eta_v(i,l),
\end{aligned}
$$

$$V_6^v(i,j) = (d_{vM} - d_{vm}) \sum_{k=-d_{vM}}^{-d_{vm}-1} \sum_{l=j+k}^{j-1} \eta_v^\top(i,l) E_v^\top R_3^v E_v \eta_v(i,l),$$

$$V_7^v(i,j) = s_{vM} \sum_{k=s_{vm}}^{s_{vM}} \sum_{l=1}^{k} \sum_{q=j-l}^{j-1} x^{v\top}(i,q) Z^v x^v(i,q).$$

In comparison to existing results, for example, [26, 27], the functional (10.5) has some special features. First, the functional (10.5) contains both the state vector in singular form and its double, triple summations. This allows us to integrate extra information of time delays into relevant augmented vectors. Second, the functionals $V_k = V_k^h + V_k^v$ $(k = 4, 5, 6, 7)$ are introduced to deal with different types of interval delays in combination with free zero-type equations which contain singular vectors. By using the functional (10.5), and by utilizing suitable 2-D summation inequalities, less conservative delay-range-dependent admissibility conditions for system (10.2) can be derived.

Our main result in this section is presented in the following theorem.

Theorem 10.1 *For given positive integers involving bounds of delays specified in* (10.3)*, assume that there exist symmetric positive definite matrices* $P = \mathrm{diag}(P^h, P^v)$*,* $Z = \mathrm{diag}(Z^h, Z^v)$*,* $Q_k = \mathrm{diag}(Q_k^h, Q_k^v)$ $(k = 1, 2)$*,* $R_l = \mathrm{diag}(R_l^h, R_l^v)$ *and real matrices* $S_l \in \mathbb{R}^{n\times(n-r)}$ $(l = 1, 2, 3)$ *which satisfy the following condition*

$$\mathscr{M} = \begin{bmatrix} \Phi & \mathscr{A}^\top P & \mathscr{D}^\top \Psi \\ * & -P & 0 \\ * & * & -\Theta \end{bmatrix} < 0, \tag{10.6}$$

where $\Psi = \begin{bmatrix} \bar{R}_1 & \bar{R}_2 & \bar{R}_3 \end{bmatrix}$*,* $\Theta = \mathrm{diag}(\bar{R}_1, \bar{R}_2, \bar{R}_3)$*,* $\Phi = (\Phi_{kl})_{5\times 5}$ *and*

$$\begin{aligned}
\Phi_{11} &= -E^\top R_1 E - E^\top R_2 E - E^\top P E + Q_1 + \bar{Q}_2 + \bar{Z} + \mathrm{Sym}\{S_1 L^\top A\},\\
\Phi_{12} &= S_1 L^\top A_d + A^\top L S_2^\top, \quad \Phi_{13} = E^\top R_1 E,\\
\Phi_{14} &= E^\top R_2 E, \quad \Gamma_{15} = S_1 L^\top A_s + A^\top L S_3^\top,\\
\Phi_{22} &= -Q_2 - 2E^\top R_3 E + \mathrm{Sym}\{S_2 L^\top A_d\},\\
\Phi_{23} &= E^\top R_3 E, \quad \Gamma_{24} = E^\top R_3 E, \quad \Phi_{25} = S_2 L^\top A_s + A_d^\top L S_3^\top,\\
\Phi_{33} &= -Q_1 - E^\top R_1 E - E^\top R_3 E, \quad \Phi_{34} = 0, \quad \Phi_{35} = 0,\\
\Phi_{44} &= -Q_2 - E^\top R_2 E - E^\top R_3 E, \quad \Phi_{45} = 0,\\
\Phi_{55} &= -Z + S_3 L^\top A_s + A_s^\top L S_3^\top,\\
\bar{Z} &= \mathbb{I}(s_{dh}, s_{dv}) Z, \quad \bar{Q}_2 = \mathbb{I}(r_{dh}+2, r_{dv}+2) Q_2,\\
r_{dh} &= d_{hM} - d_{hm}, \quad r_{dv} = d_{vM} - d_{vm},\\
s_{dh} &= \frac{s_{hM}(s_{hM} + s_{hm})(s_{hM} - s_{hm} + 1)}{2},
\end{aligned}$$

$$s_{dv} = \frac{s_{vM}(s_{vM} + s_{vm})(s_{vM} - s_{vm} + 1)}{2},$$
$$\bar{R}_1 = \mathbb{I}(d_{hm}^2, d_{vm}^2)R_1, \quad \bar{R}_2 = \mathbb{I}(d_{hM}^2, d_{vM}^2)R_2, \quad \bar{R}_3 = \mathbb{I}(r_{dh}^2, r_{dv}^2)R_2,$$
$$\mathscr{A} = \begin{bmatrix} A \ A_d \ 0 \ 0 \ A_s \end{bmatrix}, \quad \mathscr{D} = \begin{bmatrix} A - E \ A_d \ 0 \ 0 \ A_s \end{bmatrix}, \quad L = (E^\top)^\perp.$$

Then, the 2-D singular system (10.2) *is admissible*

Proof As the first step, we utilize the decompositions given in (10.1) to prove that LMI condition (10.6) guarantees the regularity and causality of system (10.2). Indeed, let $\bar{E} = MEN = \begin{bmatrix} I_r & 0 \\ 0 & 0 \end{bmatrix}$. Then, it is clear that $N^\top E^\top = \bar{E}M^{-\top}$. By this, we can parameterize the matrix $L = (E^\top)^\perp$ as

$$L = M^\top \begin{bmatrix} 0 \\ \hdashline G \end{bmatrix}.$$

We define the following matrices

$$\begin{aligned} \hat{S}_1 &= N^\top S_1 = \begin{bmatrix} S_{11} \\ S_{21} \end{bmatrix}, \quad \hat{P} = M^{-\top} P M^{-1} = \begin{bmatrix} P_{11} & P_{12} \\ P_{21} & P_{22} \end{bmatrix}, \\ \hat{R}_k &= M^{-\top} P_k M^{-1} = \begin{bmatrix} R_{k11} & R_{k12} \\ R_{k21} & R_{k22} \end{bmatrix}, \quad k = 1, 2. \end{aligned} \tag{10.7}$$

By the Schur complement lemma [see, Chap. 2, Lemma 2.7], it can be deduced from condition (10.6) that $\Phi_{11} < 0$, which yields

$$\Upsilon = -E^\top R_1 E - E^\top R_2 E - E^\top P E + \mathrm{Sym}\{S_1 L^\top A\} < 0. \tag{10.8}$$

By pre- and post-multiplying with $N^\top$ and N, from (10.8) we obtain

$$\begin{aligned} \begin{bmatrix} \hat{\Upsilon}_{11} & \hat{\Upsilon}_{12} \\ \hat{\Upsilon}_{12}^\top & \hat{\Upsilon}_{22} \end{bmatrix} &\triangleq -\bar{E}^\top \hat{R}_1 \bar{E} - \bar{E}^\top \hat{R}_2 \bar{E} - \bar{E}^\top \hat{P} \bar{E} \\ &\quad + \hat{S}_1 \begin{bmatrix} 0 & G^\top \end{bmatrix} \bar{A} + \bar{A}^\top \begin{bmatrix} 0 \\ \hdashline G \end{bmatrix} \hat{S}_1^\top < 0, \end{aligned} \tag{10.9}$$

where

$$\begin{aligned} \hat{\Upsilon}_{11} &= -R_{111} - R_{211} - P_{11} + \mathrm{Sym}\{S_{11} G^\top A_{21}\}, \\ \hat{\Upsilon}_{12} &= S_{11} G^\top A_{22} + A_{21}^\top G S_{21}^\top, \\ \hat{\Upsilon}_{22} &= S_{21} G^\top A_{22} + A_{22}^\top G S_{21}^\top. \end{aligned}$$

As a consequence of (10.9), we have

$$S_{21}G^\top A_{22} + A_{22}^\top G S_{21}^\top < 0$$

which ensures that the matrix A_{22} is nonsingular. By Lemma 10.1 and Definition 10.2, the 2-D system (10.2) is acceptable.

Next, we compute and manipulate the directional difference $\Delta V(i, j)$ along state trajectories of system (10.2)

$$\Delta V(i, j) \triangleq \underbrace{V_h(i+1, j) - V_h(i, j)}_{\Delta V_h(i,j)} + \underbrace{V_v(i, j+1) - V_v(i, j)}_{\Delta V_v(i,j)}.$$

According to (10.5) and system (10.2), the differences of V_1^h and V_2^h are given by

$$\begin{aligned}
\Delta V_1^h(i, j) &= x^{h\top}(i+1, j)E_h^\top P^h E_h x(i+1, j) - x^{h\top}(i, j)E_h^\top P^h E_h x^h(i, j), \\
\Delta V_2^h(i, j) &= x^{h\top}(i, j)Q_1^h x(i, j) - x^{h\top}(i - d_{hm}, j)Q_1^h x^h(i - d_{hm}, j).
\end{aligned} \tag{10.10}$$

We now compute the difference $\Delta V_3^h(i, j)$. Note that

$$\begin{aligned}
&\sum_{l=i+1-d_h(i+1)}^{i} x^{h\top}(l, j)Q_2^h x^h(l, j) - \sum_{l=i-d_h(i)}^{i-1} x^{h\top}(l, j)Q_2^h x^h(l, j) \\
&\quad = x^{h\top}(i, j)Q_2^h x^h(i, j) - x^{h\top}(i - d_h(i), j)Q_2^h x^h(i - d_h(i), j) \\
&\qquad + \sum_{l=i+1-d_h(i+1)}^{i-1} x^{h\top}(l, j)Q_2^h x^h(l, j) - \sum_{l=i+1-d_h(i)}^{i-1} x^{h\top}(l, j)Q_2^h x^h(l, j) \\
&\quad \le x^{h\top}(i, j)Q_2^h x^h(i, j) - x^{h\top}(i - d_h(i), j)Q_2^h x^h(i - d_h(i), j) \\
&\qquad + \sum_{l=i+1-d_{hM}}^{i-d_{hm}} x^{h\top}(l, j)Q_2^h x^h(l, j).
\end{aligned}$$

Based on this observation, we have

$$\begin{aligned}
\Delta V_3^h(i, j) \le{}& x^{h\top}(i, j)Q_2^h x^h(i, j) - x^{h\top}(i - d_h(i), j)Q_2^h x^h(i - d_h(i), j) \\
&+ \sum_{l=i+1-d_{hM}}^{i-d_{hm}} x^{h\top}(l, j)Q_2^h x^h(l, j) \\
&+ \sum_{k=-d_{hM}}^{-d_{hm}} \left[\sum_{l=i+1+k}^{i} x^{h\top}(l, j)Q_2^h x^h(l, j) - \sum_{l=i+k}^{i-1} x^{h\top}(l, j)Q_2^h x^h(l, j) \right]
\end{aligned}$$

and hence

$$
\begin{aligned}
&\le (r_{dh}+2)x^{h\top}(i,j)Q_2^h x(i,j) - x^{h\top}(i-d_h(i),j)Q_2^h x(i-d_h(i),j)\\
&\quad - x^{h\top}(i-d_{hM},j)Q_2^h x^h(i-d_{hM},j).
\end{aligned} \tag{10.11}
$$

The differences of $V_k^v(i,j)$ $(k=1,2,3)$ will be similarly estimated. In combination with (10.10) and (10.11), we have

$$
\begin{aligned}
&\sum_{k=1}^{3}\Delta(V_k^h(i,j)+V_k^v(i,j))\\
&\le x_{1,1}^\top(i,j)E^\top P E^\top x_{1,1}(i,j) - x^\top(i,j)E^\top P E x(i,j)\\
&\quad + x^\top(i,j)Q_1 x(i,j) - x_{dm}^\top(i,j)Q_1 x_{dm}(i,j)\\
&\quad + x^\top(i,j)\bar{Q}_2 x(i,j) - x_d^\top(i,j)Q_2 x_d(i,j) - x_{dM}^\top(i,j)Q_2 x_{dM}(i,j)\\
&\le \chi^\top(i,j)\mathscr{A}^\top P\mathscr{A}\chi(i,j) + x^\top(i,j)\left(-E^\top P E + Q_1 + \bar{Q}_2\right)x(i,j)\\
&\quad - x_{dm}^\top(i,j)Q_1 x_{dm}(i,j) - x_{dM}^\top(i,j)Q_2 x_{dM}(i,j) - x_d^\top(i,j)Q_2 x_d(i,j),
\end{aligned} \tag{10.12}
$$

where $Ex_{1,1}(i,j) = \underbrace{\begin{bmatrix} A & A_d & 0 & 0 & A_s \end{bmatrix}}_{\mathscr{A}}\chi(i,j)$ and $\bar{Q}_2 = \mathbb{I}(r_{dh}+2, r_{dv}+2)Q_2$.

Similar to (10.10) and (10.11), the differences ΔV_k^h $(k=4,5,6)$ are given by

$$
\begin{aligned}
\Delta V_4^h(i,j) &= d_{hm}^2\eta_h^\top(i,j)E_h^\top R_1^h E_h\eta(i,j)\\
&\quad - d_{hm}\sum_{l=i-d_{hm}}^{i-1}\eta_h^\top(l,j)E_h^\top R_1^h E_h\eta_h(l,j),\\
\Delta V_5^h(i,j) &= d_{hM}^2\eta_h^\top(i,j)E_h^\top R_2^h E_h\eta(i,j)\\
&\quad - d_{hM}\sum_{l=i-d_{hM}}^{i-1}\eta_h^\top(l,j)E_h^\top R_2^h E_h\eta_h(l,j),\\
\Delta V_6^h(i,j) &= r_{dh}^2\eta_h^\top(i,j)E_h^\top R_3^h E_h\eta(i,j)\\
&\quad - r_{dh}\sum_{l=i-d_{hM}}^{i-d_{hm}-1}\eta_h^\top(l,j)E_h^\top R_3^h E_h\eta_h(l,j).
\end{aligned}
$$

By the Jensen summation inequalities [see, Chap. 2, Corollary 2.1], we have

$$
\begin{aligned}
&-d_{hm}\sum_{l=i-d_{hm}}^{i-1}\eta_h^\top(l,j)E_h^\top R_1^h E_h\eta_h(l,j)\\
&\le -[x^h(i,j)-x^h(i-d_{hm},j)]^\top\left(E_h^\top R_1^h E_h\right)[x^h(i,j)-x^h(i-d_{hm},j)]\\
&= \begin{bmatrix} x^h(i,j)\\ x^h(i-d_{hm},j)\end{bmatrix}^\top \begin{bmatrix} -E_h^\top R_1^h E_h & E_h^\top R_1^h E_h\\ * & -E_h^\top R_1^h E_h\end{bmatrix}\begin{bmatrix} x^h(i,j)\\ x^h(i-d_{hm},j)\end{bmatrix}
\end{aligned} \tag{10.13}
$$

and

$$
\begin{aligned}
-d_{hM} \sum_{l=i-d_{hM}}^{i-1} & \eta_h^\top(l,j) E_h^\top R_2^h E_h \eta_h(l,j) \\
& \leq -[x^h(i,j) - x^h(i-d_{hM},j)]^\top \left(E_h^\top R_2^h E_h\right)[x^h(i,j) - x^h(i-\tau_{hM},j)] \\
& = \begin{bmatrix} x^h(i,j) \\ x^h(i-d_{hM},j) \end{bmatrix}^\top \begin{bmatrix} -E_h^\top R_2^h E_h & E_h^\top R_2^h E_h \\ * & -E_h^\top R_2^h E_h \end{bmatrix} \begin{bmatrix} x^h(i,j) \\ x^h(i-d_{hM},j) \end{bmatrix}.
\end{aligned} \tag{10.14}
$$

Moreover, we also have

$$
\begin{aligned}
-r_{dh} \sum_{l=i-d_{hM}}^{i-d_{hm}-1} & \eta_h^\top(l,j) E_h^\top R_3^h E_h \eta_h(l,j) \\
& \leq -(d_h(i) - d_{hm}) \sum_{l=i-d_h(i)}^{i-d_{hm}-1} \eta_h^\top(l,j) E_h^\top R_3^h E_h \eta_h(l,j) \\
& \quad - (d_{hM} - d_h(i)) \sum_{l=i-d_{hM}}^{i-d_h(i)-1} \eta_h^\top(l,j) E_h^\top R_3^h E_h \eta_h(l,j) \\
& \leq -\left[\sum_{l=i-d_h(i)}^{i-d_{hm}-1} \eta_h^\top(l,j) E_h^\top\right] R_3^h \left[E_h \sum_{l=i-d_h(i)}^{i-d_{hm}-1} z^h(l,j)\right] \\
& \quad - \left[\sum_{l=i-d_{hM}}^{i-d_h(i)-1} \eta_h^\top(l,j) E_h^\top\right] R_3^h \left[E_h \sum_{l=i-d_{hM}}^{i-d_h(i)-1} \eta_h(l,j)\right] \\
& \leq \begin{bmatrix} x^h(i-d_h(i),j) \\ x^h(i-d_{hm},j) \\ x^h(i-d_{hM},j) \end{bmatrix}^\top \mathscr{W} \otimes \left(E_h^\top R_3^h E_h\right) \begin{bmatrix} x^h(i-d_h(i),j) \\ x^h(i-d_{hm},j) \\ x^h(i-d_{hM},j) \end{bmatrix},
\end{aligned} \tag{10.15}
$$

where the coefficient matrix $\mathscr{W} = \begin{bmatrix} -2 & 1 & 1 \\ * & -1 & 0 \\ * & * & -1 \end{bmatrix}$ and the symbol $\otimes$ denotes the Kronecker product of two matrices. It follows from (10.13)–(10.15) that

$$
\begin{aligned}
\sum_{k=4}^{6} \Delta V_k^h(i,j) \leq \eta_h^\top(i,j) E_h^\top & \left(\bar{R}_1^h + \bar{R}_2^h + \bar{R}_3^h\right) E_h \eta_h(i,j) \\
& + \chi^{h\top}(i,j) \Phi_h \chi^h(i,j),
\end{aligned} \tag{10.16}
$$

where $\bar{R}_1^h = d_{hm}^2 R_1^h$, $\bar{R}_2^h = d_{hM}^2 R_2^h$, $\bar{R}_3^h = r_{dh}^2 R_3^h$ and $\Phi_h = \mathrm{diag}(\Phi_{11}^h, 0_{n_h})$ with

$$
\chi^h(i,j) = \mathrm{col}\left\{x^h(i,j), x^h(i-d_h(i),j), x^h(i-d_{hm},j), x^h(i-d_{hM},j), \sum_{l=1}^{s_h(i)} x^h(i-l,j)\right\},
$$

$$\Phi_{11}^h = \begin{bmatrix} -E_h^\top(R_1^h + R_2^h)E_h & 0 & E_h^\top R_1^h E_h & E_h^\top R_2^h E_h \\ * & -2E_h^\top R_3^h E_h & E_h^\top R_3^h E_h & E_h^\top R_3^h E_h \\ * & * & -E_h^\top(R_1^h + R_3^h)E_h & 0 \\ * & * & * & -E_h^\top(R_2^h + R_3^h)E_h \end{bmatrix}.$$

By the same arguments used to derive (10.16) we obtain

$$\begin{aligned}\sum_{k=4}^{6} \Delta V_k^v(i,j) &\le \eta_v^\top(i,j)E_v^\top\left(\bar{R}_1^v + \bar{R}_2^v + \bar{R}_3^v\right)E_v\eta_v(i,j) \\ &\quad + \chi^{v\top}(i,j)\Phi_v\chi^v(i,j), \end{aligned} \tag{10.17}$$

where $\bar{R}_1^v = d_{vm}^2 R_1^v$, $\bar{R}_2^v = d_{vM}^2 R_2^v$, $\bar{R}_3^v = r_{dv}^2 R_3^v$ and $\Phi^v = \mathrm{diag}(\Phi_{11}^v, 0_{n_v})$ with

$$\chi^v(i,j) = \mathrm{col}\left\{x^v(i,j), x^v(i,j-d_v(j)), x^v(i,j-d_{vm}), x^v(i,j-d_{vM}), \sum_{l=1}^{s_v(j)} x^v(i,j-l)\right\},$$

$$\Phi_{11}^v = \begin{bmatrix} -E_v^\top(R_1^v + R_2^v)E_v & 0 & E_v^\top R_1^v E_v & E_v^\top R_2^v E_v \\ * & -2E_v^\top R_3^v E_v & E_v^\top R_3^v E_v & E_v^\top R_3^v E_v \\ * & * & -E_v^\top(R_1^v + R_3^v)E_v & 0 \\ * & * & * & -E_v^\top(R_2^v + R_3^v)E_v \end{bmatrix}.$$

We now manipulate the difference $\Delta V_7(i,j)$. According to (10.5), we have

$$\begin{aligned}\Delta V_7^h(i,j) &= s_{hM}\sum_{k=s_{hm}}^{s_{hM}}\sum_{l=1}^{k}\left(\sum_{q=i+1-l}^{i} x^{h\top}(q,j)Z^h x^h(q,j) - \sum_{q=i-l}^{i-1} x^{h\top}(q,j)Z^h x^h(q,j)\right) \\ &= r_{sh}x^{h\top}(i,j)Z^h x^h(i,j) - s_{hM}\sum_{k=s_{hm}}^{s_{hM}}\sum_{l=1}^{k} x^{h\top}(i-l,j)Z^h x^h(i-l,j) \\ &\le s_{dh}x^{h\top}(i,j)Z^h x^h(i,j) - s_{hM}\sum_{l=1}^{s_h(i)} x^{h\top}(i-l,j)Z^h x^h(i-l,j) \\ &\le s_{dh}x^{h\top}(i,j)Z^h x^h(i,j) - s_h(i)\sum_{l=1}^{s_h(i)} x^{h\top}(i-l,j)Z^h x^h(i-l,j). \end{aligned}$$

Therefore,

$$\begin{aligned}\Delta V_7^h(i,j) &\le s_{dh}x^{h\top}(i,j)Z^h x^h(i,j) \\ &\quad - \left(\sum_{l=1}^{s_h(i)} x^h(i-l,j)\right)^\top Z^h \left(\sum_{l=1}^{s_h(i)} x^h(i-l,j)\right). \end{aligned} \tag{10.18}$$

Similarly, we also have

$$\Delta V_7^v(i,j) \le s_{dv} x^{v\top}(i,j) Z^v x^v(i,j) - \Big(\sum_{l=1}^{s_v(j)} x^v(i,j-l)\Big)^\top Z^v \Big(\sum_{l=1}^{s_v(j)} x^v(i,j-l)\Big). \tag{10.19}$$

A combination of (10.18) and (10.19) gives

$$\Delta V_7^h(i,j) + \Delta V_7^v(i,j) \le x^\top(i,j)\bar{Z}x(i,j) - x_s^\top(i,j) Z x_s(i,j). \tag{10.20}$$

On the other hand, since $L = (E^\top)^\perp$, we have $E^\top L = 0$. Thus, for any matrices S_1, S_2 and S_3 of appropriate dimension, the following zero-type equation holds

$$2\left[x^\top(i,j)S_1L^\top + x_d^\top(i,j)S_2L^\top + x_s^\top(i,j)S_3L^\top\right] E x_{1,1}(i,j) = 0. \tag{10.21}$$

Taking equation (10.2) into account, from (10.21), we have

$$\begin{aligned} &2\left[x^\top(i,j)S_1L^\top + x_d^\top(i,j)S_2L^\top + x_s^\top(i,j)S_3L^\top\right] \\ &\quad \times \left[Ax(i,j) + A_d x_d(i,j) + A_s x_s(i,j)\right] = 0. \end{aligned} \tag{10.22}$$

Finally, from the above derivation, we finally obtain

$$\begin{aligned} \Delta V(i,j) &= \sum_{k=1}^{7}\left(\Delta V_k^h(i,j) + \Delta V_k^v(i,j)\right) \\ &\le \chi^\top(i,j)\left[\Phi + \mathscr{A}^\top P \mathscr{A} + \mathscr{D}^\top\left(\bar{R}_1 + \bar{R}_2 + \bar{R}_3\right)\mathscr{D}\right]\chi(i,j). \end{aligned} \tag{10.23}$$

By Schur complement lemma [Chap. 2, Lemma 2.7], it is found that the inequality

$$\Phi + \mathscr{A}^\top P \mathscr{A} + \mathscr{D}^\top\left(\bar{R}_1 + \bar{R}_2 + \bar{R}_3\right)\mathscr{D} < 0$$

holds if and only if

$$\mathscr{M} = \begin{bmatrix} \Phi & \mathscr{A}^\top P & \mathscr{D}^\top \Psi \\ * & -P & 0 \\ * & * & -\Theta \end{bmatrix} < 0.$$

Thus, by condition (10.6), there exists a positive scalar λ_0 such that

$$\Delta V(i,j) = \Delta V^h(i,j) + \Delta V^v(i,j) \le -c_0\|x(i,j)\|^2. \tag{10.24}$$

For a positive integer q, let $E_V(q) = \sum_{(i,j)\in\Gamma(q)} V(i,j)$ denote the total energy of the functional $V(i,j)$ stored in the diagonal line $\Gamma_q = \{(i,j) \in \mathbb{N}_0^2 : i + j = q\}$. It can be deduced from (10.24) that

$$
\begin{aligned}
\sum_{(i,j)\in\Gamma_{q+1}} V(i,j) &= \sum_{(i,j)\in\Gamma_{q+1}} \left(V^h(i,j)+V^v(i,j)\right) \\
&= V^h(1,q)+V^h(2,q-1)+\cdots+V^h(q+1,0) \\
&\quad + V^v(q,1)+V^v(q-1,2)+\cdots+V^v(0,q+1) \\
&\le V^h(0,q)+V^h(1,q-1)+\cdots+V^h(q,0) \\
&\quad + V^v(q,0)+V^v(q-1,1)+\cdots+V^v(0,q)-c_0\sum_{(i,j)\in\Gamma_q}\|x(i,j)\|^2 \\
&= \sum_{(i,j)\in\Gamma_q}\left(V^h(i,j)+V^v(i,j)\right)-c_0\sum_{(i,j)\in\Gamma_q}\|x(i,j)\|^2.
\end{aligned}
$$

This shows that, under condition (10.6), the energy $E_V(q)$ stored in Γ_q decreases as q increases. More specifically, we have

$$
E_V(q+1)\le E_V(q)-c_0\sum_{(i,j)\in\Gamma_q}\|x(i,j)\|^2. \tag{10.25}
$$

It follows from (10.25) that there exists a finite limit $E_V(\infty)=\lim_{q\to\infty}E_V(q)$. Furthermore, we also have

$$
c_0\sum_{(i,j)\in\Gamma_q}\|x(i,j)\|^2\le E_V(q)-E_V(q+1)\to 0
$$

as q tends to infinity. This shows that system (10.2) is internally stable. The proof is completed. □

Remark 10.1 Theorem 10.1 provides delay-range-dependent stability conditions for 2-D singular systems in the presence of two types of delays simultaneously. As a special case, when $A_s=0$, system (10.2) is reduced to 2-D singular systems with state delays, which are typically considered in existing works [26, 27, 29, 30]. To this special case, the functional $V_7(x)$ in (10.6) is omitted and the result of Theorem 10.1 is reduced to the following corollary.

Corollary 10.1 *The 2-D singular system* (10.2) *with $A_s=0$ is admissible if there exist positive definite matrices $P=\mathrm{diag}(P^h,P^v)$, $Q_k=\mathrm{diag}(Q_k^h,Q_k^v)$ $(k=1,2)$, $R_l=\mathrm{diag}(R_l^h,R_l^v)$ $(l=1,2,3)$ and real matrices $S_k\in\mathbb{R}^{n\times(n-r)}$ $(k=1,2)$ that satisfy the following condition*

$$
\mathscr{N}=\begin{bmatrix}\hat{\Phi} & \hat{\mathscr{A}}^\top P & \hat{\mathscr{D}}^\top\Psi \\ * & -P & 0 \\ * & * & -\Theta\end{bmatrix}<0, \tag{10.26}
$$

where $\hat{\Phi}=(\hat{\Phi}_{kl})_{4\times4}$ and

$$
\begin{aligned}
&\hat{\Phi}_{11}=-E^\top R_1E-E^\top R_2E-E^\top PE+Q_1+\bar{Q}_2+\mathrm{Sym}\{S_1L^\top A\},\\
&\hat{\Phi}_{12}=S_1L^\top A_d+A^\top LS_2^\top,\quad \hat{\Phi}_{13}=E^\top R_1E,\quad \hat{\Phi}_{14}=E^\top R_2E,
\end{aligned}
$$

$$
\begin{aligned}
\hat{\Phi}_{22} &= -Q_2 - 2E^\top R_3 E + \mathrm{Sym}\{S_2 L^\top A_d\},\\
\hat{\Phi}_{23} &= E^\top R_3 E, \quad \hat{\Phi}_{24} = E^\top R_3 E,\\
\hat{\Phi}_{33} &= -Q_1 - E^\top R_1 E - E^\top R_3 E, \quad \hat{\Phi}_{34} = 0,\\
\hat{\Phi}_{44} &= -Q_2 - E^\top R_2 E - E^\top R_3 E,\\
\hat{\mathscr{A}} &= \begin{bmatrix} A & A_d & 0 & 0 \end{bmatrix}, \quad \hat{\mathscr{D}} = \begin{bmatrix} A - E & A_d & 0 & 0 \end{bmatrix},
\end{aligned}
$$

and some other matrices are defined as in (10.6).

Remark 10.2 For 2-D singular systems without delays, by specifying $A_d = 0$ and $A_s = 0$, and by similar arguments used in the proof of Theorem 10.1, it is found that system (10.2) without delay is admissible if there exist a positive definite matrix $P = \mathrm{diag}(P^h, P^v)$ and a matrix $S \in \mathbb{R}^{n\times(n-r)}$ that satisfy the following condition

$$
\begin{bmatrix} -E^\top P E + \mathrm{Sym}\{S L^\top A\} & A^\top P \\ * & -P \end{bmatrix} < 0, \tag{10.27}
$$

where $L = (E^\top)^\perp$.

Remark 10.3 In [24], stability and design conditions of a state-feedback controller were derived for 2-D singular Roesser systems without delay. It should be pointed out that the derived conditions in [24] are dependent upon selected matrices which produce a decomposition as (10.1). This approach may result more complicated conditions for testing stability and controller design. Different from that, condition (10.27) is derived in terms of tractable LMI, which does not depend on the selected decomposition. The proposed method in this chapter can also be adapted for various problems in the systems and control theory of 2-D singular systems.

10.3.3 Numerical Examples

In this section, we present some numerical examples to illustrate the effectiveness of the obtained stability conditions.

Example 10.1 Consider a thermal process which is described by the following Darboux-type equation [42].

$$
\begin{aligned}
\frac{\partial w(s,t)}{\partial x} + \frac{\partial w(x,t)}{\partial t} &= a_0 w(x,t) + a_1 w(x - d_x, t) + a_2 w(x, t - d_t)\\
&\quad + c_1 \int_0^{x_f} w(x - s, t) ds + c_2 \int_0^{t_f} w(x, t - \theta) d\theta,
\end{aligned} \tag{10.28}
$$

where $w(x, t)$ represents the unknown function at space x and time t, a_0, a_1, a_2, and c_1, c_2 are real scalars, d_x, d_t, and x_f, t_f represent delay terms.

For given increments Δx and Δt, denote $w(i, j) = w(i\Delta x, j\Delta t)$. The derivatives $\frac{\partial w(x,t)}{\partial x}$ and $\frac{\partial w(x,t)}{\partial t}$ will be approximated using a backward difference scheme as

$$\frac{\partial w(x,t)}{\partial x} \simeq \frac{w(i,j) - w(i-1,j)}{\Delta x},$$
$$\frac{\partial w(x,t)}{\partial t} \simeq \frac{w(i,j) - w(i,j-1)}{\Delta t}.$$

Define $x^h(i, j) = w(i-1, j)$ and $x^v(i, j) = w(i, j)$. Then, the equation (10.28) can be transformed into the following 2-D system

$$\begin{bmatrix} 1 & 0 \\ 0 & 0 \end{bmatrix} \begin{bmatrix} x^h(i+1,j) \\ x^v(i,j+1) \end{bmatrix} = A \begin{bmatrix} x^h(i,i) \\ x^v(i,j) \end{bmatrix} + A_d \begin{bmatrix} x^h(i-d_h(i),j) \\ x^v(i,j-d_v(j)) \end{bmatrix} + A_s \begin{bmatrix} \sum_{l=1}^{s_h} x^h(i-l,j) \\ \sum_{l=1}^{s_v} x^v(i,j-l) \end{bmatrix}, \tag{10.29}$$

where $d_h(i) = \lfloor d_x/\Delta x \rfloor$, $d_v(j) = \lfloor d_t/\Delta t \rfloor$, $s_h = \lfloor x_f/\Delta x \rfloor$ and $s_v = \lfloor t_f/\Delta t \rfloor$.

For illustrative purpose, we give the system matrices of (10.29) as

$$A = \begin{bmatrix} -0.084 & -0.7186 \\ 0.084 & 1.7186 \end{bmatrix}, \quad A_d = \begin{bmatrix} 0 & 0.4286 \\ 0 & -0.4286 \end{bmatrix}, \quad A_s = \begin{bmatrix} 0.1195 & -0.1536 \\ 0 & -0.1089 \end{bmatrix}.$$

With time-varying delays $1 \le d_h(i) \le 6$, $1 \le d_v(j) \le 6$ and $s_h = 5$, $s_v = 4$, by solving condition (10.6) using Matlab LMI toolbox, we obtain

$$P = \left[\begin{array}{c|c} 44.1235 & 0 \\ \hline 0 & 9.2954 \end{array}\right], \quad Q_1 = \left[\begin{array}{c|c} 2.0402 & 0 \\ \hline 0 & 6.7049 \end{array}\right],$$
$$Q_2 = \left[\begin{array}{c|c} 0.5405 & 0 \\ \hline 0 & 39.4883 \end{array}\right], \quad R_1 = \left[\begin{array}{c|c} 0.5417 & 0 \\ \hline 0 & 9.2954 \end{array}\right],$$
$$R_2 = \left[\begin{array}{c|c} 0.0215 & 0 \\ \hline 0 & 9.2954 \end{array}\right], \quad R_3 = \left[\begin{array}{c|c} 0.0277 & 0 \\ \hline 0 & 9.2954 \end{array}\right],$$
$$S_1 = \begin{bmatrix} -21.9586 \\ -625.2429 \end{bmatrix}, \quad S_2 = \begin{bmatrix} -0.0002 \\ 103.2363 \end{bmatrix},$$
$$S_3 = \begin{bmatrix} 2.4675 \\ 21.9502 \end{bmatrix}, \quad Z = \left[\begin{array}{c|c} 1.1479 & 0 \\ \hline 0 & 6.4587 \end{array}\right].$$

By Theorem 10.1, the 2-D singular system (10.29) is admissible. Simulation results presented in Figs. 10.1 and 10.2 are conducted with delays $d_h(i) = 1 + 5|\sin(\pi i/8)|$, $d_v(j) = 1 + 5|\sin(\pi j/8)|$, $s_h = 5$, $s_v = 4$ and finite support initial $x^h(k, j) = 1$ for $k \in \mathbb{Z}[-6, 0]$, $j \le 40$, and $x^v(i, l) = 1$ for $i \le 40$, $l \in \mathbb{Z}[-6, 0]$. It can be seen that the corresponding state trajectories $x^h(i, j)$ and $x^v(i, j)$ of (10.29) are convergent as $i + j \to \infty$ as revealed by the result of Theorem 10.1.

Example 10.2 To further illustrate the effectiveness of the derived conditions, we consider system (10.2) without summation delays (i.e. $A_s = 0$). Let $E = \text{diag}(1, 0, 1, 0)$ and the matrices A, A_d are given by

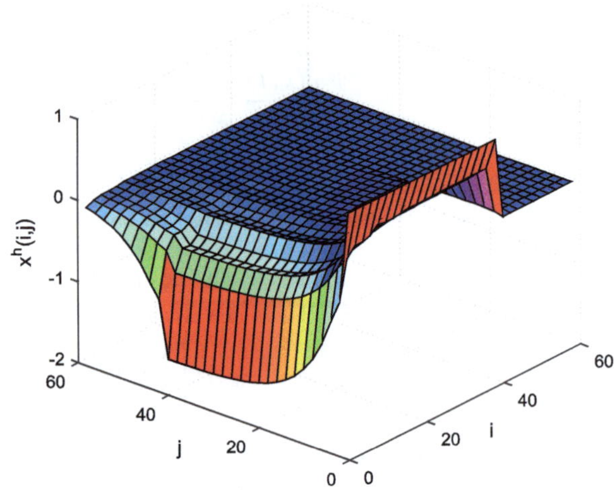

Fig. 10.1 A state trajectory of x^h with delays $d_h(i) = 1 + 5|\sin(\pi i/8)|$, $d_v(j) = 1 + 5|\sin(\pi j/8)|$, $s_h = 5$, $s_v = 4$

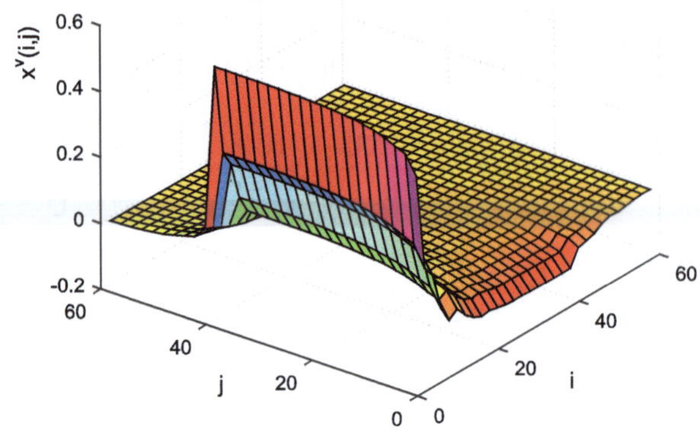

Fig. 10.2 A state trajectory of x^v with delays $d_h(i) = 1 + 5|\sin(\pi i/8)|$, $d_v(j) = 1 + 5|\sin(\pi j/8)|$, $s_h = 5$, $s_v = 4$

$$A = \left[\begin{array}{cc:cc} 0.3397 & -0.7281 & 0.2136 & 0 \\ -0.3397 & 1.7281 & -0.1692 & 0.3218 \\ \hdashline -0.1366 & 0 & 0.3961 & -5.9593 \\ 0.1598 & -0.2219 & -0.3961 & 6.9593 \end{array}\right],$$

$$A_d = \left[\begin{array}{cc:cc} 0 & 0.4286 & -0.4286 & 0.2163 \\ 0 & -0.4286 & 0.4286 & -0.1259 \\ \hdashline 0 & 0.1062 & 0.4287 & 0.1562 \\ 0.0518 & -0.2105 & -0.3865 & -0.2317 \end{array}\right].$$

Table 10.1 MAB of d_{hM} for various d_{hm} and $d_{vm} = 1$, $d_{vM} = 3$

d_{hm}	1	2	5	7	10	20
Corollary 10.1	4	6	9	11	14	24

For this system, it is found that the delay-independent stability conditions proposed in [26, Theorem 1] and [27, Lemma 2.7] are not feasible. Also, the delay-dependent condition given in [43] does not deliver any feasible solution. Thus, the derived stability conditions in [26, 27, 43] cannot access the stability of this system.

We now apply Corollary 10.1 to this example. Let $d_v(j) \in [1, 3]$. The maximum allowable bounds (MABs) of d_{hM} for various value of d_{hm} subject to condition (10.26) are given in Table 10.1. For example, with $d_{hm} = 2$, $d_{hM} = 6$, a feasible solution of (10.26) is given by

$$P = \operatorname{diag}\left(\left[\begin{array}{c:c} 0.6044 & 0 \\ \hdashline 0 & 0.7178 \end{array}\right], \left[\begin{array}{c:c} 7.5285 & 0 \\ \hdashline 0 & 0.7178 \end{array}\right]\right),$$

$$R_1 = \operatorname{diag}\left(\left[\begin{array}{c:c} 0.0280 & 0 \\ \hdashline 0 & 0.7178 \end{array}\right], \left[\begin{array}{c:c} 0.0972 & 0 \\ \hdashline 0 & 0.7178 \end{array}\right]\right),$$

$$R_2 = \operatorname{diag}\left(\left[\begin{array}{c:c} 0.0034 & 0 \\ \hdashline 0 & 0.7178 \end{array}\right], \left[\begin{array}{c:c} 0.0097 & 0 \\ \hdashline 0 & 0.7178 \end{array}\right]\right),$$

$$R_3 = \operatorname{diag}\left(\left[\begin{array}{c:c} 0.0114 & 0 \\ \hdashline 0 & 0.7178 \end{array}\right], \left[\begin{array}{c:c} 0.0438 & 0 \\ \hdashline 0 & 0.7178 \end{array}\right]\right),$$

$$Q_1 = \operatorname{diag}\left(\begin{bmatrix} 0.1269 & -0.0867 \\ -0.0867 & 0.1439 \end{bmatrix}, \begin{bmatrix} 0.2517 & -0.0469 \\ -0.0469 & 0.7102 \end{bmatrix}\right),$$

$$Q_2 = \operatorname{diag}\left(\begin{bmatrix} 0.0408 & -0.0918 \\ -0.0918 & 0.6072 \end{bmatrix}, \begin{bmatrix} 1.6265 & -0.5989 \\ -0.5989 & 0.8857 \end{bmatrix}\right),$$

$$S_1 = \begin{bmatrix} 24.9889 & 36.7921 \\ -61.7633 & -193.2811 \\ -40.9425 & 22.6660 \\ 774.2601 & -94.0574 \end{bmatrix}, \quad S_2 = \begin{bmatrix} 5.8482 & -0.1281 \\ -14.9714 & 49.3266 \\ -51.8389 & -44.2690 \\ -23.5090 & 15.7538 \end{bmatrix}.$$

Note in addition that the derived conditions of Corollary 10.1 can also produce less computational burden than [43, Theorem 1]. More specifically, the number of decision variables (NoDVs) of Corollary 10.1 is $3(n_h^2 + n_v^2) + 3n + 2n(n - r)$, which is clearly fewer than NoDV $= 8.5(n_h^2 + n_v^2) + 3.5(n_h + n_v)$ of [43, Theorem 1], where $n = n_h + n_v$ and $r = \operatorname{rank}(E)$. For this particular example, the corresponding numbers are NoDV $= 52$ and NoDV $= 82$. The obtained results illustrate that our method can deliver less conservative stability conditions than existing results in [26, 27, 43] both in terms of feasibility and computational complexity.

Example 10.3 We consider a delay-free 2-D singular system with $n_h = n_v = 2$, $E = \text{diag}(1, 1, 0, 0)$ and

$$A = \left[\begin{array}{cc:cc} 0.3397 & -0.7281 & 0.2136 & 0.0157 \\ -0.3397 & 1.7281 & -0.1692 & 0.3218 \\ \hdashline -0.1366 & -0.1173 & 0.3961 & -5.9593 \\ 0.1598 & -0.2219 & -0.3961 & 6.9593 \end{array}\right].$$

For this system, the derived conditions of [24, Theorem 6] do not provide any feasible solution. Whereas, the derived condition (10.26) is feasible with

$$P = \left[\begin{array}{cc:cc} 3.0977 & -0.7321 & \multicolumn{2}{c}{\multirow{2}{*}{0}} \\ -0.7321 & 2.1992 & & \\ \hdashline \multicolumn{2}{c:}{\multirow{2}{*}{0}} & 2.6995 & 2.2257 \\ & & 2.2257 & 2.0282 \end{array}\right].$$

This result again shows that the method presented in this section produces less conservative stability conditions.

10.4 Stabilization of 2-D Singular Roesser Systems with Delays: An Event-Triggered Control Approach

10.4.1 Event-Triggered State-Feedback Control

In this section, we address the stabilization problem via ETCs for a class of 2-D singular Roesser systems with state delays given as

$$\begin{bmatrix} E_h x^h(i+1, j) \\ E_v x^v(i, j+1) \end{bmatrix} = A \begin{bmatrix} x^h(i, j) \\ x^v(i, j) \end{bmatrix} + A_d \begin{bmatrix} x^h(i-d_h, j) \\ x^v(i, j-d_v) \end{bmatrix} + Bu(i, j), \tag{10.30}$$

where $u(i, j) \in \mathbb{R}^m$ is the control input. Other vectors and matrices are described similarly in system (10.2).

As discussed in Chap. 2, in practice, it is important to operate a system with minimum used resources. In particular, to use limited networks resources more effectively, it is not necessary to transmit the data at every 2-D time instant (i, j) to enforce the stability with desired performance of the closed-loop system. In the following, we demonstrate a design scheme of an ETC to stabilize system (10.30). Assume that the vertical and horizontal state vectors are updated at instants i_p, j_q, and the latest states will be denoted as $x^h(i_p, j)$ and $x^v(i, j_q)$. The errors $e^h(i, j)$, $e^v(i, j)$ between states $x^h(i, j)$, $x^v(i, j)$ and the latest transmitted states $x^h(i_p, j)$, $x^v(i, j_q)$ are defined by

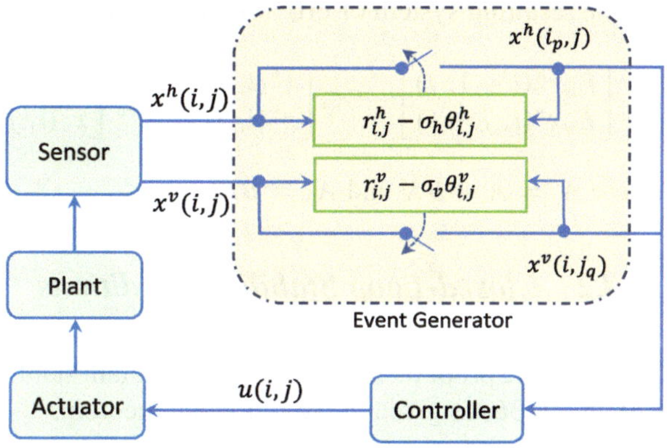

Fig. 10.3 A schematic event-triggered control

$$\begin{bmatrix} e^h(i,j) \\ e^v(i,j) \end{bmatrix} = \begin{bmatrix} x^h(i_p,j) \\ x^v(i,j_q) \end{bmatrix} - \begin{bmatrix} x^h(i,j) \\ x^v(i,j) \end{bmatrix}. \tag{10.31}$$

From the latest transmitted instants i_p and j_q, the next horizontal and vertical time instants i_{p+1}, j_{q+1} are determined respectively by

$$i_{p+1} = i_p + \min_{s>0} \left\{ s = i - i_p \middle| e^{h\top}(i,j)\Pi_h e^h(i,j) > \sigma_h x^{h\top}(i,j)\Pi_h x^h(i,j) \right\} \tag{10.32}$$

and

$$j_{q+1} = j_q + \min_{t>0} \left\{ t = j - j_q \middle| e^{v\top}(i,j)\Pi_v e^v(i,j) > \sigma_v x^{v\top}(i,j)\Pi_v x^v(i,j) \right\}, \tag{10.33}$$

where σ_h and σ_v are positive scalars related to triggering parameters and Π_h, Π_v are positive definite matrices, which will be determined. This means that the states $x^h(i,j)$ and $x^v(i,j)$ will not be transmitted within the rectangle region $[i_p, i_{p+1}) \times [j_q, j_{q+1})$. Between two consecutive transmission instants, it holds that

$$\begin{cases} e^{h\top}(i,j)\Pi_h e^h(i,j) \le \sigma_h x^{h\top}(i,j)\Pi_h x^h(i,j), \\ e^{v\top}(i,j)\Pi_v e^v(i,j) \le \sigma_v x^{v\top}(i,j)\Pi_v x^v(i,j. \end{cases} \tag{10.34}$$

An ETC for system (10.30) will be taken in the form

$$u(i,j) = K \begin{bmatrix} x^h(i_p,j) \\ x^v(i,j_q) \end{bmatrix}, \tag{10.35}$$

where $x^h(i_p,j), x^v(i,j_q)$ are the latest transmitted states and triggering instants i_p, j_q will be determined by an event generator. Figure 10.3 demonstrates a diagram of the ETC scheme (10.35), where $r^h_{i,j} = e^{h\top}(i,j)\Pi_h e^h(i,j)$, $\theta^h_{i,j} = x^{h\top}(i,j)\Pi_h x^h(i,j)$ and $r^v_{i,j} = e^{v\top}(i,j)\Pi_v e^v(i,j)$, $\theta^v_{i,j} = x^{v\top}(i,j)\Pi_v x^v(i,j)$.

The resulting system of (10.30) under ETC (10.35) is represented by

$$\begin{bmatrix} E_h x^h(i+1,j) \\ E_v x^v(i,j+1) \end{bmatrix} = A_c \begin{bmatrix} x^h(i,j) \\ x^v(i,j) \end{bmatrix} + A_d \begin{bmatrix} x^h(i-d_h,j) \\ x^v(i,j-d_v) \end{bmatrix} + A_e \begin{bmatrix} e^h(i,j) \\ e^v(i,j) \end{bmatrix}, \tag{10.36}$$

where $A_c = A + BK$ and $A_e = BK$.

10.4.2 Closed-Loop Stability Conditions

Similar to the proof of Theorem 10.1, we obtain stability conditions for 2-D singular system (10.36) as given in the following theorem.

Theorem 10.2 *For given scalars $\sigma_h > 0$, $\sigma_v > 0$, the closed-loop system* (10.36) *is admissible if there exist positive definite matrices $P = \mathrm{diag}(P^h, P^v)$, $Q = \mathrm{diag}(Q^h, Q^v)$, $R = \mathrm{diag}(R^h, R^v)$, $\Pi = \mathrm{diag}(\Pi_h, \Pi_v)$, and a matrix S of appropriate dimension that satisfy the following condition*

$$\begin{bmatrix} \Omega & \mathscr{A}_c^\top P & (\mathscr{A}_c - Ee_1)^\top \widehat{R} \\ * & -P & 0 \\ * & * & -R \end{bmatrix} < 0, \tag{10.37}$$

where $\mathscr{A}_c = A_c e_1 + A_d e_2 + A_e e_3$, $\Omega = (\Omega_{kl})_{3\times 3}$, $\widehat{R} = \mathbb{I}(d_h, d_v)R$, and

$$\begin{aligned} e_k &= \begin{bmatrix} 0_{n\times(k-1)n} & I_n & 0_{n\times(3-k)n} \end{bmatrix}, \quad k = 1,2,3, \\ \Omega_{11} &= -E^\top PE - E^\top RE - \mathrm{Sym}\{S^\top \mathscr{N}_E^\top A_c\} + Q + \mathbb{I}(\sigma_h, \sigma_v)\Pi, \\ \Omega_{12} &= E^\top RE - S^\top \mathscr{N}_E^\top A_d, \quad \Omega_{13} = -S^\top \mathscr{N}_E^\top A_e, \\ \Omega_{22} &= -Q - E^\top RE, \quad \Omega_{23} = 0, \quad \Omega_{33} = -\Pi, \end{aligned}$$

with $\mathscr{N}_E = (E^\top)^\perp$ is the null space of $E^\top$.

Proof Similar to the proof of Theorem 10.1, we first prove that subject to condition (10.37), the closed-loop system (10.36) is acceptable. Indeed, let M, N be nonsingular matrices defined by the decomposition (10.1) with the matrices $E = \mathrm{diag}(E_h, E_v)$ and A_c. We define the following matrices

$$\mathscr{S} = SN = \begin{bmatrix} S_{11} & S_{12} \end{bmatrix}, \quad \mathscr{L} = M^{-\top}\mathscr{N}_E = \begin{bmatrix} L_{11} \\ L_{21} \end{bmatrix},$$

$$\mathscr{P} = M^{-\top} P M^{-1} = \begin{bmatrix} P_{11} & P_{12} \\ P_{21} & P_{22} \end{bmatrix}, \quad \mathscr{R} = M^{-\top} R M^{-1} = \begin{bmatrix} R_{11} & R_{12} \\ R_{21} & R_{22} \end{bmatrix}.$$

Since $\mathrm{rank}(\mathscr{N}_E) = n - r$ and $E^\top \mathscr{N}_E = 0$, we have

$$E^\top \mathscr{N}_E = N^\top E^\top M^\top M^{-\top} \mathscr{N}_E = \bar{E}^\top \mathscr{L} = 0.$$

Thus $L_{11} = 0$ and we can represent the matrix $\mathscr{N}_E$ as

$$\mathscr{N}_E = M^\top \begin{bmatrix} 0 \\ L_{21} \end{bmatrix}.$$

By the Schur complement lemma, condition (10.37) induces that $\Omega_{11} < 0$ and consequently,

$$-E^\top PE - E^\top RE - \mathrm{Sym}\{S^\top \mathscr{N}_E^\top A_c\} < 0. \tag{10.38}$$

By pre- and post-multiplying with $N^\top$ and N, from (10.38), we obtain

$$-\bar{E}^\top \mathscr{P}\bar{E} - \bar{E}^\top \mathscr{R}\bar{E} - \mathrm{Sym}\{\overline{A}_c^\top \mathscr{L}\mathscr{S}\} = \left[\begin{array}{c:c} \Upsilon_{11} & \Upsilon_{12} \\ \hdashline \Upsilon_{21} & \Upsilon_{22} \end{array}\right] < 0, \tag{10.39}$$

where $\Upsilon_{22} = -S_{12}^\top L_{21}^\top A_{22}^c - A_{22}^{c\top} L_{21} S_{12}$. Again, by the Schur complement lemma, inequality (10.39) implies that $\Upsilon_{22} < 0$, and hence

$$S_{12}^\top L_{21}^\top A_{22}^c + A_{22}^{c\top} L_{21} S_{12} > 0.$$

If the matrix A_{22}^c is singular then there exists a nonzero vector v of appropriate dimension such that $A_{22}^c v = 0$ and $v^\top A_{22}^{c\top} = 0$. Therefore,

$$0 < -v^\top \Upsilon_{22} v = v^\top \left(S_{12}^\top L_{21}^\top A_{22}^c + A_{22}^{c\top} L_{21} S_{12}\right) v = 0,$$

which clearly yields a contradiction. This shows that the matrix A_{22}^c is nonsingular. By Lemma 10.1, the closed-loop system (10.36) is acceptable.

We denote the vectors

$$\begin{aligned}
x_d(i,j) &= \begin{bmatrix} x^h(i-d_h, j) \\ x^v(i, j-d_v) \end{bmatrix}, \quad e(i,j) = \begin{bmatrix} e^h(i,j) \\ e^v(i,j) \end{bmatrix}, \\
\eta_d(i,j) &= \begin{bmatrix} \sum_{k=i-d_h}^{i-1} \eta_h(k,j) \\ \sum_{l=j-d_v}^{j-1} \eta_v(i,l) \end{bmatrix}, \quad \eta(i,j) = \begin{bmatrix} \eta_h(i,j) \\ \eta_v(i,j) \end{bmatrix}, \\
\eta_h(i,j) &= x^h(i+1,j) - x^h(i,j), \quad \eta_v(i,j) = x^v(i,j+1) - x^v(i,j), \\
\chi(i,j) &= \mathrm{col}\{x(i,j), x_d(i,j), e(i,j)\}, \quad \zeta(i,j) = \mathrm{col}\{x(i,j), e(i,j)\},
\end{aligned}$$

and consider the LKF candidate

$$\underbrace{V(x(i,j))}_{V} = \sum_{k=1}^{3} \underbrace{V_k(x(i,j))}_{V_k}, \tag{10.40}$$

where

$$V_1 = x^{h\top}(i, j)E_h^\top P^h E_h x^h(i, j) + x^{v\top}(i, j)E_v^\top P^v E_v x^v(i, j),$$

$$V_2 = \sum_{k=i-d_h}^{i-1} x^{h\top}(k, j)Q^h x^h(k, j) + \sum_{l=j-d_v}^{j-1} x^{v\top}(i, l)Q^v x^v(i, l),$$

$$V_3 = d_h \sum_{k=-d_h}^{-1} \sum_{s=i+k}^{i-1} \eta_h^\top(s, j)E_h^\top R^h E_h \eta_h(s, j) + d_v \sum_{l=-d_v}^{-1} \sum_{s=j+l}^{j-1} \eta_v^\top(i, s)E_v^\top R^v E_v \eta_v(i, s).$$

Similar to the proof of Theorem 10.1, we have

$$\Delta V_1 + \Delta V_2 = \chi^\top(i, j)\mathscr{A}_c^\top P \mathscr{A}_c \chi(i, j) - x_d^\top(i, j)Q x_d(i, j) + x^\top(i, j)\left(Q - E^\top P E\right)x(i, j), \tag{10.41}$$

and

$$\Delta V_3 \le \eta^\top(i, j)\mathbb{I}\left(d_h^2, d_v^2\right)E^\top R E \eta(i, j) - (x(i, j) - x_d(i, j))^\top E^\top R E\,(x(i, j) - x_d(i, j)).$$

It follows from (10.36) that

$$\eta^\top(i, j)E^\top \widehat{R} E \eta(i, j) = \chi^\top(i, j)\left(\mathscr{A}_c - E e_1\right)^\top \widehat{R}\left(\mathscr{A}_c - E e_1\right)\chi(i, j).$$

Therefore,

$$\Delta V_3 \le \chi^\top(i, j)\left(\mathscr{A}_c - E e_1\right)^\top \widehat{R}\left(\mathscr{A}_c - E e_1\right)\chi(i, j) - (x(i, j) - x_d(i, j))^\top E^\top R E\,(x(i, j) - x_d(i, j)). \tag{10.42}$$

On the other hand, for any real matrix S of appropriate dimension, the following zero-type equation holds

$$-2x^\top(i, j)S^\top \mathscr{N}_E^\top \mathscr{A}_c \chi(i, j) = 0. \tag{10.43}$$

In addition, according to triggering condition (10.34), we have

$$x^\top(i, j)\mathbb{I}(\sigma_h, \sigma_v)\Pi x(i, j) - e^\top(i, j)\Pi e(i, j) \ge 0.$$

Therefore, from (10.41) to (10.43), we get

$$\Delta V \le \chi^\top(i, j)\left(\Omega + \mathscr{A}_c^\top P \mathscr{A}_c + \left(\mathscr{A}_c - E e_1\right)^\top \widehat{R}\left(\mathscr{A}_c - E e_1\right)\right)\chi(i, j). \tag{10.44}$$

By the Schur complement lemma [Chap. 2, Lemma 2.7], condition (10.37) holds if and only if

$$\Omega + \mathscr{A}_c^\top P \mathscr{A}_c + (\mathscr{A}_c - Ee_1)^\top \widehat{R} (\mathscr{A}_c - Ee_1) < 0.$$

Thus, under condition (10.37), there exists a $\tilde{c}_0 > 0$ such that $\Delta V \leq -\tilde{c}_0 \|x(i, j)\|^2$. Similar to the proof Theorem 10.1, we can conclude that the closed-loop system (10.36) is admissible. The proof is completed. □

10.4.3 Controller Design Under ETC Scheme

In this section, based on the stability conditions proposed in Theorem 10.2, we derive design conditions for a stabilizing ETC in the form of (10.35). Note at first that, by the SVD procedure, there exist orthogonal matrices U_h, U_v and V_h, V_v such that

$$E_h = U_h \begin{bmatrix} \Sigma_h & 0 \\ 0 & 0 \end{bmatrix} V_h^\top, \quad E_v = U_v \begin{bmatrix} \Sigma_v & 0 \\ 0 & 0 \end{bmatrix} V_v^\top, \tag{10.45}$$

where $U_h^\top U_h = I_h$, $V_h^\top V_h = I_h$, $U_v^\top U_v = I_v$, $V_v^\top V_v = I_v$ and Σ_h, Σ_v are invertible matrices. Let $X = \text{diag}(X^h, X^v)$ be an invertible matrix. We decompose the matrices X^h, X^v as

$$X^h = V_h \begin{bmatrix} X_{11}^h & X_{12}^h \\ X_{21}^h & X_{22}^h \end{bmatrix} V_h^\top, \quad X^v = V_v \begin{bmatrix} X_{11}^v & X_{12}^v \\ X_{21}^v & X_{22}^v \end{bmatrix} V_v^\top.$$

Then, it follows from (10.45) that

$$\begin{aligned} E_h X^h &= U_h \begin{bmatrix} \Sigma_h & 0 \\ 0 & 0 \end{bmatrix} V_h^\top V_h \begin{bmatrix} X_{11}^h & X_{12}^h \\ X_{21}^h & X_{22}^h \end{bmatrix} V_h^\top \\ &= U_h \begin{bmatrix} \Sigma_h X_{11}^h & \Sigma_h X_{12}^h \\ 0 & 0 \end{bmatrix} V_h^\top. \end{aligned} \tag{10.46}$$

On the other hand, let $Z = \text{diag}(Z^h, Z^v)$ be an invertible matrix. We also decompose the matrices Z^h, Z^v as

$$Z^h = U_h \begin{bmatrix} Z_{11}^h & Z_{12}^h \\ Z_{21}^h & Z_{22}^h \end{bmatrix} U_h^\top, \quad Z^v = U_v \begin{bmatrix} Z_{11}^v & Z_{12}^v \\ Z_{21}^v & Z_{22}^v \end{bmatrix} U_v^\top.$$

Then, we have

$$\begin{aligned} Z^h E_h &= U_h \begin{bmatrix} Z_{11}^h & Z_{12}^h \\ Z_{21}^h & Z_{22}^h \end{bmatrix} U_h^\top U_h \begin{bmatrix} \Sigma_h & 0 \\ 0 & 0 \end{bmatrix} V_h^\top \\ &= U_h \begin{bmatrix} Z_{11}^h \Sigma_h & 0 \\ Z_{21}^h \Sigma_h & 0 \end{bmatrix} V_h^\top. \end{aligned} \tag{10.47}$$

It can be deduced from (10.46) and (10.47) that

$$E_h X^h = Z^h E_h \tag{10.48}$$

if and only if $X_{12}^h = 0$, $Z_{21}^h = 0$ and $Z_{11}^h = \Sigma_h X_{11}^h \Sigma_h^{-1}$. The derivation for X^v and Z^v will be proceeded similarly. In summary, based on the representation (10.45), if $X = \mathrm{diag}(X^h, X^v)$ and $Z = \mathrm{diag}(Z^h, Z^v)$ are invertible matrices formulated by

$$\begin{aligned} X &= \mathrm{diag}\left(V_h \begin{bmatrix} X_{11}^h & 0 \\ X_{21}^h & X_{22}^h \end{bmatrix} V_h^\top, V_v \begin{bmatrix} X_{11}^v & 0 \\ X_{21}^v & X_{22}^v \end{bmatrix} V_v^\top \right), \\ Z &= \mathrm{diag}\left(U_h \begin{bmatrix} \Sigma_h X_{11}^h \Sigma_h^{-1} & Z_{12}^h \\ 0 & Z_{22}^h \end{bmatrix} U_h^\top, U_v \begin{bmatrix} \Sigma_v X_{11}^v \Sigma_v^{-1} & Z_{12}^v \\ 0 & Z_{22}^v \end{bmatrix} U_v^\top \right), \end{aligned} \tag{10.49}$$

then $EX = ZE$ and, consequently, we have

$$X^\top E^\top P E X = E^\top Z^\top P Z E. \tag{10.50}$$

Based on the transformation derived in (10.5), we obtain the following design conditions for an ETC as given in (10.35).

Theorem 10.3 *Given positive scalars σ_h, σ_v. Assume that there exist symmetric positive definite matrices $\widetilde{P} = \mathrm{diag}(\widetilde{P}^h, \widetilde{P}^v)$, $\widetilde{Q} = \mathrm{diag}(\widetilde{Q}^h, \widetilde{Q}^v)$, $\widetilde{R} = \mathrm{diag}(\widetilde{R}^h, \widetilde{R}^v)$, $\widetilde{\Pi} = \mathrm{diag}(\widetilde{\Pi}_h, \widetilde{\Pi}_v)$, invertible matrices X, Z as presented in* (10.49) *and a matrix Y of appropriate dimension that satisfy the following condition*

$$\begin{bmatrix} \widetilde{\Omega} & \mathscr{O}_{X,Y}^\top & \widetilde{\mathscr{O}}_{X,Y}^\top \mathbb{I}(d_h, d_v) \\ * & \widetilde{P} - Z - Z^\top & 0 \\ * & * & \widetilde{R} - Z - Z^\top \end{bmatrix} < 0, \tag{10.51}$$

where $\widetilde{\Omega} = (\widetilde{\Omega}_{kl})_{3\times 3}$ and

$$\begin{aligned} \widetilde{\Omega}_{11} &= -E^\top \widetilde{P} E - E^\top \widetilde{R} E - \mathrm{Sym}\left\{ \mathscr{N}_E^\top (AX + BY) \right\} + \widetilde{Q} + \mathbb{I}(\sigma_h, \sigma_v) \widetilde{\Pi}, \\ \widetilde{\Omega}_{12} &= E^\top \widetilde{R} E - \mathscr{N}_E^\top A_d X, \quad \widetilde{\Omega}_{13} = -\mathscr{N}_E^\top BY, \\ \widetilde{\Omega}_{22} &= -\widetilde{Q} - E^\top \widetilde{R} E, \quad \widetilde{\Omega}_{23} = 0, \quad \widetilde{\Omega}_{33} = -\widetilde{\Pi}, \\ \mathscr{O}_{X,Y} &= (AX + BY) e_1 + A_d X e_2 + BY e_3, \\ \widetilde{\mathscr{O}}_{X,Y} &= (AX - EX + BY) e_1 + A_d X e_2 + BY e_3. \end{aligned}$$

Then, there exists an ETC in the form of (10.35) *that makes the closed-loop system* (10.36) *admissible. The controller gain matrix is obtained as*

$$K = Y X^{-1}.$$

Proof Let $S = X^{-1}$, $\mathscr{D}_X = \mathrm{diag}(X, X, X)$, and $\mathscr{D} = \mathrm{diag}(\mathscr{D}_X, P^{-1}, R^{-1})$. By pre- and post-multiplying with $\mathscr{D}^\top$ and $\mathscr{D}$, condition (10.37) is feasible if and only if it holds that

$$\begin{bmatrix} \mathscr{D}_X^\top \Omega \mathscr{D}_X & \mathscr{D}_X^\top \mathscr{A}_c^\top & \mathscr{D}_X^\top (\mathscr{A}_c - Ee_1)^\top \mathbb{I}(d_h, d_v) \\ * & -P^{-1} & 0 \\ * & * & -R^{-1} \end{bmatrix} < 0. \tag{10.52}$$

By using the variable $Y = KX$, we have

$$\begin{aligned} \mathscr{D}_X^\top \mathscr{A}_c^\top &= \begin{bmatrix} X^\top (A^\top + K^\top B^\top) \\ X^\top A_d^\top \\ X^\top K^\top B^\top \end{bmatrix} = \begin{bmatrix} (AX + BY)^\top \\ (A_d X)^\top \\ (BY)^\top \end{bmatrix} = \mathscr{O}_{X,Y}^\top, \\ \mathscr{D}_X^\top (\mathscr{A}_c - Ee_1)^\top &= \begin{bmatrix} (AX - EX + BY)^\top \\ (A_d X)^\top \\ (BY)^\top \end{bmatrix} = \widetilde{\mathscr{O}}_{X,Y}^\top \end{aligned} \tag{10.53}$$

In addition, let us define the matrices

$$\begin{aligned} \widetilde{P} &= Z^\top P Z, \quad \widetilde{R} = Z^\top R Z, \\ \widetilde{Q} &= X^\top Q X, \quad \widetilde{\Pi} = X^\top \Pi X. \end{aligned}$$

Then, it is clear that P, Q, R, and Π are symmetric positive definite matrices if and only if so are the matrices $\widetilde{P}$, $\widetilde{Q}$, $\widetilde{R}$, and $\widetilde{\Pi}$. Moreover, by taking (10.50) into account we then obtain

$$\mathscr{D}_X^\top \Omega \mathscr{D}_X = \widetilde{\Omega}. \tag{10.54}$$

On the other hand, by utilizing the Cauchy matrix inequality [see, Chap. 2, Lemma 2.6], we have

$$\begin{aligned} -P^{-1} &\le Z^\top P Z - Z^\top - Z \\ &= \widetilde{P} - Z^\top - Z, \\ -R^{-1} &\le \widetilde{R} - Z^\top - Z. \end{aligned} \tag{10.55}$$

Thus, it can be be deduced from (10.52)–(10.55) that if condition (10.51) holds then the stability condition derived in (10.37) is satisfied with $K = YX^{-1}$. By Theorem 10.2, the closed-loop system (10.36) is admissible under the event-triggered controller (10.35) with the controller gain $K = YX^{-1}$. The proof is completed. □

Remark 10.4 For 2-D singular delay-free systems

$$\begin{bmatrix} E_h x^h(i+1, j) \\ E_v x^v(i, j+1) \end{bmatrix} = A \begin{bmatrix} x^h(i, j) \\ x^v(i, j) \end{bmatrix} + Bu(i, j), \tag{10.56}$$

condition (10.51) is reduced to the following LMI

$$\begin{bmatrix} \check{\Omega} & -\mathscr{N}_E^\top BY & (AX+BY)^\top \\ * & -\widetilde{\Pi} & (BY)^\top \\ * & * & \widetilde{P}-Z-Z^\top \end{bmatrix} < 0, \tag{10.57}$$

where

$$\check{\Omega} = -E^\top \widetilde{P} E - \mathrm{Sym}\left\{\mathscr{N}_E^\top (AX+BY)\right\} + \mathbb{I}(\sigma_h, \sigma_v)\widetilde{\Pi}.$$

Based on this observation, we have the following corollary.

Corollary 10.2 *Given positive scalars σ_h, σ_v and assume that there exist symmetric positive definite matrices $\widetilde{P} = \mathrm{diag}(\widetilde{P}^h, \widetilde{P}^v)$, $\widetilde{\Pi} = \mathrm{diag}(\widetilde{\Pi}_h, \widetilde{\Pi}_v)$, invertible matrices X, Z prescribed in* (10.49) *and a matrix Y of appropriate dimension that satisfy condition* (10.57). *Then, the closed-loop system of* (10.56) *is admissible by the event-triggered controller*

$$u(i,j) = YX^{-1}\begin{bmatrix} x^h(i_p, j) \\ x^v(i, j_q) \end{bmatrix}.$$

10.4.4 An Illustrative Example

In this section, we give an example with simulations to illustrate the effectiveness of the design conditions proposed in Theorem 10.3.

Example 10.4 Consider system (10.30) with the following data

$$E = \begin{bmatrix} 1 & 0 \\ 0 & 0 \end{bmatrix}, \quad A = \begin{bmatrix} 0.8 & 0.05 \\ 0.65 & -1.1 \end{bmatrix}, \quad A_d = \begin{bmatrix} 0.12 & 0.09 \\ 0.01 & 0.15 \end{bmatrix}, \quad B = \begin{bmatrix} 0.6 \\ -0.25 \end{bmatrix}.$$

Note at first that the spectral radius $\rho(A) = 1.117$ and thus the matrix A is unstable. Simulation results presented in Figs. 10.4 and 10.5 are taken with delays $d_h = d_v = 8$. It can be seen that the conducted open-loop state trajectories do not converge to the origin, which illustrates the open system is unstable.

We now apply the design method proposed in Theorem 10.3 to obtain a desired ETC. Given triggering parameters $\sigma_h = \sigma_v = 0.9$, with delays $d_h = d_v = 8$, by using the Matlab LMI toolbox, it is found that condition (10.51) is feasible with the matrices

$$\begin{aligned} \widetilde{P} &= \mathrm{diag}(44.5708, 44.1462), \quad \widetilde{Q} = \mathrm{diag}(5.3425, 2.9148), \\ \widetilde{R} &= \mathrm{diag}(0.5350, 1.1472), \quad \widetilde{\Pi} = \mathrm{diag}(8.0071, 4.8189), \\ X &= \mathrm{diag}(41.3242, -25.6878), \quad Z = \mathrm{diag}(41.3242, 57.2681), \\ Y &= \begin{bmatrix} -11.7289 & 8.1520 \end{bmatrix}. \end{aligned}$$

By Theorem 10.3, the closed-loop system of (10.30) is admissible under the ETC (10.35) with the controller gain

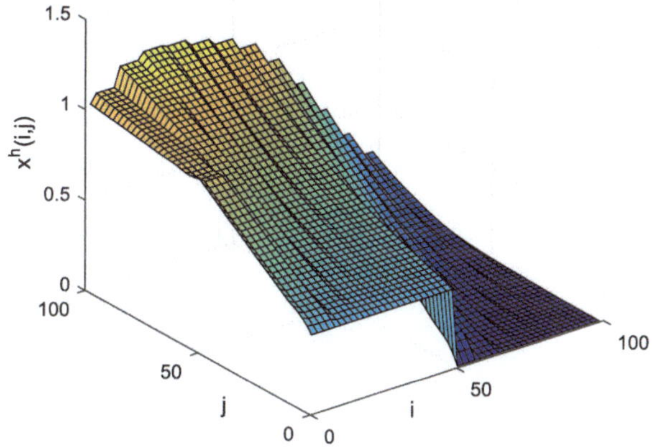

Fig. 10.4 A state trajectory x^h of the open-loop system

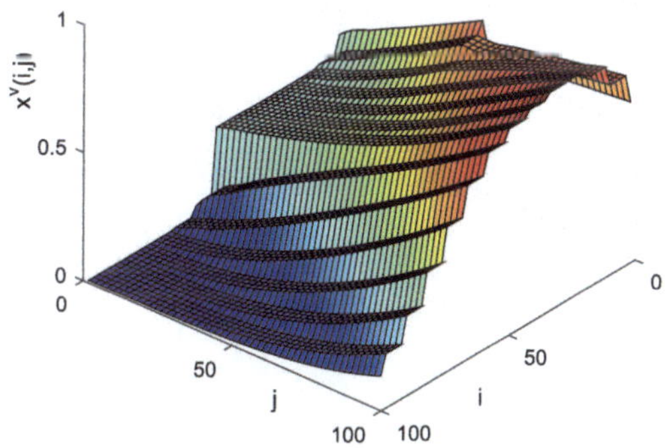

Fig. 10.5 A state trajectory x^v of the open-loop system

$$K = YX^{-1} = \begin{bmatrix} -0.2838 & -0.3173 \end{bmatrix}.$$

Simulation results for the closed-loop system (10.36) with the ETC (10.35) are presented in Figs. 10.6, 10.7, 10.8, 10.9, 10.10 and 10.11. More specifically, directional ranges of triggering times are given in Figs. 10.6 and 10.7, respectively, and 2-D triggering instants are plotted in Figs. 10.8 and 10.9. Closed-loop state trajectories with $d_h = d_v = 8$ and $\sigma_h = \sigma_v = 0.8$ are presented in Figs. 10.10 and 10.11. It can be seen that the conducted closed-loop trajectories converge to the origin as $i + j \to \infty$. This illustrates the effectiveness of the design method proposed in Theorem 10.3.

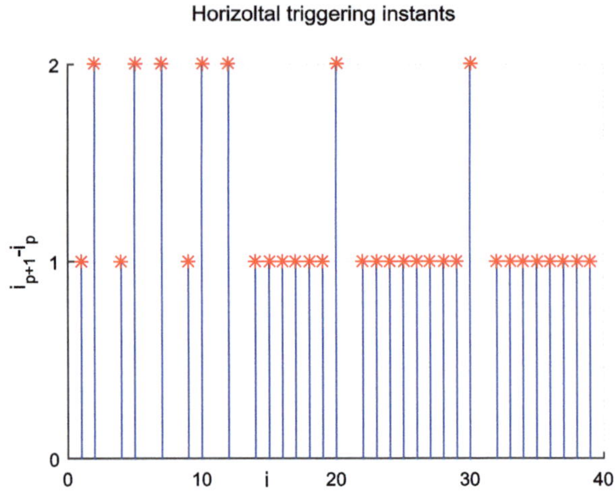

Fig. 10.6 Horizontal triggering instants $i_{p+1} - i_p$

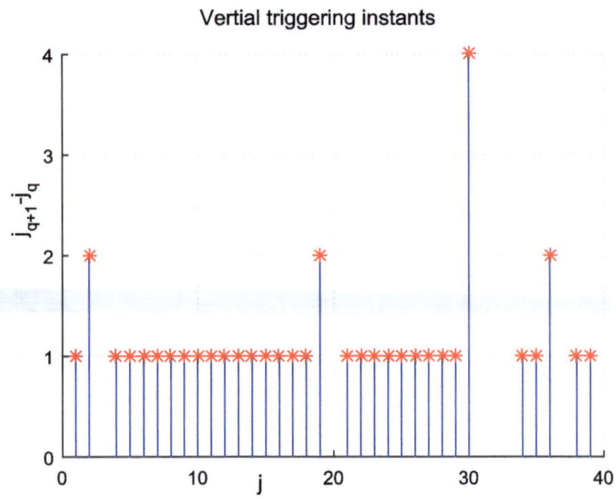

Fig. 10.7 Vertical triggering instants $j_{q+1} - j_q$

10.5 Conclusion

In this chapter, the problems of stability and stabilization via event-triggered state-feedback controllers have been addressed for 2-D singular Roesser systems with state delays. Delay-dependent conditions have been derived using a functional-based method and zero-type matrix equations to ensure the regularity, causality and internal stability of the system. The derived stability conditions have been employed to derive tractable design conditions for a stabilizing event-triggered state-feedback controller. Numerical examples have been presented to illustrate the effectiveness of the proposed analysis and design conditions.

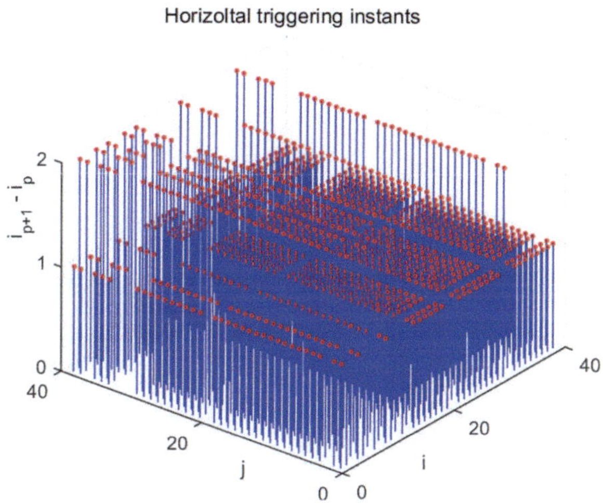

Fig. 10.8 2-D triggering instants $i_{p+1} - i_p$

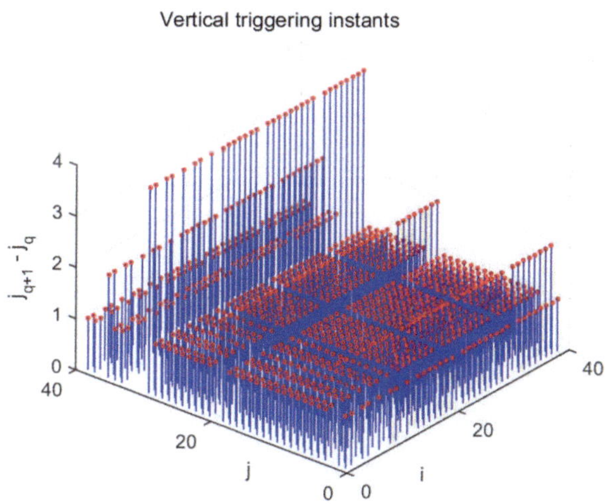

Fig. 10.9 2-D triggering instants $j_{q+1} - j_q$

The analysis and design methods presented in this chapter can be effectively utilized for various problems in the systems and control theory of 2-D singular systems with state delays such as event-triggered H_∞ control, filtering or state estimation. The proposed methods in this chapter can also be extended to complex 2-D singular systems with jumping parameters both deterministic and stochastic models. We leave it to reader more details on specific applications to such problems.

Fig. 10.10 A state trajectory x^h of the closed-loop system

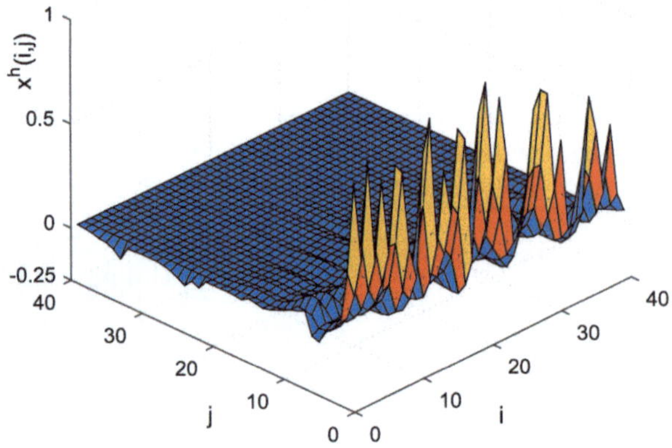

Fig. 10.11 A state trajectory x^v of the closed-loop system

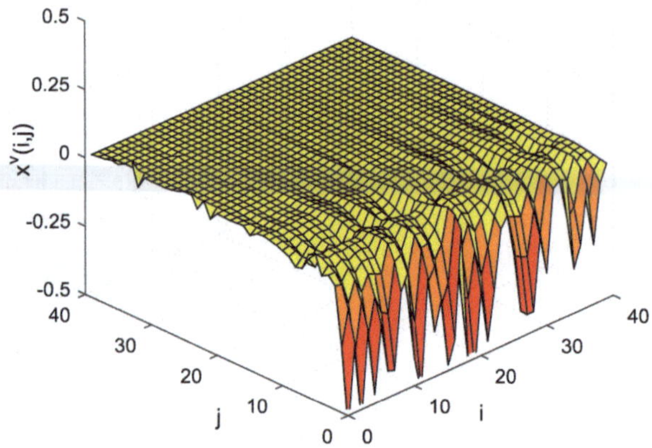

References

1. Stevens, B.L., Lewis, F.L.: Aircraft Modeling: Dynamics and Control. Wiley, New York (1991)
2. Aplevich, J.D.: Implicit Linear Systems. Springer-Varlag, Berlin (1991)
3. Duan, G.R.: Analysis and Design of Descriptor Linear Systems. Springer, New York (2010)
4. Xu, S., Lam, L.: Robust Control and Filtering of Singular Systems. Springer, New York (2006)
5. Kunkel, P., Mehrmann, V.: Differential-Algebraic Equations: Analysis and numerical solution. EMS Publishing House, Zürich (2006)
6. Fridman, E.: Stability of linear descriptor systems with delay: a Lyapunov-based approach. J. Math. Anal. Appl. **273**, 24–44 (2002)
7. Malyshev, A., Sadkane, M.: On the complex stability radius for time-delay differential-algebraic systems. Linear Alg. Appl. **708**, 355–371 (2025)

8. Witrant, E., Fridman, E., Sename, O., Dugard, L.: Recent Results on Time-Delay Systems: Analysis and Control. Springer, Basel (2016)
9. Gao, Q., Karimi, H.R.: Stability. Control and Application of Time-Delay Systems. Butterworth-Heinemann, Oxford (2019)
10. Zhang, X.M., Han, Q.L., Seuret, A., Gouaisbaut, F., He, Y.: Overview of recent advances in stability of linear systems with time-varying delays. IET Control Theory Appl. **13**, 1–16 (2019)
11. Lee, S.M., Park, M.J., Know, O.M., Cao, J.: Less conservative results for stability of sampled-data systems with constant delay. J. Frank. Inst. **357**, 10960–10976 (2020)
12. Zeng, H.B., He, Y., Teo, K.L.: Monotone-delay-interval-based Lyapunov functionals for stability analysis of systems with a periodically varying delay. Automatica **138**, 110030 (2022)
13. Wang, X., Sun, J., Berberich, J., Wang, G., Allgöwer, R., Chen, J.: Data-driven control of dynamic event-triggered systems with delays. Int. J. Robust Nonlinear Control **33**, 7071–7093 (2023)
14. Zhang, C.K., Chen, W.H., Zhu, C., He, Y., Wu, M.: Stability analysis of discrete-time systems with time-varying delay via a delay-dependent matrix- separation-based inequality. Automatica **156**, 111192 (2023)
15. Ge, X., Wang, Z., Lei, S.L., Vong, S.: Exponential stability of system with time delay based on matrix-refined weighted functions. European J. Control **79**, 101076 (2024)
16. Sha, H., Park, J.H., Chen, J., Zhu, M., Nan, C.: Stability analysis of discrete-time systems with a time-varying delay via improved methods. IET Control Theory Appl. **18**, 951–959 (2024)
17. He, J., Han, Y., Yang, F.: Relaxed mixed convex combination lemmas: Application to stability analysis of time-varying delay systems. J. Frankl. Inst. **361**, 107125 (2024)
18. Huang, Y.B., An, J., He, Y., Wu, M.: Quasi-convex combination method and its application to the stability analysis of 2D discrete-time Roesser systems with time-varying delays. IET Control Theory Appl. **12**, 718–727 (2018)
19. Hien, L.V., Trinh, H.M., Pathirana, P.N.: On l_1-gain control of 2-D positive Roesser systems with directional delays: Necessary and sufficient conditions. Automatica **112**, 108720 (2020)
20. Peng, D., Nie, H.: Abel lemma-based finite-sum inequality approach to stabilization for 2-D time-varying delay systems. Asian J. Control **23**, 1394–1406 (2021)
21. Zhang, Y., Ji, Y., Jiang, T., Zhou, J.: Event-triggered control for Roesser model-based 2D Markov jump systems under stochastic communication protocol. Circuit Syst. Sig. Process **43**, 6953–6976 (2024)
22. Xue, X., Xu, J., Zhang, H.: Linear quadratic optimal control of stochastic 2-D Roesser models. Math. Comput. Simul. **227**, 500–510 (2025)
23. Zhu, K., Wang, Z., Ding, D., Dong, H., Wei, G.: Encryption-decryption-based set-membership filtering for two-dimensional systems: on security and boundedness. Automatica **173**, 112091 (2025)
24. Chen, S.F.: Stability analysis and stabilization of 2-D singular Roesser models. Appl. Math. Comput. **250**, 779–91 (2010)
25. Dai, L.: Singular Control Systems. Springer-Verlag, Berlin (1989)
26. Xu, H., Zou, Y.: H_∞ control for 2-D singular delayed systems. Int. J. Syst. Sci. **42**, 609–619 (2011)
27. Kririm, S., Hmamed, A., Tadeo, F.: Analysis and design of H_∞ controllers for 2D singular systems with delays. Circuit Syst. Sig. Process **35**, 1579–1592 (2016)
28. Hien, L.V., Vu, L.H., Trinh, H.: Stability of two-dimensional descriptor systems with generalized directional delays. Syst. Control Lett. **112**, 42–50 (2018)
29. Chelliq, E.H., Alfidi, M., Chalh, Z.: Admissibility and robust H_∞ controller design for uncertain 2D singular discrete systems with interval time-varying delays. Optim. Control Appl. Meth. **44**, 2476–2503 (2023)
30. Vu LH (2025) Exponential stability and H_∞ controller design for 2-D discrete-state systems in Roesser model with directional delays. Optimal Control Appl Method. https://doi.org/10.1002/oca.3226
31. Hien, L.V., Vu, L.H.: Finite-region boundedness and (Q, S, R)-dissipativity of 2-D singular Roesser systems with mixed time-varying delays. J. Frankl. Inst. **361**, 1416–1435 (2024)

32. Shu, F., Zhai, J.: Global event-triggered output feedback stabilization for a class of nonlinear time-delay systems. IEEE Trans Circuit Syst I: Reg Paper **68**, 4371–4380 (2021)
33. Liu, B., Liu, T., Xiao, P.: Dynamic event-triggered intermittent control for stabilization of delayed dynamical systems. Automatica **149**, 110847 (2023)
34. Wu, C., Zhao, X.: A hybrid framework of dynamic periodic event-triggered networked control systems subject to time-varying delays. J. Frank. Inst. **361**, 1–11 (2024)
35. Ferdinando, M., Gennaro, S., Borri, A., Pola, G., Pepe, P.: On the digital event-based control for nonlinear time-delay systems with exogenous disturbances. Automatica **163**, 111567 (2024)
36. Zhi, Y., Liu, X., Wang, Y., Fan, Y.: Network-based H_∞ control for descriptor systems with time delays under a novel event-triggered strategy. IEEE Trans. Control Netw. Syst. (2024). https://doi.org/10.1109/TCNS.2024.3431412
37. Yang, R., Yu, Y.: Event-triggered control of discrete-time 2-D switched Fornasini-Marchesini systems. European J. Control **48**, 42–51 (2019)
38. Li, L., Yang, R., Feng, Z., Wu, L.: Event-triggered dissipative control for 2-D switched systems. Info. Sci. **589**, 802–812 (2022)
39. Yang, R., Zheng, W.X., Yu, Y.: Event-triggered sliding mode control of discrete-time two-dimensional systems in Roesser model. Automatica **114**, 108813 (2020)
40. Lv, Xinyu, Niu, Y., Park, J.H., Song, J.: Sliding mode control for 2D FMII systems: A bidirectional dynamic event-triggered strategy. Automatica **147**, 110727 (2023)
41. Ma, X., Dong, J., Tai, W., Zhou, J., Paszke, W.: Asynchronous event-triggered H_∞ control for 2D Markov jump systems subject to networked random packet losses. Commun. Nonlinear Sci. Numer Simul. **126**, 107453 (2023)
42. Kaczorek, T.: Two-Dimensional Linear Systems. Springer, Berlin (1985)
43. Lanning W, Weiqun W, Weimin C (2014) Delay-dependent stability for 2-D singular systems with state-varying delay in Roesser model: an LMI approach. In: Proc 33rd CCC:6074–6079, Nanjing, China, July 28–30

Chapter 11
Finite-Region Stability and Dissipativity of Two-Dimensional Singular Systems with Delays

Abstract In this chapter, the problems of stability and dissipativity analysis in finite-region are developed for two-dimensional (2-D) singular Roesser systems with mixed time-varying delays. Finite-region boundedness and dissipativity concepts are proposed and taken into consideration. Based on a 2-D Lyapunov–Krasovskii functional (LKF) method via weighted LKF candidates, and by utilizing zero-type free matrix equations, delay-dependent conditions are derived in the form of linear matrix inequalities (LMIs) to guarantee the finite-region stability and dissipativity of the system with respect to quadratic supplied rate functions. Numerical examples are given to illustrate the effectiveness of the obtained analysis results.

Keywords 2-D singular systems · Roesser model · Finite-region stability · Dissipativity

11.1 Introduction

Singular systems (also referred as descriptor systems or differential-algebraic equations) form a very important class of dynamical systems, which can be used to model various practical phenomena in circuits networks, power systems, aerospace engineering or multibody mechanics [1–4]. Unlike nominal/regular systems, states and outputs of a singular system are typically get involved to algebraic constraints that are difficult to analyze as well as lead to complicated dynamic behaviors such as impulsive state-response, non-properness of the transfer matrix or non-causality between states and inputs [5, 6]. Besides that, as mentioned in previous chapters, time-delay phenomena are frequently encountered in systems and control engineering [7, 8]. Thus, the study on analysis and control of time-delay singular systems has attracted remarkable research attention in the past decades [9–17].

In the context of long-time behavior, the theory of 2-D systems has been extensively studied due to various applications in practical models, for example, in electrical circuits, image processing, data transmission or multi-dimensional digital filtering. A very rich literature concerning various problems in systems and control theory of 2-D systems with and without delays has been reported [18–22]. However,

L. V. Hien, *Two-Dimensional Time-Delay Systems*, Studies in Systems, Decision and Control 605, https://doi.org/10.1007/978-3-031-95678-2_11

very little attempt of scholars has been successfully extended such studies to 2-D singular systems with delays. For example, in [23, 24], the H_∞ control problem was investigated for 2-D singular Roesser systems with constant delays using a bounded real lemma and delay-independent LMI-based approach. Extended results on stability and H_∞ stabilization were also developed for 2-D singular Roesser systems with time-varying delays in [25–27]. However, as discussed in many existing works, the asymptotic behavior of dynamical systems only deals with qualitative properties at sufficiently large time [28]. Nevertheless, in numerous practical processes, the transient performance is only subject to constraints that involve to specified bounds and thresholds on the system states over some prescribed time domain. To such phenomena, classical control schemes that are based on the conventional Lyapunov stability theory are not utilizable [29]. Therefore, the concepts of finite-time stability (FTS), finite-time boundedness (FTB) and their variants have been proposed and developed for many types of 1-D dynamical systems [28]. Roughly speaking, a dynamical system is said to be finite-time stability if for a given bound on initial states and inputs, the system response (states and outputs) does not exceed a prescribed threshold in a fixed time interval. This concept has been recognized to have practical meaning and is suitable for dealing with the transient performance in short time intervals. It is noted also that, due to the nature of their differences in the dynamic propagation, the existing results on FTS and FTB of 1-D systems are not directly adaptable for 2-D systems. More specifically, to cope with the transient performance of 2-D systems, it is necessary and challenging to bound a two-time state trajectory $x(k, l)$ over a finite-region. Despite of remarkable effort from researchers, only a few results concerning the finite-region analysis and control of 2-D systems have been established. For example, the problems of finite-region stability, finite-region boundedness, and input-output finite-region stability were investigated for various classes of 2-D systems in [30, 31] using the induction method and constrained LMIs approach. Finite-region H_∞ stabilization problem was studied in [32] for 2-D switched systems based on mode-dependent average dwell time switchings.

As another important issue, the theory of dissipativity of dynamical systems, introduced in the earlier of 1970s [33], provides meaningful concepts for physical processes. This theory and its variants have been intensively studied for 1-D systems in the past few years. For some references that dealt with dissipativity theory in combination with important control issues, we refer the reader to recent works [34–39]. This area has also received growing research attention in the area of 2-D systems. Preliminary results on analysis and control [40, 41], fault detection [42] or filtering [43–45] subject to dissipativity of 2-D systems have been reported. In particular, the problem of finite-region dissipative control was considered in [46, 47] via the oblique line approach. Co-design problem of 2-D event generators and dissipativity-based output feedback controllers was discussed in [48] for 2-D Roesser systems without delay. Analysis and design conditions were derived in terms of tractable LMIs using a 2-D Lyapunov function method. It should be emphasized here that existing works so far have only been developed for 2-D nominal systems without delays.

In this chapter, we study the problems of finite-region boundedness and (Q, S, R)-dissipativity for 2-D singular Roesser systems with mixed interval time-varying delays. Finite-region boundedness and dissipativity concepts are first proposed. Then, by using an improved weighted 2-D Lyapunov–Krasovskii functional and zero-type matrix equations, delay-dependent LMI-based conditions that guarantee the finite-region stability, boundedness and dissipativity with respect to quadratic supplied rate functions were derived. Numerical examples with simulations are given to illustrate the effectiveness of the obtained results.

11.2 Preliminaries

11.2.1 Model Description

Consider a class of 2-D singular systems with mixed time-varying delays given by

$$\begin{aligned}
\begin{bmatrix} E_h x^h(i+1,j) \\ E_v x^v(i,j+1) \end{bmatrix} &= A \begin{bmatrix} x^h(i,j) \\ x^v(i,j) \end{bmatrix} + A_d \begin{bmatrix} x^h(i-d_h(i),j) \\ x^v(i,j-d_v(j)) \end{bmatrix} \\
&\quad + A_s \begin{bmatrix} \sum_{l=1}^{s_h(i)} x^h(i-l,j) \\ \sum_{l=1}^{s_v(j)} x^v(i,j-l) \end{bmatrix} + B_w w(i,j), \\
z(i,j) &= C \begin{bmatrix} x^h(i,j) \\ x^v(i,j) \end{bmatrix} + C_d \begin{bmatrix} x^h(i-d_h(i),j) \\ x^v(i,j-d_v(j)) \end{bmatrix} \\
&\quad + C_s \begin{bmatrix} \sum_{l=1}^{s_h(i)} x^h(i-l,j) \\ \sum_{l=1}^{s_v(j)} x^v(i,j-l) \end{bmatrix} + D_w w(i,j),
\end{aligned} \tag{11.1}$$

where $x^h(i,j) \in \mathbb{R}^{n_h}$ and $x^v \in \mathbb{R}^{n_v}$ denote the horizontal and vertical state vectors, $w(i,j) \in \mathbb{R}^{n_w}$ and $z(i,j) \in \mathbb{R}^{n_z}$ represent the exogenous disturbance and regulated output, respectively. The system matrices $E_h \in \mathbb{R}^{n_h \times n_h}$, $E_v \in \mathbb{R}^{n_v \times n_v}$ and $A, A_d, A_s \in \mathbb{R}^{n \times n}$ $(n = n_h + n_v)$, $B_w \in \mathbb{R}^{n \times n_w}$, $C, C_d, C_s \in \mathbb{R}^{n_z \times n}$ and $D_w \in \mathbb{R}^{n_z \times n_w}$ are given real matrices of appropriate dimensions. It is assumed that the matrix $E = \text{diag}(E_h, E_v)$ is singular with $\text{rank}(E_h) = r_h \leq n_h$, $\text{rank}(E_v) = r_v \leq n_v$ and $r = r_h + r_v < n_h + n_v = n$. In model (11.1), the functions $d_h(i)$, $s_h(i)$ and $d_v(j)$, $s_v(j)$ represent the time-varying delays along the horizontal and vertical directions, which are assumed to be confined in the ranges specified by

$$\begin{aligned}
d_{hm} \leq d_h(i) \leq d_{hM}, \quad d_{vm} \leq d_v(j) \leq d_{vM}, \\
s_{hm} \leq s_h(i) \leq s_{hM}, \quad s_{vm} \leq s_v(j) \leq s_{vM},
\end{aligned} \tag{11.2}$$

where d_{hm}, d_{hM}, d_{vm}, d_{vM} and s_{hm}, s_{hM}, s_{vm}, s_{vM} are given nonnegative integers involving the upper and the lower bounds of delays. Denote $\sigma_h = \max(d_{hM}, s_{hM})$ and $\sigma_v = \max(d_{vM}, s_{vM})$ as the upper bounds of delays. The overall state vector of (11.1) is defined as $x(i,j) = \begin{bmatrix} x^h(i,j) \\ x^v(i,j) \end{bmatrix} \in \mathbb{R}^n$.

The initial condition of (11.1) is assumed to have finite support and specified by

$$\begin{aligned} x^h(k,j) &= \varphi(k,j), \quad k \in \mathbb{Z}[-\sigma_h, 0], \quad 0 \le j \le D_2, \\ x^v(i,l) &= \psi(i,l), \quad l \in \mathbb{Z}[-\sigma_v, 0], \quad 0 \le i \le D_1, \\ x^h(.,j) &= 0 \text{ for } j > D_2 \text{ and } x^v(i,.) = 0 \text{ for } i > D_1, \end{aligned} \tag{11.3}$$

where D_1, D_2 are given positive integers. We define the region

$$\mathscr{R}(D_1, D_2) = \left\{(i,j) \in \mathbb{N}_0^2 \middle| \, 0 \le i \le D_1, \; 0 \le j \le D_2\right\}$$

and the space

$$\mathscr{W}_2^{\rho} = \left\{ w : \mathscr{R}(D_1, D_2) \to \mathbb{R}^{n_w} \middle| \sum_{(i,j)\in\mathscr{R}(D_1,D_2)} w^\top(i,j)w(i,j) \le \rho^2 \right\}.$$

In model (11.1), the exogenous disturbance w is assumed to belong to the space $\mathscr{W}_2^{\rho}$ for some scalar $\rho > 0$.

11.2.2 Problem Formulation

We recall here that [see, Chap. 10, Definition 10.1], the pair of matrices (E, A) is regular if the polynomial det $(E\mathbb{I}(\lambda,\mu) - A)$ is not identically zero and (E, A) is causal if $\deg(\det(sE - A)) = \text{rank}(E)$, where $\mathbb{I}(\lambda,\mu) = \text{diag}(\lambda I_{n_h}, \mu I_{n_v})$. The 2-D system (11.1) is acceptable if the pair (E, A) is regular and causal. In addition, for given positive scalars c_h, c_v and a matrix $\Theta = \text{diag}(\Theta_h, \Theta_v) \in \mathbb{S}_n^+$, we define the following regions

$$\begin{aligned} \Omega(\Theta_h, c_h) &= \left\{x \in \mathbb{R}^{n_h} \middle| x^\top E_h^\top \Theta_h E_h x < c_h^2\right\} \cap \left\{x \in \mathbb{R}^{n_h} \middle| x^\top \Theta_h x < c_h^2\right\}, \\ \Omega(\Theta_v, c_v) &= \left\{x \in \mathbb{R}^{n_v} \middle| x^\top E_v^\top \Theta_v E_v x < c_v^2\right\} \cap \left\{x \in \mathbb{R}^{n_v} \middle| x^\top \Theta_v x < c_v^2\right\}. \end{aligned}$$

For an illustration of the regions $\Omega(\Theta_h, c_h)$ and $\Omega(\Theta_v, c_v)$, let

$$\Theta_h = \text{diag}\left(\begin{bmatrix} 2 & 1 \\ 1 & 1 \end{bmatrix}, 1\right), \quad E_h = \text{diag}\left(\begin{bmatrix} 1 & 0 \\ 0 & 0 \end{bmatrix}, 1\right)$$

and $c_h = 1$. Then, the set $\left\{x \in \mathbb{R}^{n_h} \middle| x^\top \Theta_h x < c_h^2\right\}$ is an ellipsoid determined by

$$\mathscr{E} = \{x \in \mathbb{R}^3 : 2x_1^2 + 2x_1x_2 + x_2^2 + x_3^2 < 1\}$$

whereas the set $\{x \in \mathbb{R}^{n_h} | x^\top E_h^\top \Theta_h E_h x < c_h^2\}$ is a cylinder determined by

$$\mathscr{C} = \{x \in \mathbb{R}^3 : 2x_1^2 + x_3^2 < 1\}.$$

As shown in Fig. 11.1, the region $\Omega(\Theta_h, c_h)$ is determined by the part of cylinder $\mathscr{C}$ that is confined within the ellipsoid $\mathscr{E}$.

The following definition is extended from the concept of finite-time boundedness developed for one-dimensional (1-D) systems [49].

Definition 11.1 (Singular finite-region boundedness (SFRB)) Given positive scalars c, c_h, c_v and a matrix $\Theta = \text{diag}(\Theta_h, \Theta_v) \in \mathbb{S}_n^+$. System (11.1) is said to be singularly finite-region bounded with respect to (c, c_h, c_v, Θ) if (i) system (11.1) is acceptable; and (ii) for any external disturbance $w \in \mathscr{W}_2^\rho$ and initial states $\varphi(k, j) \in \Omega(\Theta_h, c_h)$ for all $k \in \mathbb{Z}[-\sigma_h, 0]$, $0 \le j \le D_2$, $\psi(i, l) \in \Omega(\Theta_v, c_v)$ for all $0 \le i \le D_1$, $l \in \mathbb{Z}[-\sigma_v, 0]$, the corresponding state trajectory $x(i, j)$ of (11.1) satisfies

$$x^\top(i, j) E^\top \Theta E x(i, j) < c^2 \text{ for all } (i, j) \in \mathscr{R}(D_1, D_2). \tag{11.4}$$

Remark 11.1 For 2-D nominal systems (i.e. E_h and E_v are identity matrices), condition (11.4) is reduced to the normal condition that the state trajectory $x(i, j)$ of the system is always confined within the ellipsoid $\{x \in \mathbb{R}^n : x^\top \Theta x < c^2\}$. In this regard, the concept of SFRB for 2-D singular Roesser systems presented in Definition 11.1 extends both the concepts of finite-time boundedness (FTB) proposed for 1-D systems [28] and finite-region boundedness proposed for 2-D nominal systems (see, e.g., [31, 32]).

Remark 11.2 It can be deduced from Definition 11.1 that, for 2-D singular systems, the components of initial states that fall within the range space ImE of the matrix E are taken into account to define the property of SFRB as indicated in (11.4). This is a special feature between 2-D nominal systems and singular systems.

Fig. 11.1 An illustration of the region $\Omega(\Theta_h, c_h)$.

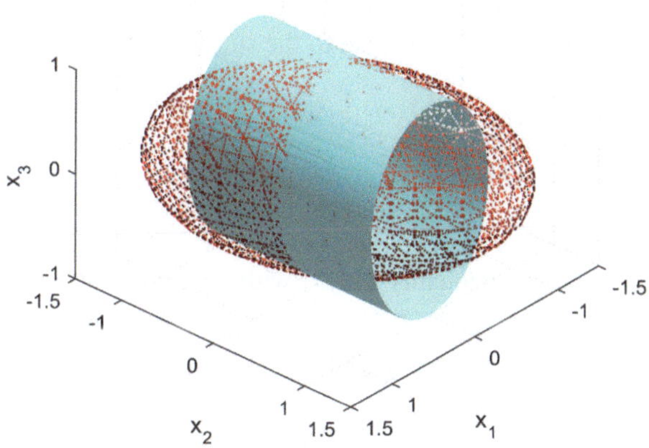

We now introduce the concept of finite-region dissipativity for 2-D singular systems. Typically, we say that a dynamical system is dissipative if the energy stored in the system does not exceed the total energy that is supplied to the system from external sources. To give a more detail description, we define the input-output operator of system (11.1) as $S : (\varphi, \psi, w) \to z$. Consider a real-valued function $E(z, w)$ associated with S, which is called the *supplied rate function*. For given positive integers T, K, the supplied energy in the time domain $\mathbb{Z}[0, T) \times \mathbb{Z}[0, K)$ is defined as

$$\sum_{i=0}^{T-1}\sum_{j=0}^{K-1} E(z(i, j), w(i, j)).$$

Similar to [48, 50], we introduce the following definition.

Definition 11.2 System (11.1) is said to be dissipative with respect to the supplied rate function $E(w, z)$ if there exists a storage function $V(x) = V^h(x^h) + V^v(x^v)$ with $V(0) = 0$ such that for any positive integers T, K, and external input w, it holds that

$$\begin{aligned}\sum_{j=0}^{K-1}\left[V^h(x^h(T, j)) - V^h(x^h(0, j))\right] + \sum_{i=0}^{T-1}\left[V^v(x^v(i, K)) - V^v(x^v(i, 0))\right]\\ \leq \sum_{i=0}^{T-1}\sum_{j=0}^{K-1} E(z(i, j), w(i, j)).\end{aligned} \tag{11.5}$$

Inequality (11.5) is referred to the dissipation inequality. It reveals that the total energy stored in the system at the boundary $C(T, K) = \{(T, j) : 0 \leq j < K\} \cup \{(i, K) : 0 \leq i < T\}$ is no more than the addition of initial stored energy and supplied energy to the system in the domain $\mathbb{Z}[0, T) \times \mathbb{Z}[0, K)$.

In the literature, quadratic supplied rate functions in w and z are widely used [40–42]. More specifically, for given real matrices of appropriate dimensions $\mathscr{Q} = \mathscr{Q}^\top$, $\mathscr{S}$ and $\mathscr{R} = \mathscr{R}^\top$, system (11.1) is said to be $(\mathscr{Q}, \mathscr{S}, \mathscr{R})$-dissipative if it is dissipative with respect to the supplied rate function

$$E(z, w) = z^\top \mathscr{Q} z + 2z^\top \mathscr{S} w + w^\top \mathscr{R} w. \tag{11.6}$$

Definition 11.3 For given positive scalars c, c_h, c_v, and a matrix $\Theta = \mathrm{diag}(\Theta_h, \Theta_v) \in \mathbb{S}_n^+$, system (11.1) is said to be finite-region $(\mathscr{Q}, \mathscr{S}, \mathscr{R})$-dissipative with respect to (c, c_h, c_v, Θ) if it is SFRB with respect to (c, c_h, c_v, Θ) and the dissipation inequality (11.5) with supplied rate function (11.6) is satisfied, where $\mathscr{Q} = \mathscr{Q}^\top$, $\mathscr{S}$, and $\mathscr{R} = \mathscr{R}^\top$ are given real matrices of appropriate dimensions.

Remark 11.3 To the supplied rate function (11.6), if the kernel $\mathscr{R}$ is of the form $\mathscr{R} = \mathscr{R}_0 - \mu I_n$, where $\mu > 0$ is some scalar, Definition 11.3 is reduced to the concept of finite-region $(\mathscr{Q}, \mathscr{S}, \mathscr{R}_0)$-$\mu$-dissipativity of 2-D singular systems. In addition, subject to the dissipation inequality (11.5), nonzero initial conditions can also be taken into account. To this regard, Definition 11.3 extends existing dissipativity concepts proposed for both 1-D and 2-D systems in the existing literature.

Remark 11.4 Some other important performance indices in the systems and control theory of 2-D systems can be obtained from (11.6) by specifying particular kernel matrices. For example, if system (11.1) is finite-region $(\mathscr{Q}, \mathscr{S}, \mathscr{R})$-dissipative for $\mathscr{Q} = -I_n, \mathscr{S} = 0$ and $\mathscr{R} = \gamma^2 I_n$, then it is finite-region stable with H_∞ attenuation level γ. If $\mathscr{S} = I_n$ and $\mathscr{R} = -\nu I_n$ or $\mathscr{Q} = -\rho I_n$, then we obtain finite-region input or output passivity indices, respectively [50]. Moreover, if $\mathscr{Q} = -\theta I_n, \mathscr{S} = (1-\theta)I_n$ and $\mathscr{R} = [\theta\gamma^2 + (\theta - 1)\nu]I$, then we obtain a finite-region mixed H_∞ and passivity performance index, where $\theta \in [0, 1]$ is a parameter that defines the trade-off between the H_∞ performance and the passivity.

The main objective of this chapter is to derive tractable conditions that ensure simultaneously the properties of SFRB and finite-region $(\mathscr{Q}, \mathscr{S}, \mathscr{R})$-dissipativity.

11.3 Finite-Region Boundedness of 2-D Singular Roesser Systems with Mixed Interval Delays

In this section, we derive tractable LMI-based conditions that guarantee the SFRB of system (11.1). For the brevity, we denote the following augmented state vectors

$$x_{1,1}(i,j) = \begin{bmatrix} x^h(i+1,j) \\ x^v(i,j+1) \end{bmatrix}, \quad x_d(i,j) = \begin{bmatrix} x^h(i-d_h(i),j) \\ x^v(i,j-d_v(j)) \end{bmatrix},$$
$$x_{dM}(i,j) = \begin{bmatrix} x^h(i-d_{hM},j) \\ x^v(i,j-d_{vM}) \end{bmatrix}, \quad x_{dm}(i,j) = \begin{bmatrix} x^h(i-d_{hm},j) \\ x^v(i,j-d_{vm}) \end{bmatrix},$$
$$x_s(i,j) = \begin{bmatrix} \sum_{l=1}^{s_h(i)} x^h(i-l,j) \\ \sum_{l=1}^{s_v(j)} x^v(i,j-l) \end{bmatrix}, \quad \eta(i,j) = \begin{bmatrix} \eta_h(i,j) \\ \eta_v(i,j) \end{bmatrix},$$
$$\eta_h(i,j) = x^h(i+1,j) - x^h(i,j), \quad \eta_v(i,j) = x^v(i,j+1) - x^v(i,j),$$
$$\chi(i,j) = \mathrm{col}\{x(i,j), x_d(i,j), x_{dm}(i,j), x_{dM}(i,j), x_s(i,j), w(i,j)\}.$$

Theorem 11.1 *Given positive scalars c_h, c_v, c, and a matrix $\Theta = \mathrm{diag}(\Theta_h, \Theta_v) \in \mathbb{S}_n^+$. The 2-D singular system (11.1) is SFRB with respect to (c, c_h, c_v, Θ) if there exist a scalar $\alpha > 0$, positive definite matrices $P = \mathrm{diag}(P^h, P^v)$, $Q = \mathrm{diag}(Q^h, Q^v)$, $R = \mathrm{diag}(R^h, R^v)$, $Q_k = \mathrm{diag}(Q_k^h, Q_k^v)$, $Z_k = \mathrm{diag}(Z_k^h, Z_k^v)$ $(k = 1, 2)$, Ψ, and real matrices X_l $(l = 1, 2, \ldots, 6)$ of appropriate dimensions that satisfy the following conditions*

$$\begin{bmatrix} \Xi & \mathscr{A}^\top P & \hat{\mathscr{A}}^\top [\overline{Z}_1 \ \overline{Z}_2] \\ * & -P & 0 \\ * & * & -\mathrm{diag}(\overline{Z}_1, \overline{Z}_2) \end{bmatrix} < 0, \tag{11.7}$$

$$\alpha_{1s}(D+1)\frac{\lambda_h c_h^2 + \lambda_v c_v^2 + \rho^2 \lambda_{\max}(\Psi)}{c^2} < \lambda_{\min}(\Theta^{-1/2} P \Theta^{-1/2}), \tag{11.8}$$

where $D = D_1 + D_2$, $L = (E^\top)^\perp$ is the null space of $E^\top$, $\Xi = (\Xi_{kl})_{6\times 6}$ and other notations are defined as

$$\begin{aligned}
&\Xi_{11} = Q_1 + Q_2 + \overline{Q} + \overline{R} - \alpha E^\top P E - E^\top(\hat{Z}_1 + \hat{Z}_2)E - \mathrm{Sym}\{A^\top L X_1\},\\
&\Xi_{12} = E^\top \hat{Z}_1 E - A^\top L X_2, \quad \Xi_{13} = -A^\top L X_3 - X_1^\top L^\top A_d,\\
&\Xi_{14} = E^\top \hat{Z}_2 E - A^\top L X_4, \quad \Xi_{15} = -A^\top L X_5 - X_1^\top L^\top A_s,\\
&\Xi_{16} = -A^\top L X_6 - X_1^\top L^\top B_w, \quad \Xi_{22} = -\widehat{Q}_1 - E^\top \hat{Z}_1 E,\\
&\Xi_{23} = -X_2^\top L^\top A_d, \quad \Xi_{24} = 0, \quad \Xi_{25} = -X_2^\top L^\top A_s,\\
&\Xi_{26} = -X_2^\top L^\top B_w, \quad \Xi_{33} = -\widehat{Q} - \mathrm{Sym}\{A_d^\top L X_3\}, \quad \Xi_{34} = -A_d^\top L X_4,\\
&\Xi_{35} = -X_3^\top L^\top A_s - A_d^\top L X_5, \quad \Xi_{36} = -X_3^\top L^\top B_w - A_d^\top L X_6,\\
&\Xi_{44} = -\widehat{Q}_2 - E^\top \hat{Z}_2 E, \quad \Xi_{45} = -X_4^\top L^\top A_s, \quad \Xi_{46} = -X_4^\top L^\top B_w,\\
&\Xi_{55} = -\widehat{R} - \mathrm{Sym}\{A_s^\top L X_5\}, \quad \Xi_{56} = -X_5^\top L^\top B_w - A_s^\top L X_6,\\
&\Xi_{66} = -\Psi - \mathrm{Sym}\{B_w^\top L X_6\}, \quad \overline{Z}_1 = \mathbb{I}(d_{hm}^2, d_{vm}^2) Z_1,\\
&\overline{Z}_2 = \mathbb{I}(d_{hM}^2, d_{vM}^2) Z_2, \quad \overline{Q} = \mathbb{I}(r_{d_h}+1, r_{d_v}+1) Q, \quad \overline{R} = \mathbb{I}(\varrho_h, \varrho_v) R,\\
&\widehat{Z}_1 = \mathbb{I}\big(\alpha^{d_{hm}}, \alpha^{d_{vm}}\big) Z_1, \quad \widehat{Z}_2 = \mathbb{I}\big(\alpha^{d_{hM}}, \alpha^{d_{vM}}\big) Z_2, \quad \widehat{R} = \mathbb{I}\big(\alpha^{s_{hM}}, \alpha^{s_{vM}}\big) R,\\
&\widehat{Q}_1 = \mathbb{I}\big(\alpha^{d_{hm}}, \alpha^{d_{vm}}\big) Q_1, \quad \widehat{Q}_2 = \mathbb{I}\big(\alpha^{d_{hM}}, \alpha^{d_{vM}}\big) Q_2, \quad \widehat{Q} = \mathbb{I}\big(\alpha^{d_{hM}}, \alpha^{d_{vM}}\big) Q,\\
&\mathscr{A} = \big[A \; 0 \; A_d \; 0 \; A_s \; B_w\big], \quad \hat{\mathscr{A}} = \big[A - E \; 0 \; A_d \; 0 \; A_s \; B_w\big],\\
&r_{d_h} = d_{hM} - d_{hm}, \quad \varrho_h = \frac{s_{hM}(s_{hM} + s_{hm})(s_{hM} - s_{hm} + 1)}{2},\\
&r_{d_v} = d_{vM} - d_{vm}, \quad \varrho_v = \frac{s_{vM}(s_{vM} + s_{vm})(s_{vM} - s_{vm} + 1)}{2},\\
&\lambda_h = \lambda_{\max}(\mathscr{P}^h) + \lambda_{\max}(\mathscr{Q}_1^h)\alpha_{1s}(d_{hm}) + \big(\lambda_{\max}(\mathscr{Q}_2^h) + \lambda_{\max}(\mathscr{Q}^h)\big)\alpha_{1s}(d_{hM})\\
&\qquad + 4(d_{hm}\lambda_{\max}(\mathscr{Z}_1^h)\alpha_{2s}(d_{hm}) + d_{hM}\lambda_{\max}(\mathscr{Z}_2^h)\alpha_{2s}(d_{hM}))\\
&\qquad + \lambda_{\max}(\mathscr{Q}^h)\alpha_{3s}(d_{hm}, d_{hM}) + \lambda_{\max}(\mathscr{R}^h)\alpha_{4s}(s_{hm}, s_{hM}),\\
&\lambda_v = \lambda_{\max}(\mathscr{P}^v) + \lambda_{\max}(\mathscr{Q}_1^v)\alpha_{1s}(d_{vm}) + \big(\lambda_{\max}(\mathscr{Q}_2^v) + \lambda_{\max}(\mathscr{Q}^v)\big)\alpha_{1s}(d_{vM})\\
&\qquad + 4(d_{vm}\lambda_{\max}(\mathscr{Z}_1^v)\alpha_{2s}(d_{vm}) + d_{vM}\lambda_{\max}(\mathscr{Z}_2^v)\alpha_{2s}(d_{vM}))\\
&\qquad + \lambda_{\max}(\mathscr{Q}^v)\alpha_{3s}(d_{vm}, d_{vM}) + \lambda_{\max}(\mathscr{R}^v)\alpha_{4s}(s_{vm}, s_{vM}),\\
&\mathscr{P} = \Theta^{-\frac{1}{2}} P \Theta^{-\frac{1}{2}}, \quad \mathscr{Q} = \Theta^{-\frac{1}{2}} Q \Theta^{-\frac{1}{2}}, \quad \mathscr{R} = \Theta^{-\frac{1}{2}} R \Theta^{-\frac{1}{2}},\\
&\mathscr{Q}_k = \Theta^{-\frac{1}{2}} Q_k \Theta^{-\frac{1}{2}}, \quad \mathscr{Z}_k = \Theta^{-\frac{1}{2}} Z_k \Theta^{-\frac{1}{2}} \; (k = 1, 2),\\
&\alpha_{1s}(d) = \sum_{k=0}^{d-1} \alpha^k = \frac{1-\alpha^d}{1-\alpha}, \quad d \in \{d_{hm}, d_{vm}, d_{hM}, d_{vM}, D+1\},
\end{aligned}$$

$$\alpha_{2s}(d) = \sum_{k=0}^{d-1}\sum_{j=0}^{k}\alpha^j = \frac{d(1-\alpha)-\alpha(1-\alpha^d)}{(1-\alpha)^2}, \quad d \in \{d_{hm}, d_{vm}, d_{hM}, d_{vM}\},$$

$$\alpha_{3s}(p,q) = \sum_{k=p}^{q-1}\sum_{j=0}^{k-1}\alpha^j = \frac{(q-p)(1-\alpha)-(\alpha^p-\alpha^q)}{(1-\alpha)^2},$$

$$(p,q) \in \{(d_{hm}, d_{hM}), (d_{vm}, d_{vM})\},$$

$$\alpha_{4s}(p,q) = \sum_{s=p}^{q}\sum_{k=1}^{s}\sum_{j=1}^{k}\alpha^{j-1} = \frac{q-p+1}{1-\alpha}\left(\frac{p+q}{2}-\frac{\alpha}{1-\alpha}\right)+\frac{\alpha^{p+1}-\alpha^{q+2}}{(1-\alpha)^3},$$

$$(p,q) \in \{(s_{hm}, s_{hM}), (s_{vm}, s_{vM})\}.$$

Proof We first prove that the derived conditions in (11.7) and (11.8) ensure the regularity and causality of system (11.1). Similar to the proof of Theorem 10.1 [see, Chap. 10, Sect. 10.3], since $\text{rank}(E) = r < n$, these exist two nonsingular matrices M and N by which we have the decomposition (10.1). We now define the matrices

$$\boldsymbol{X}_1^\top = N^\top X_1^\top = \begin{bmatrix} X_{11} \\ X_{21} \end{bmatrix}, \quad \boldsymbol{P} = M^{-\top}(\alpha P)M^{-1} = \begin{bmatrix} P_{11} & P_{12} \\ P_{21} & P_{22} \end{bmatrix},$$

$$\boldsymbol{L} = M^{-\top}L = \begin{bmatrix} L_{11} \\ L_{21} \end{bmatrix}, \quad \boldsymbol{Z}_1 = M^{-\top}\hat{Z}_1 M^{-1} = \begin{bmatrix} Z_{111} & Z_{112} \\ Z_{121} & Z_{122} \end{bmatrix},$$

$$\boldsymbol{Z}_2 = M^{-\top}\hat{Z}_2 M^{-1} = \begin{bmatrix} Z_{211} & Z_{212} \\ Z_{221} & Z_{222} \end{bmatrix}.$$

Then, it can be verified using the fact $L = (E^\top)^\perp$ that

$$E^\top L = N^\top E^\top M^\top M^{-\top} L = \overline{E}^\top \boldsymbol{L} = 0,$$

where $\overline{E} = \begin{bmatrix} I_r & 0 \\ 0 & 0 \end{bmatrix}$. Thus, $L_{11} = 0$ and we have

$$L = M^\top \begin{bmatrix} 0 \\ L_{21} \end{bmatrix}.$$

By the Schur lemma [see, Chap. 2, Lemma 2.7], condition (11.7) induces that $\Xi_{11} < 0$ and therefore,

$$-\alpha E^\top P E - X_1^\top L^\top A - A^\top L X_1 + \text{Sym}\{E^\top(H_1+T_1)\} < 0. \tag{11.9}$$

By pre- and post-multiplying both sides of (11.9) with $N^\top$ and N, we obtain

$$
\begin{aligned}
-\overline{E}^\top \boldsymbol{P}\overline{E} - \overline{E}^\top \mathbf{Z}_1\overline{E} - \overline{E}^\top \mathbf{Z}_1\overline{E} - \overline{A}^\top \begin{bmatrix} 0 \\ L_{21} \end{bmatrix} \mathbf{X}_1 - \mathbf{X}_1^\top \begin{bmatrix} 0 & L_{21}^\top \end{bmatrix} \overline{A} \\
= \begin{bmatrix} \mathscr{O}_{11} & \mathscr{O}_{12} \\ * & \mathscr{O}_{22} \end{bmatrix} < 0,
\end{aligned} \tag{11.10}
$$

where

$$
\mathscr{O}_{22} = -X_{21}L_{21}^\top A_{22} - A_{22}^\top L_{21}X_{21}^\top.
$$

Again, by the Schur complement lemma, inequality (11.10) implies that

$$
\left(X_{21}L_{21}^\top\right)A_{22} + A_{22}^\top\left(L_{21}X_{21}^\top\right) > 0,
$$

which guarantees that the matrix A_{22} is nonsingular. Therefore, the pair (E, A) is regular and causal [see, Chap. 10, Lemma 10.2] and as a consequence system (11.1) is regular and causal.

Next, we now prove the 2-D system (11.1) is SFRB. By virtue of the LKF method, we construct the following weighted LKF candidate

$$
\mathscr{V}(i,j) \triangleq V(x(i,j)) = \underbrace{\sum_{k=1}^{8} V_k^h(x^h(i,j))}_{\mathscr{V}^h(i,j)} + \underbrace{\sum_{k=1}^{8} V_k^v(x^v(i,j))}_{\mathscr{V}^v(i,j)}, \tag{11.11}
$$

where

$$
\begin{aligned}
V_1^h(x^h(i,j)) &= x^{h\top}(i,j)E_h^\top P^h E_h x^h(i,j), \\
V_2^h(x^h(i,j)) &= \sum_{k=i-d_{hm}}^{i-1} \alpha^{i-k-1} x^{h\top}(k,j)Q_1^h x^h(k,j), \\
V_3^h(x^h(i,j)) &= \sum_{k=i-d_{hM}}^{i-1} \alpha^{i-k-1} x^{h\top}(k,j)Q_2^h x^h(k,j), \\
V_4^h(x^h(i,j)) &= \sum_{k=i-d_h(i)}^{i-1} \alpha^{i-k-1} x^{h\top}(k,j)Q^h x^h(k,j), \\
V_5^h(x^h(i,j)) &= \sum_{k=-d_{hM}+1}^{-d_{hm}} \sum_{l=i+k}^{i-1} \alpha^{i-l-1} x^{h\top}(l,j)Q^h x^h(l,j), \\
V_6^h(x^h(i,j)) &= d_{hm} \sum_{k=-d_{hm}}^{-1} \sum_{l=i+k}^{i-1} \alpha^{i-l-1} \eta_h^\top(l,j)E_h^\top Z_1^h E_h \eta_h(l,j),
\end{aligned}
$$

$$V_7^h(x^h(i,j)) = d_{hM} \sum_{k=-d_{hM}}^{-1} \sum_{l=i+k}^{i-1} \alpha^{i-l-1} \eta_h^\top(l,j) E_h^\top Z_2^h E_h \eta_h(l,j),$$

$$V_8^h(x^h(i,j)) = \sum_{p=-s_{hM}}^{-s_{hm}} \sum_{k=p}^{-1} \sum_{l=i+k}^{i-1} \alpha^{i-l-1} x^{h\top}(l,j) R^h x^h(l,j),$$

and

$$V_1^v(x^v(i,j)) = x^{v\top}(i,j) E_v^\top P^v E_v x^v(i,j),$$

$$V_2^v(x^v(i,j)) = \sum_{l=j-d_{vm}}^{j-1} \alpha^{j-l-1} x^{v\top}(i,l) Q_1^v x^v(i,l),$$

$$V_3^v(x^v(i,j)) = \sum_{l=j-d_{vM}}^{j-1} \alpha^{j-l-1} x^{v\top}(i,l) Q_2^v x^v(i,l),$$

$$V_4^v(x^v(i,j)) = \sum_{l=j-d_v(j)}^{j-1} \alpha^{j-l-1} x^{v\top}(i,l) Q^v x^v(i,l),$$

$$V_5^v(x^v(i,j)) = \sum_{t=-d_{vM}+1}^{-d_{vm}} \sum_{l=j+t}^{j-1} \alpha^{j-l-1} x^{v\top}(i,l) Q^v x^v(i,l),$$

$$V_6^v(x^v(i,j)) = d_{vm} \sum_{t=-d_{vm}}^{-1} \sum_{l=j+t}^{j-1} \alpha^{j-l-1} \eta_v^\top(i,l) E_v^\top Z_1^v E_v \eta_v(i,l),$$

$$V_7^v(x^v(i,j)) = d_{vM} \sum_{t=-d_{hM}}^{-1} \sum_{l=j+t}^{j-1} \alpha^{j-l-1} \eta_v^\top(i,l) E_v^\top Z_2^v E_v \eta_v(i,l),$$

$$V_8^v(x^h(i,j)) = \sum_{q=-s_{vM}}^{-s_{vm}} \sum_{t=q}^{-1} \sum_{l=j+t}^{j-1} \alpha^{j-l-1} x^{v\top}(i,l) R^v x^v(i,l).$$

The weighted directional difference $\Delta_\alpha \mathscr{V}(i,j)$ of the functional $V(x)$ along state trajectory of system (11.1) is defined as

$$\Delta_\alpha \mathscr{V}(i,j) \triangleq \underbrace{V^h(i+1,j) - \alpha V^h(i,j)}_{\Delta_\alpha \mathscr{V}^h(i,j)} + \underbrace{V^v(i,j+1) - \alpha V^v(i,j)}_{\Delta_\alpha \mathscr{V}^v(i,j)}.$$

For the simplicity, we denote $V_k^h(i,j) = V_k^h(x^h(i,j)$ and $V_k^v(i,j) = V_k^v(x^v(i,j)$. The difference $\Delta_\alpha \mathscr{V}(i,j)$ will be computed and manipulated as follows. First, for the functional $V_k^h(i,j)$ $(k = 1, 2, \ldots, 5)$, we have

$$
\begin{aligned}
\Delta_\alpha V_1^h(i,j) &= x^{h\top}(i+1,j)E_h^\top P^h E_h x^h(i+1,j) - \alpha x^{h\top}(i,j)E_h^\top P^h E_h x^h(i,j),\\
\Delta_\alpha V_2^h(i,j) &= x^{h\top}(i,j)Q_1^h x^h(i,j) - \alpha^{d_{hm}} x^{h\top}(i-d_{hm},j)Q_1^h x^h(i-d_{hm},j),\\
\Delta_\alpha V_3^h(i,j) &= x^{h\top}(i,j)Q_2^h x^h(i,j) - \alpha^{d_{hM}} x^{h\top}(i-d_{hM},j)Q_2^h x^h(i-d_{hM},j),
\end{aligned}
\tag{11.12}
$$

and

$$
\begin{aligned}
\Delta_\alpha V_4^h(i,j) = & \sum_{k=i+1-d_h(i+1)}^{i} \alpha^{i-k} x^{h\top}(k,j)Q^h x^h(k,j)\\
& - \sum_{k=i-d_h(i)}^{i-1} \alpha^{i-k} x^{h\top}(k,j)Q^h x^h(k,j)\\
= & \, x^{h\top}(i,j)Q^h x^h(i,j) - \alpha^{d_h(i)} x^{h\top}(i-d_h(i),j)Q^h x^h(i-d_h(i),j)\\
& + \sum_{k=i+1-d_h(i+1)}^{i-1} \alpha^{i-k} x^{h\top}(k,j)Q^h x^h(k,j)\\
& - \sum_{k=i+1-d_h(i)}^{i-1} \alpha^{i-k} x^{h\top}(k,j)Q^h x^h(k,j)\\
\le & \, x^{h\top}(i,j)Q^h x^h(i,j) - \alpha^{d_{hM}} x^{h\top}(i-d_h(i),j)Q^h x^h(i-d_h(i),j)\\
& + \sum_{k=i+1-d_{hM}}^{i-1} \alpha^{i-k} x^{h\top}(k,j)Q^h x^h(k,j)\\
& - \sum_{k=i+1-d_{hm}}^{i-1} \alpha^{i-k} x^{h\top}(k,j)Q^h x^h(k,j).
\end{aligned}
$$

Therefore,

$$
\begin{aligned}
\Delta_\alpha V_4^h(i,j) \le & \, x^{h\top}(i,j)Q^h x^h(i,j) - \alpha^{d_{hM}} x^{h\top}(i-d_h(i),j)Q^h x^h(i-d_h(i),j)\\
& + \sum_{k=i+1-d_{hM}}^{i-d_{hm}} \alpha^{i-k} x^{h\top}(k,j)Q^h x^h(k,j).
\end{aligned}
\tag{11.13}
$$

In addition, we have

$$\begin{aligned}
\Delta_\alpha V_5^h(i,j) &= \sum_{k=-d_{hM}+1}^{-d_{hm}} \left[\sum_{l=i+1+k}^{i} \alpha^{i-l} x^{h\top}(l,j) Q^h x^h(l,j) \right. \\
&\quad \left. - \sum_{l=i+k}^{i-1} \alpha^{i-l} x^{h\top}(l,j) Q^h x^h(l,j) \right] \\
&= r_{d_h} x^{h\top}(i,j) Q^h x^h(i,j) - \sum_{k=-d_{hM}+1}^{-d_{hm}} \alpha^{-k} x^{h\top}(i+k,j) Q^h x^h(i+k,j) \\
&= r_{d_h} x^{h\top}(i,j) Q^h x^h(i,j) - \sum_{k=i+1-d_{hM}}^{i-d_{hm}} \alpha^{i-k} x^{h\top}(k,j) Q^h x^h(k,j). \qquad (11.14)
\end{aligned}$$

By similar arguments, we also have

$$\begin{aligned}
\Delta_\alpha V_1^v(i,j) &= x^{v\top}(i,j+1) E_v^\top P^v E_v x^v(i,j+1) - \alpha x^{v\top}(i,j) E_v^\top P^v E_v x^v(i,j) \\
\Delta_\alpha V_2^v(i,j) &= x^{v\top}(i,j) Q_1^v x^v(i,j) - \alpha^{d_{vm}} x^{v\top}(i-d_{vm},j) Q_1^v x^v(i-d_{vm},j), \qquad (11.15) \\
\Delta_\alpha V_3^v(i,j) &= x^{v\top}(i,j) Q_2^v x^v(i,j) - \alpha^{d_{vM}} x^{v\top}(i-d_{vM},j) Q_2^v x^v(i-d_{vM},j),
\end{aligned}$$

the estimate

$$\begin{aligned}
\Delta_\alpha V_4^v(i,j) &\le x^{v\top}(i,j) Q^v x^v(i,j) - \alpha^{d_{vM}} x^{v\top}(i,j-d_v(j)) Q^v x^v(i,j-d_v(j)) \\
&\quad + \sum_{l=j+1-d_{vM}}^{j-d_{vm}} \alpha^{j-l} x^{v\top}(i,l) Q^v x^v(i,l), \qquad (11.16)
\end{aligned}$$

and

$$\Delta_\alpha V_5^v(i,j) = r_{d_v} x^{v\top}(i,j) Q^v x^v(i,j) - \sum_{l=j+1-d_{vM}}^{j-d_{vm}} \alpha^{j-l} x^{v\top}(i,l) Q^v x^v(i,l). \qquad (11.17)$$

Together with the fact $Ex_{1,1}(i,j) = \mathscr{A}\chi(i,j)$, from (11.12) to (11.17), we have

$$\begin{aligned}
&\sum_{k=1}^{5} \Delta_\alpha \left(V_k^h(i,j) + V_k^v(i,j) \right) \\
&\quad \le x_{1,1}^\top(i,j) E^\top P E x_{1,1}(i,j) + x^\top(i,j) \big(Q_1 + Q_2 + \overline{Q} - \alpha E^\top P E \big) x(i,j) \\
&\quad\quad - x_{dm}^\top(i,j) \widehat{Q}_1 x_{dm}(i,j) - x_{dM}^\top(i,j) \widehat{Q}_2 x_{dM}(i,j) - x_d^\top(i,j) \widehat{Q} x_d(i,j) \\
&\quad \le \chi^\top(i,j) \mathscr{A}^\top P \mathscr{A} \chi(i,j) + x^\top(i,j) \big(Q_1 + Q_2 + \overline{Q} - \alpha E^\top P E \big) x(i,j) \\
&\quad\quad - x_{dm}^\top(i,j) \widehat{Q}_1 x_{dm}(i,j) - x_{dM}^\top(i,j) \widehat{Q}_2 x_{dM}(i,j) - x_d^\top(i,j) \widehat{Q} x_d(i,j). \qquad (11.18)
\end{aligned}$$

Next, by utilizing the 2-D summation inequalities given in [Chap. 2, Corollary 2.1] to manipulate the differences $\Delta_\alpha V_6(i,j)$ and $\Delta_\alpha V_7(i,j)$, we have

$$
\begin{aligned}
\Delta_\alpha V_6^h(i,j) &= d_{hm} \sum_{k=-d_{hm}}^{-1} \Bigg(\sum_{l=i+1+k}^{i} \alpha^{i-l} \eta_h^\top(l,j) E_h^\top Z_1^h E_h \eta_h(l,j) \\
&\quad - \sum_{l=i+k}^{i-1} \alpha^{i-l} \eta_h^\top(k,j) E_h^\top Z_1^h E_h \eta_h(l,j) \Bigg) \\
&= d_{hm}^2 \eta_h^\top(i,j) E_h^\top Z_1^h E_h \eta_h(i,j) \\
&\quad - d_{hm} \sum_{k=-d_{hm}}^{-1} \alpha^{-k} \eta_h^\top(i+k,j) E_h^\top Z_1^h E_h \eta_h(i+k,j) \\
&\le d_{hm}^2 \eta_h^\top(i,j) E_h^\top Z_1^h E_h \eta_h(i,j) \\
&\quad - \alpha^{d_{hm}} \Bigg(\sum_{k=i-d_{hm}}^{i-1} \eta_h(k,j) \Bigg)^\top E_h^\top Z_1^h E_h \Bigg(\sum_{k=i-d_{hm}}^{i-1} \eta_h(k,j) \Bigg) \\
&\le d_{hm}^2 \eta_h^\top(i,j) E_h^\top Z_1^h E_h \eta_h(i,j) - \alpha^{d_{hm}} \left(x^h(i,j) - x^h(i-d_{hm},j) \right)^\top \\
&\quad \times E_h^\top Z_1^h E_h \left(x^h(i,j) - x^h(i-d_{hm},j) \right)
\end{aligned} \tag{11.19}
$$

and

$$
\begin{aligned}
\Delta_\alpha V_7^h(i,j) &= d_{hM} \sum_{k=-d_{hM}}^{-1} \Bigg(\sum_{l=i+1+k}^{i} \alpha^{i-l} \eta_h^\top(l,j) E_h^\top Z_2^h E_h \eta_h(l,j) \\
&\quad - \sum_{l=i+k}^{i-1} \alpha^{i-l} \eta_h^\top(l,j) E_h^\top Z_2^h E_h \eta_h(l,j) \Bigg) \\
&= d_{hM}^2 \eta_h^\top(i,j) E_h^\top Z_2^h E_h \eta_h(i,j) \\
&\quad - d_{hM} \sum_{k=i-d_{hM}}^{i-1} \alpha^{i-k} \eta_h^\top(k,j) E_h^\top Z_2^h E_h \eta_h(k,j) \\
&\le d_{hM}^2 \eta_h^\top(i,j) E_h^\top Z_2^h E_h \eta_h(i,j) \\
&\quad - \alpha^{d_{hM}} \Bigg(\sum_{k=i-d_{hM}}^{i-1} \eta_h(k,j) \Bigg)^\top E_h^\top Z_2^h E_h \Bigg(\sum_{k=i-d_{hM}}^{i-1} \eta_h(k,j) \Bigg) \\
&\le d_{hM}^2 \eta_h^\top(i,j) E_h^\top Z_2^h E_h \eta_h(i,j) - \alpha^{d_{hM}} \left(x^h(i,j) - x^h(i-d_{hM},j) \right)^\top \\
&\quad \times E_h^\top Z_2^h E_h \left(x^h(i,j) - x^h(i-d_{hM},j) \right).
\end{aligned} \tag{11.20}
$$

By similar arguments used to derive (11.19) and (11.20), we also have

$$
\begin{aligned}
\Delta_\alpha V_6^v(i,j) &\le d_{vm}^2 \eta_v^\top(i,j) E_v^\top Z_1^v E_v \eta_v(i,j) \\
&\quad - \alpha^{d_{vm}} \left(x^v(i,j) - x^v(i,j-d_{vm}) \right)^\top E_v^\top Z_1^v E_v \\
&\quad \times \left(x^v(i,j) - x^v(i,j-d_{vm}) \right)
\end{aligned} \tag{11.21}
$$

and

$$\begin{aligned}\Delta_\alpha V_7^v(i,j) \le\ & d_{vM}^2 \eta_v^\top(i,j) E_v^\top Z_2^v E_v \eta_v(i,j) \\ & - \alpha^{d_{vM}} \left(x^v(i,j) - x^v(i,j-d_{vM})\right)^\top E_v^\top Z_2^v E_v \\ & \times \left(x^v(i,j) - x^v(i,j-d_{vM})\right). \end{aligned} \tag{11.22}$$

By the results of (11.19) to (11.22), we get

$$\begin{aligned}&\sum_{k=6}^{7} \Delta_\alpha (V_k^h(i,j) + V_k^v(i,j)) \\ &\quad \le \eta^\top(i,j) E^\top \left(\overline{Z}_1 + \overline{Z}_2\right) E \eta(i,j) - x^\top(i,j) E^\top (\widehat{Z}_1 + \widehat{Z}_2) E x(i,j) \\ &\quad - x_{dm}^\top(i,j) E^\top \widehat{Z}_1 E x_{dm}(i,j) - x_{dM}^\top(i,j) E^\top \widehat{Z}_2 E x_{dM}(i,j) \\ &\quad + 2x^\top(i,j) E^\top \widehat{Z}_1 E x_{dm}(i,j) + 2x^\top(i,j) E^\top \widehat{Z}_2 E x_{dM}(i,j). \end{aligned} \tag{11.23}$$

Next, the difference $\Delta_\alpha V_8^h(i,j)$ along state trajectory of system (11.1) is given by

$$\begin{aligned}\Delta_\alpha V_8^h(i,j) &= s_{hM} \sum_{k=s_{hm}}^{s_{hM}} \sum_{k=1}^{p} \Bigg(\sum_{l=i+1-k}^{i} \alpha^{i-l} x^{h\top}(l,j) R^h x^h(l,j) \\ &\qquad - \sum_{l=i-k}^{i-1} \alpha^{i-l} x^{h\top}(l,j) R^h x^h(l,j) \Bigg) \\ &= \varrho_h x^{h\top}(i,j) R^h x^h(i,j) - s_{hM} \sum_{p=s_{hm}}^{s_{hM}} \sum_{k=1}^{p} \alpha^k x^{h\top}(i-k,j) R^h x^h(i-k,j). \end{aligned}$$

Therefore,

$$\begin{aligned}\Delta_\alpha V_8^h(i,j) &\le \varrho_h x^{h\top}(i,j) R^h x^h(i,j) - \alpha^{s_{hM}} s_{hM} \sum_{k=1}^{s_h(i)} x^{h\top}(i-k,j) R^h x^h(i-k,j) \\ &\le \varrho_h x^{h\top}(i,j) R^h x^h(i,j) - \alpha^{s_{hM}} \left(\sum_{k=1}^{s_h(i)} x^h(i-k,j) \right)^\top R^h \left(\sum_{k=1}^{s_h(i)} x^h(i-k,j) \right). \end{aligned} \tag{11.24}$$

Similarly, we can get

$$\begin{aligned}\Delta_\alpha V_8^v(i,j) \le\ & \varrho_v x^{v\top}(i,j) R^v x^v(i,j) \\ & - \alpha^{s_{vM}} \left(\sum_{l=1}^{s_v(j)} x^v(i,j-l) \right)^\top R^v \left(\sum_{l=1}^{s_v(j)} x^v(i,j-l) \right). \end{aligned} \tag{11.25}$$

Note also that

$$\begin{aligned}E\eta(i,j) &= Ex_{1,1}(i,j) - Ex(i,j)\\ &= (A-E)x(i,j) + A_d x_d(i,j) + A_s x_s(i,j) + B_w w(i,j)\\ &= \hat{\mathscr{A}}\chi(i,j).\end{aligned}$$

By this, we readily obtain

$$\eta^\top(i,j)E^\top\left(\overline{Z}_1+\overline{Z}_2\right)E\eta(i,j) = \chi^\top(i,j)\hat{\mathscr{A}}^\top\left(\overline{Z}_1+\overline{Z}_2\right)\hat{\mathscr{A}}\chi(i,j). \quad (11.26)$$

In addition, since $E^\top L = 0$, for any block matrix $\mathscr{X} = \begin{bmatrix}X_1 \ X_2 \ \cdots \ X_6\end{bmatrix}$, where X_k $(k = 1, 2, \ldots, 6)$ are real matrices of appropriate dimensions, the following zero-type equation holds

$$-2\chi^\top(i,j)\mathscr{X}^\top L^\top E x_{1,1}(i,j) = 0.$$

This identity, together with (11.1), yields

$$-2\chi^\top(i,j)\mathscr{X}^\top L^\top \mathscr{A}\chi(i,j) = 0. \quad (11.27)$$

From the estimations in (11.18)–(11.27), we can get

$$\begin{aligned}\Delta_\alpha\mathscr{V}(i,j) &= \sum_{k=1}^{8}\Delta_\alpha(V_k^h(i,j)+V_k^v(i,j))\\ &\le \chi^\top(i,j)\left[\Xi + \mathscr{A}^\top P\mathscr{A} + \hat{\mathscr{A}}^\top\left(\overline{Z}_1+\overline{Z}_2\right)\hat{\mathscr{A}}\right]\chi(i,j)\\ &\quad + w^\top(i,j)\Psi w(i,j). \end{aligned} \quad (11.28)$$

By the Schur lemma, condition (11.7) holds if and only if

$$\Xi + \mathscr{A}^\top P\mathscr{A} + \hat{\mathscr{A}}^\top\begin{bmatrix}\overline{Z}_1 \ \overline{Z}_2\end{bmatrix}\mathrm{diag}(\overline{Z}_1,\overline{Z}_2)^{-1}\begin{bmatrix}\overline{Z}_1^\top\\ \overline{Z}_2^\top\end{bmatrix}\hat{\mathscr{A}} < 0$$

which can be recast as

$$\Xi + \mathscr{A}^\top P\mathscr{A} + \hat{\mathscr{A}}^\top\left(\overline{Z}_1+\overline{Z}_2\right)\hat{\mathscr{A}} < 0. \quad (11.29)$$

It follows from (11.28) and (11.29) that

$$\Delta_\alpha\mathscr{V}(i,j) \le w^\top(i,j)\Psi w(i,j),$$

which clearly gives

$$V^h(x^h(i+1,j)) + V^v(x^v(i,j+1)) \le \alpha \mathscr{V}(i,j) + \lambda_{\max}(\Psi) w^\top(i,j) w(i,j). \tag{11.30}$$

Denote

$$E_S(D) = \sum_{i+j=D} \mathscr{V}(i,j), \quad D \in \mathbb{Z}[0, D_1 + D_2],$$

as the energy stored in the system. From (11.30), we have

$$\begin{aligned} E_S(D+1) &= \sum_{i+j=D+1} \left(\mathscr{V}^h(i,j) + \mathscr{V}^v(i,j)\right) \\ &= \sum_{i+j=D} \left[V^h(x^h(i+1,j)) + V^v(x^v(i,j+1))\right] \\ &\quad + V^h(x^h(0,D+1)) + V^v(x^v(D+1,0)) \\ &\le \alpha E_S(D) + \lambda_{\max}(\Psi) \sum_{i+j=D} w^\top(i,j) w(i,j) \\ &\quad + V^h(x^h(0,D+1)) + V^v(x^v(D+1,0)). \end{aligned} \tag{11.31}$$

By the constant variation formula for linear difference equations, from (11.31) we get

$$\begin{aligned} E_S(D) &\le \alpha^D E_S(0) + \lambda_{\max}(\Psi) \sum_{k=0}^{D-1} \alpha^{D-1-k} \sum_{i+j=k} w^\top(i,j) w(i,j) \\ &\quad + \sum_{j=1}^{D} \alpha^{D-j} V^h(x^h(0,j)) + \sum_{i=1}^{D} \alpha^{D-i} V^v(x^v(i,0)) \\ &\le \sum_{j=0}^{D} \alpha^{D-j} V^h(x^h(0,j)) + \sum_{i=0}^{D} \alpha^{D-i} V^v(x^v(i,0)) \\ &\quad + \lambda_{\max}(\Psi) \sum_{k=0}^{D-1} \alpha^{D-1-k} \sum_{i+j=k} w^\top(i,j) w(i,j). \end{aligned} \tag{11.32}$$

For any $w \in \mathscr{W}_2^\rho$, from (11.32), we have

$$\begin{aligned} E_S(D) &\le \sum_{j=0}^{D} \alpha^{D-j} V^h(x^h(0,j)) + \sum_{i=0}^{D} \alpha^{D-i} V^v(x^v(i,0)) \\ &\quad + \rho^2 \lambda_{\max}(\Psi) \sum_{k=0}^{D-1} \alpha^{D-1-k}. \end{aligned} \tag{11.33}$$

To get a growth bounding for $E_S(D)$, we need to estimate the value of $V^h(x^h(0, j))$ and $V^v(x^v(i, 0))$. For this purpose, we notice that

$$
\begin{aligned}
V^h\big(x^h(0, j)\big) = {} & x^{h\top}(0, j)E_h^\top \Theta_h^{\frac{1}{2}} \mathscr{P}^h \Theta_h^{\frac{1}{2}} E_h x^h(0, j) \\
& + \sum_{k=-d_{hm}}^{-1} \alpha^{-k-1} x^{h\top}(k, j)\Theta_h^{\frac{1}{2}} \mathscr{Q}_1^h \Theta_h^{\frac{1}{2}} x^h(k, j) \\
& + \sum_{k=-d_{hM}}^{-1} \alpha^{-k-1} x^{h\top}(k, j)\Theta_h^{\frac{1}{2}} \mathscr{Q}_2^h \Theta_h^{\frac{1}{2}} x^h(k, j) \\
& + \sum_{k=-d_h(0)}^{-1} \alpha^{-k-1} x^{h\top}(k, j)\Theta_h^{\frac{1}{2}} \mathscr{Q}^h \Theta_h^{\frac{1}{2}} x^h(k, j) \\
& + \sum_{k=-d_{hM}+1}^{-d_{hm}} \sum_{l=k}^{-1} \alpha^{-l-1} x^{h\top}(l, j)\Theta_h^{\frac{1}{2}} \mathscr{Q}^h \Theta_h^{\frac{1}{2}} x^h(k, j) \\
& + d_{hm} \sum_{k=-d_{hm}}^{-1} \sum_{l=k}^{-1} \alpha^{-l-1} \eta_h^\top(l, j)E_h^\top \Theta_h^{\frac{1}{2}} \mathscr{Z}_1^h \Theta_h^{\frac{1}{2}} E_h \eta_h(l, j) \\
& + d_{hM} \sum_{k=-d_{hM}}^{-1} \sum_{l=k}^{-1} \alpha^{-l-1} \eta_h^\top(l, j)E_h^\top \Theta_h^{\frac{1}{2}} \mathscr{Z}_2^h \Theta_h^{\frac{1}{2}} E_h \eta_h(l, j) \\
& + \sum_{p=s_{hm}}^{s_{hM}} \sum_{k=1}^{p} \sum_{l=-k}^{-1} \alpha^{-l-1} x^{h\top}(l, j)\Theta_h^{\frac{1}{2}} \mathscr{R}^h \Theta_h^{\frac{1}{2}} x^h(k, j).
\end{aligned}
$$

Therefore,

$$
\begin{aligned}
& V^h(x^h(0, j)) \le \lambda_{\max}(\mathscr{P}^h) x^{h\top}(0, j)E_h^\top \Theta_h E_h x^h(0, j) \\
& + \lambda_{\max}(\mathscr{Q}_1^h) \sum_{k=-d_{hm}}^{-1} \alpha^{-k-1} x^{h\top}(k, j)\Theta_h x^h(k, j) \\
& + \Big(\lambda_{\max}(\mathscr{Q}_2^h) + \lambda_{\max}(\mathscr{Q}^h)\Big) \sum_{k=-d_{hM}}^{-1} \alpha^{-k-1} x^{h\top}(k, j)\Theta_h x^h(k, j) \\
& + \lambda_{\max}(\mathscr{Q}^h) \sum_{k=-d_{hM}+1}^{-d_{hm}} \sum_{l=k}^{-1} \alpha^{-l-1} x^{h\top}(k, j)\Theta_h x^h(l, j) \\
& + \lambda_{\max}(\mathscr{Z}_1^h) d_{hm} \sum_{k=-d_{hm}}^{-1} \sum_{l=k}^{-1} \alpha^{-l-1} \eta_h^\top(l, j)E_h^\top \Theta_h E_h \eta_h(l, j) \\
& + \lambda_{\max}(\mathscr{Z}_2^h) d_{hM} \sum_{k=-d_{hM}}^{-1} \sum_{l=k}^{-1} \alpha^{-l-1} \eta_h^\top(k, j)E_h^\top \Theta_h E_h \eta_h(l, j) \\
& + \lambda_{\max}(\mathscr{R}^h) \sum_{p=s_{hm}}^{s_{hM}} \sum_{k=1}^{p} \sum_{l=-k}^{-1} \alpha^{-l-1} x^{h\top}(k, j)\Theta_h x^h(k, j).
\end{aligned}
\tag{11.34}
$$

Note also that

$$
\begin{aligned}
&\sum_{k=-d_{hm}}^{-1}\sum_{l=k}^{-1}\alpha^{-l-1}\eta_h^\top(l,j)E_h^\top\Theta_hE_h\eta_h(l,j)\\
&\le 2\sum_{k=-d_{hm}}^{-1}\sum_{l=k}^{-1}\alpha^{-l-1}\Big(x^{h\top}(l+1,j)E_h^\top\Theta_hE_hx^{h\top}(l+1,j)\\
&+x^h(l,j)E_h^\top\Theta_hE_hx^h(l,j)\Big). \qquad (11.35)
\end{aligned}
$$

From the estimations (11.34) and (11.35), we obtain

$$
\begin{aligned}
V^h(x^h(0,j)) \le\ & \Big(\lambda_{\max}(\mathscr{P}^h)+4d_{hm}\lambda_{\max}(\mathscr{Z}_1^h)\alpha_{2s}(d_{hm})+4d_{hM}\lambda_{\max}(\mathscr{Z}_2^h)\alpha_{2s}(d_{hM})\Big)\\
&\times\sup_{\mathbb{Z}[-\sigma_h,0]\times[0,D_1]}\varphi^\top(k,j)E_h^\top\Theta_hE_h\varphi(k,j)\\
&+\Big(\lambda_{\max}(\mathscr{Q}_1^h)\alpha_{1s}(d_{hm})+(\lambda_{\max}(\mathscr{Q}_2^h)+\lambda_{\max}(\mathscr{Q}^h))\alpha_{1s}(d_{hM})\\
&+\lambda_{\max}(\mathscr{Q}^h)\alpha_{3s}(d_{hm},d_{hM})+\lambda_{\max}(\mathscr{R}^h)\alpha_{4s}(\tau_{hm},\tau_{hM})\Big)\\
&\times\sup_{\mathbb{Z}[-\sigma_h,0]\times[0,D_1]}\varphi^\top(k,j)\Theta_h\varphi(k,j). \qquad (11.36)
\end{aligned}
$$

For initial states $\varphi(k,j)\in\Omega(\Theta_h,c_h)$, according to (11.36), we have

$$V^h(x^h(0,j))\le\lambda_hc_h^2. \qquad (11.37)$$

Similarly, if $\psi(k,j)\in\Omega(\Theta_v,c_v)$ then

$$V^v(x^v(i,0))\le\lambda_vc_v^2. \qquad (11.38)$$

From (11.33), (11.37) and (11.38), we have

$$
\begin{aligned}
E_S(D) &\le (\lambda_hc_h^2+\lambda_vc_v^2)\sum_{k=0}^{D}\alpha^{D-k}+\rho^2\lambda_{\max}(\Psi)\sum_{k=0}^{D-1}\alpha^{D-1-k}\\
&<(\lambda_hc_h^2+\lambda_vc_v^2+\rho^2\lambda_{\max}(\Psi))\sum_{k=0}^{D}\alpha^{D-k}\\
&=\alpha_{1s}(D+1)\left(\lambda_hc_h^2+\lambda_vc_v^2+\rho^2\lambda_{\max}(\Psi)\right). \qquad (11.39)
\end{aligned}
$$

On the other hand, it can be verified by (11.11) that, for any $(i, j) \in \mathscr{R}(D_1, D_2)$,

$$\begin{aligned}V(x(i,j)) &\geq x^{h\top}(i,j)E_h^\top P^h E_h x(i,j) + x^{v\top}(i,j)E_v^\top P^v E_v x(i,j)\\ &= x^\top(i,j)E^\top \Theta^{\frac{1}{2}} \mathscr{P} \Theta^{\frac{1}{2}} E x(i,j)\\ &\geq \lambda_{\min}(\mathscr{P}) x^\top(i,j)E^\top \Theta E x(i,j). \end{aligned} \tag{11.40}$$

According to (11.39) and (11.40), condition (11.8) ensures that

$$x^\top(i,j)E^\top \Theta E x(i,j) \leq \alpha_{1s}(D+1)\frac{\lambda_h c_h^2 + \lambda_v c_v^2 + \rho^2 \lambda_{\max}(\Psi)}{\lambda_{\min}(\Theta^{-1/2} P \Theta^{-1/2})} < c^2 \tag{11.41}$$

for all $(i, j) \in \mathscr{R}(D_1, D_2)$. The last inequality indicates that the 2-D singular system (11.1) is SFRB with respect to (c, c_h, c_v, Θ). The proof is now completed. □

Remark 11.5 It is noticing that in the proof Theorem 11.1, the expressions of $\alpha_{1s}(d)$, $\alpha_{2s}(d)$, and $\alpha_{3s}(p, q)$, $\alpha_{4s}(p, q)$ are defined for $0 < \alpha \neq 1$. When $\alpha = 1$, these constants are determined by trivial sums as $\alpha_{1s}(d) = d$, $\alpha_{2s}(d) = \frac{d(d+1)}{2}$, $\alpha_{3s}(p, q) = \frac{(q-p)(q+p-1)}{2}$ and $\alpha_{4s}(p, q) = \frac{q(q+1)(q+2)-(p-1)p(p+1)}{6}$.

Remark 11.6 Theorem 11.1 provides a delay-dependent criterion for finite-region stability (when the exogenous disturbance w is ignored) and finite-region boundedness of 2-D singular systems extended time-varying delays. To tackle with the feasibility checking of nonlinear condition (11.8), we consider the following LMI-based conditions

$$\begin{aligned} &\gamma_1 \Theta < P < \gamma_2 \Theta,\\ Q < \gamma_3 \Theta, \quad &Q_1 < \gamma_3 \Theta, \quad Q_2 < \gamma_3 \Theta,\\ R < \gamma_4 \Theta, \quad &Z_1 < \gamma_5 \Theta, \quad Z_2 < \gamma_5 \Theta,\\ \Psi < \gamma_6 I, \quad &\gamma_k > 0, \quad k = 1, 2, \ldots, 6. \end{aligned} \tag{11.42}$$

Then, subject to (11.42), condition (11.8) is reduced to the following condition

$$\gamma_h c_h^2 + \gamma_v c_v^2 + \rho^2 \gamma_6 < \frac{\gamma_1 c^2}{\alpha_{1s}(D+1)}, \tag{11.43}$$

where

$$\begin{aligned}\gamma_h &= \gamma_2 + \gamma_3\left(\alpha_{1s}(d_{hm}) + 2\alpha_{1s}(d_{hM})\right) + 4\left(d_{hm}\alpha_{2s}(d_{hm}) + d_{hM}\alpha_{2s}(d_{hM})\right)\gamma_5\\ &\quad + \gamma_3\alpha_{3s}(d_{hm}, d_{hM}) + \gamma_4\alpha_{4s}(s_{hm}, s_{hM}),\\ \gamma_v &= \gamma_2 + \gamma_3\left(\alpha_{1s}(d_{vm}) + 2\alpha_{1s}(d_{vM})\right) + 4\left(d_{vm}\alpha_{2s}(d_{vm}) + d_{vM}\alpha_{2s}(d_{vM})\right)\gamma_5\\ &\quad + \gamma_3\alpha_{3s}(d_{vm}, d_{vM}) + \gamma_4\alpha_{4s}(s_{vm}, s_{vM}).\end{aligned}$$

On the basis of the above derivation, we have the following result.

Corollary 11.1 *Given positive scalars c, c_h, c_v, and a matrix $\Theta = \mathrm{diag}(\Theta_h, \Theta_v) \in \mathbb{S}_n^+$. The 2-D singular system (11.1) is SFRB with respect to (c, c_h, c_v, Θ) if there* *exist scalars* $\alpha > 0$, $\gamma_k > 0$ $(k = 1, 2, \ldots, 6)$ *such that the LMI-based conditions governed by* (11.7), (11.42) *and* (11.43) *are feasible for positive definite matrices* $P = \mathrm{diag}(P^h, P^v)$, $Q = \mathrm{diag}(Q^h, Q^v)$, $Q_k = \mathrm{diag}(Q_k^h, Q_k^v)$, $R = \mathrm{diag}(R^h, R^v)$, $Z_k = \mathrm{diag}(Z_k^h, Z_k^v)$ $(k = 1, 2)$, Ψ, *and real matrices* X_l $(l = 1, 2, \ldots, 6)$ *of appropriate dimensions, where the matrix notations are mentioned in Theorem 11.1.*

Remark 11.7 For given scalars c_h, c_v and c involving the thresholds of initial states and response states, the total time $D = D_1 + D_2$, which determines the target region, can be induced from (11.8) as

$$D + 1 \le \left\lfloor \frac{\ln\left(1 + (\alpha - 1)\frac{c^2\lambda_{\min}(\Theta^{-1/2}P\Theta^{-1/2})}{\lambda_h c_h^2 + \lambda_v c_v^2 + \rho^2\lambda_{\max}(\Psi)}\right)}{\ln \alpha} \right\rfloor \quad \text{if } 0 < \alpha \neq 1 \tag{11.44}$$

and

$$D + 1 \le \left\lfloor \frac{c^2\lambda_{\min}(\Theta^{-1/2}P\Theta^{-1/2})}{\lambda_h c_h^2 + \lambda_v c_v^2 + \rho^2\lambda_{\max}(\Psi)} \right\rfloor \quad \text{if } \alpha = 1, \tag{11.45}$$

where $\lfloor x \rfloor$ denotes the floor value of x.

Consider the following delay-free singular system

$$\begin{aligned} \begin{bmatrix} E_h x^h(i+1, j) \\ E_v x^v(i, j+1) \end{bmatrix} &= A \begin{bmatrix} x^h(i, j) \\ x^v(i, j) \end{bmatrix} + B_w w(i, j), \\ z(i, j) &= C \begin{bmatrix} x^h(i, j) \\ x^v(i, j) \end{bmatrix} + D_w w(i, j). \end{aligned} \tag{11.46}$$

By simplifying the proof of Theorem 11.1, we obtain the following result.

Corollary 11.2 *Given positive scalars* c, c_h, c_v, *and a matrix* $\Theta = \mathrm{diag}(\Theta_h, \Theta_v) \in \mathbb{S}_n^+$. *The delay-free singular 2-D system* (11.46) *is SFRB with respect to* (c, c_h, c_v, Θ) *if there exist scalars* $\alpha > 0$, $\gamma_k > 0$ $(k = 1, 2, 3)$, *positive definite matrices* $P = \mathrm{diag}(P^h, P^v)$, Ψ, *and a matrix* $S \in \mathbb{R}^{n\times(n-r)}$ *that satisfy the following conditions*

$$\begin{aligned} \begin{bmatrix} A^\top PA - \alpha E^\top PE + \mathrm{Sym}\{SL^\top A\} & A^\top PB_w + SL^\top B_w \\ * & -\Psi + B_w^\top PB_w \end{bmatrix} &< 0, \\ \gamma_1\Theta < P < \gamma_2\Theta, \quad \Psi &< \gamma_3 I, \\ \left(\gamma_2(c_h^2 + c_v^2) + \gamma_3\rho^2\right)\frac{1 - \alpha^{D+1}}{1 - \alpha} &< \gamma_1 c^2, \end{aligned} \tag{11.47}$$

where $L = (E^\top)^\perp$.

Remark 11.8 For 2-D nominal systems, we have $E = I_n$ and $L = 0$. By changing the variable $\Psi = \alpha M$, the derived conditions in (11.47) are reduced to the FRB conditions derived in [46, Theorem 3.1]. Thus, Corollary 11.2 is an extension of Theorem 3.1 in [46] for 2-D singular Roesser systems.

11.4 Finite-Region Dissipativity of 2-D Singular Roesser Systems with Delays

In this section, we consider the problem of finite-region $(\mathscr{Q}, \mathscr{S}, \mathscr{R})$-dissipativity of 2-D singular systems with extended delays as presented in (11.1). By employing the analysis result obtained in Theorem 11.1, we derive delay-dependent conditions that ensure the SFRB and finite-region $(\mathscr{Q}, \mathscr{S}, \mathscr{R})$-dissipativity of system (11.1).

Theorem 11.2 *Given positive numbers c_h, c_v, c, and symmetric matrices $\Theta = \mathrm{diag}(\Theta_h, \Theta_v)$, $\mathscr{Q}$, $\mathscr{S}$, and $\mathscr{R}$ with $\Theta > 0$, $\mathscr{Q} \leq 0$ and $\mathscr{R} > 0$. Then, the 2-D singular system* (11.1) *is finite-region $(\mathscr{Q}, \mathscr{S}, \mathscr{R})$-dissipative with respect to (c, c_h, c_v, Θ) if there exist scalars $\alpha > 0$, $\beta \in (0, 1)$, positive definite matrices $P = \mathrm{diag}(P^h, P^v)$, $Q = \mathrm{diag}(Q^h, Q^v)$, $Q_k = \mathrm{diag}(Q_k^h, Q_k^v)$, $R = \mathrm{diag}(R^h, R^v)$, $Z_k = \mathrm{diag}(Z_k^h, Z_k^v)$ $(k = 1, 2)$, and real matrices X_l $(l = 1, 2, \ldots, 6)$ of appropriate dimensions that satisfy condition* (11.8) *and the following LMI-based condition*

$$\begin{bmatrix} \Xi - \mathscr{C}^\top \mathscr{Q} \mathscr{C} & \mathscr{A}^\top P & \hat{\mathscr{A}}^\top [\overline{Z}_1 \;\; \overline{Z}_2] & \mathscr{C}^\top \mathscr{S} \\ * & -P & 0 & 0 \\ * & * & -\mathrm{diag}(\overline{Z}_1, \overline{Z}_2) & 0 \\ * & * & * & -\beta \mathscr{R} \end{bmatrix} < 0, \tag{11.48}$$

where $\mathscr{C} = \begin{bmatrix} C & 0 & C_d & 0 & C_s & D_w \end{bmatrix}$, $\Psi = (1 - \beta)\mathscr{R}$ and the other matrix notations are defined as in Theorem 11.1.

Proof By the Schur complement lemma, condition (11.48) with $\Psi = (1 - \beta)\mathscr{R}$ holds if and only if

$$\begin{bmatrix} \Xi + \hat{\Xi} & \mathscr{A}^\top P & \hat{\mathscr{A}}^\top [\overline{Z}_1 \;\; \overline{Z}_2] \\ * & -P & 0 \\ * & * & -\mathrm{diag}(\overline{Z}_1, \overline{Z}_2) \end{bmatrix} < 0,$$

where

$$\hat{\Xi} = \mathscr{C}^\top \mathscr{S} (\beta \mathscr{R})^{-1} \mathscr{S}^\top \mathscr{C} - \mathscr{C}^\top \mathscr{Q} \mathscr{C} \geq 0$$

according to the assumption $\mathscr{Q} \leq 0$. Thus, the feasibility of condition (11.48) induces that of condition (11.7). By Theorem 11.1, the 2-D singular system (11.1) is SFRB with respect to (c, c_h, c_v, Θ).

We now prove the dissipation inequality (11.5). For this, we define the supplied rate function

$$E(z, w) = z^\top \mathscr{Q} z + 2 z^\top \mathscr{S} w + w^\top \mathscr{R} w \tag{11.49}$$

and consider a storage function $V(x) = V^h(x^h) + V^v(x^v)$ as presented in (11.11). Similar to the proof of Theorem 11.1, we have

$$\begin{aligned}
&\Delta_\alpha \mathscr{V}(i,j) - E(z(i,j),w(i,j)) \\
&\qquad = \Delta_\alpha \mathscr{V}(i,j) - \chi^\top(i,j)\mathscr{C}^\top \mathscr{Q}\mathscr{C}\chi(i,j) \\
&\qquad - 2\chi^\top(i,j)\mathscr{C}^\top \mathscr{S} w(i,j) - w^\top(i,j)\mathscr{R} w(i,j).
\end{aligned} \tag{11.50}$$

By utilizing the Cauchy matrix inequality [see, Chap. 2, Lemma 2.6], we have

$$\begin{aligned}
-2\chi^\top(i,j)\mathscr{C}^\top \mathscr{S} w(i,j) \le \chi^\top(i,j)\mathscr{C}^\top \mathscr{S}(\beta\mathscr{R})^{-1}\mathscr{S}^\top \mathscr{C}\chi(i,j) \\
+ \beta w^\top(i,j)\mathscr{R} w(i,j).
\end{aligned} \tag{11.51}$$

Together with (11.28), from (11.50) and (11.51), we have

$$\begin{aligned}
\Delta_\alpha \mathscr{V}(i,j) - E(z(i,j),w(i,j)) \le \chi^\top(i,j)\Big[\Xi - \mathscr{C}^\top \mathscr{Q}\mathscr{C} + \mathscr{A}^\top P \mathscr{A} \\
+ \hat{\mathscr{A}}^\top \left(\overline{Z}_1 + \overline{Z}_2\right)\hat{\mathscr{A}} + \mathscr{C}^\top \mathscr{S}(\beta\mathscr{R})^{-1}\mathscr{S}^\top \mathscr{C}\Big]\chi(i,j) \\
+ w^\top(i,j)\left(\Psi - (1-\beta)\mathscr{R}\right)w(i,j).
\end{aligned} \tag{11.52}$$

Again, by the Schur complement lemma, condition (11.48) holds if and only if

$$\begin{aligned}
\Xi - \mathscr{C}^\top \mathscr{Q}\mathscr{C} + \mathscr{A}^\top P \mathscr{A} + \hat{\mathscr{A}}^\top \left(\overline{Z}_1 + \overline{Z}_2\right)\hat{\mathscr{A}} \\
+ \mathscr{C}^\top \mathscr{S}(\beta\mathscr{R})^{-1}\mathscr{S}^\top \mathscr{C} < 0.
\end{aligned}$$

Therefore, with $\Psi = (1-\beta)\mathscr{R}$, it follows from condition (11.48) and inequality (11.52) that

$$\Delta_\alpha \mathscr{V}(i,j) - E(z(i,j),w(i,j)) \le 0. \tag{11.53}$$

If the scalar α in (11.48) satisfies $0 < \alpha \le 1$, then we have

$$\Delta_\alpha \mathscr{V}(i,j) = \Delta \mathscr{V}(i,j) + (1-\alpha)\mathscr{V}(i,j) \ge \Delta \mathscr{V}(i,j),$$

where the difference $\Delta \mathscr{V}(i,j)$ is defined as

$$\Delta \mathscr{V}(i,j) = V^h(x^h(i+1,j)) - V^h(x^h(i,j)) + V^v(x^v(i,j+1)) - V^v(x^v(i,j)).$$

For any positive integers T, K, by summing up both sides of the inequality

$$\Delta \mathscr{V}(i,j) \le E(z(i,j),w(i,j)),$$

we get

$$\sum_{i=0}^{T-1}\sum_{j=0}^{K-1} E(z(i,j),w(i,j)) \geq \sum_{i=0}^{T-1}\sum_{j=0}^{K-1}\Big[V^h(x^h(i+1,j)) - V^h(x^h(i,j)) + V^v(x^v(i,j+1)) - V^v(x^v(i,j))\Big]$$
$$= \sum_{j=0}^{K-1}\big(V^h(x^h(T,j)) - V^h(x^h(0,j))\big) + \sum_{i=0}^{T-1}\big(V^v(x^v(i,K)) - V^v(x^v(i,0))\big). \tag{11.54}$$

The estimate in (11.54) indicates that the dissipation inequality (11.5) is satisfied with the supplied rate function $E(z,w) = z^\top \mathscr{Q} z + 2z^\top \mathscr{S} w + w^\top \mathscr{R} w$ and storage function $V(x) = V^h(x^h) + V^v(x^v)$ as given in (11.11).

If $\alpha > 1$, then we define the storage function

$$\widehat{V}(x) = \widehat{V}^h(x^h) + \widehat{V}^v(x^v),$$

where

$$\widehat{V}^h(x^h(i,j)) = \frac{V^h(x^h(i,j))}{\alpha^{i+j-1}}, \quad \widehat{V}^v h(x^v(i,j)) = \frac{V^v(x^v(i,j))}{\alpha^{i+j-1}}.$$

It follows from (11.53) that

$$E(z(i,j),w(i,j)) \geq \big(V^h(x^h(i+1,j)) - \alpha V^h(x^h(i,j))\big) + \big(V^v(x^v(i,j+1)) - \alpha V^v(x^v(i,j))\big)$$
$$= \alpha^{i+j}\Big[\big(\widehat{V}^h(x^h(i+1,j)) - \widehat{V}^h(x^h(i,j))\big) + \big(\widehat{V}^v(x^v(i,j+1)) - \widehat{V}^v(x^v(i,j))\big)\Big]$$
$$\geq \Delta\widehat{V}(x(i,j)).$$

Similar to (11.54), we can get

$$\sum_{i=0}^{T-1}\sum_{j=0}^{K-1} E(z(i,j),w(i,j)) \geq \sum_{j=0}^{K-1}\big(\widehat{V}^h(x^h(T,j)) - \widehat{V}^h(x^h(0,j))\big) + \sum_{i=0}^{T-1}\big(\widehat{V}^v(x^v(i,K)) - \widehat{V}^v(x^v(i,0))\big). \tag{11.55}$$

It can be deduced from (11.54) and (11.55) that the 2-D singular system (11.1) is finite-region $(\mathcal{Q}, \mathcal{S}, \mathcal{R})$-dissipative. The proof is completed. □

As a consequence of Theorem 11.2 and Corollary 11.2, the finite-region dissipativity of delay-free singular system (11.46) is presented in the following corollary.

Corollary 11.3 *For given positive numbers c_h, c_v, c, and symmetric matrices $\Theta = \text{diag}(\Theta_h, \Theta_v)$, $\mathcal{Q}$, $\mathcal{S}$, $\mathcal{R}$ with $\Theta > 0$, $\mathcal{Q} \leq 0$ and $\mathcal{R} > 0$, the delay-free singular system* (11.46) *is finite-region $(\mathcal{Q}, \mathcal{S}, \mathcal{R})$-dissipative with respect to (c, c_h, c_v, Θ) if there exist scalars $\alpha > 0$, $\beta \in (0, 1)$, $\gamma_1 > 0$, $\gamma_2 > 0$, a positive definite matrix $P = \text{diag}(P^h, P^v)$ and a matrix $S \in \mathbb{R}^{n\times(n-r)}$ that satisfy the following conditions*

$$\begin{aligned}
&\left[\begin{array}{cc|c} \Phi_{11} & \Phi_{12} & C^\top \mathcal{S} \\ * & \Phi_{22} & D_w^\top \mathcal{S} \\ \hline * & * & -\beta\mathcal{R} \end{array}\right] < 0, \\
&\gamma_1\Theta < P < \gamma_2\Theta, \\
&\left(\gamma_2(c_h + c_v) + (1-\beta)\lambda_{\max}(\mathcal{R})\rho^2\right)\frac{1-\alpha^{D+1}}{1-\alpha} < \gamma_1 c,
\end{aligned} \tag{11.56}$$

where $L = (E^\top)^\perp$ and

$$\begin{aligned}
\Phi_{11} &= A^\top PA - \alpha E^\top PE + \text{Sym}\{SL^\top A\} - C^\top \mathcal{Q} C, \\
\Phi_{12} &= A^\top PB_w + SL^\top B_w - C^\top \mathcal{Q} D_w, \\
\Phi_{22} &= -(1-\beta)\mathcal{R} + B_w^\top PB_w - D_w^\top \mathcal{Q} D_w.
\end{aligned}$$

11.4.1 Numerical Examples

In this section, we present some numerical examples to illustrate the effectiveness of the obtained stability conditions.

Example 11.1 Consider a thermal process in reactors, heat exchanger or pipe furnaces described by the following Darboux-type equation [51]

$$\begin{aligned}
\frac{\partial v(s,t)}{\partial x} + \frac{\partial v(x,t)}{\partial t} &= a_0 v(x,t) + a_1 v(x-d_x,t) + a_2 v(x,t-d_t) \\
&\quad + c_1\int_0^{s_x} v(x-\xi,t)d\xi + c_2\int_0^{s_t} v(x,t-\tau)d\tau,
\end{aligned} \tag{11.57}$$

where $v(x,t)$ represents the unknown function at space x and time t, a_0, a_1, a_2, c_1, c_2 are real numbers and d_x, d_t, s_x, s_t are time-varying delays.

Given increments Δx and Δt. We denote $v(i,j) = v(i\Delta x, j\Delta t)$. By forward Euler approximation, the derivatives $\frac{\partial v(x,t)}{\partial x}$ and $\frac{\partial v(x,t)}{\partial t}$ will be approximated as

$$\frac{\partial v(x,t)}{\partial x} \simeq \frac{v(i,j)-v(i-1,j)}{\Delta x},$$
$$\frac{\partial v(x,t)}{\partial t} \simeq \frac{v(i,j)-v(i,j-1)}{\Delta t}.$$

Define $x^h(i,j) = v(i-1,j)$ and $x^v(i,j) = v(i,j)$. Then,

$$v(i,j) = x^h(i+1,j) = \begin{bmatrix} 1 & 0 \\ 0 & 0 \end{bmatrix} \begin{bmatrix} x^h(i+1,j) \\ x^v(i,j+1) \end{bmatrix}$$

and Eq. (11.57) can be recast into a 2-D discrete-time system of the form

$$\begin{bmatrix} 1 & 0 \\ 0 & 0 \end{bmatrix} \begin{bmatrix} x^h(i+1,j) \\ x^v(i,j+1) \end{bmatrix} = A \begin{bmatrix} x^h(i,i) \\ x^v(i,j) \end{bmatrix} + A_d \begin{bmatrix} x^h(i-d_h(i),j) \\ x^v(i,j-d_v(j)) \end{bmatrix}$$
$$+ A_s \begin{bmatrix} \sum_{l=1}^{s_h(i)} x^h(i-l,j) \\ \sum_{l=1}^{s_v(j)} x^v(i,j-l) \end{bmatrix} + B_w w(i,j), \tag{11.58}$$

where $d_h(i) = \lfloor d_x/\Delta x \rfloor$, $d_v(j) = \lfloor d_t/\Delta t \rfloor$, $s_h(i) = \lfloor x_f/\Delta x \rfloor$, $s_v(j) = \lfloor t_f/\Delta t \rfloor$ and $w(i,j)$ is introduced as approximation errors.

To demonstrate the efficacy of the analysis results, we select the system matrices of (11.58) as

$$A = \begin{bmatrix} 0.25 & 0.5 \\ 0.1 & 1.5 \end{bmatrix}, \quad A_d = \begin{bmatrix} 0.1 & 0.12 \\ 0 & -0.21 \end{bmatrix},$$
$$A_s = \begin{bmatrix} 0.1 & -0.1 \\ 0.05 & 0.1 \end{bmatrix}, \quad B_w = \begin{bmatrix} 0.1 \\ 0.1 \end{bmatrix}.$$

Time delays $d_h(i)$, $d_v(j)$, and $s_h(i)$, $s_v(j)$ are assumed to be confined within the ranges defined by

$$d_{hm} = 1, \quad d_{hM} = 6, \quad d_{vm} = 1, \quad d_{vM} = 6,$$
$$s_{hm} = 0, \quad s_{hM} = 4, \quad s_{vm} = 0, \quad s_{vM} = 4.$$

Let $\Theta = I_2$. By solving the LMI condition (11.7) using Matlab LMI toolbox, it is found that condition (11.7) is feasible with $\alpha = 1.11$ and the following matrix variables (we omit some other matrices here)

$$P = \left[\begin{array}{c:c} 6.4652 & 0 \\ \hdashline 0 & 0.5253 \end{array}\right], \quad Q = \left[\begin{array}{c:c} 0.2025 & 0 \\ \hdashline 0 & 1.3361 \end{array}\right],$$
$$Q_1 = \left[\begin{array}{c:c} 0.2853 & 0 \\ \hdashline 0 & 1.3013 \end{array}\right], \quad Q_2 = \left[\begin{array}{c:c} 0.3762 & 0 \\ \hdashline 0 & 0.8307 \end{array}\right],$$
$$R = \left[\begin{array}{c:c} 0.0924 & 0 \\ \hdashline 0 & 0.4426 \end{array}\right], \quad Z_1 = \left[\begin{array}{c:c} 0.5434 & 0 \\ \hdashline 0 & 0.5253 \end{array}\right],$$

Fig. 11.2 State trajectory $x^h(i, j)$

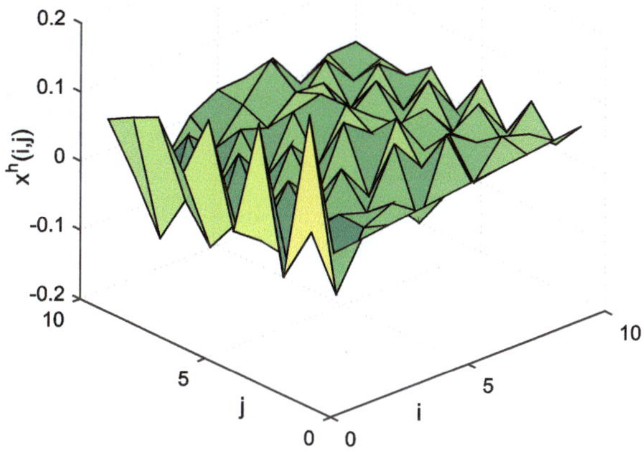

$$Z_2 = \left[\begin{array}{c|c} 0.0142 & 0 \\ \hline 0 & 0.0146 \end{array}\right], \quad \Psi = 2.3355.$$

Moreover, the scalars λ_h, λ_v in condition (11.8) are obtained as

$$\lambda_h = 14.7735, \quad \lambda_v = 1.1505.$$

Let $\rho = 0.2$, $c_h = c_v = 0.1$ and $c = 6$, then by Theorem 11.1, system (11.58) is SFRB with respect to (c_h, c_v, c, I_2) for finite times D_1, D_2 such that $D + 1 \leq \lfloor 21.2996 \rfloor = 21$.

To simulate the results, we take $D_1 = D_2 = 10$, time-varying delays $d_h(i) = 1 + 5|\cos(\pi i/8)|$, $d_v(j) = 1 + 5|\sin(\pi j/8)|$, $s_h(i) = 4|\sin(i\pi/2)|$, $s_v(j) = 4|\cos(j\pi/2)|$, initial sequences $\varphi(k, j) = 0.1\sin(0.1j\pi/6)$, $\psi(i, l) = 0.1\cos(0.1i\pi/4)$ and disturbance $w(i, j) = 0.2(-1)^{(i+j)}\frac{\sin(0.1i\pi+0.1j\pi)}{\sqrt{i+j}}$. State trajectories $x^h(i, j)$ and $x^v(i, j)$ are presented in Figs. 11.2 and 11.3, respectively, while the trajectory of disturbance $w(i, j)$ is depicted in Fig. 11.4. It can be seen that, for $(i, j) \in \mathscr{R}(D_1, D_2)$, the norm-state trajectory of $x^\top(i, j)x(i, j)$ does not exceed the given threshold. This illustrates the obtained analysis result of Theorem 11.1.

Example 11.2 Consider a delay-free system in the form of (11.46). The system matrices are given by

$$E = \mathrm{diag}\left(\begin{bmatrix} 1 & 0 \\ 0 & 0 \end{bmatrix}, \begin{bmatrix} 1 & 0 \\ 0 & 0 \end{bmatrix}\right), \quad A = \left[\begin{array}{c|c} A_{11} & A_{12} \\ \hline 0 & A_{22} \end{array}\right],$$

$$A_{11} = \begin{bmatrix} -0.5 & 0.8 \\ 0.2 & -0.1 \end{bmatrix}, \quad A_{12} = \begin{bmatrix} 0.2 & 0.15 \\ -0.2 & 0.3 \end{bmatrix},$$

$$A_{22} = \begin{bmatrix} -0.05 & 0.5 \\ 0.05 & 1.5 \end{bmatrix}, \quad B_w^\top = [0.1\ 0.2\ 0\ 0.1].$$

Fig. 11.3 State trajectory $x^v(i, j)$

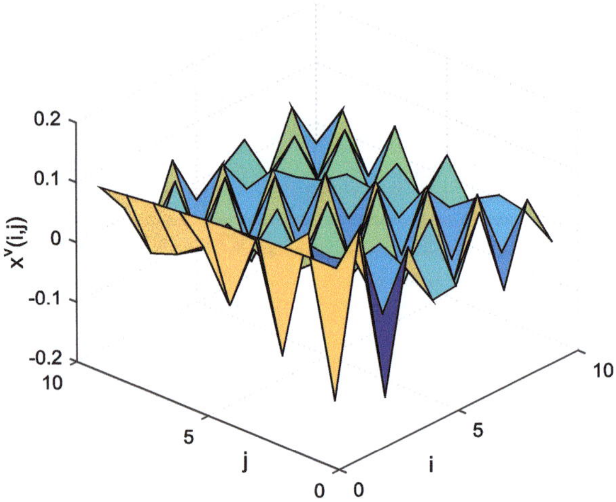

Fig. 11.4 Disturbance $w(i, j) = 0.2(-1)^{(i+j)} \frac{\sin(0.1i\pi+0.1j\pi)}{\sqrt{i+j}}$

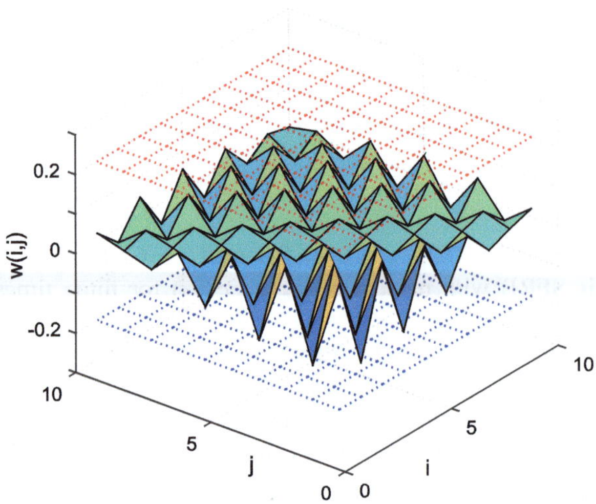

This example is to show that a 2-D singular system can be stable in any finite-region despite it is not asymptotically stable in the sense of Lyapunov. Indeed, let $P(z_1, z_2) = \det(E - A\mathbb{I}(z_1, z_2))$ denote the characteristic polynomial of the system. Then, system (11.46) is admissible (regular, causal and asymptotically stable) if and only if (see, [52], Lemma 1)

$$P(z, z) \neq 0 \text{ for } 0 < |z| \leq 1 \text{ and } P(z_1, z_2) \neq 0 \text{ for } |z_1| = |z_2| = 1.$$

Since $P(z_1, z_2) = \det(E_h - z_1 A_{11}) \det(E_v - z_2 A_{22})$ and $\det(E_h - z_1 A_{11}) = 0$ for $z_1 = \frac{10}{11}$, according to the aforementioned stability criterion, we can conclude that the system is not asymptotically stable.

We now apply Corollary 11.2. By using the Matlab LMI toolbox, it is found that condition (11.47) is feasible with $\alpha = \sqrt{e}$, $\gamma_2/\gamma_1 = 14.2944$, $\gamma_3/\gamma_2 = 16.0161$ and the matrix variables

$$P = \operatorname{diag}\left(\begin{bmatrix} 0.2368 & 0.001 \\ 0.001 & 2.9404 \end{bmatrix}, \begin{bmatrix} 3.3846 & 0 \\ 0 & 2.9189 \end{bmatrix}\right),$$

$$S^{\top} = \begin{bmatrix} -4.6084 & 2.8323 & -4.4775 & -1.3013 \\ 1.0149 & -0.6287 & 0.9003 & -3.2729 \end{bmatrix}.$$

By Corollary 11.2, the system is SFRB with respect to (c, c_h, c_v, I_4) with the time-region $\mathscr{R}(D_1, D_2)$ subject to

$$D_1 + D_2 < 2\ln\left(1 + \frac{c^2}{22.0347(c_h^2 + c_v^2) + 24.6887\rho^2}\right) - 1.$$

Example 11.3 This example is to demonstrate the effectiveness of the dissipativity analysis result obtained in Theorem 11.2. Consider a 2-D singular system as given in (11.1) with the system matrices

$$A = \left[\begin{array}{cc:cc} -0.0261 & -0.4806 & 0.2231 & 0.1482 \\ 0.0338 & 1.4806 & -0.1962 & 0.3318 \\ \hdashline -0.1366 & 0.2301 & -0.0458 & -0.5282 \\ 0.1598 & -0.2519 & 0.0535 & 1.5282 \end{array}\right], \quad B_w = \begin{bmatrix} 0.1 \\ 0.1 \\ 0.1 \\ 0.2 \end{bmatrix},$$

$$A_d = \left[\begin{array}{cc:cc} 0 & -0.1 & 0 & 0 \\ -0.05 & 0.12 & 0 & 0 \\ \hdashline 0 & 0 & -0.1 & 0.05 \\ 0 & 0 & 0.09 & -0.2 \end{array}\right], \quad A_s = \left[\begin{array}{cc:cc} 0.1 & 0.05 & 0 & 0 \\ 0.04 & 0 & 0 & 0 \\ \hdashline 0 & 0 & 0.05 & 0.1 \\ 0 & 0 & -0.1 & 0.12 \end{array}\right],$$

$$C = \begin{bmatrix} 0.01 & -0.02 & 0.05 & 0.1 \\ 0.05 & 0.1 & 0 & -0.12 \end{bmatrix}, \quad C_d = \begin{bmatrix} 0 & 0.1 & -0.08 & 0.02 \\ -0.05 & 0.1 & 0 & 0.14 \end{bmatrix},$$

$$C_s = \begin{bmatrix} 0.08 & 0.11 & 0.06 & 0 \\ 0.15 & 0.12 & 0.1 & 0.12 \end{bmatrix}, \quad D_w = \begin{bmatrix} 0.1 \\ 0.1 \end{bmatrix}, \quad E = \operatorname{diag}\left(\begin{bmatrix} 1 & 0 \\ 0 & 0 \end{bmatrix}, \begin{bmatrix} 1 & 0 \\ 0 & 0 \end{bmatrix}\right).$$

Time delays $d_h(i)$, $d_v(j)$, and $s_h(i)$, $s_v(j)$ are assumed to be confined within the ranges defined by upper and lower bounds

$$d_{hm} = 1, \quad d_{hM} = 5, \quad d_{vm} = 1, \quad d_{vM} = 5,$$
$$s_{hm} = 0, \quad s_{hM} = 4, \quad s_{vm} = 0, \quad s_{vM} = 4.$$

Let $\Theta = I_4$, $\mathscr{Q} = -0.1I_2$, $\mathscr{R} = 10$ and $\mathscr{S} = \begin{bmatrix} 0.1 \\ -0.1 \end{bmatrix}$. It is found that condition (11.48) is satisfied with $\alpha = 1.23$, $\beta = 0.95$ and the following matrix (we omit some other matrices)

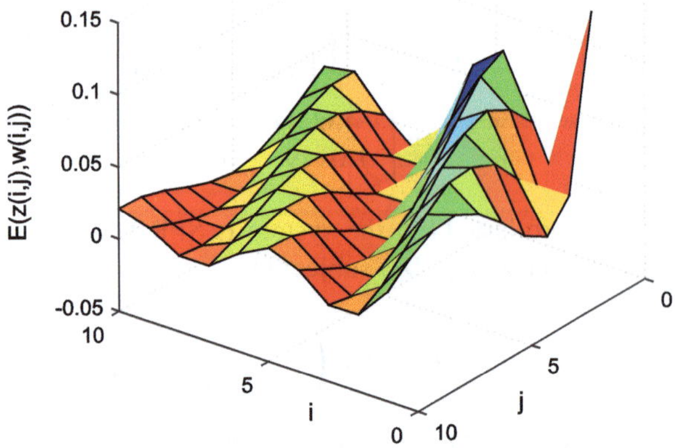

Fig. 11.5 Supplied rate $E(z(i, j), w(i, j))$

$$P = \text{diag}\left(\begin{bmatrix} 4.6285 & 0.578 \\ 0.578 & 0.6827 \end{bmatrix}, \begin{bmatrix} 3.5335 & 0.4706 \\ 0.4706 & 0.6137 \end{bmatrix}\right).$$

In addition, the constants λ_h, λ_v are obtained as $\lambda_h = 29.318$ and $\lambda_v = 22.6343$. By Theorem 11.2, system (11.1) is finite-region $(\mathscr{Q}, \mathscr{S}, \mathscr{R})$-dissipative with respect to (c, c_h, c_v, I_4) for positive scalars c, c_h, c_v and time-region $\mathscr{R}(D_1, D_2)$ subject to

$$D_1 + D_2 < 4.8306 \ln\left(1 + \frac{0.1241c^2}{29.318c_h^2 + 22.6343c_v^2 + 0.5\rho^2}\right) - 1.$$

To simulate the result, we take $D_1 = D_2 = 10$, $d_h(i) = 1 + 4|\sin(i\pi/2)|$, $d_v(j) = 1 + 4|\sin(j\pi/2)|$, $s_h(i) = 4$, $s_v(j) = 4$ and disturbance $w(i, j) = 0.2\frac{\cos(0.1i\pi + 0.2j\pi)}{\sqrt{i+j}}$.

The supplied rate $E(z(i, j), w(i, j))$ and exogenous disturbance $w(i, j)$ are presented in Figs. 11.5 and 11.6, respectively. The state trajectory of $x^\top(i, j)x(i, j)$ is given in Fig. 11.7. It can be seen that the trajectory of $x^\top(i, j)x(i, j)$ is confined by the given bound $c = 5$ in the region $\mathscr{R}(D_1, D_2)$. This illustrates the effectiveness of the obtained result in Theorem 11.2.

11.5 Conclusion

In this chapter, the problems of stability and dissipativity analysis in a finite-region have been extended for 2-D singular Roesser systems with directional time-varying delays. By virtue of the Lyapunov–Krasovskii functional method via weighted LKF candidates, delay-dependent conditions have been derived using linear matrix

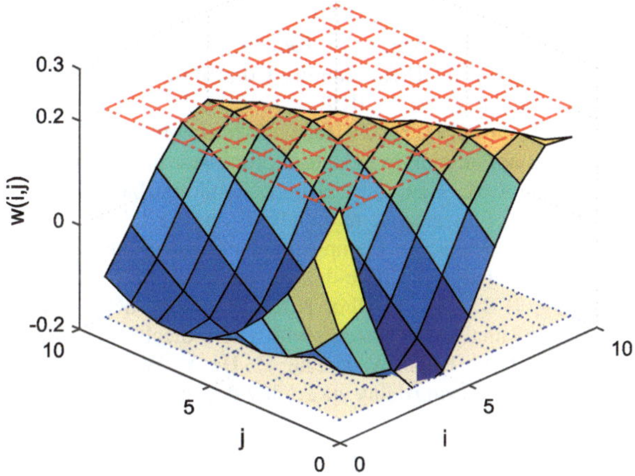

Fig. 11.6 Disturbance $w(i, j) = 0.2\frac{\sin(0.1i\pi+0.2j\pi)}{\sqrt{i+j}}$

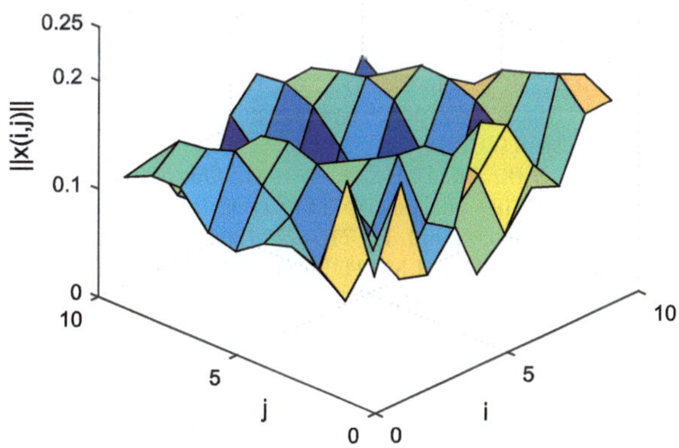

Fig. 11.7 State trajectory $\|x(i, j)\|$

inequalities. The derived conditions guarantee the finite-region stability and dissipativity of the system with respect to quadratic supplied rate functions. Numerical examples have been provided to illustrate the efficacy of the obtained analysis results.

References

1. Stevens, B.L., Lewis, F.L.: Aircraft Modeling: Dynamics and Control. Wiley, New York (1991)
2. Xu, S., Lam, L.: Robust Control and Filtering of Singular Systems. Springer, New York (2006)
3. Kunkel, P., Mehrmann, V.: Differential-Algebraic Equations: Analysis and numerical solution. EMS Publishing House, Zürich (2006)
4. Duan, G.R.: Analysis and Design of Descriptor Linear Systems. Springer, New York (2010)

5. Fridman, E.: Stability of linear descriptor systems with delay: a Lyapunov-based approach. J. Math. Anal. Appl. **273**, 24–44 (2002)
6. Malyshev, A., Sadkane, M.: On the complex stability radius for time-delay differential-algebraic systems. Linear Alg. Appl. **708**, 355–371 (2025)
7. Gao, Q., Karimi, H.R.: Stability. Control and Application of Time-Delay Systems. Butterworth-Heinemann, Oxford (2019)
8. Yu, W., Song, J., Yu, J.: Robust hybrid controller design for batch processes with time delay and its application in industrial processes. Int. J. Control Autom. Syst. **17**, 2881–2894 (2019)
9. Zhang, X.M., Han, Q.L., Seuret, A., Gouaisbaut, F., He, Y.: Overview of recent advances in stability of linear systems with time-varying delays. IET Control Theory Appl. **13**, 1–16 (2019)
10. Tian, Y., Wang, Z.: Composite slack-matrix-based integral inequality and its application to stability analysis of time-delay systems. Appl. Math. Lett. **120**, 107252 (2021)
11. Zeng, H.B., He, Y., Teo, K.L.: Monotone-delay-interval-based Lyapunov functionals for stability analysis of systems with a periodically varying delay. Automatica **138**, 110030 (2022)
12. Zhang, C.K., Chen, W.H., Zhu, C., He, Y., Wu, M.: Stability analysis of discrete-time systems with time-varying delay via a delay-dependent matrix- separation-based inequality. Automatica **156**, 111192 (2023)
13. Ge, X., Wang, Z., Lei, S.L., Vong, S.: Exponential stability of system with time delay based on matrix-refined weighted functions. European J. Control **79**, 101076 (2024)
14. Sha, H., Park, J.H., Chen, J., Zhu, M., Nan, C.: Stability analysis of discrete-time systems with a time-varying delay via improved methods. IET Control Theory Appl. **18**, 951–959 (2024)
15. He, J., Han, Y., Yang, F.: Relaxed mixed convex combination lemmas: Application to stability analysis of time-varying delay systems. J. Frankl. Inst. **361**, 107125 (2024)
16. Vu, L.H.: Reachable set estimation for singular discrete-time nonlinear systems with time-varying delays. Int. J. Robust Nonlinear Control **35**, 843–857 (2025)
17. Chen, J., Ji, F., Zhuang, G., Sha, H.: Stability analysis of discrete-time delayed neural networks via delay-product-type Lyapunov-Krasovskii functionals. IET Control Theory Appl. **19**, e12778 (2025)
18. Huang, Y.B., An, J., He, Y., Wu, M.: Quasi-convex combination method and its application to the stability analysis of 2D discrete-time Roesser systems with time-varying delays. IET Control Theory Appl. **12**, 718–727 (2018)
19. Hien, L.V., Trinh, H.M., Pathirana, P.N.: On l_1-gain control of 2-D positive Roesser systems with directional delays: Necessary and sufficient conditions. Automatica **112**, 108720 (2020)
20. Zhang, Y., Ji, Y., Jiang, T., Zhou, J.: Event-triggered control for Roesser model-based 2D Markov jump systems under stochastic communication protocol. Circuit Syst. Sig. Process **43**, 6953–6976 (2024)
21. Xue, X., Xu, J., Zhang, H.: Linear quadratic optimal control of stochastic 2-D Roesser models. Math. Comput. Simul. **227**, 500–510 (2025)
22. Zhu, K., Wang, Z., Ding, D., Dong, H., Wei, G.: Encryption-decryption-based set-membership filtering for two-dimensional systems: on security and boundedness. Automatica **173**, 112091 (2025)
23. Xu, H., Zou, Y.: H_∞ control for 2-D singular delayed systems. Int. J. Syst. Sci. **42**, 609–619 (2011)
24. Kririm, S., Hmamed, A., Tadeo, F.: Analysis and design of H_∞ controllers for 2D singular systems with delays. Circuit Syst. Sig. Process **35**, 1579–1592 (2016)
25. Hien, L.V., Vu, L.H., Trinh, H.: Stability of two-dimensional descriptor systems with generalized directional delays. Syst. Control Lett. **112**, 42–50 (2018)
26. Chelliq, E.H., Alfidi, M., Chalh, Z.: Admissibility and robust H_∞ controller design for uncertain 2D singular discrete systems with interval time-varying delays. Optim. Control Appl. Meth. **44**, 2476–2503 (2023)
27. Vu, L.H.: Exponential stability and H_∞ controller design for 2-D discrete-state systems in Roesser model with directional delays. Optim. Control Appl. Method **46**, 726–736 (2025)
28. Amato, F., Ambrosino, R., Ariola, M., Cosentino, C., Tommasi, G.D.: Finite-Time Stability and Control. Springer-Verlag, Berlin (2014)

29. Hien, L.V.: An explicit criterion for finite-time stability of linear nonautonomous systems with delays. Appl. Math. Lett. **30**, 12–18 (2014)
30. Zhang, G., Wang, W.: Finite-region stability and finite-region boundedness for 2D Roesser models. Math. Meth. Appl. Sci. **39**, 5757–5769 (2016)
31. Zhang, G., Trentelman, H.L., Wang, W., Gao, J.: Input-output finite-region stability and stabilization for discrete 2-D Fornasini-Marchesini models. Syst. Control Lett. **99**, 9–16 (2017)
32. Hua, D., Wang, W., Ren, Y., Yao, J.: Finite-region stabilization and H_∞ control for 2-D FMLSS asynchronously switched system. J. Frank. Inst. **357**, 9127–9153 (2020)
33. Willems, J.C.: Dissipative dynamical systems part I: General theory. Arch. Ration Mech. Anal. **45**, 321–351 (1972)
34. Feng, Z., Li, W., Lam, J.: Dissipativity analysis for discrete singular systems with time-varying delay. ISA Trans. **64**, 86–91 (2016)
35. Fu, X., Sheng, Z., Lin, C., Chen, B.: New results on admissibility and dissipativity analysis of descriptor time-delay systems. Appl. Math. Comput. **419**, 126860 (2022)
36. Han, S., Zhong, Q., Cui, L., Shi, K., Cai, X., Kwon, O.M.: Extended dissipativity analysis for T-S fuzzy systems based on reliable memory control and aperiodic sampled-data method. J. Frankl. Inst. **359**, 2156–2175 (2022)
37. Li, Y., He, Y., Zhang, C.K., Wu, M.: Discrete-state decomposition technique of dissipativity analysis for discrete-time singular systems with time-varying delays. IEEE Trans. Cybern. **53**, 5459–5468 (2022)
38. Lima, T.A., de Sousa, Madeira D., Jungers, M.: Dissipativity-based conditions for the feedback stabilization of systems with time-varying input delays. European J. Control **74**, 100855 (2023)
39. Wang, P., Lin, P., Sang, H., Zhang, L., Dimirovski, G.M.: Bumpless transfer control for switched positive time-delay systems with linear-type dissipative inequalities. J. Frankl. Inst. **361**, 107121 (2024)
40. Ahn, C.K., Shi, P., Basin, M.V.: Two-dimensional dissipative control and filtering for Roesser model. IEEE Trans. Autom. Control **60**, 1745–1759 (2015)
41. Wang, L., Chen, W., Li, L.: Dissipative stability analysis and control of two-dimensional Fornasini-Marchesini local state-space model. Int. J. Syst. Sci. **48**, 1744–1751 (2017)
42. Shen, Y., Wu, Z.G., Shi, P., Wen, G.: Dissipativity based fault detection for 2D Markov jump systems with asynchronous modes. Automatica **106**, 8–17 (2019)
43. Xu, Z., Su, H., Lu, R., Wu, Z.G.: Non-fragile dissipative filtering for 2-D switched systems. J. Frankl. Inst. **354**, 6234–6246 (2017)
44. Yang, R., Li, L., Shi, P.: Dissipativity-based two-dimensional control and filtering for a class of switched systems. IEEE Trans. Syst. Man Cybern. **51**, 2737–2750 (2021)
45. Li, L., Yang, R., Feng, Z., Wu, L.: Dissipative filtering for two-dimensional LPV systems: A hidden Markov model approach. ISA Trans. **130**, 409–419 (2022)
46. Hua, D., Wang, W., Yao, J.: Finite-region dissipative control for 2-D systems in the Roesser model. Int. J. Syst. Sci. **49**, 3406–3417 (2018)
47. Yang, R., Li, L., Su, X.: Finite-region dissipative dynamic output feedback control for 2-D FM systems with missing measurements. Inf. Sci. **514**, 1–14 (2020)
48. Yang, R., Ding, S., Zheng, W.X.: Co-design of event-triggered mechanism and dissipativity-based output feedback controller for two-dimensional systems. Automatica **130**, 109694 (2021)
49. Ma, Y., Jia, Z., Zhang, Q.: Robust observer-based finite-time H_∞ control for discrete-time singular Markovian jumping system with time delay and actuator saturation. Nonlinear Anal. Hybrid Syst. **28**, 1–22 (2018)
50. Zakeri H, Antsaklis PJ (201) Passivity and passivity indices of nonlinear systems under operational limitations using approximations. Int. J. Control 94:1114–1124
51. Kaczorek, T.: Two-Dimensional Linear Systems. Springer, Berlin (1985)
52. Xu, H., Zou, Y., Xu, S., Lam, J.: Bounded real lemma and robust H_∞ control of 2-D singular Roesser models. Syst. Control Lett. **54**, 339–346 (2005)

Index

L. V. Hien, *Two-Dimensional Time-Delay Systems*, Studies in Systems, Decision and Control 605, https://doi.org/10.1007/978-3-031-95678-2

MIX
Papier aus verantwortungsvollen Quellen
Paper from responsible sources
FSC® C105338

If you have any concerns about our products,
you can contact us on
ProductSafety@springernature.com

In case Publisher is established outside the EU,
the EU authorized representative is:
Springer Nature Customer Service Center GmbH
Europaplatz 3, 69115 Heidelberg, Germany

Printed by Libri Plureos GmbH
in Hamburg, Germany